Medicinal Chemistry for the 21st Century

INTERNATIONAL UNION OF PURE AND APPLIED CHEMISTRY

Medicinal Chemistry for the 21st Century

A 'Chemistry for the 21st Century' monograph

EDITED BY

C.G. WERMUTH
Laboratoire de Pharmacochimie Moléculaire
Faculté de Pharmacie
Université Louis Pasteur Strasbourg
France

WITH

N. KOGA
Daiichi Pharmaceutical Co., Ltd
Tokyo 103
Japan

H. KÖNIG
BASF Aktiengeselischaft
D-6700 Ludwigshafen
Germany

B.W. METCALF
SmithKline Beecham Pharmaceuticals
King of Prussia, PA 19406-2799
USA

OXFORD

BLACKWELL SCIENTIFIC PUBLICATIONS

LONDON EDINBURGH BOSTON

MELBOURNE PARIS BERLIN VIENNA

Editorial Offices:
Osney Mead, Oxford OX2 0EL
25 John Street, London WC1N 2BL
23 Ainslie Place, Edinburgh EH3 6AJ
238 Main Street, Cambridge
Massachusetts 02142, USA
54 University Street, Carlton
Victoria 3053, Australia

Other Editorial Offices:
Librairie Arnette SA
2, rue Casimir-Delavigne
75006 Paris
France

Blackwell Wissenschafts-Verlag
Meinekestrasse 4
D-1000 Berlin 15
Germany

Blackwell MZV
Feldgasse 13
A-1238 Wien
Austria

First published 1992

Set by Semantic Graphics, Singapore
Printed in Great Britain
at The Alden Press, Oxford
and bound at the Green Street Bindery, Oxford

DISTRIBUTORS

Marston Book Services Ltd
PO Box 87
Oxford OX2 0DT
(*Orders*: Tel: 0865 791155
Fax: 0865 791927
Telex: 837515)

Australia
Blackwell Scientific Publications
(Australia) Pty Ltd
54 University Street
Carlton, Victoria 3053
(*Orders*: Tel: (03) 347–0300)

Distributed in the USA
and North America by
CRC Press, Inc.
2000 Corporate Blvd, NW
Boca Raton
Florida 33431

A catalogue record for this book
is available from the British Library

ISBN 0-632-03408-4

Library of Congress
Cataloging in Publication Data

Medicinal chemistry for the 21st century /
edited by C. G. Wermuth . . . [*et al.*].
p. cm.
At head of title: International Union
of Pure and Applied Chemistry.
ISBN 0-632-03408-4
1. Pharmaceutical chemistry.
2. Pharmaceutical chemistry—Forecasting.
I. Wermuth, C. G. II. International Union
of Pure and Applied Chemistry.
III. Title: Medicinal chemistry for the
twenty-first century.
[DNLM: 1. Chemistry, Pharmaceutical—trends.
QV 744 M4893]
RS403.M39 1992
615'.19—dc20

Contents

Contributors

K. ARAI *Department of Molecular and Developmental Biology, Institute of Medical Science, University of Tokyo, 4-6-1 Shirokanedai, Minato-ku, Tokyo 108, Japan* [119]

J. BERGERS *Department of Pharmaceutics, University of Utrecht, PO Box 80.082, 3508 TB Utrecht, The Netherlands* [351]

J. BOCKAERT *Centre CNRS-INSERM de Pharmacologie-Endocrinologie, rue de la Cardonille, 34094 Montpellier Cedex 5, France* [145]

W. BODE *Max-Planck-Institut für Biochemie, D-8033 Martinsried, Germany* [73]

H. BUNDGAARD *PhD, DSc, Professor, The Royal Danish School of Pharmacy, Department of Pharmaceutical Chemistry, 2 Universitetsparken, DK-2100 Copenhagen, Denmark* [321]

M.G. BURES *PhD, Senior Research Scientist, Computer Assisted Molecular Design Project, Pharmaceutical Products Division, Abbott Laboratories, Abbott Park, IL 60064, USA* [295]

F.E. COHEN *PhD, MD, Departments of Pharmaceutical Chemistry and Medicine, University of California, San Francisco, CA 94143, USA* [107]

C.S. CRAIK *PhD, Departments of Pharmaceutical Chemistry and Biochemistry, University of California, San Francisco, CA 94143, USA* [107]

D.J.A. CROMMELIN *PhD, Professor of Pharmaceutics, Department of Pharmaceutics, University of Utrecht, PO Box 80.082, 3508 TB Utrecht, The Netherlands* [351]

E. DE CLERCQ *PhD, MD, Professor, Rega Institute for Medical Research, Katholieke Universiteit Leuven, B-3000 Leuven, Belgium* [45]

R.M. FREIDINGER *Department of Medicinal Chemistry, Merck Research Laboratories, West Point, PA 19486, USA* [233]

C.R. GANELLIN *FRS, Professor of Medicinal Chemistry, Department of Chemistry, University College London, Christopher Ingold Laboratories, 20 Gordon Street, London WC1H 0AJ, UK* [3]

C. HANSCH *Department of Chemistry, Pomona College, Claremont, CA 91711, USA* [281]

P. HERDEWIJN *PhD, Professor, Rega Institute for Medical Research, Katholieke Universiteit Leuven, B-3000 Leuven, Belgium* [45]

H.-D. HÖLTJE *Professor of Pharmaceutical Chemistry, Free University of Berlin, Institute of Pharmacy, Königin-Luise-Strasse 2 + 4, W-1000 Berlin 33, Germany* [181]

R. HUBER *Max-Planck-Institut für Biochemie, D-8033 Martinsried, Germany* [73]

W.F. HUFFMAN *PhD, Director, Peptidomimetic Research Department, SmithKline Beecham, 709 Swedeland Road, PO Box 1539, King of Prussia, PA 19406-0939, USA* [247]

R. IGARASHI *PhD, Assistant Professor, Institute of Medical Science, St Marianna University, 2-16-1 Sugao, Miyamae-ku, Kawasaki 216, Japan* [381]

A. ITAI *PhD, Faculty of Pharmaceutical Sciences, University of Tokyo, Tokyo 113, Japan* [191]

P.A.J. JANSSEN *Janssen Research Foundation, B-2340 Beerse, Belgium* [13]

Y. KATO *PhD, Faculty of Pharmaceutical Sciences, University of Tokyo, Tokyo 113, Japan* [191]

D.S. KEMP *Department of Chemistry Room 18-582, Massachusetts Institute of Technology, Cambridge, MA 01239, USA* [259]

K.-H. KIM *PhD, Research Investigator, Computer Assisted Molecular Design Project, Pharmaceutical Products Division, Abbott Laboratories, Abbott Park, IL 60064, USA* [295]

R. KIRSH *Department of Drug Delivery, SmithKline Beecham Pharmaceuticals, 709 Swedeland Road, King of Prussia, PA 19406, USA* [367]

P.M. LADURON *PhD, MD, Research Center Rhône-Poulenc Rorer, 13 Quai Jules Guesde BP-14, F-94403 Vitry Sur Seine Cedex, France* [39]

J.H. MCKERROW *PhD, MD, Department of Pathology, University of California, San Francisco, CA 94143, and the San Francisco Veterans Administration Medical Center, CA 94121, USA* [107]

G.R. MARSHALL *PhD, Center for Molecular Design, Washington University, Box 1099, 1 Brookings Drive, St Louis, MO 63130, USA* [163]

Y.C. MARTIN *PhD, Senior Project Leader, Computer Assisted Molecular Design Project, Pharmaceutical Products Division, Abbott Laboratories, Abbott Park, IL 60064, USA* [295]

Y. MIZUSHIMA *PhD, MD, Professor, Institute of Medical Science, St Marianna University, 2-16-1 Sugao, Miyamae-ku, Kawasaki 216, Japan* [381]

J.H. MUSSER *PhD, Director, Glycomed Inc., 860 Atlantic Avenue, Alameda, CA 94501, USA* [25]

M.V. NESTEROVA *Research Center of Molecular Diagnostics, Sympheropolsky Boulevard 8, Moscow 113149, Russia* [133]

Y. NISHIBATA *PhD, Faculty of Pharmaceutical Sciences, University of Tokyo, Tokyo 113, Japan* [191]

S. SAITO *PhD, Faculty of Pharmaceutical Sciences, University of Tokyo, Tokyo 113, Japan* [191]

J.A. SAKANARI *PhD, Department of Pathology, University of California, San Francisco, CA 94143, and the San Francisco Veterans Administration Medical Center, CA 94121, USA* [107]

P.W. SCHILLER *PhD, FRSC, Laboratory of Chemical Biology and Peptide Research, Clinical Research Institute of Montreal, 110 Pine Avenue West, Montreal, Quebec, Canada H2W 1R7* [215]

J.-C. SCHWARTZ *PhD, Professor, Unité de Neurobiologie et Pharmacologie (U.109) de l'INSERM, Centre Paul Broca, 2ter rue d'Alésia, 75014 Paris, France* [63]

E.S. SEVERIN *Research Center of Molecular Diagnostics, Sympheropolsky Boulevard 8, Moscow 113149, Russia* [133]

J.P. TOLLENAERE *Department of Theoretical Medicinal Chemistry, Janssen Research Foundation, B-2340 Beerse, Belgium* [13]

N. TOMIOKA *PhD, Faculty of Pharmaceutical Sciences, University of Tokyo, Tokyo 113, Japan* [191]

S. WATANABE *Department of Molecular and Developmental Biology, Institute of Medical Science, University of Tokyo, 4-6-1 Shirokanedai, Minato-ku, Tokyo 108, Japan* [119]

G. WILSON *Department of Drug Delivery, SmithKline Beecham Pharmaceuticals, 709 Swedeland Road, King of Prussia, PA 19406, USA* [367]

J. ZUIDEMA *PhD, Department of Pharmaceutics, University of Utrecht, PO Box 80.082, 3508 TB Utrecht, The Netherlands* [351]

Preface

During the 34th IUPAC General Assembly in 1987, our colleague Kirill Zamaraev suggested that it would be beneficial for all the scientific divisions or sections represented within IUPAC to assemble their respective knowledge and edit a series of monographs devoted to the various aspects and prospects of chemistry at the dawn of the 21st century. This ambitious enterprise was accepted with enthusiasm and the idea of publishing a series of specialized monographs, under the generic name 'Chemistry for the 21st Century', was considered to be an important opportunity to increase the influence, recognition and accomplishments of the IUPAC divisions and sections.

In this context, the IUPAC Section on Medicinal Chemistry* decided to contribute in sponsoring and editing the present volume.

Hesitating between the title 'Medicinal Chemistry—Future Directions' and 'Medicinal Chemistry for the 21st Century', the second title was finally adopted. The selection of topics and authors gave rise to many passionate discussions relating to the design and development of new chemical molecules for the treatment, cure or prevention of human and animal diseases. However, it was rapidly recognized that three critical stages essentially govern the obtaining of valuable new drugs: (i) the identification and production of an original active principle (lead compound); (ii) the improvement of the lead compound in terms of potency, selectivity and safety; and (iii) the suitability of the molecule selected for clinical use, primarily that it reaches its target organ. In addition, one must render its administration acceptable to patients and doctors.

In the first stage, the real problem lies in the discovery or identification of an original line of research, since there is no laid-down formula and no means of planning a laboratory's creative work. As a result, the discovery of new lead compounds represents the most uncertain stage in the development programme of a drug. The first part of the book, therefore, illustrates the quest for active substances. Traditional approaches to the generation of new lead compounds that we think will remain valid, are presented. Rational strategies, based on a complete knowledge of biochemical and physiopathological disturbances, are also described.

Protein structure–function relationships constitute the second part of the book. Great advances in the knowledge of enzyme and receptor interactions with smaller molecules have helped to explain the mechanisms of biological effects. Signal transduction will be an absolute requisite for the medicinal chemist dealing with drug design during the next few decades.

The third part of the book constitutes the more classical structure–activity relationships in medicinal chemistry. In this field, benefits can be expected from two major advances in supramolecular chemistry: (i) computer-aided drug design, allowing modelling and quantifying of drug–receptor interactions; and (ii) design of

* Section on Medicinal Chemistry, 1987–1991 period: Camille G. Wermuth, President; Brian W. Metcalf, Vice-President; Naofumi Koga, Secretary.

peptidomimetics, in other words, non-peptide surrogates acting on receptors for peptidic ligands. In any event, from leads provided by these methods, medicinal chemists can quickly develop more active and, at the same time, more selective and less toxic molecules.

The final part of the book is devoted to chemical formulation and to specific administration devices in order to obtain optimal bioavailability of active molecules and thus to render them chemically efficient. As Molière once wrote: *La façon de donner vaut mieux que ce que l'on donne.*

DEREK H.R. BARTON *Nobel Laureate*
A & M University
College Station
Texas, USA

C.G. WERMUTH
Université Louis Pasteur Strasbourg
France

Part 1
New Lead Discovery
Historical Perspectives

1 Past Approaches to Discovering New Drugs

C.R. GANELLIN
Department of Chemistry, University College London, Christopher Ingold Laboratories, 20 Gordon Street, London WC1H 0AJ, UK

1 Introduction

The development of new medicines takes place mainly in the modern pharmaceutical industry and many of the new drugs are also discovered in the research laboratories of pharmaceutical companies.

New drug discovery depends on a complex interplay and exchange of ideas between individual researchers and research teams in the pharmaceutical industry, in universities, hospitals and research institutes. It involves interactions between various scientific disciplines drawn mainly from the chemical, biological and medical sciences, and it is a highly complicated process. It has reached an extraordinary level of sophistication in the organization and interdependency between researchers which is why most of the advances now come from the laboratories of the developed industrial nations.

To gain a historical perspective it is of interest to consider the main approaches to drug discovery which have been taken in the past, particularly during the 20th century with the rise of the modern multinational pharmaceutical industry. For an excellent account of many examples of drug discovery, see Sneader (1985).

Broadly speaking, one can discern four main sources for new drugs. These are:
- natural products;
- existing drugs;
- screens;
- physiological mechanisms.

2 Natural products as a source for new drugs

Historically, natural products provide the oldest source for new medicines. Natural selection during evolution, and competition between the species, has produced powerful biologically active natural products which can serve as chemical leads, to be refined by the chemist to give more specifically acting drugs. For example, moulds and bacteria produce substances that prevent other organisms from growing in their vicinity. The famous *Penicillium* mould led to a whole range of penicillins, and identified a concept of seeking naturally occurring antibiotics. This was taken up by the microbiologists who argued that bacteria which cause infections in humans do not survive for long in soil because they are destroyed by other soil-inhabiting microbes. Extensive soil-screening research programmes have led to a vast array of antibiotics which have given rise to very potent life-saving drugs, e.g. streptomycin, terramycin, neomycin, bacitracin and polymyxin.

A recent example of the use of microbial fermentation as a source for a drug lead was the discovery of a novel cholecystokinin (CCK) antagonist, asperlicin, from

Aspergillus alliaceus (Chang *et al.*, 1985) which served as the starting point for Merck scientists to develop the very specific and potent non-peptide antagonists at CCK-A and CCK-B receptors, respectively. Such fermentation broths contain hundreds, if not thousands, of chemicals and are a potentially rich source of novel enzyme inhibitors.

A variety of venoms and toxins are used by animals as protection or to paralyse their prey; some are extremely potent, requiring only minute doses, e.g. bungarotoxin (from snake venom) which combines with acetylcholine receptors, tetrodotoxin (from puffer fish) which blocks sodium channels, apamin (from bee venom) which blocks Ca^{2+}-activated K^{+} channels and batroxobin (from the venom of a pit viper) which is a thrombin-like enzyme. They have served as starting points for investigation of hormone receptors, ion channels and enzymes. Such natural products have provided important leads to drugs as muscle relaxants and haemostatics. Recent subjects for study include spiders and sponges. Indeed, marine life offers a vast untapped resource for future investigation (Hall & Strichartz, 1990).

Another fruitful means for identifying pharmacologically active natural products has been the folklore remedies, which are mainly plant products: alkaloids such as morphine and heroin (from the opium poppy known in ancient Egypt), atropine and scopolamine (from plants of the Solanaceae family known to the ancient Greeks) and reserpine (from *Rauwolfia serpentina*, the snakeroot, popular in India as a herbal remedy), and non-nitrogenous natural products such as salicylates from the willow tree (genus *Salix*, botanical sources known to Hippocrates) and the glycosides in digitalis from the foxglove (in folk use in England for centuries).

3 Existing drugs as a basis for new drug discovery

The most fruitful basis for the discovery of a new drug is to start with an old drug. This has been the most common and reliable route to new products. Thus, existing drugs may need to be improved, e.g. to get a better dosage form, to improve drug absorption or duration, to increase potency to reduce the daily dose, or to avoid certain side-effects. On the other hand there may be side-effects that can be exploited, arising from astute observation during pharmacological studies in animals or from clinical investigation in patients.

The discovery of sulphonamide diuretics in the 1950s followed from the observation that sulphanilamide rendered the urine alkaline through inhibition of the enzyme carbonic anhydrase (Schwartz, 1949). The phenothiazine tranquillizers, which revolutionized the treatment of psychiatric patients, resulted from astute observation of the effects of a chloro derivative of the antihistamine, promethazine, whilst it was being studied for use in preventing surgical shock (Laborit, 1954).

Unfortunately, however, many new products arise which are not real advances on the old one, or where the new version is only a minor improvement. These are the so-called 'me-too' products. 'me-too' is a term often used pejoratively by those who like to attack the pharmaceutical industry. Actually, historically, the process has provided the main route whereby a particular type of drug action has been optimized in terms both of selectivity (to avoid side-effects) and application (for a particular patient population).

Companies generally start with the intention of discovering an improved product, and sometimes the 'me-too' product is a substantial advance and may eventually gain wider use than the original product. In general, to realize the full potential of a new 'me-too' drug and to reveal its advantages it is necessary to market it in order to gain access to a sufficiently wide patient population. Also, there is always a good chance that a competitor's product may run into difficulties, e.g. side-effects, poor clinical performance and inadequate marketing. Thus there are powerful reasons for developing new 'me-too' compounds but, unfortunately, this can lead to the proliferation of products and it can take many years for clinicians to determine the most suitable drug treatment.

4 Screens for new drugs

The screening approach with natural products for new antibiotics, antimetabolites and enzyme inhibitors has had its counterpart with synthetic chemicals, the idea being to test large numbers of compounds on a relatively simple system to reveal the required activity. This is the third main source of new drugs. The background to this approach lies in the dyestuffs' industry: Paul Ehrlich, founder and prophet of chemotherapy, discovered that synthetic dyes were absorbed differentially into tissues and that they could kill parasites and bacteria without affecting mammalian cells. From this work came salvarsan in 1910, an arsenic compound for treating syphilis (Burger, 1954). Several large chemical companies followed up this discovery by establishing their own research programmes seeking drugs against venereal diseases. From then on, the key to new drug discovery was seen to be the systematic examination of hundreds of synthetic chemicals.

In 1931, Gerhard Domagk, working for I.G. Farbenindustrie, turned to screening sulphonamide derivatives of azo dyes. This work gave rise to prontosil red in 1935, the first truly effective chemotherapeutic agent for any generalized bacterial infection. This discovery led, via sulphanilamide, to the development of the sulphonamide class of antibacterials (Rose, 1964).

The success of these discoveries in chemotherapy dominated research thinking in the pharmaceutical industry for many years, but the success of the screening approach is limited. America's National Cancer Institute is reputed to screen some 10 000 synthetic compounds each year as potential anticancer agents, but very few have become marketed as drugs. Screens were also established for non-infectious diseases, e.g. for anticonvulsants (useful in epilepsy), analgesics (in the hope of being non-addictive), antihypertensives, anti-inflammatory, antiulcer, etc.

There is, however, a fundamental distinction between the anti-infective screens, and the screens which seek a treatment for a disease which we can call 'metabolically based'. In the anti-infective screens (antibacterial, antifungal, antihelminthic, antiprotozoal, antiviral) a drug is sought which is lethal to the pathogen, but leaves the host unharmed. It is a search for selective toxicity between species.

By contrast, in the metabolically based diseases, e.g. allergy, asthma, cancer, duodenal ulcers, epilepsy, gout, hypertension, inflammation and so on, the cause is usually unknown, and we seek selectivity within the same being. In these latter situations, an animal model is used as a screening test, in which a clinical condition is induced in a laboratory animal such as a rabbit or rat, and compounds are tested to see whether they alleviate it.

The model often simulates the disease by presenting similar symptoms, but it may be misleading if the underlying causes are quite different; then the procedure throws up false leads, e.g. compounds that protect the laboratory animal but when tested clinically are found to be not active in humans. Pharmaceutical companies spent some 30 years screening for antiulcer agents without much success. Not all attempts have failed, however. The non-steroidal anti-inflammatory drugs are a very good example; they were discovered by screening in animals in which various forms of inflammation had been artifically induced. Many other types of drug have been discovered by such relatively simplistic screening procedures. A major drawback is how to generate a lead compound, because the screen offers no chemical guidance.

5 Physiological mechanisms; the modern 'rational approach' to drug design

With the advent of greater understanding of physiological mechanisms it has become possible to take a more mechanistic approach to research and start from a rationally argued hypothesis to design drugs. The target diseases selected for study are generally those prevalent in Western society and progress depends largely upon the current state of understanding of physiology in relation to diseases. This is the modern 'rational approach' to drug design which is becoming increasingly important with the development of information in cell biochemistry and cell biology, especially where this is understood at the molecular level.

The modern approach to drug discovery requires very close collaboration between chemists and biologists and, reflecting this, most large pharmaceutical companies now organize their research in project teams instead of by scientific discipline.

The 'rational approach' to new drug discovery requires several essential ingredients for success:

- evidence of a physiological basis for understanding a disease, so that one may hypothesize that a drug with a particular action should be therapeutically beneficial;
- an explicit chemical starting point;
- bioassay systems which measure the desired drug activity in the laboratory;
- a test which measures the activity of the drug in humans that can be related to a potential therapeutic treatment.

One may discern, broadly, five main sites for drug action:

1 *Enzymes*—where new molecules are made in tissues (the basis of metabolic activity).

2 *Receptors*—where circulating messengers, e.g. biogenic amines and peptides, act to alter cellular activity.

3 *Transport systems*—that selectively permit access through membranes into and out of cells, e.g. ion channels, transporter molecules.

4 *Cell replication and protein synthesis*—DNA and RNA.

5 *Storage sites*—where molecules are kept in an inactive form for later use, e.g. mast cells, blood platelets, neurones.

The body is controlled by chemical messengers (physiological mediators). There is a very complex communication system and each messenger has specific functions and is recognized specifically at a specialized site where it acts. In disease, something has got out of balance, and in drug therapy the aim is to redress that balance. Thus, enzymes have active sites which specifically recognize the appropriate substrates which they can process. If we wish to interrupt this process we may design enzyme inhibitors, using the chemistry of the substrate as a starting point, e.g. antimetabolites such as 6-mercaptopurine, azathioprine, pyrimethamine, trimethoprim and allopurinol (Hitchings, 1980).

If the enzyme has been isolated, it may be possible to characterize the active site. For example, angiotensin-converting enzyme (ACE) was characterized as a metalloenzyme containing zinc at the active site, and this led to the design of the inhibitor captopril for treatment of hypertension (Ondetti *et al.*, 1977). Or, the substrate specificity may be studied using dipeptides as probes (Llorens *et al.*, 1980) which may then be converted into potent inhibitors, as was effected in the design of the enkephalinase inhibitor, thiorphan. In the future, as the structures of enzymes become known it should prove possible to design inhibitors using knowledge of the molecular structure of the active site.

Many different substances circulate as chemical messengers and combine with their own receptors, e.g. amino acids, biogenic amines, peptides, prostaglandins, purines, steroids; they are remarkably specific since the messengers are not normally recognized by other receptors, i.e. the receptors discriminate between different messengers. In the case of biogenic amines (Fig. 1.1) it has been possible to differentiate between sites for the same messenger, suggesting different subpopulations of receptors, to provide further scope for introducing selectivity of drug action. If too many messages are getting through, then we can design an antagonist to block the receptor, using the chemistry of the natural messenger as a lead (Black, 1976).

There are many potential sites for drug intervention to affect biogenic amine action especially in their role as neurotransmitters (Fig. 1.2). In addition to blocking postsynaptic receptors (R_1, R_2, etc.), there are presynaptic autoreceptors (R) which may modulate transmitter release or synthesis, storage sites and re-uptake mechanisms which modulate free transmitter concentration, and enzymes which can be blocked to alter the rate of transmitter turnover, e.g. inhibit biosynthesis to deplete the transmitter or inhibit metabolism to prolong its existence. Furthermore, the amine receptor may be coupled to a second messenger enzyme (protein kinase) or an ion channel, so that its influence may be modified by drug interference with the amplifying transducing mechanism.

Undoubtedly, some of the best examples of rational drug design have been based on using the chemical structure of the natural transmitter (i.e. the biogenic amine) as a template. Thus selective agonists for postsynaptic receptors have been made for adrenergic β-receptors, e.g. salbutamol (Hartley *et al.*, 1968), dobutamine (Tuttle & Mills, 1975), xamoterol (Barlow *et al.*, 1981), and 5-hydroxytryptamine, e.g. sumatriptan (Brittain *et al.*, 1987), by the modification and addition of appropriate substituents to the biogenic amine structure (Fig. 1.3).

Antagonists for biogenic amine receptors have been designed by retaining a partial structure of the amine (for receptor recognition) and then incorporating additional groups for increasing the binding affinity between drug and receptor, e.g. pronethalol (Black & Stephenson, 1962) for adrenergic β-receptors, burimamide

AMINE	STRUCTURE	RECEPTORS
Acetylcholine	$CH_3CO.O.CH_2CH_2\overset{+}{N}Me_3$	Nicotinic (muscle, neuronal) M_1, M_2, M_3, M_4
Dopamine	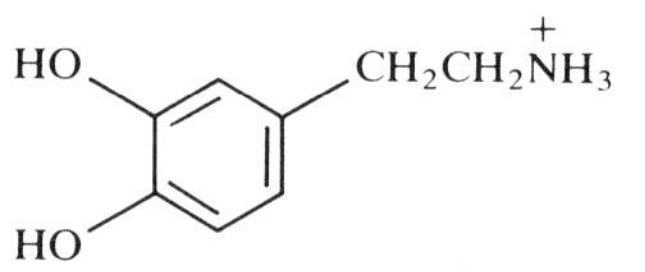	D_1, D_2, D_3, D_5
Noradrenaline R=H Adrenaline R=CH_3	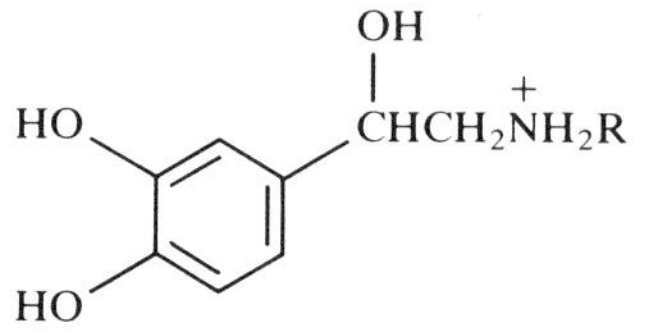	α_1, α_2, β_1, β_2, β_3
Serotonin (5-hydroxytryptamine)	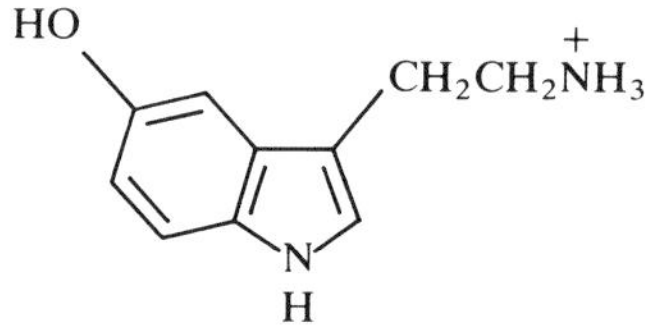	5-HT_{1A-1D} 5-HT_2, 5-HT_3, 5-HT_4
Histamine	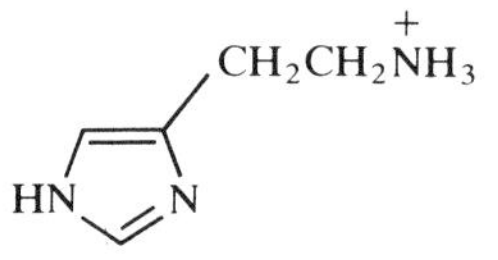	H_1, H_2, H_3
GABA (γ-aminobutyric acid)	$^{-}O_2CCH_2CH_2CH_2\overset{+}{N}H_3$	$GABA_A$, $GABA_B$

Figure 1.1. Chemical structures of the biogenic amines (shown as the most prevalent ionic species at pH 7.4) and receptor subtypes (see Watson & Abbott, 1991, for additional information on receptor subtypes).

(Black *et al.*, 1972) for histamine H_2-receptors, granisetron for 5-hydroxytryptamine $5HT_3$-receptors (Bermudez *et al.*, 1990) (Fig. 1.3).

Thus both enzymes and receptors are highly specialized sites involved in molecular recognition, whereby only the appropriate substances interact with them in a productive manner. This leads to a very high degree of selectivity of action and hence they provide very good chemical starting points for drug design. Since the key to a successful drug lies in its selectivity of action, the rational approach has been especially effective in modern drug discovery when based on enzymes or hormone receptors.

The past three decades have seen some notable drug-design successes (Table 1.1) based on the approach of using as leads enzymes and substrates or receptors for biogenic amines.

With the development of biochemistry and cell biology has come a greater understanding of cellular mechanisms and control and many of the drugs discovered serendipitously or by screening procedures of the past three decades have subsequently been shown to act by interfering with particular mechanisms; some examples are shown in Table 1.2. Such examples have provided a further stimulus to develop agents rationally, from mechanistic considerations. Thus there is a

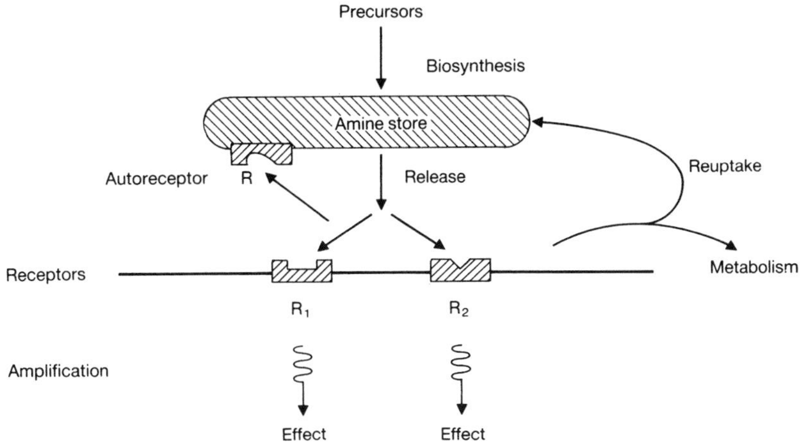

Figure 1.2. Biogenic amine transmission can be modified by drug intervention at various stages and sites using enzyme inhibitors, or selective receptor blockers or stimulants. Drugs may also affect release or uptake.

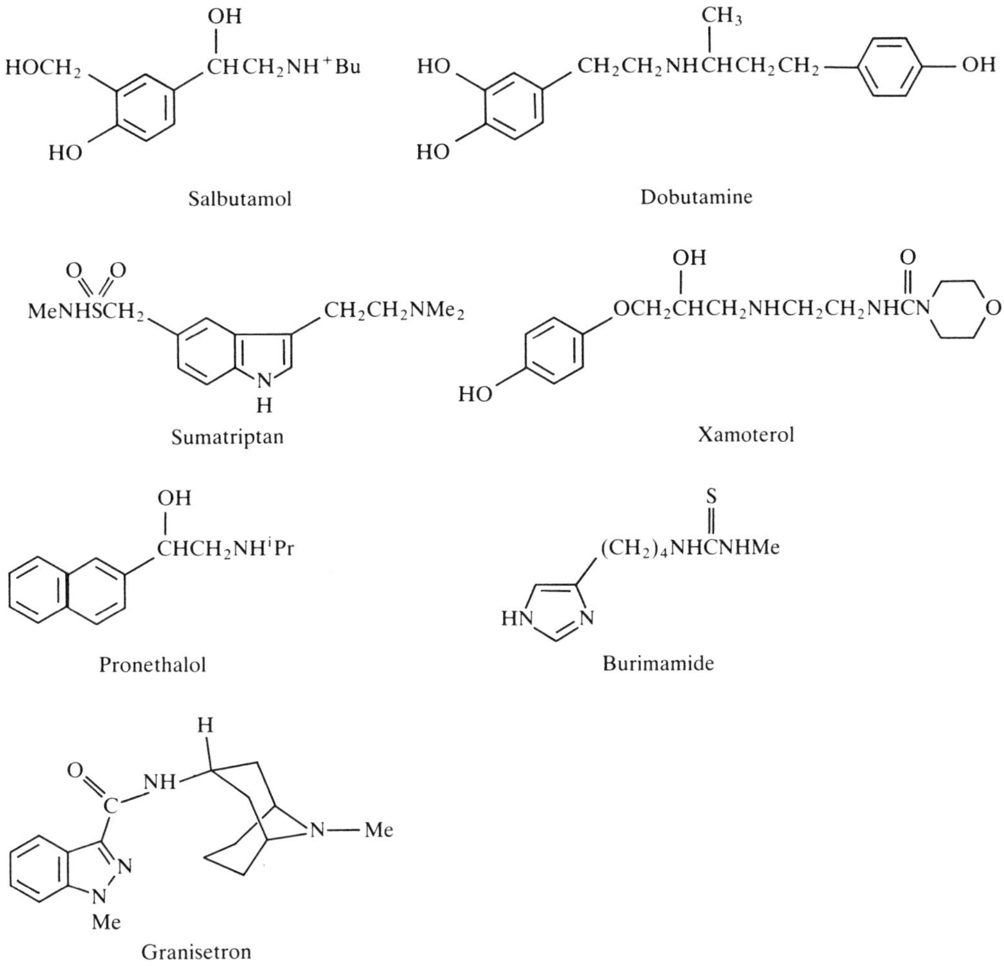

Figure 1.3. Chemical structures of some agonists and antagonists at biogenic amine receptors where the design was based on the amine structure as a template.

Table 1.1. Some examples of drugs designed to act on enzymes or receptors

Drug	Use	Mechanism of action
Enzyme inhibitors		
Pargyline	Antihypertensive	Inhibits monoamine oxidase-B
Captopril	Antihypertensive	Inhibits angiotensin-converting enzyme (ACE)
Carbidopa	Potentiates use of L-DOPA in Parkinson's disease	Inhibits DOPA-decarboxylase peripherally
Clavulanic acid	Potentiates antibiotic action of penicillins	Inhibits β-lactamase
Methotrexate	Antitumour	Inhibits dihydrofolate reductase
Cytarabine	Antileukaemia	Inhibits DNA polymerase
Lovastatin	Lowers cholesterol level	Inhibits HMG-CoA reductase
Omeprazole	Antiulcer	Inhibits H^+/K^+-ATPase ('proton pump')
Receptor stimulants or blockers		
Atracurium	Neuromuscular blockade	Blocks acetylcholine nicotinic receptors
Pirenzepine	Antiulcer	Selectively blocks acetylcholine muscarinic M_1-receptors
Fenoldopam	Congestive heart failure	Stimulates dopamine D_1-receptors
Butaclamol	Antipsychotic	Blocks dopamine D_2-receptors
Prazosine	Antihypertensive	Blocks adrenergic α_1-receptors
Labetalol	Antihypertensive	Blocks adrenergic α- and β-receptors
Propranolol	Antihypertensive	Blocks adrenergic β-receptors
Xamoterol	Heart failure	Stimulates adrenergic β_1-receptors
Salbutamol	Antiasthma	Stimulates adrenergic β_2-receptors
Ondansetron	Antiemetic	Blocks serotonin $5HT_3$-receptors
Sumatriptan	Antimigraine	Stimulates serotonin '$5HT_1$-like' receptors
Mepyramine	Antiallergy	Blocks histamine H_1-receptors
Cimetidine	Antiulcer	Blocks histamine H_2-receptors

strong interplay between drug discovery and retrospective rationalization. A semi-empirical discovery of a useful drug provides the stimulus for deeper probing into how and why it works and thus to a deeper understanding of what underlies the disease. This in turn may give rise to new concepts and new discoveries.

6 The 'rational approach' to drug design is still very speculative

There has been a phenomenal increase in recent years in the amount of information about mechanisms and therefore, as we approach the 21st century, the outlook for

Table 1.2. Some examples of old drugs and new mechanisms

Drug	Use	Mechanism of action
Aspirin	Anti-inflammatory	Inhibits the enzyme prostaglandin synthetase
Theophylline	Antiasthma (bronchodilator)	Inhibits the enzyme phosphodiesterase
Atropine	Mydriatic (dilates pupil of the eye)	Blocks the muscarinic receptors for acetylcholine
Haloperidol	Neuroleptic	Blocks the receptors for dopamine in the brain
Chlordiazepoxide	Antianxiety	Binds allosterically to the receptors for GABA (γ-aminobutyric acid) and opens the channels for the transport of Cl^- in brain neurones
Tolbutamide	Hypoglycaemic (antidiabetic)	Blocks the transport of K^+ through its channels in the insulin-secreting β-cell of the pancreas
Nifedipine	Coronary vasodilator used to treat angina	Blocks the channels for transport of Ca^{2+} in vascular smooth muscle
Aminacrine	Antibacterial	Intercalates into the DNA of bacterial nucleus to inhibit growth
Cromoglycate	Antiasthma	Prevents histamine release from its storage sites in mast cells in the lung
Guanethidine	Antihypertensive	Prevents release of noradrenaline from stores at nerve junctions

rational drug design ought to be very optimistic. Unfortunately, however, there has been a key factor missing from our understanding.

The body is under multifactorial control. There are many natural checks and balances. For any given function, there are usually several messengers and several types of receptor; there are also amplification systems, modulating systems, feedback inhibitory mechanisms, various ion fluxes, and so on; if we block one pathway by drug action, another pathway is likely to take over. The consequence is that we cannot be sure at the outset that designing a drug to act on a particular receptor or enzyme will necessarily provide treatment for a given medical condition, even though we may know that it is involved in the physiological controlling mechanisms. Furthermore, enzymes and receptors are ubiquitous and occur in many different tissue systems. Blocking them at a tissue site involved in a disease may be therapeutically effective but blockade concurrently in other tissue sites (not involved in the disease) may be thoroughly undesirable. Thus there is still a strong element of speculation in drug design, and a considerable uncertainty of success.

Another complicating feature for drug design arises because a drug has to be administered and find its way to the desired site of action, whereas the natural messenger may be generated locally or stored nearby to its required site of action. Furthermore, after it has acted, the natural transmitter is disposed of naturally by specific enzymes or re-uptake mechanisms, whereas the drug has to be disposed of by being excreted. We have to balance the desired pharmacology with the biochemical needs to achieve drug access and elimination. In altering the chemistry of a drug to achieve adequate disposition one may inadvertently introduce other pharmacological properties, thereby reducing the selectivity of action.

Finally, the biggest hurdle to new drug discovery and development is still the prediction of drug safety, i.e. to determine that a potential new drug will not cause unacceptable toxicity or side-effects unconnected with its main pharmacological action.

For the above reasons there is no doubt that new drug discovery in the 21st century will continue to be an exciting challenge in which the role of the medicinal chemist will be less empirical and even more intellectually demanding than it is today.

7 References

Barlow, J.J., Main, B.G. & Snow, H.M. (1981). *J. Med. Chem.*, **24**, 315.

Bermudez, J., Fake, C.S., Joiner, C.F., Joiner, K.A., King, F.D., Miner, W.D. & Sanger, G.J. (1990). *J. Med. Chem.*, **33**, 1924.

Black, J.W. (1976). *Pharmaceut. J.*, **217**, 303.

Black, J.W. and Stephenson, J.S. (1962). *Lancet*, **2**, 311.

Black, J.W., Duncan, W.A.M., Durant, G.J., Ganellin, C.R. & Parsons, M.E. (1972). *Nature (London)*, **236**, 385.

Brittain, R.T., Butina D., Coates, I.H., Feniuk, W., Humphrey, P.P.A., Jack, D., Oxford, A.W. & Perren, M.J. (1987). *Br. J. Pharmacol.*, **92** (Suppl.), 618P.

Burger, A. (1954). *Chem. Eng. News*, **32**, 4172.

Chang, R.S.L., Lotti, V.J., Monaghan, R.L., Birnbaum, J., Stapley E.O., Goetz, M.A., Albers-Schönberg, G., Patchett, A.A., Liesch, J.M., Hensens, O.D. & Springer, J.P. (1985). *Science*, **230**, 177.

Hall, S. & Strichartz, G. (Eds.) (1990). *ACS Symposium Series 418. Marine Toxins: Origin, Structure and Molecular Pharmacology*, American Chemical Society, Washington, DC.

Hartley, D., Jack, D., Lunts, L. & Ritchie, A.C.H. (1968). *Nature (London)*, **219**, 861.

Hitchings, G.H. (1980). *Trends Pharmacol. Sci.*, **1**, 167.

Laborit, H. (1954). *Presse Med.*, **62**, 359.

Llorens, C., Gacel, G., Swerts, J.P., Perdrisot, R., Fournie-Zaluski, M.C., Schwartz, J.C. & Roques, B.P. (1980). *Biochem. Biophys. Res. Commun.*, **96**, 1710.

Ondetti, M.A., Rubin, B. & Cushman, D.W. (1977). *Science*, **196**, 441.

Rose, F.L. (1964). *Chem. Ind.*, 858.

Schwartz, W.B. (1949). *New Engl. J. Med.*, **240**, 173.

Sneader, W. (1985). *Drug Discovery: The Evolution of Modern Medicines*, John Wiley, Chichester.

Tuttle, R.R. & Mills, J. (1975). *Circ. Res.*, **36**, 185.

Watson, S. & Abbott, A. (1991). *TIPS Receptor Nomenclature (Supplement)*, Elsevier, Amsterdam.

2 Reflections and Perspectives

P.A.J. JANSSEN and J.P. TOLLENAERE
Janssen Research Foundation, B-2340 Beerse, Belgium

1 Historical introduction

Ever since recorded history, humans have sought to correct their conditions of living. In early times, people already knew and used psychoactive plant substances to alter their moods or to alleviate pain. In the *Pursuit of Intoxication* (Malcolm, 1971), Malcolm states that 'in almost every part of the world man . . . found plants that contained narcotic and hallucinogenic principles'. The use of drugs to alter consciousness is so ubiquitous that it must represent a basic human appetite (Kramer & Merlin, 1983). It is known, for example, that hashish was used in the 8th century BC by the Assyrians during religious ceremonies. Well-known are the ritual practices of praecolumbian Central and South American peoples chewing coca leaves to experience pleasure and rare sensations. Opium, it is said, was given by the goddess Ceres so that humankind could be relieved from pain and sorrow.

Opium, however, is medicinally speaking, a concoction of several alkaloids of which morphine is the main constituent. In the early 19th century, the German pharmacist Sertürner succeeded in isolating the active ingredient which he called morphium after the Greek god of sleep. Subsequently, other constituents of opium were isolated such as codeine, noscapine, thebaine and papaverine.

Also in the 19th century, the active constituent of the willow bark, salicin, was identified. This glycoside upon hydrolysis leads to glucose and saligenin which can be oxidized to yield salicylic acid. Subsequent chemical work over many decades eventually led to the synthesis of acetylsalicylic acid which was introduced into medical practice in 1899 by the Farbenfabriken Bayer.

Extracts of barks of trees were used by many ancient societies to relieve ague and fevers. Originally imported from South America where powder of the cinchona bark was used to treat fever, the active ingredient, quinine, was isolated in the early 19th century.

All this spurred on innumerable efforts to isolate and identify new chemicals, some of which are still in medicinal use today. For vivid and sometimes witty accounts of what can be considered as the early beginnings of pharmaceutical research, the reader is referred to *Discoveries in Pharmacology* (Parnham & Bruinvels, 1983).

Interestingly enough, as the 19th century, and particularly the last quarter of it, saw a frantic activity in not-yet organized organo-pharmaceutical research, another observation proved, by hindsight, to be of major significance. In the same year as aspirin was introduced into medical practice, Meyer (1899) and Overton (1901) observed a strong correlation between the anaesthetic activity of many simple organic compounds and their oil–water partition coefficients. Although the original enthusiasm of medicinal chemists quickly subsided, the first structure–activity relationship was nevertheless established. It remained to Collander (1954) and others to generate new interest in the use of partition coefficients to correlate biological data. Following the discovery (Fujita *et al.,* 1964) that partition coefficients were additive–constitutive in nature, it was Corwin Hansch and his school who pioneered the successful quantitative structure–activity relationships (SAR) approach in medicinal chemistry in the 1970s.

At the end of the 19th century and in the early 20th century, while chemists were searching for morphine substitutes, reports were published on the merits of heroin as a safe substitute for morphine! Within a couple of years, however, the so-called cure for morphine dependence could no longer by purchased as an over-the-counter medication! The elucidation of the structure of morphine in 1925 stimulated organic chemists into synthesizing new derivatives. The search for synthetic alternatives to morphine, however, was not very successful at all until the synthesis of meperidine or pethidine (4-phenylpiperidine derivative) was reported (Eisleb & Schaumann, 1939). Meperidine, which was originally designed to be a spasmolytic, turned out to be a useful narcotic analgesic. The initial hope that this compound was non-addictive proved to be false.

During World War II, in Germany where supply of morphine was short, a derivative of diphenylpropylamine was synthesized which is now known as methadone.

Both meperidine and methadone, being the archetypes of chemically simple structures, proved to be rich sources for the further design and exploration of novel analgesics and other related compounds sometimes devoid of analgesic activity or with unexpected biological properties. Particularly telling in this respect was the systematic search for more potent meperidine derivatives, which eventually led to the discovery of the butyrophenone-type neuroleptics (Janssen & Tollenaere, 1983).

Another example where an archetypal chemical structure stimulated the subsequent discovery of many thousands of derivatives is chlorpromazine. This compound itself would have remained on the shelves of Rhône-Poulenc had a french surgeon Dr H. Laborit not asked for a compound with a stronger central effect than that of promethazine (Deniker, 1983).

Summarizing, when reading the older literature one cannot escape the following conclusions that (i) the use of drugs in the broadest sense is as old as human recorded civilization, (ii) the search for the active principle in folk medicines led to the isolation of many useful compounds which subsequently were subjected to chemical analysis and modification to find cheaper and better substitutes and (iii) modern drug research has its roots in chemistry.

2 Present-day drug research

Drug research, as we know it today, gradually emerged in the late 1940s and in the 1950s as an endeavour in its own right. It is recognized that current achievements have come about not as a result of a single branch of the sciences such as chemistry but by a wide range of scientific disciplines. Starting in the 1950s, various laboratories—chiefly industrial ones—were engaged in systematic synthesis and biological screening of many tens of thousands of compounds. Fuelled by increasingly more sophisticated organic chemical methods and analytical chemical equipment, this pace is unabated today. As more knowledge came about from synthetic and analytical efforts, structure–activity relationships based mainly on the topological or two-dimensional (2-D) properties of compounds were put forward. One of the early attempts to systematize the current body of knowledge of narcotic analgesics was the pharmacophore model that used the morphine skeleton as a basis (Beckett & Casy, 1954). By the end of the 1960s, reviews on the anatomy of analgesics, including all known structural families, began to appear (Janssen & Van der Eycken, 1968; Casy, 1973, 1978, 1989; Janssen & Tollenaere, 1979). These reviews, which span a 20-year period, reveal strikingly and significantly that, at the end of the 1980s, medicinal chemists were still examining derivatives of the basic structures known at the end of the 1960s, i.e. morphine, thebaines, morphinans, benzomorphans, diphenylpropylamines and the 4-arylpiperidine derivatives. To be historically correct, however, it should be noted that the latest review (Casy, 1989) contains a large section discussing the developments of opioid peptides that were discovered in 1975. Apart from opening up a novel class of compounds, the isolation of met- and leu-enkephalin gave a major boost to the idea of receptor multiplicity.

In the 1960s, Janssen and his collaborators systematically investigated the comparative pharmacology of the better-known neuroleptics of that time (Janssen *et al.*, 1965a, 1965b, 1966, 1967). In the 1970s, it was stated that all potent neuroleptics are tertiary or exceptionally secondary aliphatic amines containing at least one aromatic ring (Ar) linked to an amine fragment (N) by an intermediate chain obeying the general structure Ar–α,β-C–C–N where α and β are a variety of functional groups (Janssen, 1973; Janssen & Van Bever, 1975).

1.2 *X-ray crystallography*

If one recalls that the X-ray crystallographic 3-D structures of, for example, reserpine (Karle & Karle, 1968) and chlorpromazine (McDowell, 1969) only became known in the late 1960s and that of haloperidol in the 1970s (Reed & Schaefer, 1973) then it is not surprising that the systematic studies in the earliest days were based solely on a topological (2-D) classification of compounds. Of even more significance was the fact that in the 1960s and early 1970s there were almost no links, or very few organized collaborative efforts, between X-ray crystallographers on the one hand and pharmacologists and medicinal chemists on the other hand. During the 1970s, this unhappy situation was remedied. In 1983, a course on X-ray crystallography and drug action was arranged under the auspices of the International School of Crystallography at the E. Majorana Centre in Erice, Sicily. Perhaps for the first time, people working in these fields were brought together in order to try to gain a better understanding of each other's achievements, aims and problems, and also with the hope of increasing collaboration amongst the various disciplines (Horn & De Ranter, 1984). A few years earlier, the *Atlas of the Three-Dimensional Structure of Drugs* (Tollenaere *et al.*, 1979) which was intended to bridge the gap between X-ray crystallographers and pharmacologists and medicinal chemists was published.

Now, X-ray crystallography is considered to be an integral part and to play an important role in our understanding of how drugs in a given pharmacological class are related and how they might act. X-ray crystallography continues to provide valuable information about drugs and neurotransmitters and slowly but surely enlarges our knowledge about the shapes of proteins and the details of drug–receptor interactions.

1.3 *Theoretical methods*

In a pioneering series of papers, Kier introduced the use of quantum chemical methods into the field of medicinal chemistry. A few years after the appearance of semi-empirical quantum chemical methods such as Extended Hückel Theory (Hoffmann, 1963), Kier calculated the conformational profiles of acetylcholine, muscarone and muscarine (Kier, 1967, 1971). This work was followed and partly repeated by Pullman and collaborators who used the far superior PCILO (Perturbative Configuration Interaction using Localized Orbitals) molecular orbital (MO) method (Pullman, 1974). Thus in the 1970s, 'quantum pharmacology' (whatever this term may mean) or the idea that the methods of quantum chemistry could be useful to probe biological events at the molecular level became firmly established (Richards, 1977). Summarizing, one may safely state that conformational analysis using X-ray crystallographic techniques for the solid-state and theoretical methods for the gaseous or isolated state, and to a lesser extent nuclear magnetic resonance (NMR) analysis, considerably enlarged our knowledge of what drug molecules look like. It should be borne in mind that the determination of the conformation—however fascinating it may be—is not an end in itself but rather the beginning of, and the basis for, answering a more important question. That question is, at least for the medicinal chemist, how does conformation relate to biological activity (Humber *et al.*, 1979; Tollenaere *et al.*, 1980, 1986; Olson *et al.*, 1981; Rognan *et al.*, 1990).

1.4 *Computers*

The last decade saw the gradual migration of the computer as a central facility used for administrative and scientific chores alike, to the departmental computer dedicated solely to scientific calculation. With the revolution brought about by the concept of the personal computer, the existence of the workstation in the laboratory with processing powers superior to the average main frame of a decade ago is a fact. Thus, as computers became cheaper and more powerful, computational chemistry entered the (medicinal) chemical laboratory. Since high-performance workstations have come to the forefront, computational chemistry has become closely associated with molecular graphics. Both are at the heart of what is called computer-assisted molecular modelling (CAMM) which is the latest but still far-from-mature tool for visualizing, manipulating, building and storing molecular models. Because of the intimate connection with computational chemistry, molecular properties possibly relevant to a hypothetical SAR can be, in many cases, computed interactively. Instead of presenting the results of a calculation in terms of numbers, modern display systems offer the opportunity to present, and to communicate, the results in a form with which the organic chemist is familiar.

Generations of chemists have been familiar, if not satisfied, with a static picture of molecules. With the advent of molecular dynamics (MD) this situation has changed quite dramatically. MD calculations, which essentially solve Newton's equations of motion, provide a dynamic picture of a molecular ensemble over a given period of time. In doing so, MD calculation may provide estimates of solvation and entropic effects which are of fundamental importance for the adequate description of the energetics of interaction between, for example, a drug and its receptor. Thus, instead of ignoring entropy altogether, or at best assuming entropic effects to be constant as is done so often, MD calculations offer the exciting possibility of simulating the physical behaviour of molecules and proteins and their mutual interactions (McCammon & Harvey, 1987; Van Gunsteren, 1989; Beveridge & DiCapua, 1989) on a theoretically sounder basis than was hitherto possible.

1.5 *Polypeptides and proteins*

There are now approximately 60 000 protein sequences available as a result of significant developments in DNA cloning technology. In sharp contrast to this figure, are the ca. 400 X-ray structures of proteins in the Brookhaven Protein Data Bank. Despite many years of both experimental and theoretical work (Lambert & Scheraga, 1989a,b,c), progress in predicting the 3-D characteristics of proteins is rather slow and all together far from satisfactory. The challenge posed by the protein-folding problem and the difficulties involved in handling the multiple minimum energy conformations of proteins is far from trivial. While empirical force field calculations (Burkert & Allinger, 1982) are satisfactory and feasible for determining minimum energy conformations of small molecules and peptides, the brute force approach in predicting the folding of, for example, 40-residue protein is impractical with the current generation of computers. At a sampling rate of one minimum energy conformation per picosecond, the approximately $4^{40} \approx 10^{24}$ conformations would take 38 000 years. This figure not only puts into perspective the formidable problems involved in the attempts to predict the shape of proteins, it also tells something about the way proteins experimentally fold from the denatured state. Because the rate of the folding process is of the order of seconds or minutes (Kawajima *et al.*, 1985), the way proteins fold from the unfolded state to their native conformation is anything but random (Skolnick & Kolinski, 1989). Thus, the slow rate with which X-ray structures of proteins and enzymes become available, the rather poor performance in predicting the tertiary structure of proteins from their amino acid sequence, and the not-yet understood problem of globular protein folding are all major stumbling blocks in the progress towards our understanding of how proteins function and how they interact with drug molecules.

1.6 *Receptors and their multiplicity*

The mechanism of action of hormones and many drugs can be understood in terms of the mediation through the response of cell receptors. Over the years, classical pharmacological evidence has accumulated that strongly supports the idea of receptor multiplicity. Well known in this respect are the adrenergic receptors, which can be divided into the α- and β-subtypes (Ahlquist, 1948). Receptor multiplicity seems to be also well established in the case of the opioid receptor (Casy, 1989), the dopaminergic receptor (Kebabian & Calne, 1979), the histaminergic receptor (Black *et al.*, 1972) and the muscarinic receptor (Bonner *et al.*, 1988). Perhaps most conspicuous of all is the bewildering array of 5-HT receptor subtypes (Hartig, 1989). Without sides being taken in the lively debate on the classification or functional role of all these receptor subtypes, it may be worthwhile to make the analogy with the situation that prevailed at a certain time in nuclear physics. Until physicists could settle their disputes in unification theories, the discovery of so-called 'fundamental particles' designated with Greek letters was reported almost weekly up to the exhaustion of the Greek alphabet!

Whatever the outcome of the debate, there seems to be little doubt that the muscarinic, dopaminergic, adrenergic and serotonergic receptors are characterized by the putative presence of seven-membrane-spanning domains (Schofield *et al.*, 1990). Of major importance is the observation that the transmembrane domains of the GTP- or G-protein-coupled receptors are the most highly conserved regions of these proteins. Although definite conclusions can, as yet, not be reached bearing on the exact location of the neurotransmitter binding site, site-directed mutagenesis techniques (Strader *et al.*, 1987), careful SARs of receptor-binding data of closely congeneric series of drugs and the detailed analysis of the homologies and identities (Moereels *et al.*, 1990) of both the transmembrane domains and the extracellular loops of these receptors may yield interesting working hypotheses.

What is conspicuously lacking and urgently needed is solid proof of the exact tertiary structure of these receptors from high-resolution X-ray structures. Growing crystals of a membrane protein, however, is a formidable task (Deisenhofer & Michel, 1989). If the recent elucidation of the tertiary structure of the photosynthetic reaction centre from the purple bacterium *Rhodopseudomonas viridis* (Deisenhofer and Michel, 1989) portends anything for the future then it may not be far fetched to assume that the X-ray structure of other interesting membrane-bound proteins will become available.

3 Summary and conclusions

From our brief and certainly biased excursion back through the history of the quest for drugs it is very clear that drug research in the 19th century and up to the early 1950s was conspicuously dominated by chemists. Starting in the 1950s, this situation gradually changed in favour of a multidisciplinary and collaborative endeavour combining a large diversity of scientific disciplines. From the chemist, being at the helm of things, and the pharmacologist, in the role of screening the compounds, we are now in the lucky situation where chemists, pharmacologists, physicists, theoreticians, biochemists, computer scientists and clinicians work in a concerted fashion towards the discovery and the development of new and better medicines.

Gone are the days when simple chemical tests assured the purity of the compounds. Nowadays, and nothing foresees this to be different in the decades ahead, complex analytical instruments such as NMR spectrometers and X-ray spectrometers are available to guarantee the purity and to establish the 3-D structure of drugs. Indeed, in many cases, analyses obtained from these and other expensive techniques are demanded by the regulatory public authorities.

Gone also are the days of simple pharmacological screening tests. Fairly complex *in vivo* and *in vitro* tests are now used not merely to determine the biological activity of a compound but to establish a biological profile based on a whole range of tests.

We have seen how computer technology has revolutionized laboratory practices. In particular, the introduction of high-speed workstations and associated software has made it possible to look at and to analyse drug molecules and proteins at the atomic level. If one can extrapolate from the past then the prospects are bright in the decades ahead for computers, many orders of magnitude faster than the supercomputers of today, allowing far more realistic simulations of the behaviour of complex biological systems. It is hoped that computer technology will be instrumental in tackling the elucidation of the mechanism of protein folding and the prediction of the shape of proteins based on the knowledge of their primary structure. In the same way as small synthetic molecules have had a tremendous impact on our living conditions, so will protein engineering surely lead to new varieties of proteins and enzymes that can supplant in appropriate cases those from plant or animal sources. The analogy with the quest for a synthetic substitute for 'natural' morphine is obvious!

Perhaps most important of all, it was seen that, as time went on, a variety of scientific disciplines have gradually brought themselves together in an attempt to deepen and expand our knowledge.

The aetiology and pathogenesis of many chronic diseases are still poorly understood. Despite the successes of modern multidisciplinary drug research, it is a sobering thought that almost half the annual death rate in the Western world is due to vascular disease. Other areas where little progress has been made include the treatment of rheumatic disorders, senile dementia, autoimmune diseases, cancer and AIDS. Diseases such as malaria, schistosomiasis, filariasis, trypanosomiasis and leishmaniasis, which afflict many hundreds of millions of people, are still poorly controlled (Vanden Bossche, 1978). Although in many cases, acceptable cures are available, many of these plagues are still rampaging through large parts of Africa, Asia and South America because of political, economic and social factors which are not very conducive for the utilization of the existing drugs.

The possibility that new interesting drugs will emerge from the screening of molecules in outdated biological test systems is becoming less and less likely. What is needed is the development of new pharmacological models and the identification of biological targets, maybe at the molecular level, for progress in drug-based treatment of, for example, chronic diseases and other diseases in humans and animals that are not yet curable. After years of debate over the pros and cons, the mapping and sequencing of the human genome is underway (Watson, 1990). When completed through international efforts, the sequence of roughly 3 billion base pairs will be known. When finally interpreted, the genetic message encoded in our DNA will, we hope, be helpful in understanding how genetic factors play a role in a wide range of diseases such as cancers, Alzheimer's disease, Huntington's chorea and complex diseases such as hypertension and some psychiatric disorders.

For humankind to succeed in further improving its living conditions and well-being on a global scale, an economic, political and social climate must be created in which imaginative and creative powers are nurtured and encouraged to prosper.

4 References

Ahlquist, R.P. (1948). *Am. J. Physiol.*, **153**, 586.

Beckett, A. & Casy, A.F. (1954). *J. Pharm. Pharmac.*, **10**, 868.

Beveridge, D.L. & DiCapua, F.M. (1989). *Ann. Rev. Biophys. Chem.*, **18**, 431.

Black, J.W., Duncan, W.A.M., Durrant, C.J., Ganellin, C.R. & Parsons, M.E. (1972). *Nature (London)*, **236**, 385.

Bonner, T.I., Young, A.C., Brann, M.R. & Buckley, N.J. (1988). *Neurone*, **1**, 403.

Burkert, U. & Allinger, N.L. (1982). In: *Molecular Mechanics. American Chemical Society, Monograph 177*, American Chemical Society, Washington, D.C.

Casy, A.F. (1973). *Mod. Pharmac.*, **1**, 217.

Casy, A.F. (1978). *Progr. Drug Res.*, **22**, 149.

Casy, A.F. (1989). *Adv. Drug Res.*, **18**, 177.

Collander, R. (1954). *Physiol. Plant.*, **7**, 420.

Deisenhofer, J. & Michel, H. (1989). *Angew. Chemie (English Edition)*, **28**, 829.

Deniker, P. (1983). In: M.J. Parnham and J. Bruinvels (Eds.), *Discoveries in Pharmacology*, Vol. 1, p. 163, Elsevier Science Publishers, Amsterdam.

Eisleb, O. & Schaumann, O. (1939). *Dt. Med. Wschr.*, **65**, 967.

Fujita, T., Iwasa, J. & Hansch, C. (1964). *J. Am. Chem. Soc.*, **86**, 5175.

Hartig, P.R. (1989). *Trends Pharmaceut. Sci.* **10**, 64.

Hoffmann, R. (1963). *J. Chem. Phys.*, **39**, 1397.

Horn, A.S. & De Ranter, C.J. (1984). In: A.S. Horn and C.J. De Ranter (Eds.), *X-Ray Crystallography and Drug Action*, Clarendon Press, Oxford.

Humber, L.G., Bruderlein, F.T., Philipp, A.H., Götz, M. & Voith, K. (1979). *J. Med. Chem.*, **22**, 761.

Janssen, P.A.J., (1973). In: C. Peters (Ed.), *International Encyclopedia of Pharmacology and Therapeutics*, Vol. 1, p.37, Pergamon Press, Oxford.

Janssen, P.A.J. & Van der Eycken, C.A.M. (1968). In: A. Burger (Ed.), *Drugs Affecting the Central Nervous System*, Vol. 2, p. 25, Marcel Dekker, New York.

Janssen, P.A.J. & Van Bever, W.F.M. (1975). In: W.B. Essman and L. Valzelli (Eds.), *Current Developments in Psychopharmacology*, Vol. 2, p. 167, Spectrum Publications, New York.

Janssen, P.A.J. & Tollenaere, J.P. (1979). In: H.H. Loh and D.H. Ross (Eds.), *Neurochemical Mechanism of Opiates and Endorphins, Adv. Biochem. Psychopharmacol.*, Vol. 20, p. 103, Raven Press, New York.

Janssen, P.A.J. & Tollenaere, J.P. (1983). In: M.J. Parnham and J. Bruinvels (Eds.), *Discoveries in Pharmacology*, Vol. 1, p. 181, Elsevier Science Publishers, Amsterdam.

Janssen, P.A.J., Niemegeers, C.J.E. & Schellekens, K.H.L. (1965a). *Arzneimittel-Forsch.*, **15**, 104.

Janssen, P.A.J., Niemegeers, C.J.E. & Schellekens, K.H.L. (1965b). *Arzneimittel-Forsch.*, **15**, 1196.

Janssen, P.A.J., Niemegeers, C.J.E. & Schellekens, K.H.L. (1966). *Arzneimittel-Forsch.*, **16**, 339.

Janssen, P.A.J., Niemegeers, C.J.E. & Schellekens, K.H.L. (1967). *Arzneimittel-Forsch.*, **17**, 841.

Karle, I.L. & Karle, J. (1968). *Acta Cryst.*, **B24**, 81.

Kawajima, K., Hiraoka, Y., Ikeguchi, M. & Sugai, S. (1985). *Biochemistry*, **24**, 874.

Kebabian, J.W. & Calne, D.B. (1979). *Nature (London)*, **277**, 93.

Kier, L.B. (1967). *Mol. Pharmacol.*, **3**, 487.
Kier, L.B. (1971). In: G. deStevens (Ed.), *Medicinal Chemistry. A Series of Monographs*, Vol. 10, p. 162, Academic Press, New York.
Kramer, J.C. & Merlin, M.D. (1983). In: M.J. Parnham and J. Brunivels (Eds.), *Discoveries in Pharmacology*, Vol. 1, p. 23, Elsevier Science Publishers, Amsterdam.
Lambert, M.H. & Scheraga, H.A. (1989a). *J. Comput. Chem.*, **10**, 770.
Lambert, M.H. & Scheraga, H.A. (1989b). *J. Comput. Chem.*, **10**, 798.
Lambert, M.H. & Scheraga, H.A. (1989c). *J. Comput. Chem.*, **10**, 817.
McCammon, J.A. & Harvey, S.C. (1987). *Dynamics of Proteins and Nucleic Acids*, Cambridge University Press, Cambridge.
McDowell, J.J.H. (1969). *Acta Cryst.*, **B25**, 2175.
Malcolm, A. (1971). *The Pursuit of Intoxication*, p. 42, Washington Square Press, New York, N.Y.
Meyer, H. (1899). *Arch. Exp. Pathol. Pharmakol.*, **42**, 110.
Moereels, H., De Bie, L. & Tollenaere, J.P. (1990). *J. Comp. Aided Mol. Design*, **4**, 131.
Olson, G.L., Cheung, H.-C., Morgan, K.D., Blount, J.F., Todaro, L., Berger, L., Davidson, A.B. & Boff, E. (1981). *J. Med. Chem.*, **24**, 1026.
Overton, E. (1901). *Studien über die Narkose*, Fischer, Jena.
Parnham, M.J. & Bruinvels, J. (Eds.) (1983). *Discoveries in Pharmacology*, Vol. 1, Elsevier Science Publishers, Amsterdam.
Pullman, B. (1974). In: E.D. Bergmann and B. Pullman (Eds.), *Molecular and Quantum Pharmacology,* Vol. 7, p. 9 and references cited therein, Reidel Publishing, Dordrecht.
Reed, L.L. & Schaefer, J.P. (1973). *Acta Cryst.*, **B29**, 1886.
Richards, W.G. (1977). *Quantum Pharmacology*, Butterworth, London.
Rognan, D., Sokoloff, P., Mann, A., Martres, M.-P., Schwartz, J.-C., Costentin, J. & Wermuth, C.-G. (1990). *Eur. J. Pharmacol.*, **189**, 59.
Schofield, P.R., Shivers, B.D. & Seeburg, P.H. (1990). *Trends Neurosci.*, **13**, 8.
Skolnick, J. & Kolinski, A. (1989). *Ann. Rev. Phys. Chem.*, **40**, 207.
Strader, C.D., Sigal, I.S., Register, R.B., Candelore, M.R., Rands, E. & Dixon, R.A.F. (1987). *Proc. Natl. Acad. Sci. USA*, **84**, 4384.
Tollenaere, J.P., Moereels, H. & Raymaekers, L.A. (1979). In: *Atlas of the Three-Dimensional Structure of Drugs.* Elsevier/North-Holland Biomedical Press, Amsterdam.
Tollenaere, J.P., Moereels, H. & Raymaekers, L.A. (1980). In: E.J. Ariëns (Ed.), *Drug Design*, Vol. X, p. 71, Academic Press, New York.
Tollenaere, J.P., Moereels, H. & Van Loon, M. (1986). *Prog. Drug Res.*, **30**, 92.
Vanden Bossche, H. (1978). *Nature (London)*, **273**, 626.
Van Gunsteren, W.F. (1989). *Mol. Simul.*, **3**, 187.
Watson, J.D. (1990). *Science*, **248**, 44.

Strategies for the Creation of New Drugs

3 Trends in New Lead Identification: A Structural Approach

J.H. MUSSER
Glycomed Inc., 860 Atlantic Avenue, Alameda, CA 94501, USA

The problem will not be our ability to do things. The terrible problem is, what will we choose to do next? [Sir James Black, interview by T. Bass, 1990]

1 Introduction

In medicinal chemistry, the two basic approaches to new lead identification are screening and rational design, and both of these approaches are undergoing fundamental changes because of advances in biotechnology and instrumentation. The biotechnical revolution is having a major effect not only on how novel leads are generated but is influencing every aspect of new drug discovery and development. Many aspects of medicinal chemistry are impacted by the development of molecular biological techniques for the detection and sequence analysis of genes that encode enzymes or receptors, and for the cloning, site-specific mutagenesis and expression of their cDNAs in cells. For the purposes of this chapter, however, no attempt will be made to review recombinant methodologies, nor will there be a discussion on macromolecular therapeutics produced by cDNA technology. Nevertheless, medicinal chemists will incorporate DNA recombinant technology into their armamentarium of tools for new lead generation (Venuti, 1990).

The revolution in instrumentation includes structure analysis, computational methodologies and robotics. Although advances in instruments used in structural analysis have been incremental, the recent net result on structural determination of high molecular weight biomolecules and biosystems has reached a critical mass. With automated multiware area detectors and synchrotron sources, X-ray crystallographic data for proteins that once required months or even years to obtain can now be collected in a matter of weeks. Also, the new techniques of fast atom bombardment (FAB) (Barber, 1981) and electrospray are making high-quality mass spectra of polar biopolymers a routine task. Finally, multidimensional nuclear magnetic resonance (NMR) techniques allow for the first time the determination of solution conformations of biopolymer and ligand–biopolymer interactions. Computational chemistry is a new area of multidisciplinary research and involves *ab initio* molecular orbital (MO) calculations and semi-empirical methods, molecular mechanics and molecular dynamics in the design of new drug candidates. Thus, recombinant DNA technology in combination with advances in instrumentation will come together in a synergistic fashion to play a significant role in new lead identification in the 21st century.

Ironically, advances in biotechnology and instrumentation are affecting a divergence of screening and rational design with respect to the total numbers of compounds examined. Rational design will produce far fewer compounds than are currently synthesized, whereas trends in screening will require even greater numbers of compounds to be synthesized or isolated.

Screening as an approach to the identification of new leads is evolving. In the classical sense, screening involves the biological testing of randomly identified chemicals. The chemicals are generated either by isolation from natural sources or by traditional methods of synthetic organic chemistry. Often, synthetic compounds are made because their structures are unusual or the chemistry associated with their preparation is scientifically interesting. Also, unexpected chemistry which results in the generation of novel structures can provide additional test compounds. This process of lead generation is based on trial and error with its attendant critical dependence on serendipity. Thus, random screening can be costly, unreliable and slow. Nevertheless, the use of natural products as non-peptide ligands has renewed interest and is discussed in Section 2 below. Contemporary methods of screening involve selection criteria, such as a pharmacophoric relationship, a chemical class, or physical characteristics (molecular weight, log P, water solubility). For the generation of new leads, selective screening is certainly an improvement over random screening. Currently, the biological testing of compounds is being facilitated with the use of robotics; however, lagging behind in development are automated techniques for large-scale synthesis of structurally distinct compounds. In the future, the trend in screening will thus focus on very high volume both in number of assays performed and chemicals synthesized. This trend is covered in Section 6 of this chapter.

In contrast to screening, the rational design of new leads is based on an extensive knowledge of a target biopolymer involved in the control of a particular disease process and its interactions with a small molecular weight ligand. In turn, the detailed knowledge of biopolymer–ligand interaction is provided by a number of biotechnical, analytical and computational methodologies. Recombinant DNA technology, as in screening, provides the target biopolymer; however, in contrast to screening, the biopolymer is used for informational content not just an indicator of activity. The informational content is extracted from the biopolymer and biopolymer–ligand complex by X-ray crystallography, high-field multidimensional NMR spectroscopy and optical rotatory dispersion/circular dichroism (ORD/CD) studies. Ligands used in initial analysis can be the natural substrates or agonists. Once a suitable database or 'picture' is available, computer-aided molecular modelling can be used to design the new lead.

As a conceptional note, it is important in the design of new drugs to distinguish between new lead identification and lead optimization. Lead optimization is an iterative process and relies on empirical observation. There are many lead optimization methods, such as quantitative structure–activity relationship (QSAR) of hydrophobic, electronic and steric effects, Hansch analysis, Topliss decision tree, Free–Wilson model, topological, geometrical and substructural approaches, heuristic methods, receptor mapping and response profiles (Rainer, 1984). Because lead optimization is a mature science, we will not focus our attention in this area of drug discovery but concentrate on lead identification which in the 21st century will progress from an art to a science.

Since knowledge of the detailed structure of how biopolymers interact with small molecular weight ligands will play a critical role in new lead discovery in the 21st century, this chapter is organized by biopolymer class and includes peptides, oligonucleotides, lipids and carbohydrates. No attempt will be made to be inclusive of all the trends in new lead discovery in each class. Each section presents one or two new trends and, in some cases, examples of how specific ligands may generate new leads in drug discovery are given. Also, due to the length requirements for this chapter, no attempt is made to give the more negative aspects of each structural approach. It is reasoned that by the year 2001, many of the technical or resource issues which inhibit current implementations of the following trends will be obviated. Further, new approaches to lead generation not yet on the horizon will relegate some to obscurity. Thus, some of the following will fall by the side not because of inherent problems but will disappear because of the strength of science and technology behind yet unanticipated approaches.

2 The decade of peptides

In the 21st century, the versatile power of the mammalian immune system via the use of antibodies may be employed to gain information on cellular receptors, enzymes or viruses. The introduction of hybridoma technology (Seiler, 1985) now makes it possible to generate homogeneous, high-affinity antibodies or 'monoclonal antibodies' that are able to recognize peptide ligands ranging in size from proteins up to 34 Å long to smaller peptide fragments as small as 6 Å with association constants in the range of 10^4 to 10^{14} M^{-1} (Goodman, 1985). The theoretical pool of potential antibodies is enormous. Combinatorial joining of the gene segments encoding the variable light and heavy antibody genes and the combinatorial association of different polypeptide light and heavy chains are able to generate a minimum of 10^8 different antibody molecules (mutations greatly expand this number) (Tonegawa, 1983).

Since antibodies are large complex proteins themselves, they are sometimes recognized by the immune system as antigens and elicit the production of additional antibodies. Thus, each antibody or idiotype triggers the secretion of a complementary anti-idiotypic antibody which can combine with, and inactivate, the originating antibody.

In theory, the idiotype creates a 'negative image' of a certain part on the antigen surface or epitope. The structural basis for antibody binding is known through X-ray crystallography, electron spin resonance (ESR), NMR and affinity labelling experiments, and the results support the negative image theory. A number of antibody fragment–ligand complexes, such as antibody fragments with lysozyme (Amat, 1986) and viral protein (Colman, 1987), were solved by high-resolution X-ray diffraction studies and are examples of complementary fit. Thus, if the anti-idiotype faithfully generates a positive image which mimics the steric, electronic and hydrophobic characteristics of the epitope, this information can be used in direct drug screening or in drug design (Fig. 3.1). With the anti-idiotype 'image' peptide sequence in hand, all that remains to obtain an orally active agent with a reasonable biological half-life is the development of a peptidomimetic based on the 3-D structure of the epitope mimic.

Today, however, the conversion of a peptide sequence into peptidomimetic remains a significant problem (Veber & Freidinger, 1985). Although certain solutions were found (for example, the best illustration of this approach is the design of non-peptidic inhibitors of angiotensin-converting inhibitors based on a nona-

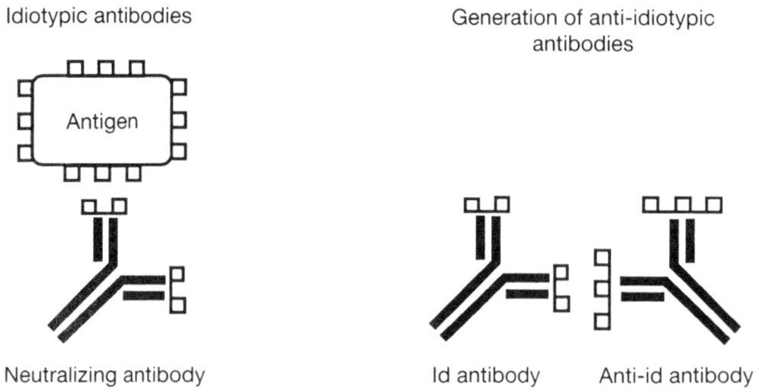

Figure 3.1. Use of antibodies as informational intermediates.

peptide; Wyvratt, 1985), the solutions will not be discussed in this chapter. The reader is referred to Chapters 15, 17 and 18 for further thoughts on the design of peptidomimetics. Nevertheless, by the 21st century, the technology should reach the state of the science currently enjoyed by lead optimization.

A current question with protein-based drug discovery is how to identify non-peptide ligands for peptide receptor or enzyme systems. An approach which has received renewed interest is to find leads from natural products. Drug discovery in the 21st century will continue to employ this strategy with one major difference—the tools of recombinant DNA technology and instrumentation mentioned in the introduction will have a major impact on natural product new lead identification. The current science stemming from the macrocyclic immunomodulatory agents (Caufield & Musser, 1990), cyclosporin-A (CsA), FK-506 and rapamycin (RPM), can serve as an example of the forthcoming synergy.

The natural product, CsA, is a cyclic polypeptide and is the subject of a review article (Wenger, 1986) and will not be discussed here. What follows is a brief summary of the discovery of CsA peptidic binding protein, its relation to enzymes involved in protein folding, its extension to the other natural products, FK-506 (Tanaka *et al.*, 1987) and RPM (Sehgal *et al.*, 1975), and questions that must be answered for the identification of new leads.

The search for a possible cellular receptor for CsA led to the isolation and characterization of an erythrocyte CsA-binding protein, cyclophilin (CyP), in 1984 (Handschumacher, 1984). Recently, CyP was identified as being identical to peptidyl-prolyl-*cis,trans*-isomerase (PPIase), an enzyme which catalyses the slow protein folding around prolyl amide bonds (Fisher *et al.*, 1989). The existence of CyP prompted the search for a similar cellular receptor for other macrocyclic immunomodulatory agents, such as FK-506 and RPM (Fig. 3.2) Both natural products are macrocyclic lactones containing the unusual hemiketal masked α,β-diketopipecolate moiety. FK-506 binding protein (FKBP) is isolated from a Jurkat T-cell line (Siekierka *et al.*, 1989). Like CyP, FKBP is a small protein (molecular weight 11 747) with prolyl isomerase activity, although it catalyses the isomerization of the substrate succinyl-Ala-Ala-Pro-Phe-*p*-nitroanilide (Fisher *et al.*, 1984) at 25-fold higher concentrations than CyP. Interestingly, FKBP also binds RPM.

Both, FK-506 and CsA act by inhibiting the production of growth-promoting lymphokines. They inhibit expression and accumulation of mRNAs for a number

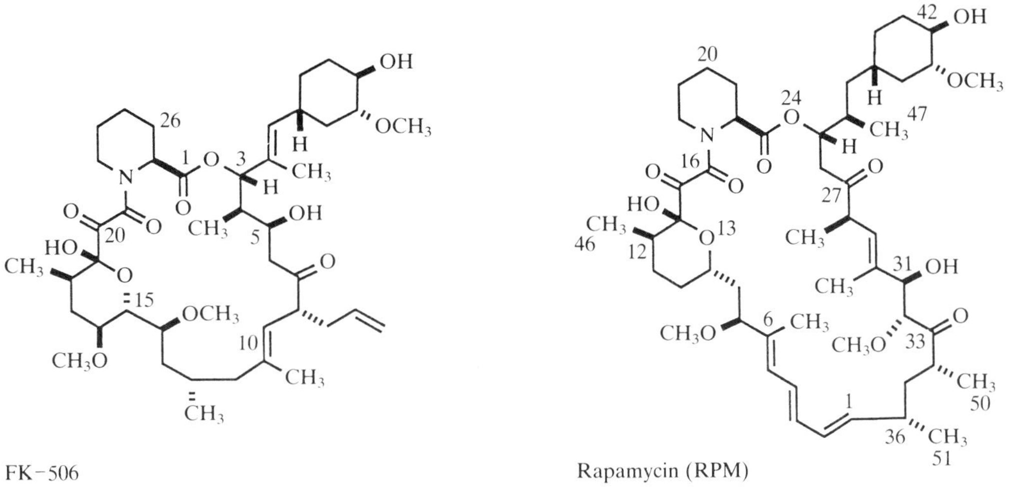

FK-506 Rapamycin (RPM)

Figure 3.2. Macrocyclic immunomodulators.

of cytokines including granulocyte/macrophage colony-stimulating factors (GM-CSF) and interleukin-2 (IL-2) (Kino *et al.*, 1987). Although structurally similar to FK-506, RPM does not suppress the expression and production of IL-2 or IL-2R. It exerts its immunosuppressive effects through impairment of the response of T-cells to lymphokines (Dumont *et al.*, 1990a). Although RPM and FK-506 appear to act by different mechanisms of action, on a cellular level both are mutually antagonistic (Dumont *et al.*, 1990b) suggesting the involvement of a common receptor.

Critical questions to the generation of new leads based on the preceding considerations include:

1 Does binding to the immunophilins, CyP and FKBP, correlate with immunosuppression?

2 Does inhibition of PPIase correlate with immunosuppression?

3 Is there an endogenous peptide ligand equivalent for CsA, FK-506 and RPM?

4 Is there a 'third' protein corresponding to the effector region?

5 Can linear analogues of FK-506 and RPM retain immunophilin binding activity?

Certainly, by the 21st century answers to the above questions will be known. Our efforts at Wyeth-Ayerst Research are addressing some of the above questions with respect to RPM. We are preparing several hundred analogues of RPM for structure–activity relationship (SAR) purposes and have used recombinant biotechnology to clone and express FKBP for use both in binding and PPIase assays. Correlation studies of immunophilin binding and PPIase inhibition with antiproliferative activity in lymphocytes and immunosuppressive activity *in vivo* are underway. We are collaborating to isolate the RPM receptor and the photoaffinity label FKBP with RPM. Protein NMR using 2-D techniques and ORD/CD studies are planned in order to understand on an atomic basis the interaction between RPM and its immunophilin. Finally, we have investigated the last question by computational methodology and have come up with a possible answer.

The single-crystal X-ray structures of FK-506 and RPM show that the amide carbonyl and the α-ketone are approximately orthogonal in both compounds. The

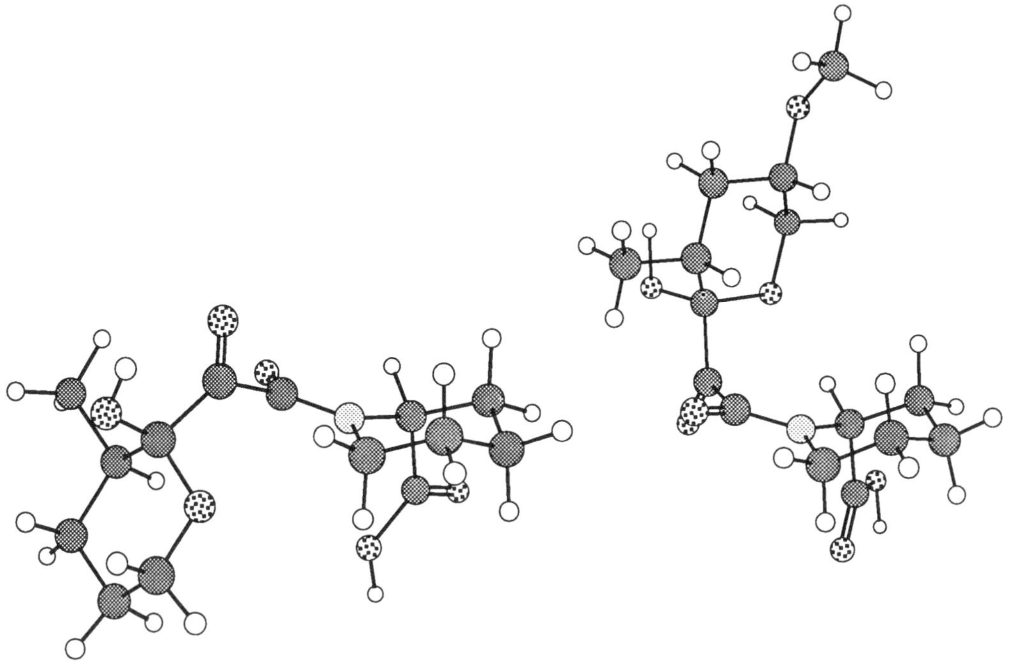

Figure 3.3. α-Keto amide segments of RPM and FK-506.

significance of the orthogonality is brought out in Schreiber's postulate that these α-keto amides serve as surrogates for the twisted amide of a PPIase-bound peptide substrate and that this binding is responsible, in part, for these compounds immunosuppressive activity (Schreiber, 1991) (Fig. 3.3).

Thus, a critical question is whether FK-506 and RPM adopt their conformations because of an intrinsic orthogonal relationship of the α-keto amide or the scaffolding effect of the macrolide that forces the amide carbonyl and the α-ketone to adopt an approximate 90° relationship. To address this question, semi-empirical molecular orbital calculations were applied to model α-keto amides with MOPAC 5.0 using the AM1 Hamiltonian (Mobilio & Musser, 1991). The results indicate that the orthogonality of RPM is due to natural tendencies of *N*-disubstituted-α-keto amides and that acyclic RPM will readily adopt conformations similar to their parent structures.

Taken together, the above information and speculation on macrocyclic immunomodulators is suggestive of how the new synergy of natural products chemistry and biotechnology can be applied to the discovery of non-peptide ligands for peptide receptors or enzymes.

3 The age of nucleotides

In contrast to protein structure-based drug discovery which has focused primarily on enzymatic inhibition and antagonism of cell-surface receptors, medicinal chemists in the 21st century will employ small oligonucleotide sequences or 'nucleotidomimetics' to intercept single-stranded messenger RNA (mRNA) in cells before the target enzyme or receptor involved in a disease process can be expressed. As rational drug design, this approach is particularly attractive because it takes advantage of the base-pair specific binding or 'duplexing' of RNA nucleotide sequences (duplexing of single-stranded DNA is the normal state of DNA) as the mode of inhibition or genetic code blocking. In double-stranded DNA, the sequence which contains the base pairs that code for a gene is called the sense strand, whereas the antisense strand contains the nucleotide sequence that is complementary. This nomenclature is directly applicable to RNA. (Fig. 3.4). Thus, if the nucleotide sequence of the target protein is known, it is possible directly to derive the chemical formula of the inhibitor for the corresponding base sequence of the antisense RNA. In theory, complementary synthetic RNA sequences to mRNA molecules transcribed from disease-related gene action are targets of choice as potential drugs. Nevertheless, in practice, DNA oligonucleotides or deoxyribose intranucleotides are designed as potential antisense drugs because of their greater stability relative to RNA or ribose intranucleotides.

Size is an important consideration in antisense oligonucleotide drug design. In order to be highly selective in its ability to recognize and bind to only one gene, a synthetic antisense sequence should be at least 12 nucleotides long. At this length a single mismatch in base-pairing can reduce affinity for duplexing by several orders of magnitude. Typically, a 21-member oligomer (21-mer) has an association constant in the order of 10^{24} M^{-1}, whereas, a 12-mer has an association constant in the order of 10^{12} M^{-1}. Thus, a high degree of selectivity is expected with larger oligonucleotides which should result in lower toxicity and other side-effects that are noted with more traditional drugs. Statistically, a 17-mer occurs only once in the human genome. Although the greater selectivity is expected with even longer

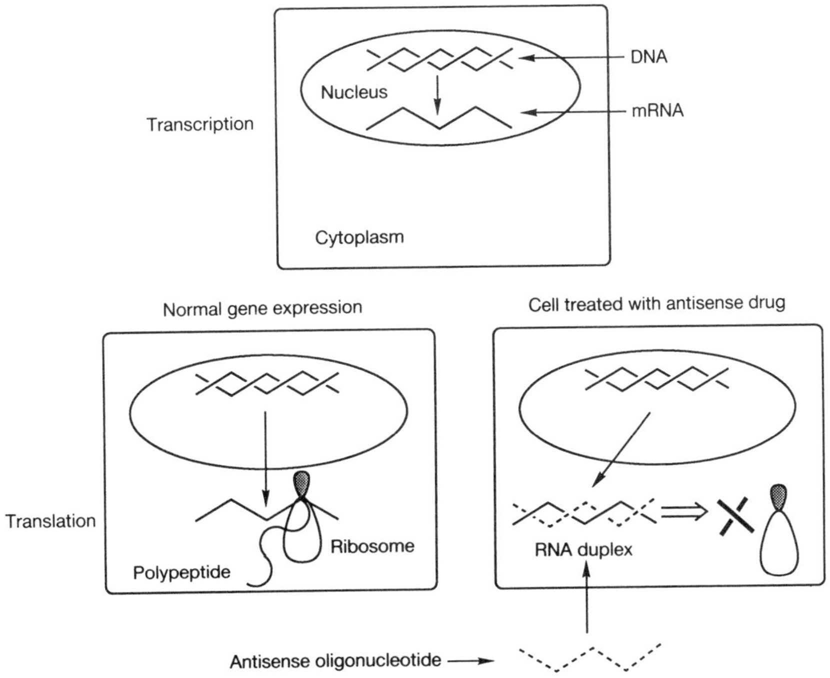

Figure 3.4. Possible mechanism of antisense inhibition.

strands, oligonucleotides larger than 25 bases have difficulty in penetrating cell membranes.

Over the years, synthesis of chemically modified oligonucleotides or 'nucleotidomimetics' has progressed (Uhlman, 1990) until it is now possible to think of antisense oligonucleotides as a viable approach to control gene expression and to design new drugs. Advanced chemical modifications have focused on enhancing structural stability, resistance to enzymatic degradation and affinity for binding. Several types of analogues are known and include modified phosphodiesters, phosphodiester surrogates, modified deoxyriboses, base analogues and terminal chain conjugates (Fig. 3.5). Methyl phosphonates and alkylphosphotriester analogues result in neutral nucleotidomimetics which contrast significantly with the natural phosphodiester oligonucleotides that are negatively charged (Miller *et al.*, 1981). These analogues have an increased permeability for cell membranes since non-charged molecules are more permeable than charged molecules. In another subclass, the sulphur atom of phosphothiolate analogues interferes with nucleotidase binding which slows enzymatic degradation (Eckstein, 1989). With intranucleotide analogues, substitution of sulphur for phosphate is poor according to SAR studies. Nevertheless, computer-aided molecular modelling suggests that a carbonate surrogate may be better although this group has chemical stability problems. Changes in stereochemistry may offer another strategy. The α-epimer of the β-configured riboses provide oligonucleotides which are resistant to degradative enzymes. Also, oppositely configured phosphodiesters and achiral phosphodiester substitutes, such as carbonates, may be employed. Finally, terminal-chain analogues involve linking the oligonucleotide to non-related molecules to enhance absorption, distribution or metabolism. Polyamine (L-lysine) (Schell, 1971) and bovine serum albumin (Mukhopadhyay *et al.*, 1989) terminal-chain modifiers are examples. There are a number of other terminal-chain modifiers, such as

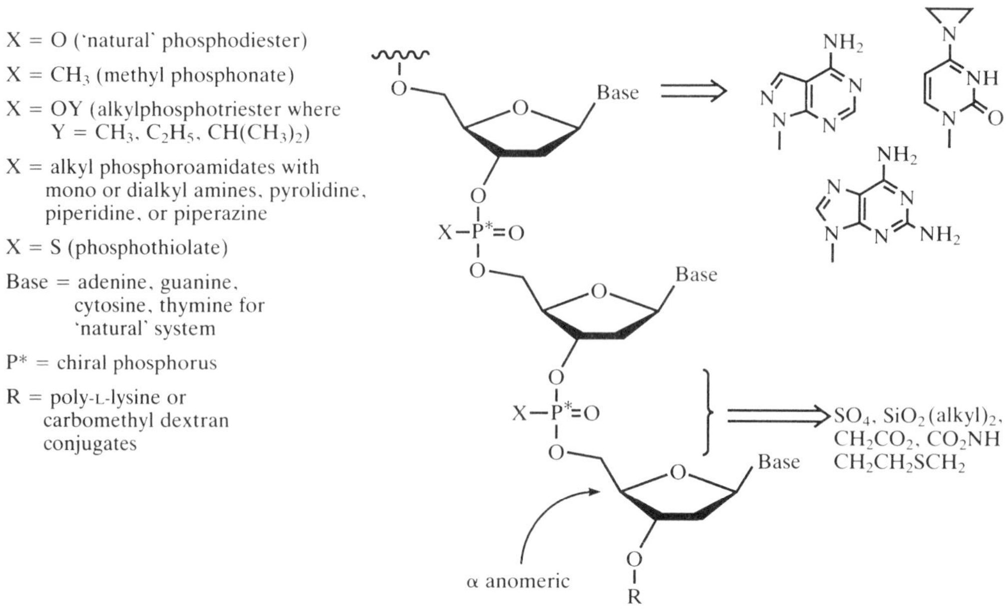

Figure 3.5. Analogues of oligonucleotides.

8-methoxypsoralen (Pieles, 1989), a photoactivating cross-linking agent, and ethylenediamine tetra-acetic acid/iron complex (Moser & Dervan, 1987), a mRNA cleavage system. However, these compounds are more likely research tools than potential antisense oligonucleotide drug candidates.

In addition to mRNA, another target for oligonucleotide drug candidates is double-stranded DNA involved in gene regulation. The resulting 'triplex' DNA responsible for signal initiation of transcription would have the same net inhibitory effect on protein expression as intercepting mRNA. In addition, triplex-forming oligonucleotides would have distinct theoretical advantages over mRNA interception including the need for less drug and less frequent dosing. These advantages stem from the consideration that oligonucleotide drugs would be required to bind to only one DNA molecule and not to the many copies of mRNA that is normally transcribed during gene activation, and the observation that triplex formation with regulatory sequences is irreversible, respectively.

Thus, both single-stranded mRNA and double-stranded DNA are targets for antisense oligonucleotides. The oligonucleotide sequences, in turn, can serve as new leads for nucleotidomimetics which can act as synthetic chemical signals for the inhibition of protein expression.

4 The emergence of lipids

Receptors on the outer surface of cell membranes receive chemical signals or 'first messengers' from circulating factors, such as mediators, hormones and neurotransmitters. Upon binding to a receptor by a first messenger, a cascade of biochemical events is triggered starting on the inner surface of the membrane and eventually ending within the nucleus of the cell. The transduction of a signal from a cell-surface receptor to the nucleus via key intercellular enzymes is naturally regulated by 'lipid second messengers'. Some of the key intracellular enzymes, in order of their appearance in the cascade, are phospholipase C (PLC), protein kinase C (PKC) and the *ras* protein. The lipid second messengers include diacylglycerol (DAG) and inositol triphosphate (IP_3) (Fig. 3.6).

In the 21st century, knowledge of lipid–protein interaction will be used to generate drugs that modulate enzymes, such as PKC and *ras*, their subtypes, and enzymes that control the production and breakdown of lipid second messengers. For example, over 100 different protein kinases have been identified using molecular cloning techniques (Hunter, 1987).

PKC is a Ca^{2+}–phospholipid-dependent protein kinase and catalyses phosphorylation at serine and threonine. It is involved in the regulation of many cellular process pivotal for growth, differentation and tumour promotion. Apparently, PKC is the only cellular target for the tumour-promoting phorbol esters. Although there are many different isoforms of PKC, there is a common primary structure consisting of two functional domains: catalytic and regulatory (Nishizuka, 1988). Potent inhibitors which interact with the catalytic domain, such as staurasporin (Tamaoki *et al.*, 1986), are not selective for PKC presumably because this domain is highly conserved among the various isozymes. In contrast, agents that target the regulatory domain of PKC should be much more selective since this domain varies

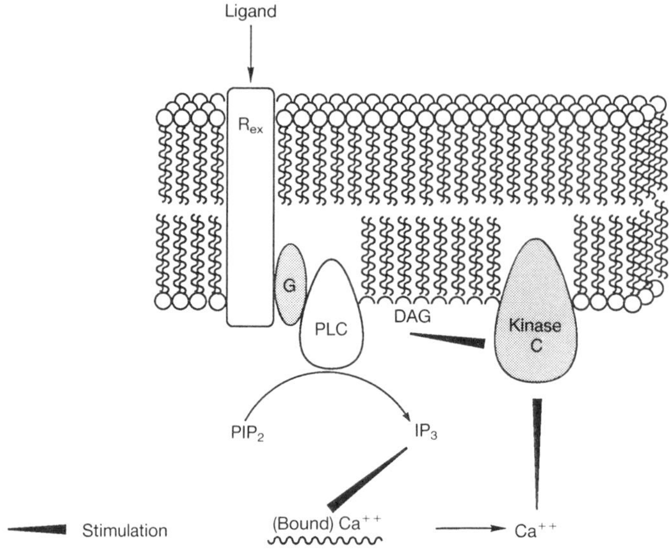

Figure 3.6. Lipid second messengers.

greatly from isozyme to isozyme. An example of a selective inhibitor of PKC is sphingosine, a naturally occurring lipid (Hannun & Bell, 1987).

Lipid hormones, such as the adrenal steroids (cortisol and aldosterol), the sex steroids (oestrogen, progesterone and testosterone), vitamin D_3 and retinoic acid are of profound importance for mammalian function and development. Recent work on steroid receptors (Evans, 1989) indicates that this important class of lipids function through binding with intracellular receptors. The resulting hormone–receptor complex undergoes a structural alteration or transformation and, thus, becomes an 'activated' binary complex. This activated complex then can bind to specific sites or hormone response elements (HREs) on the cell's DNA. The HREs regulate transcriptional activity of genes which ultimately modulates protein synthesis (Fig. 3.7).

Lipid-induced, intracellular, binary complex modulation of gene action and modulation of lipid second messengers will emerge in the 21st century as significant new ways to identify new leads for drug discovery.

5 The rise of carbohydrate chemistry

Carbohydrates have not been a major approach to new lead discovery in the pharmaceutical industry primarily for two reasons: the biological impact of this class was underestimated, and the chemistry is theoretically simple but, in practice, very difficult to perform. The pharmacologists considered these molecules as energy stores or intracellular support substances. In addition, the apparent random distribution and lack of research tools made the localization and functional characterization of carbohydrates difficult, whereas the medicinal chemists considered carbohydrate synthesis as involving one basic reaction, carbon–oxygen or ether bond formation, and thus simple. However, the practical aspects of purification

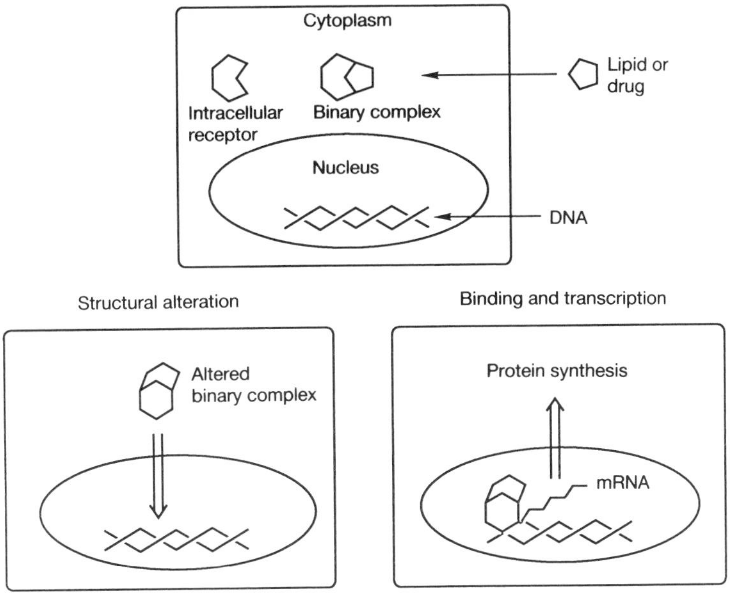

Figure 3.7. Lipid-induced gene regulation.

and characterization of carbohydrates being tedious at best and often intractable, turned away potential research efforts.

By the 21st century, however, research on carbohydrates will become a major focus for medicinal chemistry. Today there is an opinion among some scientists that whenever two or more cells interact in a specific way, cell-surface carbohydrates play a major role. For example, carbohydrates are critical in the operation of two fundamental but opposing cellular processes: the proper maintenance of the body's immune defence, and the initiation of bacterial and viral infections. Also, when cells go astray, as in autoimmune disease or cancer, cell-surface carbohydrates change in structure and composition. In parallel, mainstream chemists are recognizing the fact that carbohydrates are innately exquisite forms of concise informational packages. In comparison to polypeptides or oligonucleotides on a unit basis, carbohydrates have the potential for greater complexity. Two identical amino acids or nucleotides when joined together yield only one dipeptide or one dinucleotide; however, when two identical saccharides are linked, 11 different disaccharides are possible. Furthermore, three different nucleotides can be arranged in a trinucleotide in only six distinct ways, whereas three different saccharides can be arranged in 1056 different ways. There are only 120 ways to link five selected amino acids, whereas five different monosaccharides can be assembled in more than 31 million pentasaccharide arrays. Thus, of all the structural types, carbohydrates have the greatest theoretical potential for specificity and new lead generation.

Recent findings on the interaction of leukocytes with platelets or vascular endothelium (Stoolman, 1989; Osborn, 1990) support the postulate that cell-surface carbohydrates are involved in specific cell–cell recognition. The vascular endothelium participates in the recruitment of leukocytes by the expression of cell-adhesion molecules (CAMs) in response to cytokines (Pober, 1986). Specifically, leukocyte–endothelial cell CAM (LECCAM2) is produced within 2–4 h in

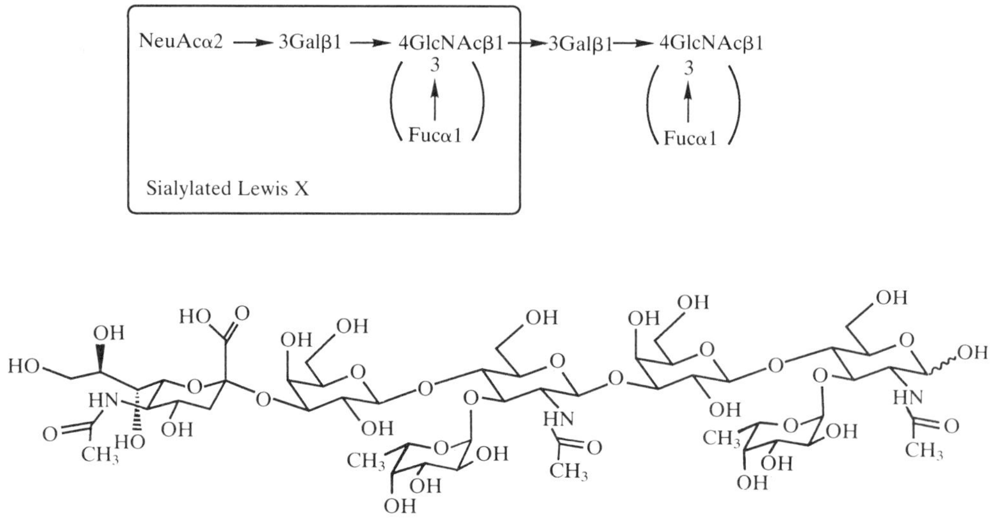

Figure 3.8. Structure of sialyl-Lewis X. Fucoses in parentheses in LECCAM2 ligand can be either mono- or disubstituted. NeuAc = sialic acid, Gal = galactose, GlcNAc = *N*-acetylglucosamine, Fuc = fucose. The structure at the bottom is disubstituted with Fuc.

response to induction by either IL-1 or tumour necrosis factor (TNF) and mediates the binding of leukocytes at the endothelium surface (Bevilacqua *et al.*, 1987). Indeed, cell adhesion by LECCAM2 is mediated by the carbohydrate ligand, sialyl-Lewis X (SLe^x) (Fig. 3.8), a terminal structure found on cell-surface glycoproteins and glycolipids (Phillips *et al.*, 1990). Because LECCAM2 is thought to be involved in both the recruitment of leukocytes to sites of inflammation and the establishment of metastasis of certain carcinomas (Rice & Bevilacqua, 1989), the identification of SLe^x is a major advance toward the discovery of carbohydrates as new leads in drug discovery.

6 A renaissance of screening

All the above structural approaches are amenable to lead generation via screening assays which are automated with advanced robotics. However, high throughput bioassays have the potential quickly to deplete the chemical libraries found in the largest of companies in the pharmaceutical industry. In the 21st century, what will be needed for the renaissance of screening is ways to generate new chemical libraries on a very large scale. A recent report of the successful combination of three established technologies, solid-phase chemical synthesis, photolabile protecting groups and photolithography, represents one possible approach to the synthesis of unique chemical entities on a very large scale (Forder *et al.*, 1991). Apparently, the treatment of a glass plate with an activated amino acid (AA) protected with a photolabile group results in a covalent tethering of AA across the entire surface. Using methodology developed in the semiconductor industry to print electrical circuits or photolithography, laser light is passed through a mask which specifically removes a designated array of the AA-protecting groups. The unmasked AAs are then able to react with a second protected AA giving a dipeptide. More selected

AA-protecting groups are then removed and through a series of light exposures followed by additional AAs, a third AA, fourth AA, *n*th AA is added giving 3, 4, . . . , *n* different dipeptides. By changing masks and AAs, thousands of peptides varying in length are possible. Once a plate filled with peptides is made, they are then exposed to specific monoclonal antibodies or soluble receptors that are fluorescently tagged. A fluorescent scan of the plate would then identify the AA sequence or peptide of interest.

Not only are peptides but all of the biopolymeric types discussed above are amenable to light-directed, spatially addressable parallel chemical synthesis. In addition, totally novel sequences composed of unusual, completely synthetic units or building blocks are possible.

Other methods to generate new chemical libraries on a very large scale are possible. For example, randomly synthesized oligonucleotides inserted in the genomes of phage can express millions of different peptides. Each phage clone displays a different peptide and those that bind to a receptor can be identified by panning, isolation of individual clones and DNA sequencing. Chemical libraries produced on a very large scale will make the renaissance of screening as important as rational design in the generation of new chemical leads in the 21st century.

7 Summary

Modern drug discovery has focused on agents which interact with peptidic enzymes and cell-surface receptors and in the decade of the 1990s there will be a maturation of this trend in new lead generation with respect to peptidomimetics. As a follow on to this approach, the emerging discoveries based on lipid second messengers and their effect on proteins will lead to a new generation of therapeutic compounds that will modulate cellular growth. Also, in light of the Human Genome Project where the ultimate goal is to know the blueprint for humans and genetically induced disease, nucleotide research may hold the key to promising new drugs. However, if informational content inherent in structure would predict specificity of ligand–biopolymer interaction, drugs based on carbohydrate chemistry may appear as the most promising avenue for pursuit.

The intention of this chapter was to give the reader an overview of some of the trends in new lead identification for the 21st century. In the spirit of the opening quote of this chapter, it appears that many of the mentioned approaches will be successful in the discovery of new drugs. The only real problem for medicinal chemists is, which trend to choose?

8 References

Amat, A.G., Mariuzza, R.A., Phillips, S.E. & Poljak, R. (1986). *Science*, **233**, 747.

Barber, M., Bordoli, R.S., Sedgwick, R.D. & Tyler, A.N. (1981). *J. Chem. Soc., Chem. Commun.*, 325.

Bass, T. (1990). *Omni*, **13**, 78.

Bevilacqua, M.P., Prober, J.S., Mendrick, D.L., Cotran, R.S. & Gimbrone, M.A. (1987). *Proc. Natl. Acad. Sci. USA*, **84**, 9238.

Caufield, C. & Musser, J.H., (1990). *Ann. Rep. Med. Chem.*, **25**, 195.

Colman, P.M. *et al.* (1987). *Nature* (*London*), **302**, 358.
Dumont, F.J., Staruch, M.J., Koprak, S.L., Melino, M.R. & Sigal, N.H. (1990a). *J. Immunol.,* **144**, 251.
Dumont, F.J., Melino, M.R., Staruch, M.J., Koprak, S. L., Fisher, P.A. & Sigal, N. H. (1990b). *J. Immunol.,* **144**, 1418.
Eckstein, F. (1989). *Trends Biochem. Sci.,* **14**, 97.
Evans, R.M. (1989). *Science,* **240**, 889.
Fisher, G., Bang, H. & Mech, C. (1984). *Biomed. Biochim. Acta,* **43**, 1101.
Fisher, G., Wittmann-Liebold, B., Lang, K., Kiefhaber, T. & Schmid, F.X. (1989). *Nature* (*London*), **337**, 476.
Fordor, S.P.A., Read, J.L., Pirrung, M.C., Stryer, L., Lu, A.T. & Solas, D. (1991). *Science*, **251**, 767.
Goodman, J.W. (1985). In: M. Sela (Ed.), *The Antigens, 3,* Academic Press, New York.
Handschumacher, R.E., Harding, M.W., Rice, J., Drugger, R.J. & Speicher, D.W. (1984). *Science*, **226**, 544.
Hannun, Y.A. & Bell, R.A. (1987). *Science,* **235**, 670.
Hunter, T. (1987). *Cell,* **50**, 823.
Kino, T., Inamura, N., Sakai, F., Nakahara, K., Goto, T., Okuhara, M., Kohsaka, M., Aoki, H. & Ochiai, T. (1987). *Transplant. Proc.,* **19**, 36.
Lipkow, K.B. & Boyd, D.B. (1990). *Reviews in Computational Chemistry,* VCH Publishers, New York.
Miller, P.S., MacFarland, K.P., Jayaraman, K. & Ts'O, P.O.P. (1981). *Biochemistry*, **20**, 1874.
Mobilio, D. & Musser, J.H. (1991). *Med. Chem. Res.*, **1**, 166.
Moser, H.E. & Dervan, P.B. (1987). *Science,* **238**, 645.
Mukhopadhyay, A., Cahndhuri, G., Arora, S.K., Sehgal, S. & Basu, S.K. (1989). *Science*, **244**, 705.
Nishizuka, Y. (1988). *Nature* (*London*), **334**, 661.
Osborn, L. (1990). *Cell*, **62**, 3.
Phillips, M.L., Nudleman, E., Gaeta, F.C.A., Perez, M., Singhal, A.K., Hakomori, S. & Paulson, J.C. (1990). *Science,* **250**, 1130.
Pieles, U. & Englisch, U. (1989). *Nucleic Acids Res.*, **17**, 285.
Pober, J.S. *et al.* (1986). *J. Immunol.*, **136**, 1680.
Rainer, F. (1984). In: W.T. Nauta and R.F. Rekker (Eds)., *Theoretical Drug Design Methods, 7,* Elsevier, Amsterdam.
Rice, G.E. & Bevilacqua, M.P. (1989). *Science*, **246**, 1303.
Schell, P.L. (1971). *Biochim. Biophys. Acta*, **340**, 472.
Schreiber, S.L. (1991). *Science*, **251**, 283.
Sehgal, S.N., Baker, H. & Vezina, C. (1975). *J. Antibiot.*, **28**, 727.
Seiler, F. *et al.* (1985). *Angew Chem.* (*English Edition*), **24**, 139.
Siekierka, J.J., Hung S.H.Y., Poe, M., Lin, C.S. & Sigal, N. H. (1989). *Nature* (*London*), **341**, 755.
Stoolman, L.M. (1989). *Cell*, **56**, 907.
Tamaoki, T., Nomoto, H., Takahashi, I., Kato, Y., Morimoto, M. & Tomita, F. (1986). *Biochem. Biophys. Res. Commun.,* **135**, 397.
Tanaka, H., Kuroda, A., Marusawa, H., Hatanaka, H., Kino, T., Goto, T., Hashimoto, M. & Taga, T. (1987). *J. Am. Chem. Soc.,* **109**, 503.
Tonegawa, S. (1983). *Nature* (*London*), **302**, 575.
Uhlman, E. & Peyman, A. (1990). *Chem. Rev.*, **90**, 543.
Veber, D.F. & Freidinger, R.M. (1985). *Trends Neurosci.,* **8**, 392.
Venuti, M.C. (1990). *Ann. Rep. Med. Chem.,* **25**, 289.
Wenger, R.M. (1986). *Fortschr. Chem. Org. Naturst.*, **50**, 123.
Wyvratt, M.J. & Prachett, A.A. (1985). *Med. Chem. Rev.*, **5**, 483.

4 From Membrane to Genomic Pharmacology or from Short-term to Long-term Effects

P.M. LADURON

Research Center Rhône-Poulenc Rorer, 13 Quai Jules Guesde BP-14, F-94403 Vitry sur Seine Cedex, France

1 Introduction

Most drugs operate at the level of cell membranes—their targets are receptors, ion channels and membrane-bound enzymes. Some drugs, for example those used in hormone and cancer therapy, are aimed at nuclear targets. For drugs that interact with cell membranes, pharmacologists study the short-term effects. The acute experiments devised to test these effects, however, are not always appropriate for drugs, such as antidepressants, that have to be administered to patients over a period of at least several weeks. The discrepancy between the type of experiment used to test the drug and the fact that eventually it will have to be administered to patients for 3 or 4 weeks frustrates the pharmacologist.

The most important problems that remain to be solved today are the treatment and prevention of degenerative and proliferative diseases. Such diseases require not only chronic treatment of patients, but also an experimental mirror in which the effects of chronic treatment can be observed in animals. The necessity for long-term treatment suggests that the mood-elevating properties of antidepressant drugs, for instance the tricyclics, are not directly related to the amine re-uptake blockade, which is only their immediate effect. Whatever is responsible for the long-term effects of these drugs remains to be discovered.

The gap between the pharmacological activity of certain drugs and their clinical efficacy may partly be ascribed to the lack of appropriate animal models. When such models exist, as they do for certain pathologies, for example arthritis, atherosclerosis and cancer, long-term treatment in animals is usually evaluated. The gap is further widened by how little is known of the consequences of chronic treatment on gene expression. A molecular dissection of the precise mode of action of antidepressants and other drugs is needed, if new pharmacological targets are to be found. Study of the mRNA changes observed in the course of chronic treatment may constitute a new branch of *in vivo* pharmacology.

The present chapter discusses the reasons why pharmacology in the future will require chronic, or genomic, approaches and why the targets of new drugs are likely to be nuclear ones, such as hormone receptors.

2 Signal molecules and long-term effects

Most drugs interact with signal molecule receptors, not only in the brain but also in blood platelets, lymphocytes, smooth muscle and many other cells. Once signal molecules have bound to receptors, their action is of varying duration: the effects of noradrenaline and of acetylcholine, for example, do not last as long as those of neuropeptide Y (NPY), interleukin-1 (IL-1) or leukotriene B_4.

Today, our therapeutic arsenal is dominated by drugs interacting with classical neurotransmitters (noradrenaline, dopamine, γ-aminobutyric acid, acetylcholine, 5-hydroxytryptamine). While the possibilities in this field have not yet been exhausted, particularly in terms of multitarget drugs, it could be suggested that in the future the main efforts should be devoted to the development of compounds interfering with signal molecules that have long-term effects, such as neuropeptides, interleukins, third messengers and transcription factors. The non-peptide compound opium, one of the oldest therapeutic agents, is known to interact with a neuropeptide receptor. The synthesis of non-peptide compounds is difficult. Nevertheless, non-peptide antagonists were recently synthesized for cholecystokinin (Evans *et al.*, 1986) and substance P (Garret *et al.*, 1991; Snider *et al.*, 1991). Perhaps, in the future, compounds that interfere with other neuropeptides, such as NPY, CGRP, neurotensin, galanin, etc., will also be synthesized.

In numerous cases, the physiological role of neuropeptides and their implications in pathological disorders are not clear or totally unknown. Neuropeptide antagonists will not all necessarily become therapeutic agents; nevertheless, they are useful for both acute and chronic studies, which are necessary if the physiological roles of the corresponding neuropeptides are to be elucidated. However, we already know a lot about the function of some of these peptides, and this knowledge can lead to therapeutic applications. For example, substance P is definitely involved in pain and inflammation. Furthermore, NPY is released more when the organism is under stress than when it is in the unstressed state (Lundberg *et al.*, 1986), whereas noradrenaline is constantly released at a low-stimulation frequency. This means that NPY antagonists will be potential antihypertensive drugs and may even become the drugs of choice for treating hypertension.

The major difficulty in the next decades is how to treat animals for several days or weeks with agonists or antagonists in order to reveal the long-term effects of these compounds. Of course, the field of neuropeptides is not the only domain in which a methodology for chronic pharmacology needs to be developed—interleukins (IL-1, IL-2, IL-6), leukotrienes and growth factors are also signal molecules, the agonists or antagonists of which represent potential therapeutic agents. Such molecules have to be studied in chronic experiments. IL-1 antagonists may turn out to be more beneficial for treating arthritic patients than non-steroidal anti-inflammatory drugs (NSAIDs), which have been reported to accelerate the progression of osteoarthritis (Rasbad *et al.*, 1989).

The synthesis of signal molecule antagonists represents an important goal, not only for obtaining new therapeutic agents but also for elucidating the precise role of neuropeptides and immunoregulators in neuronal communication and in cell–cell interaction. Classical neurotransimtters are often co-expressed with one or more neuropeptides (Hökfelt *et al.*, 1980), and cytokines may regulate neuropeptide expression through specific interactions, such as those between substance P and IL-1 (Freidin & Kessler, 1991).

As a rule, a clear-cut distinction has to be made between the short-term and the long-term effects of various neuropeptides and growth factors. For example, the short-term effect of NGF is to increase Ca^{2+} concentrations in nerve endings, while its long-term effect is to change the genetic expression of numerous enzymes, receptors and ion channels (Laduron, 1987). Similarly, the short-term effect of the opioids is to diminish the release of substance P and acetylcholine at the nerve terminals, whereas the long-term, or delayed, effects of these compounds are physical dependence and addiction. The latter effects require protein synthesis and are therefore sensitive to protein synthesis inhibitors, whereas the former are not. Recently, we found that opiate and chronic ethanol treatment reduced anterograde axonal transport of substance P in rat vagus nerves (De Witte *et al.*, 1990). Another example of a neuropeptide that elicits different short- and long-term effects is that of neurotensin, which increases dopamine release at the synapses in the striatum (short-term) (Nemeroff & Cain, 1985) and increases tyrosine hydroxylase mRNA in the substantia nigra (long-term) (Burgevin *et al.*, 1992). These examples suggest that neuropeptides may have more than one site of action, unless the same site is operating in two different ways. When the mechanism of action of a drug is being considered, both sites must be investigated in order to determine whether the clinical efficacy of the drug is directly related to its interaction with both of these sites or with only one.

The same transduction system often operates for different receptors; transduction mechanisms are not, therefore, highly specific. Drug interactions with transduction mechanisms will not be discussed simply because these mechanisms are not good targets for drugs.

3 Nuclear receptors

Even though genomic pharmacology appears to be a new concept, it has existed for a long time, in fact, since the introduction of hormone therapy. In this field, nuclear receptors are of particular importance. The classical model is the steroid hormone receptor, which belongs to the superfamily of ligand-activated enhancer binding factors. Besides this class of nuclear receptors, the existence of which has been firmly established (cf. Green & Chambon, 1988), there is a second class of receptors, still considered putative because only indirect evidence supports their occurrence in the nucleus (cf. Laduron, 1992).

The first group comprises steroid (deoxycorticosterone, progesterone, testosterone, cortisol and oestradiol), thyroxine, retinoic acid and vitamin D_3 receptors. These receptors consist of two domains, the DNA-binding domain and the ligand-binding domain. Binding of the ligand to the receptor and of the ligand–receptor complex to DNA is a prerequisite for inducing a broad spectrum of effects, such as changes in enzymatic activities or in levels of secretory proteins, as well as more complex changes in tissue development, growth and differentiation. In fact, these hormone receptors constitute a class of transcription factors which regulate gene expression.

The putative second class of nuclear receptors, that is, the non-hormone receptors, includes interleukin (IL-1, IL-2, IL-6), cytokine (TNF), growth factors (epidermal growth factor (EGF), fibroblast growth factor (FGF), nerve growth factor (NGF) and platelet derived growth factor (PDGF)) and peptide (thymosine, vasoactive intestinal peptide, neurotensin) receptors. Those most clearly established as nuclear receptors are the interleukin receptors, which have been localized in the nucleus by microscopic autoradiography and immunohistochemistry. These receptors were found to be translocated from the cell membrane to the nucleus after treatment with the ligand (cf. Laduron, 1992). This suggests that in the absence of the ligand, the majority of receptors remain at the cell surface. If such a process of translocation really occurs, the difficulty of revealing a nuclear localization for these receptors becomes clear: they are usually present in much greater amounts at the cell membrane than in the nucleus. Bishr Omary & Kagnoff (1987) report the nuclear localization of VIP (vasoactive intestinal peptide). Recent data from our laboratory suggest the presence of neurotensin receptors in the nucleus of dopaminergic cell bodies in the substantia nigra following retrograde axonal transport of labelled neurotensin injected in the striatum (Castel *et al.*, 1990; Burgevin *et al.*, 1992). It is tempting to speculate that nuclear receptors might be identified for other neuropeptides in the near future. In any case, the number of well-defined nuclear receptors is already sufficiently large so that the nucleus can be considered an important potential target for new drugs.

Of the potential sites for drug action in the nucleus, the most appropriate are the nuclear proteins, such as transcription factors, receptors or those proteins that have been called 'third messengers'. DNA itself is not an appropriate drug target. The nuclear proteins represent the target of choice for genomic, or long-term, pharmacology. Synthetic drugs have the advantage over neuropeptides in that they can reach either nuclear or membrane receptors. When neuropeptides are released in the synaptic cleft, they bind only to membrane receptors. The neuropeptide–receptor complex is then internalized and transported to the nucleus. This succession of events was recently demonstrated for neurotensin (Castel *et al.*, 1990). That the internalization process does not operate in the presence of antagonist indicates that antagonists act principally by blocking the initial step of internalization, which occurs at the cell membrane. However, it is also possible that some antagonists act on the nucleus. In pharmacology, interactions between drugs and cell membranes will continue to be important; however, to determine the long-term effects of a particular compound, the study of what occurs at the nuclear level will become vital.

4 Drugs and the regulation of gene expression

A considerable number of environmental phenomena, such as hyperalgesia, inflammation, growth factors, neuropeptides, denervation and nerve stimulation, may alter gene expression. Little is known about the changes in gene expression elicited by drug treatment, simply because this field of research is just beginning to develop. Nevertheless, this approach appears to be promising, especially for examining the long-term effects of drugs and for developing new classes of therapeutic agents. Today, the effects of drug treatment on mRNA can be studied by molecular biology. Whether the effects observed are the consequence of interactions between drug and membrane receptors (or of other mechanisms involving nuclear proteins) remains to be elucidated.

One of the major goals in research on nuclear targets should be to try to block the defence reactions of the organism that decrease drug efficacy. Indeed, nature often finds a way to overcome the effects of drugs, particularly in the brain. For instance, one consequence of blocking dopamine D_2 receptors by neuroleptics is that dopamine synthesis increases, counteracting the effects of the neuroleptics. It is of prime importance to interrupt such a vicious circle by antagonizing the signal responsible for the increase of dopamine synthesis. Recently, chronic haloperidol treatment was found to increase the levels both of pro-enkephalin mRNA in the striatum and the nucleus accumbens and of pro-neurotensin mRNA in neurones of the striatum and of the septal nuclei (cf. Laduron, 1992). Augmenting mRNA levels may be related to the efficacy and side-effects of antipsychotic drugs. It is important that such potential consequences of long-term treatment be brought to light.

In the areas of inflammation and arthritis, new concepts can be derived from investigating drugs that change gene expression. The roles of IL-1 and the neuropeptides, in particular substance P, are of great importance for understanding the mechanisms of inflammation and degradation of articular cartilage. Data are already available on the interactions between IL-1 and substance P. Of course, pain and arthritis are multifactorial and therefore require multitarget therapy. Recently, indomethacin was shown to increase the mRNA of type I and type III pro-collagen and fibronectin, as well as that of collagen (Mauviel *et al.*, 1988), an example of how molecular biology can contribute to our awareness of the undesirable effects of drugs.

To analyse changes in gene expression induced by chronic treatment with drugs will require a major effort, involving systemic studies of different classes of drugs. For drugs that act in the CNS, such studies will be laborious and time-consuming, because a great variety of mRNAs in several brain regions must be tested. However, such a pragmatic approach is more rewarding than starting from a gene in which one or several mutations have been found that cause disease.

5 Conclusions

If pharmacology is to evolve in the next century, it will have to encompass a genomic aspect. Finding treatments for proliferative and degenerative diseases requires a genomic approach, which will be most effective if it is associated with chronic, or long-term, pharmacology. If more nuclear receptors are found, they will provide additional drug targets. In this regard, the field of signal molecules (including neuropeptide, interleukin, leukotriene, etc.) will be of prime importance for generating new therapeutic agents. However, chemists have learned by experience that it is not necessary to mimic nature. The chemotherapy of the future will still be based on synthetic non-peptide compounds and multitarget drugs. It is possible today, through binding studies, to design multitarget drugs with appropriate affinity for each of their targets.

6 References

Bishr Omary, M. & Kagnoff, M.F. (1987). Identification of nuclear receptors for VIP on a human colonic adenocarcinoma cell line. *Science*, **238**, 1578–81.

Burgevin, M.C., Castel, M.N., Quarteronet, D., Chevet, T. & Laduron, P.M. (1992). Neurotensin increases tyrosine hydroxylase messenger RNA-positive neurons in sulstantia nigra after retrograde axonal transport. *Neuroscience*. In press.

Castel, M.N., Malgouris, C., Blanchard, J.C. & Laduron, P.M. (1990). Retrograde axonal transport of neurotensin in the dopaminergic nigrostriatal pathway in the rat. *Neuroscience*, **36**, 425–30.

De Witte, P.A., Hamon, M., Mauborgne, A., Cesselin, F., Levy, C. & Laduron, P.M. (1990). Ethanol and opiate decrease the axonal substance-P like immuno-reactive material in rat vagus nerves. *Neuropeptides*, **16**, 15–20.

Evans, B.E., Bock, M.G., Rittle, K.E., Dipardo, R.M., Whitter, W.L., Veber, D.F., Anderson, P.S. & Freidinger, R.F. (1986). Design of potent orally effective, non-peptidal antagonists of the peptide hormone cholecystokinin. *Proc. Natl. Acad. Sci. USA*, **83**, 4918–22.

Freidin, M. & Kessler, J.A. (1991). Cytokine regulation of substance P expression in sympathetic neurons. *Proc. Natl. Acad. Sci. USA*, **88**, 3200–3.

Garret, C., Caruette, A., Fardin, V., Moussaoui, S., Peyronel, J.F., Blanchard, J.C. & Laduron, P.M. (1991). Pharmacological properties of a potent and selective substance P antagonist. *Proc. Natl. Acad. Sci. USA*, **88**, 10208–11.

Green, S. & Chambon, P. (1988). Nuclear receptors enhance our understanding of transcription regulation. *Trends Genet.*, **4**, 309–14.

Hökfelt, T., Johansson, O., Ljungdahl, A. & Lundberg, J.M. (1980). Peptidergic neurones. *Nature (London)*, **284**, 515–21.

Laduron, P.M. (1987). Axonal transport of neuroceptors: possible involvement in long-term memory. *Neuroscience*, **22**, 767–79.

Laduron, P.M. (1992). Towards genomic pharmacology: from membranal to nuclear receptors. *Adv. Drug Res.*, **22**, 107–148.

Lundberg, J.M., Rudehill, A., Sollevi, A., Theodorsson-Norheim, E. & Hamberger, B. (1986). Frequency and reserpine dependent chemical coding of sympathetic transmission: differential release of noradrenaline and neuropeptide Y from pig spleen. *Neurosci. Lett.*, **63**, 96–100.

Mauviel, A., Käbäri, V.M., Heino, J., Daireaux, M., Hartmann, D.J., Loyau, G. & Pujol, J.P. (1988). Gene expression of fibroblast matrix protein is altered by indomethacin. *FEBS Lett.* **231**, 125–9.

Nemeroff, C.B. & Cain, S.T. (1985). Neurotensin dopamine interactions in the CNS. *Trends Pharmacol. Sci.*, **6**, 201–5.

Rashad, S., Hemingway, A., Rainsford, R., Revell, P., Low, F. & Walker, F. (1989). Effect of non-steroidal antiinflammatory drugs on the course of osteoarthritis. *Lancet*, **2**, 519–22.

Snider, R.M., Constantine, J.N., Lowe, III J.A., Longo, K.P., Lebel, W.S., Woody, H.A., Drozdar, S.E., Desai, M.C., Vinick, F.J., Spencer, R.W. & Hess, H.J. (1991). A potent nonpeptide antagonist of the substance P (NK1) receptor. *Science*, **251**, 435–7.

5 Future Applications of Oligonucleotides in Antiviral and Antitumoral Chemotherapy

P. HERDEWIJN and E. DE CLERCQ
Rega Institute for Medical Research, Katholieke Universiteit Leuven, B-3000 Leuven, Belgium

1 Introduction

The master control molecule which determines the cell's morphology and function is DNA. For the inheritance of genetic information, DNA must be replicated through the aid of a DNA polymerase. Otherwise, DNA is normally transcribed to a premessenger RNA which undergoes splicing and maturation reactions to afford the mature messenger RNA (mRNA). The mRNA is then translated to proteins (Fig. 5.1).

For a healthy individual to function normally, these processes have to proceed continuously during his or her lifetime without irreversible mistakes. Many diseases find their origin in a defect of the genome itself or in an alteration of the transcription/translation process. Most of these diseases are now treated symptomatically, however. Actual cure of such a disease can only be accomplished by eradicating directly its origin. Diseases where the relationship between an alteration of the DNA function and the clinical consequences are quite clear are cancers and viral diseases. Cancers can be induced by several different factors. However, the principal change which occurs within the cell during carcinogenesis is an alteration of pre-existing genes to oncogenes. The products of these genes are then responsible for inappropriate cell growth. Also, incorporation of the genomic material of retroviruses as occurs in the DNA of the host cell in AIDS is an irreversible process which can lead to uncontrolled proliferation or, alternatively, cell death. Attempts

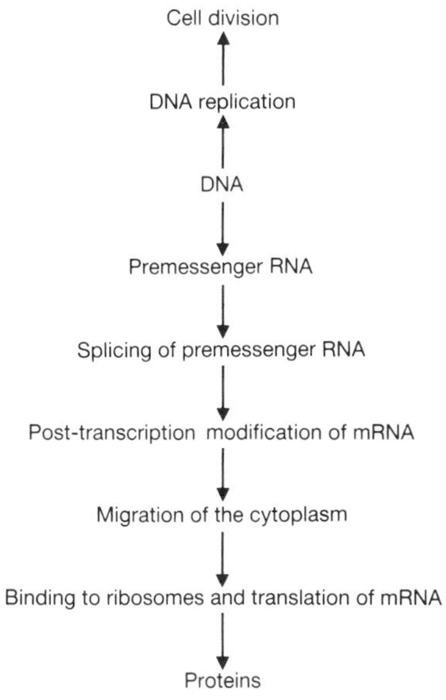

Figure 5.1. The actions of DNA.

to control these diseases should open new insights in the control of all illnesses that originate from genomic defects. Therefore, we should aim to create molecules that are directly targeted at the DNA or RNA material in a cell. Such compounds should, in principle, be able to eliminate the cause(s), and hence reverse the consequences, of viral infections and cancer.

2 Strategies to control gene expression

It is commonly accepted that the nucleic acids (DNA and RNA) are responsible for the transfer of genetic information, while the proteins, whether acting as enzymes or not, are responsible for the execution of these orders. In this viewpoint, the classical concept of a drug requires that it should interact selectively with a protein so as to modify these orders. The proteins, which function as targets for drugs, are either enzymes or receptors. The structure of the drug is generally well known from nuclear magnetic resonance (NMR) and X-ray studies. The elucidation of the protein structure is a much greater challenge, especially for membrane-embedded receptors. Recently it has become increasingly clear that, in addition to proteins, nucleic acids can also function as either enzymes (i.e. ribozymes) or receptors (i.e. for antitumoral agents).

Attempts to regulate gene expression which have been based mostly on interaction with the regulatory proteins, could in principle also be targeted at the nucleic acids. The advantage of targeting nucleic acids, in their quality of receptors, is that the structure of these receptors is well established and that, theoretically, it should be easier to design antagonists at the nucleic acid than at the protein level. Another advantage of targeting nucleic acids is that gene expression could be blocked at the earliest possible stage from its initiation.

The fundamental property of nucleic acids, involved in DNA replication to DNA, DNA transcription to RNA, and RNA translation to protein, is the Watson–Crick base-pairing (Fig. 5.2). The adenine bases of one strand of the DNA double helix are coupled with thymine bases in the other DNA strand (or uracil bases in RNA), via two hydrogen bonds [(A)N_1 . . . HN_3T; (A)NH_2 . . . O(T)]. The guanine bases are bound to the cytosine bases via three hydrogen bounds [(G)N_1H . . . N_1C); (G)NH_2 . . . O(C); (G)O . . . H_2N(C)]. The specificity of the DNA–DNA, DNA–RNA and RNA–RNA interactions is dependent on the specific base sequences.

Oligodeoxynucleotides with base sequences that are complementary to a seg-

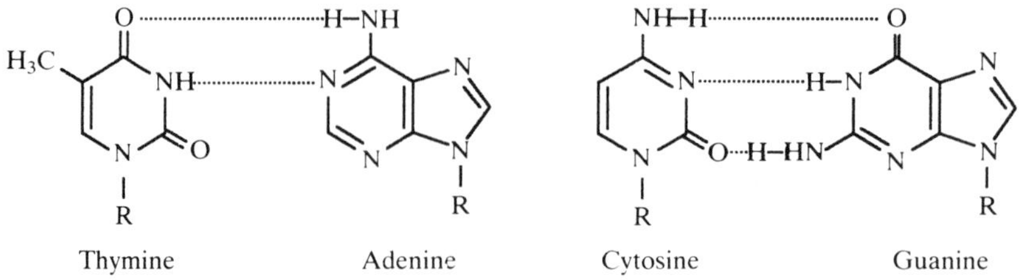

Figure 5.2. Watson–Crick base-pairing.

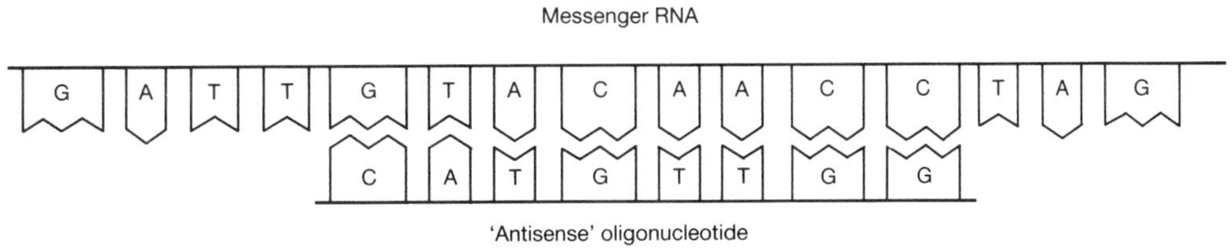

Figure 5.3. The 'antisense' oligonucleotide.

ment of DNA or mRNA could bind to these targets, thus leading to the formation of a DNA–DNA or RNA–DNA hybrid and selective inhibition of gene expression (Fig. 5.3). Such oligonucleotides have been named 'antisense' oligodeoxynucleotides (or 'antigene' oligodeoxynucleotides, if specifically directed at genes). The term 'antisense' is used for each compound whose mechanism of action is based on hybridization with a complementary target sense sequence. That this approach for drug design is not unrealistic should be evident from the fact that antisense RNA occurs in nature as a repressor of gene expression.

Two main targets can be envisaged for the antisense oligodeoxynucleotides (ASOs): DNA and mRNA. However, DNA targeting is not as simple as mRNA targeting because antisense oligonucleotides more easily hybridize with single-stranded polynucleotides than with double-stranded polynucleotides. Possible sites and modalities for interaction with DNA could then be at unpaired regions, at opened regions during supercoiling, replication or transcription, or through the formation of triple helices. However, exogenously added oligonucleotides are most likely to be concentrated in the cytoplasm, which would make translation arrest (based on interaction with mRNA) a more attainable goal than transcription arrest (based on interaction with the DNA genome).

Oligonucleotides are naturally occurring macromolecules and the problems in converting these macromolecules to drugs are of the same magnitude as the problems encountered in trying to use peptides as drugs. During the next decades we will learn not only to design peptidomimetics starting with the amino acid sequences of proteins as model but also to design nucleomimetics starting with the base sequences of nucleic acids as model. These two approaches are perhaps the most exciting new areas in medicinal chemistry for the 21st century.

3 Factors influencing the activity of natural antisense oligodeoxynucleotides

Natural ASOs are those formed from the four natural 2′-deoxynucleotides connected to each other by a phosphodiester linkage (Fig. 5.4). These antisense oligonucleotides are polyanionic substances which bind to the target RNA through a simple physicochemical process, i.e. formation of hydrogen bonds stabilized by base stacking. This hybridization should prevent the interaction of the target RNA with proteins (i.e. polymerases, ribosomes) or other nucleic acids (i.e. tRNA), or prevent its transport from the nucleus to the cytoplasm. The efficiency of the translation arrest by exogenously applied ASO is dependent on several factors (Fig. 5.5).

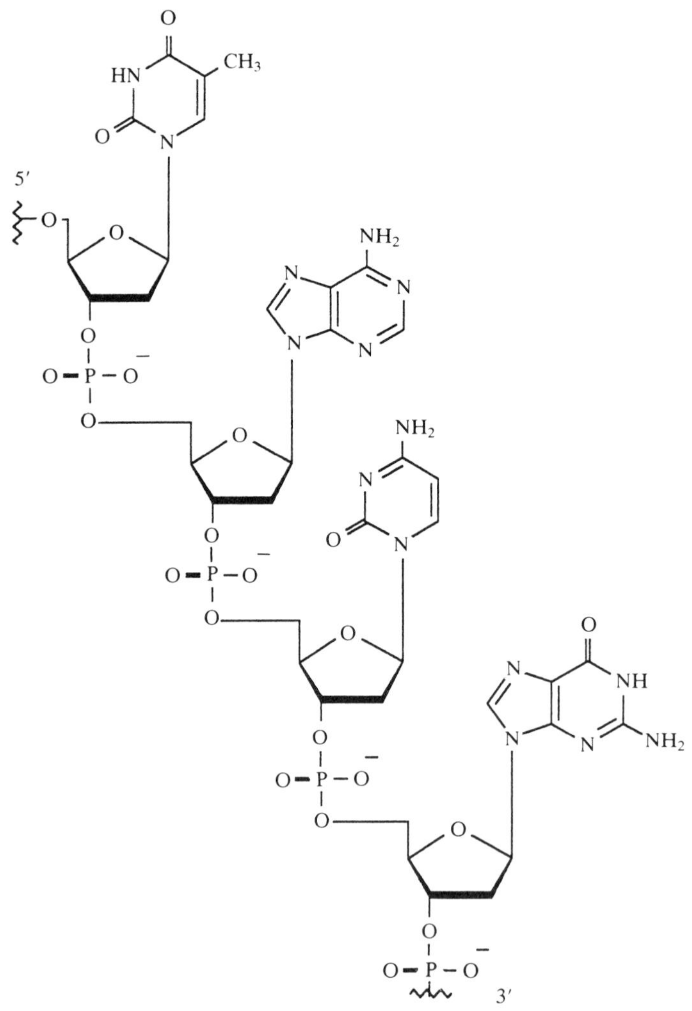

Figure 5.4. Oligodeoxynucleotide structure.

3.1 *Chain length*

The chain length of the ASO is not only important for the specificity of interaction, but also for the stability of the hybrid. The minimum chain length for a unique sequence within the human genome should be about 17 bases. As the length of the oligomer should, to some extent, also depend on the base composition (as the four different bases—A, T, G, C—are not equally distributed in the human genome) the chain length may vary from 15 (high G–C content) to about 20 (high A–T content). The necessary chain length for oligonucleotides aimed at translation arrest would also depend on the target, since only a fraction of the whole DNA is transcribed in a certain cell at a certain moment, so that specificity can be reached with less than 17 bases.

However, the specificity of ASO decreases with decreasing chain length, which means that with shorter chains, higher ASO concentrations are required. Of additional importance is that oligonucleotides have to compete with the secondary structure of the target RNA. RNA normally adopts a folded conformation, and ASO must disrupt these secondary structural regions to be able to hybridize with the target sequence.

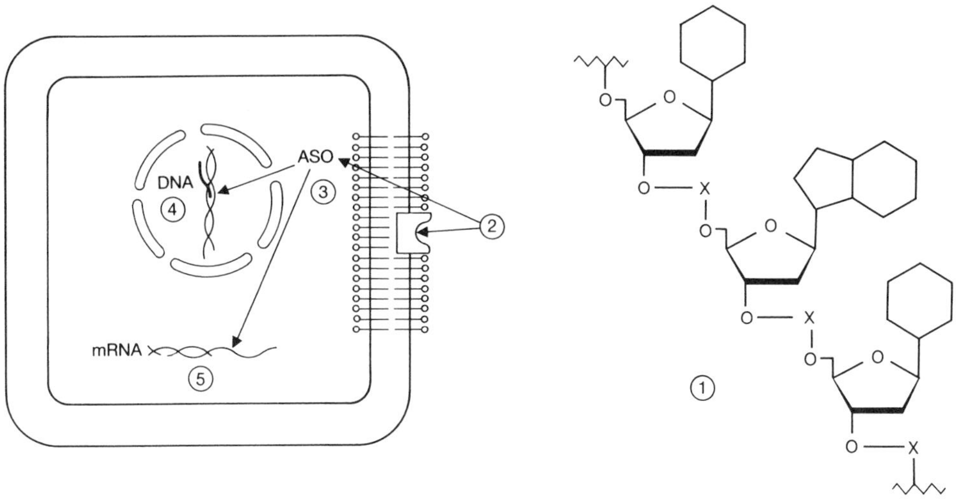

Figure 5.5. Mechanism of action of ASO. 1. nuclease resistance; 2, cellular uptake; 3, diffusion; 4, transcription arrest; 5, translation arrest.

More stable complexes can also be obtained with oligonucleotides of high C–G content or by covalently binding an intercalating agent to the ASO. Otherwise, the chain has to be elongated. However, this elongation cannot be unlimited, as it increases the risk for creating intramolecular secondary structures in the ASO itself. In this context, the use of a mixture of two oligonucleotides complementary to consecutive sequences in the target polynucleotide (the so-called 'tandem oligonucleotides') may be advantageous to the use of an oligonucleotide of double length. Also, cellular penetration would be easier for tandem oligonucleotides than for double-length oligonucleotides.

3.2 *Nuclease resistance*

Nucleases are enzymes that degrade polynucleotides by cleavage of the phosphodiester bond. They are present both inside and outside the cells. Nucleases can degrade nucleic acids from the *exo*-sites (5′- or 3′-exonucleases) or *endo*-sites (endonucleases). These enzymes may also have some sequence specificity. The presence of nucleases limits the use of normal oligodeoxynucleotides as 'antisense' because of their rapid breakdown in the extracellular medium. This could be prevented by locking the oligonucleotides into liposomes which protect them from extracellular degradation, or, alternatively, the oligonucleotides could be made resistant to nucleases by chemical modification.

An increase in stability towards degradation by 5′-exonucleases has been observed upon modification of the 5′-end, i.e. following substitution of the 5′-hydroxyl group by an amino group, or replacement of the first internucleotide linkage by an alkylphosphonate, or addition of an intercalating agent at the 5′-end. Binding of an intercalating agent at the 3′-end also protects the oligonucleotide against degradation by 3′-exonucleases. Protection of a phosphodiester bond against endonucleases can be achieved by modifying the adjacent internucleotide bond. An oligonucleotide with alternating phosphodiester and phosphotriester linkages is more resistant to endonucleases than the unmodified oligonucleotide.

3.3 *Cellular uptake of oligonucleotides*

The antisense approach is based on a specific translation arrest by sequence-specific oligonucleotides. The oligonucleotide responsible for this has to be delivered into the cell. This could be done theoretically by plasmid constructs which produce antisense RNA intracellularly or by transfection with antisense genes which are stably inserted into the host DNA (viral vector). Such constructs could control the transcription process as it is located in the nucleus. When oligonucleotides are directly delivered to the cells, the ability of the oligonucleotide to penetrate cell membranes becomes a crucial factor.

Oligonucleotides with normal phosphodiester bonds can be internalized by endocytosis. However, this process does not seem to be very efficient. Cellular uptake of normal oligonucleotides can be increased by binding to lipophilic substituents, by lipofusion or by removing the negative charges from the internucleotide phosphates. Such neutral oligonucleotides can enter cells by passive diffusion.

Modifications of the phosphodiester linkages may at the same time increase resistance to enzymatic degradation and facilitate intracellular delivery. For example, methylphosphonates are both resistant to nucleases and efficiently taken up by cells. Conjugation of oligonucleotides with poly(L-lysine) also seems to increase their cellular uptake and resistance to 3′-exonucleases. However, the effect of poly-(L-lysine) conjugation is dependent on the cell type used. Also, pH-sensitive liposomes could be employed to protect oligonucleotides against premature degradation, increase their cellular uptake and target the oligonucelotide to a particular cell type (if the liposomes have been coated with cell-specific antibodies—immunoliposomes). Liposomes can be further modified so as to prolong their half-life in the circulation. Also, coupling to intercalating agent can increase penetration across cell membranes and increase resistance towards exonucleases.

3.4 *Role of ribonuclease H*

Ribonuclease H, which is part of the reverse transcriptase complex of retroviruses, cleaves the RNA component of a DNA–RNA complex. Its role in the life cycle of retroviruses is to free the DNA formed in the first round (reverse transcription) to another round (DNA replication), thus leading to the formation of duplex DNA. Its function in the normal cell is to remove the RNA primers during DNA replication. The enzyme also occurs in the cytoplasm. When acting on an RNA–ASO complex, ribonuclease H can degrade the ASO target sequence and thus make the ASO available for hybridization to the next target sequence. In this sense, ribonuclease H can function as a kind of a catalyst. The enzyme recognizes DNA–RNA duplexes as short as four base-pairs long.

The role of ribonuclease H in the activity of unmodified antisense oligonucleotides seems to be very important. In some systems it plays a dominant role, in others not. Chemical modification of the antisense oligonucleotide (i.e. through introduction of methylphosphonate groups) may diminish the susceptibility (recognition or cleavage reaction) to ribonuclease H, which may, in turn, reduce the activity of the antisense oligonucleotides. In those cases where ribonuclease H is essential for activity, the methylphosphonate analogues may not be very active.

3.5 *The target sequence*

Ideally, the target RNA sequence should be free of secondary structures and easily attainable. Inhibition of mRNA translation can be achieved most efficiently by targeting the 5′-non-coding region or the translation start region. The likelihood of translation arrest may be enhanced if the antisense oligonucleotides are targeted at the initiation codon so that binding of the mRNA to the ribosomes is also prevented. Ribonuclease H activity and inhibition of binding to ribosomes may then result in a synergistic effect. Of additional importance is the ASO concentration. One can imagine that, at a 1 : 1 ratio of ASO to mRNA, ribosomes can easily displace the oligonucleotides from the mRNA.

3.6 *Triple helix formation*

A promising approach for targeting DNA is based on triple helix formation. After formation of a Watson–Crick base-pair with thymine, adenine is still able to bind a second thymine base via two hydrogen bonds. Also the guanine of a guanine–cytosine base-pair can still bind a second, protonated cytosine via two hydrogen bonds (Fig. 5.6). If thus a triple helix would be formed, the second thymine and protonated cytosine should fit in the major groove of the original double helix

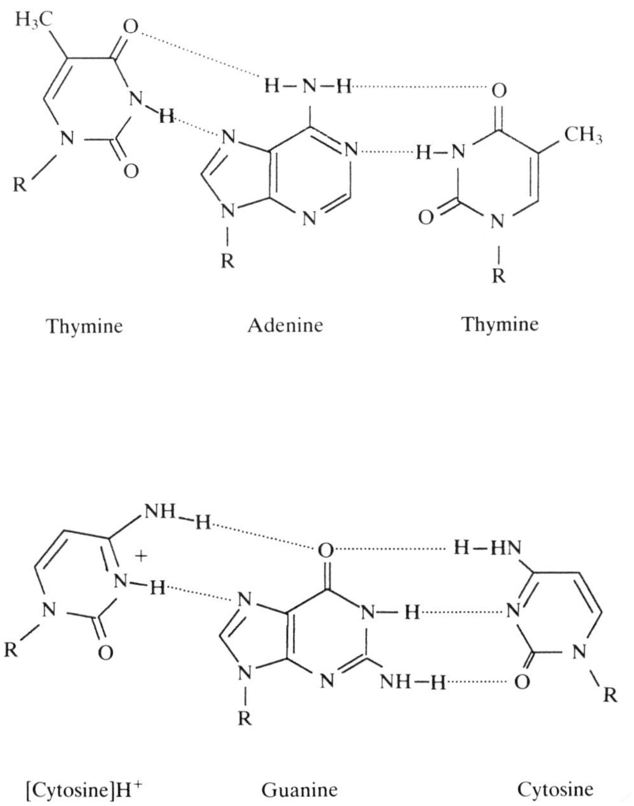

Figure 5.6. Hoogsteen and Watson–Crick base-pairing.

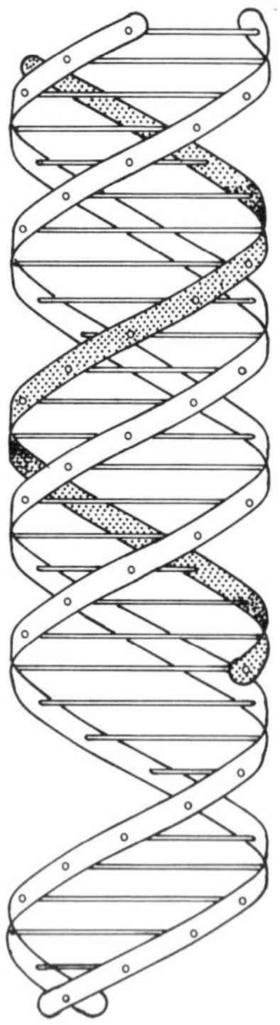

Figure 5.7. Triple helix formation.

(Fig. 5.7). This means that a homopyrimidine oligonucleotide should be able to bind by Hoogsteen base-pairing to the major groove of a duplex DNA consisting of homopyrimidine–homopurine strands. This has proved to be the case. Homopyrimidine–homopurine tracts in DNA form triple helixes with the corresponding oligodeoxyribopyrimidines. These homopyrimidine oligonucleotides bind in a parallel orientation with respect to the homopurine strand.

Oligonucleotides of specific sequence that transform duplex DNA to triple helices may permit the design of artificial sequence-specific endonucleases. Indeed, covalent attachment of DNA cleavage agents (EDTA–iron) should give site-specific double-strand breaks. Such 'restriction-like' artificial endonucleases may have an infinite choice of recognition sites (depending only on the base sequence). More tight binding could be achieved with modified pyrimidine bases. The affinity of the binding can be increased by substituting 5-methylcytosine for cytosine and 5-bromouracil for thymine.

Also, the binding of these oligonucleotides could be strengthened by covalently linking intercalating agents to the homopyrimidine strand targeted at the homopurine–homopyrimidine sequences of the duplex DNA. Theoretically, it should be possible to design more sequence-specific major groove-binding oligonucleotides composed of all four types of bases. These could regulate expression of genes of any base composition at the transcriptional level, and, depending on the target sequence, both inhibition and activation of transcription could be accomplished.

4 Modification of 'antisense oligonucleotides'

4.1 *Modification at the internucleotide linkage*

In designing modified oligonucleotides so as to increase cellular uptake and resistance to nucleases, the question arises, which modifications would not jeopardize the affinity of the ASO for its target sequence?

• not to loose specificity modification in the base part of the nucleotide should not disrupt normal Watson–Crick base-pairing;

• the five-membered furanose ring is essential for helix formation—only small modifications to this five-membered ring seem to be permissible (e.g. change in configuration from β to α);

• modification of the phosphate backbone seems to be the most logical strategy. As long as the internucleotide linkages still contain a pentavalent phosphorus atom, analogous geometry can be expected as with the normal phosphodiester bonds.

Some of the modifications which have been carried out till now, together with their characteristics, are summarized in Table 5.1 and Fig. 5.8.

α-Deoxyoligonucleotides form stable duplexes with a complementary β-strand RNA, with parallel orientation of the two chains. α-Deoxyoligonucleotides are poor substrates for nucleases. However, they do not cause sequence specific inhibition of translation in biological systems, probably due to the fact that RNA duplexes with α-4oligonucleotides are not substrates for ribonuclease H.

Table 5.1. Modification of oligonucleotides

	Modification			
Effect	Phosphodiesters with α-deoxynucleosides	Methylphosphonates	Phosphorothioates	Phosphotriesters
Retain H_2O solubility of phosphodiesters	+	–	+	–
Increased nuclease resistance, relative to natural phosphodiesters	+	+	+	+
Increased cellular uptake		+	+	+
Capability of hybridization and formation of stable duplexes	+	+	+	+
Main problem	RNA hybrids not substrates for RNase H	Chirality; RNA hybrids not substrates for RNase H	Chirality; poor internalization; sequence non-specific action	Chirality; tendency to undergo self-association by hydrophobic interaction

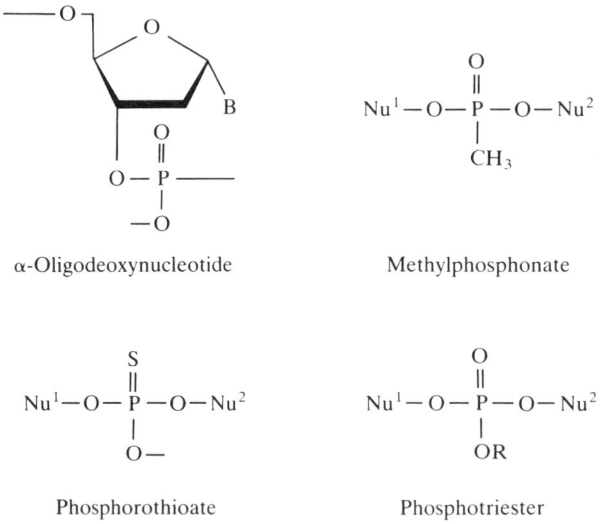

Figure 5.8. Modification of 'antisense oligonucleotides'.

Phosphorothioate oligodeoxynucleotides have a chiral centre at the phosphorus atom which means that n phosphodiester linkages represent 2^n stereoisomers, of which presumably only one has the ideal configuration for activity. Phosphorothioate oligodeoxynucleotides have a higher affinity for the receptor regulating cellular uptake than normal oligonucleotides. However, this does not mean that phosphorothioate oligodeoxynucleotides are taken up better than phosphodiester oligodeoxynucleotides. In fact, they seem to stick so tightly to the receptor that they are poorly delivered intracellularly. Also, oligothioates behave as specific antisense inhibitors at low concentration and as sequence non-specific inhibitors at high concentration. This sequence non-specific inhibition of translation is highly dependent on the length of the oligonucleotides, since, for example, a homopolymer of 28 deoxycytidines (SdC_{28}) is a potent inhibitor of translation. These molecules clearly act by a mechanism other than hybridization arrest. This could be explained by the tendency of the polyanionic thiophosphates to bind non-specifically to several proteins, thus blocking the normal functioning of these proteins. As an example, SdC_{28} is an inhibitor of HIV reverse transcriptase because it directly binds to the enzyme. Phosphorothioate–mRNA hybrids are good substrates for ribonuclease H. It even appears that ribonuclease H prefers internucleotide phosphorothioate linkages over unmodified phosphates.

None of the three oligonucleotides (α-oligonucleotides or β-oligonucleotides with normal phosphodiester linkages or phosphorothioate linkages) is ideal to function as an ASO. However, through combination of the different internucleotide bonds, oligonucleotides could be obtained that combine the advantages of the individual oligonucleotides. Also, they could be conceived with modified phosphodiester linkages only at certain critical sites. Within a single ASO, normal phosphodiester bonds could be combined with phosphorothioate or methylphosphonate bonds, as well as α-anomers substituted for the β-anomers. Such constructs may be resistant to nucleases, sufficiently soluble in water, internalize readily, hybridize well, form efficient substrates for ribonuclease H, and exert the expected activity without toxicity for the host cell.

Oligodeoxynucleotides connected by non-ionic phosphate internucleotide bonds (n) also consit of 2^n isomers. They could be phosphotriesters, alkylphosphonates, alkylphosphonothioates, alkylphosphoramidates, alkylphosphothioamidates, and so on. Duplexes of RNA with such compounds should show increased stability because of the reduced charge repulsion between the non-ionic ASO backbone and the negatively charged RNA backbone. In fact, oligonucleotide methylphosphonates have been shown specifically to inhibit viral expression in cell cultures. However, oligonucleotide methylphosphonates do not bind well to RNA regions with a particular secondary structure. In addition, methylphosphonates do not activate ribonuclease H. This problem is partially compensated by their better cellular uptake and could also be overcome by using longer oligomers.

Oligonucleotides with non-ionic non-phosphate internucleotide linkages (e.g. carbonates, carbamate) may be expected to behave more similarly to the oligonucleotide methylphosphonates than to those with the regular phosphodiester groups. However, the geometry of such internucleoside linkage may also be drastically changed.

4.2 *Oligodeoxynucleotides with a covalently bound intercalating agent*

All aforementioned oligonucleotides bind to their mRNA target sequences according to an equilibrium process, so that rather high concentrations are required for inhibition of translation. Sufficiently tight binding to the mRNA is needed for inhibition of protein synthesis. ASO showing a melting point below 36 °C cannot be expected to inhibit the translation process *in vivo*. Here, a melting point means the temperature at which 50% of the duplex is dissociated into the two components (mRNA and ASO).

The stability of the duplex between an antisense oligonucleotide and its target sequence can be increased by covalent linking of an intercalating agent (Fig. 5.9). This procedure should increase the efficiency of translation of inhibition. The intercalating agent could be a polyaromatic or polyheteroaromatic compound

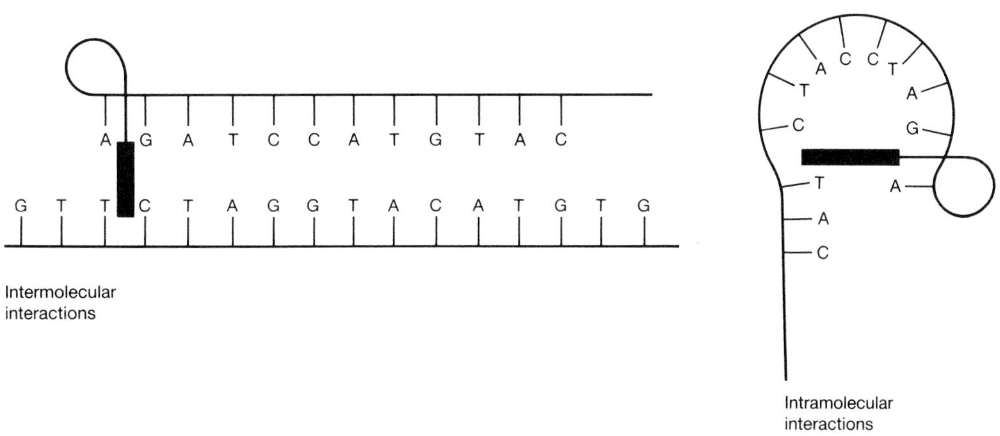

Figure 5.9. Intercalating agent covalently bound to an ASO.

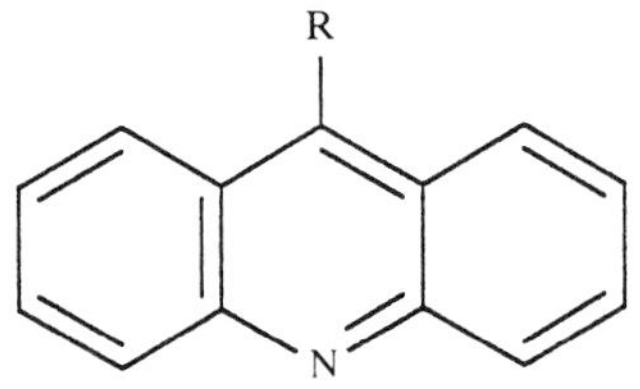

Figure 5.10. Acridine (dibenzo[*b*,*e*]pyridine).

which is inserted between two consecutive base-pairs in the double helix. The structure of the linker by which the intercalating agent is bound to the antisense oligonucleotide and the site of attachment is crucial for ensuring a strong interaction with the target sequence.

It has been proven that with an acridine derivative (Fig. 5.10) bound to the end of the oligonucleotide, a construct is obtained with a higher melting point than without the intercalating agent. This means that the affinity of the ASO for the target sequence is increased. The complex formed between the complementary sequences is stabilized by additional binding energy.

As mentioned above, normal oligomers of 15–20 bases long are efficient inhibitors of translation. This chain length can be reduced to 11 by binding an intercalating agent. Such constructs may be quite effective even in the absence of ribonuclease H activity. Moreover, an acridine moiety linked to the end of an oligonucleotide does not abolish its affinity for ribonuclease H.

However, binding of intercalating agent is not without disadvantages. Indeed, intramolecular interactions between the intercalating agent and the nucleic acid bases and non-specific intermolecular interactions may occur. The additional binding energy provided by the intercalating agent could stabilize secondary structures or compensate for mismatches. These interactions have to be disrupted before the antisense oligonucleotide could interact with its target molecule. Also, at high concentrations of the intercalator-linked oligonucleotides, non-specific inhibition of gene expression may occur due to non-specific interactions with other cellular components.

4.3 *Oligonucleotides covalently bound to a reactive group*

As mentioned above, the process of translation arrest can be potentiated by the presence of ribonuclease H. As a rule, antisense oligonucloeotides do not inhibit protein synthesis in cell-free extracts in the absence of ribonuclease H activity. However, the relative importance of this enzyme in the mechanism of action of ASO is not completely clear and seems to depend on the given situation. Therefore, it would seem desirable to induce irreversible changes in the target nucleic acid so that even in the absence of ribonuclease H translation arrest occurs. This can be achieved by covalently attaching a reactive group at the end of the oligonucleotide. Other possibilities include the attachment of nucleases to the oligonucleotides or the insertion of sequences responsible for ribozyme activity.

Oligonucleotides can be substituted by reactive groups which induce irreversible changes in the target sequence, such as single- or double-strand cleavage. This should prevent translation of RNA at the site where the oligonucleotide is bound.

Cross-linking can be obtained with alkylating agents bound to the ASO. Typical cross-linking agents are molecules containing an alkylating group such as *N*-chloroethyl (Fig. 5.11). Other reagents can be activated in a chemical or photochemical reaction, and then induce the irreversible damage at the target sequence. The photochemical activation approach might be useful in the treatment of skin cancers, as well as haematological disorders provided that extracorporeal irradiation is feasible.

Chemical damage can be induced by the activation of oxygen or by the formation of hydroxyl radicals with the aid of transition-metal complexes (i.e. phenanthroline–copper complexes or EDTA–iron complexes). Azido derivatives can be used as photoactive cross-linking groups to generate nitrene radicals. The reactive intermediate would be generated by light. Also, psoralene derivatives could cause cross-linking after irradiation. Proflavine can be used to form singlet oxygen upon irradiation.

The advantage of using reactive oligonucleotides is that this should reduce the ASO concentration needed and that the target mRNA sequence would be inactivated irreversibly. Once it has been bound, the ASO cannot be dissociated from the target sequence or displaced by ribosomes or polymerases. Efficiency of ASO binding could even be increased by coupling the ASO at one end with an intercalating agent and at the other end with a reactive group.

However, problems may be encountered from groups that are too reactive. They could react non-specifically with proteins or other cellular components thus leading to toxicity, or they could react with one another, thus leading to self-destruction. Cross-linking via covalent bond formation is not a good approach when a catalytic process is envisaged whereby one antisense molecule must inactivate several target sequences one after the other. In this case, the modified ASO should be recoverable after each reaction.

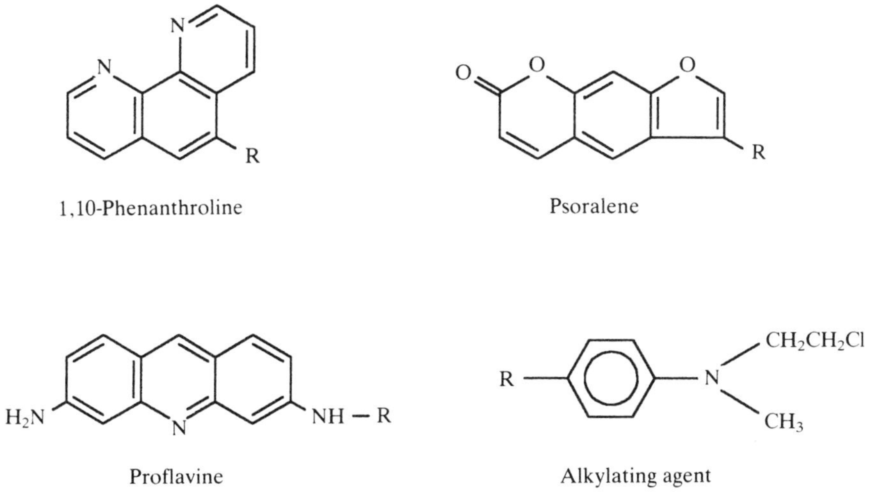

Figure 5.11. Examples of reactive groups. The alkylating agent contains an alkylating group in the form of *N*-chloroethyl.

5 Potential therapeutic applications

5.1 *Toxicity*

Before any therapeutic applications can be envisaged, the potential toxicity of the antisense oligonucleotides should be taken into consideration. Acute toxicity studies with normal oligonucleotides and some modified oligonucleotides (phosphorothioates, methylphosphonates) have demonstrated that these molecules are well tolerated. However, nothing is known about the long-term toxicity of these oligonucleotides. Nor have acute toxicity studies been carried out with most of the modified ASOs.

The normal oligonucleotides are hydrolysed to nucleotides and nucleosides by the action of nucleases. Although most modified oligonucleotides do show increased resistance to nucleases, some are still digested very slowly and the potential toxicological consequences of large pools of modified nucleotides within the cells should not be ignored. Oligonucleotide phosphorothioates could be hydrolysed to nucleoside monothiophosphates which could then be converted to their α-thiophosphate-β,γ-diphosphates and interact with DNA synthesis.

Degradation of oligonucleotides with an enzymatically stable internucleotide linkage by glycosylases may release the free purine and pyrimidine bases. Degradation of oligonucleotides with a covalently bound intercalating agent may release the free intercalating agents, known to be cytotoxic. A chemically and enzymatically stable linker may therefore be required to prevent the release of the intercalating agents.

Questions about detoxification and elimination of the ASO have to be resolved before the compounds could be considered for therapeutic use.

5.2 *Therapeutic application*

The ASO approach towards the control of gene expression should, in the first place, be useful in the treatment of parasitic and viral infections, and cancer. In addition, antisense oligonucleotides should be useful in elucidating the functioning of the individual genes, as could be deduced from a selective inhibition of the synthesis of proteins that are normally encoded by these genes. ASOs could also prove useful in the study of the targets for drug action, such as receptors, enzymes and ion channels.

As has been demonstrated with several viruses (i.e. influenza virus, herpes simplex virus, HIV and vesicular stomatitis virus), ASO can suppress virus replication through translation arrest. ASOs have also been shown to possess antiparasitic activity (*Trypanosoma, Leishmania).* Antisense oligonucleotides may also be expected not to lead as easily to drug resistance as do the classical chemotherapeutic agents.

An important target for ASO is the expression of oncogenes. Distinct oncogene products which have been identified are tyrosine-specific protein kinases (e.g. *abl*), guanine nucleotide-binding proteins with GTP-ase activity (e.g. *ras*), growth factors (e.g. *sis*) and nuclear proteins (e.g. *myc*) which show altered expression following chromosomal translocation, mutations or gene amplification. ASOs are able to reduce the expression of several of these proteins (i.e. *myc, ras, fos, mos*). The use of ASO could be envisaged in both the prevention and therapy of cancer.

Before ASOs could be used to control the expression of these or any other proteins, much has to be learned about the intracellular and extracellular factors that can influence the activity of the oligonucleotide in the body, i.e. absorption, metabolism, distribution, compartmentalization, elimination and toxicity. We still do not know whether these oligonucleotides would survive long enough *in vivo* to exert their action. We know little about their exact mechanism of action, their hybridization capacities in physiological conditions and, exactly, how important is the role of ribonuclease H. The concept of triple helix formation with its potential to target any sequence in the human genome has to be further elaborated. We still do not know whether the sequence-specific binding of ASO to double-helical DNA would be sufficiently effective completely to suppress gene expression.

6 Conclusion

Antisense oligonucleotides are aimed at hybridizing with a target DNA or RNA sequence and inactivating its expression. This approach could be followed to control selectively the expression of virtually all genes of the human genome. Thus, ASOs could theoretically be used to treat all illnesses that originate from genomic defects. However, much improvement is required before we can enter the clinic with these synthetic constructs. Cellular uptake, nuclease resistance and site-specific delivery remain problems inherent to the ASO approach, but besides these problems, the pharmacokinetics also need to be improved considerably. ASO constructs are needed that bind to and inhibit the target sequence with greater efficiency so that the ratio between the amount necessary for translation arrest and the administered dose can be decreased. Various new chemically modified derivatives have to be synthesized, since we still do not know the optimum structural characteristics. Limiting factors in assessing the therapeutic efficacy of ASO *in vivo* are the very high production costs and the difficulties in scaling up the chemical synthesis. Regardless of these problems, antisense oligonucleotide may be predicted to be one of the major tools of pharmacology in the 21st century.

7 Further reading

Cohen, J.S. (Ed.) (1989). Oligonucleotides. Antisense inhibitors of gene expression. *Topics Molec. Struct. Biol.*, **12**, 1–255, Macmillan Press, (London).

Dolnick, B.J. (1990). Antisense agents in pharmacology. *Biochem. Pharmacol.*, **40**, 671–5.

Green, P.J., Pines, O. & Inouye, M. (1986). The role of antisense RNA in gene regulation. *Ann. Rev. Biochem.*, **55**, 569–97.

Stein, C.A. & Cohen J.S. (1988). Oligodeoxynucleotides as inhibitors of gene expression: a review. *Cancer Res.*, **48**, 2659–68.

Thuong, N.T. & Asseline, U. (1985). Chemical synthesis of natural and modified oligodeoxynucleotides. *Biochimie*, **67**, 673–84.

Toulmé, J.-J. & Hélène, C. (1988). Antimessenger oligodeoxyribonucleotides: an alternative to antisense RNA for artificial regulation of gene expression—a review. *Gene*, **72**, 51–8.

Uhlmann, E. & Peyman A. (1990). Antisense oligonucleotides: a new therapeutic principle. *Chem. Rev.*, **90**, 543–84.

Zon, G. (1987). Synthesis of backbone-modified DNA analogues for biological applications. *J. Protein Chem.*, **6**, 131–45.

Zon, G. (1988). Oligonucleotide analogues as potential chemotherapeutic agents. *Pharmaceut. Res.*, **5**, 539–49.

Part 2
Protein Structure–Function Relationships

Cloning and Structural Studies on Receptors and Enzymes

6 Molecular Biology of Drug Receptors and the Advent of Reverse Pharmacology

J.-C. SCHWARTZ

Unité de Neurobiologie et Pharmacologie (U. 109) de l'INSERM, Centre Paul Broca, 2ter rue d'Alésia, 75014 Paris, France

1 Introduction

A large part of pharmacology and drug design deals with studies of, as well as attempts at, mimicking or inhibiting the interaction of small molecules, e.g. amino acids and amines, and oligo- or polypeptides (messengers or substrates) with large proteins, e.g. enzymes or receptors. For a long time these studies were mainly performed using crude tissues or tissue preparations and analysing the interaction of the small, natural or chemically synthesized ligands with the protein indirectly, e.g. via quantification of a reaction product, a biological response or, more recently, competition with a radioactive probe.

In the best cases a 'working model' of the target, i.e. the active site of an enzyme or the ligand-binding domain of a receptor, has been derived from structure–activity analysis of the apparent affinities of diverse synthetic ligands.

Whereas these models have often been extremely useful in operational terms, the process of drug design could never be an entirely rational one, like one that would have been based upon the precise knowledge of the mechanisms of interaction of small ligands with large proteins. However, this situation is rapidly changing, mainly as a result of the development of cDNA technologies and their wide application to the cloning of a number of genes of high pharmacological interest, particularly those of receptors. It seems, therefore, quite likely that this evolution in biology will be soon accompanied by a parallel evolution in pharmacology and drug design. In this chapter, the evolution is analysed in the field of receptors, one in which molecular biology has provided dramatic advances during recent years.

2 Classical pharmacology: from drug to receptor and from receptor to gene

From the time of Ehrlich until the 1960s, the receptor remained a necessary but essentially theoretical entity: *corpora non agunt nisi fixata*. During this period the 'discovery' of a novel receptor (or receptor subtype) relied upon the identification or chemical design of ligands able to interact more or less selectively with the novel entity. Thus Ahlquist's suggestion of the existence of distinct α- and β-adrenoreceptors was mainly based upon the observation that a series of catecholamines had different rank order potencies on various biological responses: the existence of the two adrenergic receptors was considered as proven with the design by Sir J.W. Black and colleagues of the first selective β-adrenergic receptor antagonists.

A similar process can be described for the histamine H_2-receptor, the existence of which was suspected from the use of a series of agonists by Ash and Schild and definitively demonstrated through the design by Black and colleagues of the first H_2-receptor antagonists.

In this traditional process of receptor subtype identification, two critical conditions have to be satisfied: a biological response selectively mediated by the novel receptor should be identified and several ligands with some degree of selectivity toward the various receptor subtypes should be available. This is well illustrated in the case of the discovery of the H_3-histamine receptor subtype. Thus, the two key issues in this process were (i) the design of a biological test system to monitor histamine release from brain slices, and (ii) the availability of a series of histamine agonists and antagonists with established potencies at H_1- and H_2-receptors (Arrang *et al.*, 1983). The definitive demonstration of the existence of the H_3-receptor was provided through the design of highly selective agonists and antagonists (Arrang *et al.*, 1987).

Chemical ligands in the form of reagents for affinity chromatography were also key tools in the isolation of the first receptor proteins, e.g. the nicotinic acetylcholine receptor (reviewed by Changeux, 1981) or the β_2-adrenergic receptor (reviewed by Lefkowitz *et al.*, 1989). In both instances, partial peptide sequences were obtained from the purified receptor proteins. This information was then used to construct oligonucleotide probes and the specific genes were isolated from cDNA or genomic libraries. Another approach consists of screening cDNA or genomic libraries using an expression system to identify the encoded receptor protein via serial electrophysiological or biochemical tests.

These 'normal' processes, requiring a knowledge of partial sequences or, at least, of the pharmacology of the receptors, were used to identify the genes (and thereby deduce the entire amino acid sequences) of numerous receptors belonging to various superfamilies. These comprise the superfamilies of ligand-regulated channels (nicotinic, γ-aminobutyric, glycine and glutamate receptors), G-protein-linked receptors (various adrenergic, muscarinic and luteinizing hormone receptors), ligand-regulated enzymes (receptors with tyrosine kinase activity such as the insulin receptor) or steroid receptors (receptors for not only steroids but also thyroid hormone, vitamin D_3 and retinoic acid).

From a practical viewpoint, the discovery of this procedure meant a rather tedious method since it required a selective biological model and several 'preliminary' ligands for receptor identification; for receptor and then gene isolation, selective ligands and powerful analytical methods are generally required.

From the mid-1970s onwards, the progressive development of radioligand binding methodology suggested it would provide great help in defining novel receptor subtypes and designing novel classes of drugs (Snyder, 1984). Yet, although this methodology has largely facilitated drug screening in the pharmaceutical industry, it has also brought some confusion as to the definition of novel receptor subtypes when used alone, and it has met with rather limited success in respect of the design of original drug classes.

Assuming that the definition of biological targets is the initial and, therefore, rate-limiting step for the design of novel drugs, the landscape has changed rapidly during the last few years with the advent of 'reverse pharmacology'.

3 Reverse pharmacology: from gene to receptor and from receptor to drug

The isolation of the first genes in the various receptor superfamilies revealed that the clustering of members (according to their mode of signal transmission) was accompanied by large structural similarities and even similarities in amino acid sequence. Thus the genes for many members of a particular receptor family, e.g. the catecholamine or serotonin receptors, are sufficiently close to one another in protein—and therefore in DNA—sequences that they can be used to isolate one another at the gene level using various molecular biological techniques. Two main approaches are currently being used: (i) screening of cDNA or DNA libraries using a series of degenerated probes, i.e. radioactive nucleotide sequences resembling those of already known (cloned) receptor genes; and (ii) amplification of cDNAs via the polymerase chain reaction (PCR), using degenerated primers based upon sequences of already known receptor genes. Both approaches lead to the cloning of genes for receptors of unknown identity which have, thereafter, to be expressed in mammalian cells for the purpose of identification. In many cases this identification, through binding or electrophysiological techniques as well as through monitoring of biochemical responses (e.g. assay of intracellular second messengers), constitutes a difficult task and receptors remain 'orphan', i.e. in search of a function or a ligand for varying periods. As an example, the group of Vassart and co-workers, who initiated the use of the PCR methodology in receptor gene cloning, described the sequence of a number of receptors belonging to the superfamily of G-protein-linked

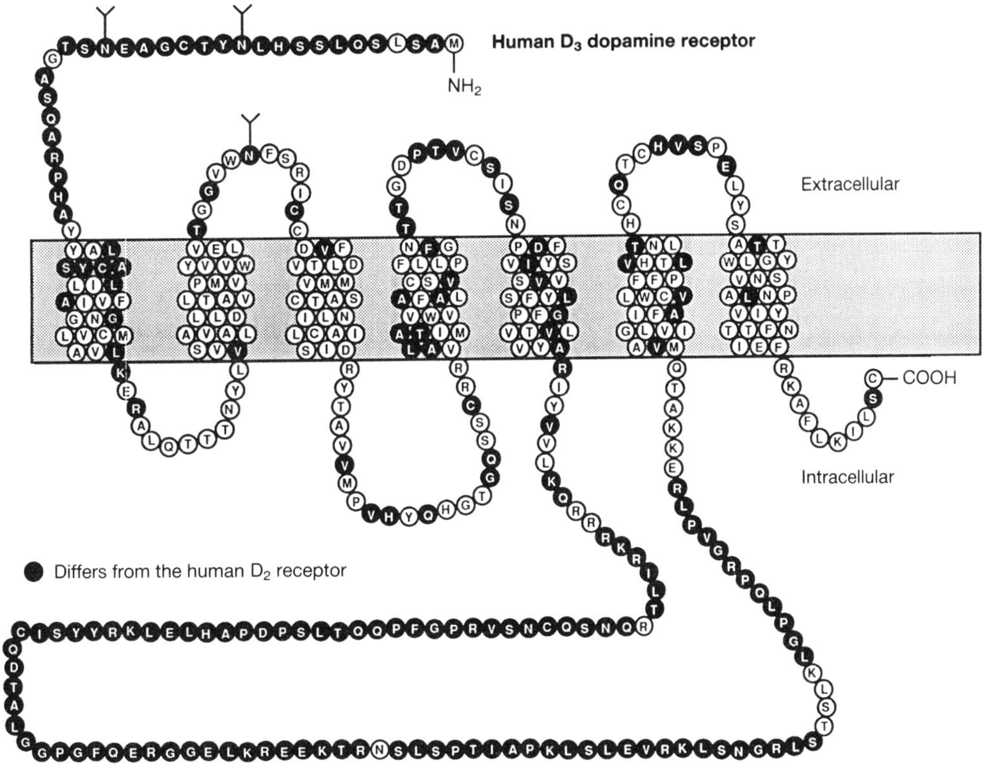

Figure 6.1. Amino acid sequence of the human dopamine D_3-receptor and comparison with that of the D_2-receptor.

receptors but whose identity remained unknown for a while (Libert *et al.*, 1989); one of them was recently identified as an adenosine receptor (Libert *et al.*, 1991). A similar case is the cloning of the cannabinoid receptor which belongs to the same superfamily (Matsuda *et al.*, 1990) and orphan receptors are also to be found in the steroid receptor superfamily (O'Malley, 1990).

In several other instances, cloning by serendipity leads to the isolation of a receptor whose close sequence homology with an already known receptor leaves little doubt as to its identity. As an example, the cloning of the rat and human dopamine D_3-receptors, using probes derived from the dopamine D_2-receptor sequence, led to a sequence displaying with the latter a homology as high as ~80% in the transmembrane domains which determine the pharmacological profile (Fig. 6.1). In many such instances, cloning reveals the existence of receptors which were previously unexpected or, at least, a matter for controversy: this is illustrated in the case of the multiple muscarinic (Burgen, 1984) or dopamine D_2-receptor subtypes (Schwartz *et al.*, 1984), the existence of which—mainly suggested by binding approaches—was denied by a number of pharmacologists.

The molecular biological approach not only proves definitively the existence of receptors by revealing their sequence and tissue localization but provides powerful biological assay systems to design novel classes of drugs.

4 Reverse pharmacology reveals multiple targets for drug design

The most fascinating observation to evolve from the cloning of receptors is the unexpectedly large size of the various superfamilies. It is sufficient to quote the cases of the five muscarinic receptors or the five dopaminergic receptors among G-protein-linked receptors; in this superfamily, diversity arises essentially from the existence of distinct, generally intronless, genes, although diversity can also occur, as in the case of D_2- and D_3-receptors, through alternative splicing of a pro-mRNA derived from a single gene (Giros *et al.*, 1989, 1991). In the case of ligand-regulated channels, e.g. nicotinic or $GABA_A$-receptors, an unexpected and even larger diversity also arises, this time as a result of assembly of diverse homologous subunits by distinct genes (Galzi *et al.*, 1991).

From a theoretical viewpoint, it is clear that this multiplicity means a greater diversity, flexibility and complexity within processes of cellular communication than has been envisaged so far. From a therapeutic viewpoint, one obvious practical consequence of this diversity is that it offers the opportunity of markedly improving the specificity of drugs. Indeed the multiple dopamine or GABA-receptor subtypes are diversely expressed among cell types where they presumably control diverse functions. As a consequence, a great deal of tissue and functional selectivity can reasonably be expected from drugs interacting with a single receptor subtype. Hence, among the five dopamine receptor subtypes known so far, only D_1- and D_2-receptors seem to be abundantly expressed in all dopaminoceptive areas of the brain, whereas the other subtypes seem much less abundant and, more importantly, expressed only in restricted cerebral areas. For instance, the dopamine D_3-receptor is absent or poorly expressed in the neostriatum or pituitary (which both express very high D_2-receptor density), whereas it seems to be highly expressed in several limbic areas (Sokoloff *et al.*, 1990). From their pharmacological profile, both D_2- and D_3-receptors are good candidates for responsibility for the antipsychotic activity of neuroleptics, whereas the motor and endocrine side-effects of these drugs could well arise from blockade of neostriatal and pituitary D_2-receptors. It is widely believed that antipsychotic activity results from dopamine antagonism in the limbic system, so that selective blockade of D_3-receptors through the design of novel compounds constitutes a promising novel approach to the treatment of schizophrenics without many of the troublesome side-effects of presently used treatments. In other words, the existence of multiple, differentially expressed receptor subtypes for a single cell-to-cell messenger provides a unique means to design tissue-selective drugs, a goal traditional for pharmacochemists but which has proved difficult to reach by other approaches.

In addition, knowledge of the detailed structure and metabolism of receptors is likely to reveal novel targets for drug design. This is well established in the case of $GABA_A$-receptors whose benzodiazepine and barbiturate sites have been the targets of large numbers of drugs in the past, in as much as they seem to mediate more subtle effects than the GABA binding site. Such allosteric regulatory sites, distinct from the main ligand recognition site, are apparently to be found in other types of channel-associated receptors: this is illustrated with the chlorpromazine and glycine-regulatory sites of the nicotinic (Galzi *et al.*, 1991) and glutamate-NMDA receptor, respectively (Johnson & Ascher, 1987). There is no evidence for such allosteric regulatory sites so far among G-protein-linked receptors but, in this case, novel drug targets were recently revealed: intracellular enzymes involved in receptor modification. Homologous β-adrenergic receptor desensitization involves phosphorylation of hydroxyl amino acids at the C-terminus tail by a unique, non-cAMP-dependent kinase, the so-called β-adrenergic receptor kinase (Dohlman *et al.*, 1991). The specificity profile of this enzyme toward various substrates remains to be established but it could well represent an interesting target for novel drugs modifying β- adrenergic transmission via its inhibition. As the details of receptor metabolism, including regulated biosynthesis, translocation, membrane insertion, aggregation and degradation, are progressively unravelled, it seems likely that novel drug targets will appear through this process.

Expression of cloned receptors through transfection into a well-defined recipient cell allows more detailed analysis of the complex protein–protein interactions leading to signal transduction. Thus recent studies with G-protein-linked receptors have shown that a single receptor can potentially interact with several distinct G-proteins and thereby use several intracellular signalling systems. The number of such G-proteins appears to be much higher than was previously thought, as a result of the existence of multiple genes (mainly for the α-subunit interacting with receptors) as well as from the generation of various gene transcripts via alternative splicing (Simon *et al.*, 1991). A definite pool of G-proteins is present in a given cell type and each may interact with different membrane receptors expressed by this cell; this presumably accounts for the receptor–receptor interactions that could be exploited pharmacologically. Among various such examples, we have recently shown that transfected dopamine D_2-receptors, which were only known to interact negatively with adenylate cylase and phospholipase C, could also trigger the release of arachidonic acid in transfected Chinese hamster ovary (CHO) cells under certain conditions: Ca^{2+} influx has to be simultaneously triggered by stimulation of another receptor inducing such influx; furthermore, the effect is enhanced in a markedly synergistic manner by co-stimulation of dopamine D_1-receptors transfected in the same cell (Piomelli *et al.*, 1991). It seems clear that a therapeutic strategy, in, for example, Parkinson's disease, could derive from this knowledge.

5 Drug screening with cloned receptors

In some instances, assessing the biological activity of drugs on cells (or cell membranes) transfected with a cloned receptor gene represents the only possibility of assessing such activity. This is the case for recently identified receptor subtypes for which no fully selective ligand or biological response is available for assays with crude tissue preparations, particularly when their abundance is extremely low (e.g. some of the muscarinic- or dopamine-receptor subtypes). However, even in the case of receptors for which natural cells or tissue can be used, the transfected cell often offers several advantages:

1 The most important advantage is *selectivity*, in as much as a wild-type cell can be chosen for transfection which does not express any potentially interfering receptor subtype. The recent discovery of an unexpected multiplicity of receptor subtypes shows that this might rarely be the case in tissues. This difficulty is again well illustrated with dopamine receptor subtypes: even a ligand such as [^{125}I]iodosulpiride, previously thought to label a homogeneous population of D_2-receptors in the brain (Martres *et al.*, 1985), was presumably also interacting with D_3-receptors (Sokoloff *et al.*, 1990) and, possibly, other subtypes yet to be discovered.

A major advantage of transfected cells, particularly for the design of drugs for clinical use, is that they allow assessment of the activity of drugs on cloned human receptors: pharmacological effect may differ among species (e.g. histamine H_1-receptors).

2 Another important advantage of transfected cells is that of *technical facility*: the detection of binding sites or signals over low background is made easy by the high level of receptor expression. Also, this level being generally constant among batches of cloned cells, a much higher degree of reproducibility than with animal tissues can be reached. In addition, fibroblasts or tumoral cells that are generally selected for this purpose are easy, and not very expensive, to culture.

3 Finally, the model of cells transfected with cloned receptors offers much *flexibility*, allowing testing of the effects of ligands, particularly full or partial agonists, under a variety of conditions. For instance, the density of receptors per cell, which is known to influence the intrinsic activity of agonists, can be varied in different clones. Also, in the superfamily of G-protein-linked receptors, it is becoming more apparent that the nature of the final biological response they mediate may vary among cells and, in a given cell, is specified by the association of three proteins: the receptor, the G-protein and the effector, each of which exists under multiple subtypes. As soon as the kind of association prevailing in a natural tissue or cell type selected as a drug target is established (by visualization through *in situ* hybridization), it is feasible to reproduce the association artificially by co-transfection of an appropriate recipient cell. As an example, CHO cells co-transfected with dopamine D_1- and D_2-receptors can be considered as models for a subpopulation of striatal neurones on which antiparkinsonian dopamine agonists can easily be studied (Piomelli *et al.*, 1991).

Whereas it is clear that cells expressing cloned receptors constitute useful tools for drug screening and establishing precise structure–activity relationships, the next step to expect is a fully rational drug design based upon physicochemical studies of ligand–receptor interactions.

6 Rational drug design with cloned receptors

Until now, the design of receptor ligands has not been an entirely rational process based upon knowledge of the chemical groups in the active-site amino acid residues responsible for the binding. This situation is likely to be progressively modified, however, as the structure of receptor proteins is unravelled in greater detail experimentally, e.g. irreversible labelling, construction of chimeras and site-directed mutagenesis. These studies have been extensive in the case of nicotinic (Galzi *et al.*, 1991) and β-adrenergic (Strader *et al.*, 1989; Dohlman *et al.*, 1991) receptors. In the latter case, analysis of chimeras and site-directed mutagenesis, in conjunction with structural alterations of ligands, has revealed areas of the trans-membrane domains responsible for the recognition of ligands. Binding sites for agonists and antagonists have both distinct and shared determinants within β-adrenergic receptors and major determinants of antagonist binding specificity lie in the 7th transmembrane domain (TM7). Three amino acid residues appear to be mainly responsible for the binding of catecholamines within the β-adrenergic receptor as well as, presumably, in other catecholamine receptors (e.g. in the dopamine D_3-receptor; see Fig. 6.2), where they are also present. The first is an acidic aspartic residue within TM3, conserved among all receptors of the super-

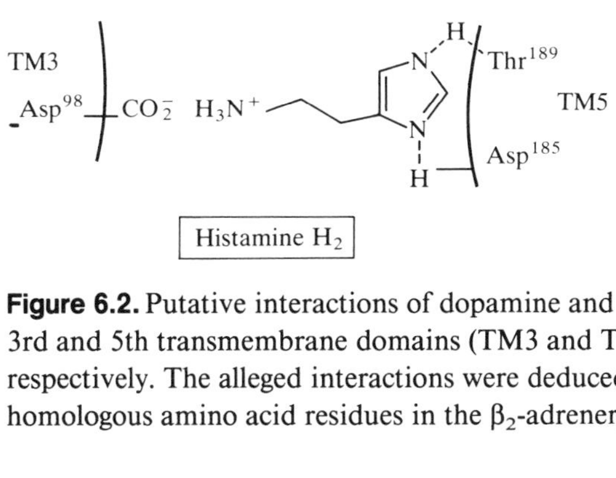

Figure 6.2. Putative interactions of dopamine and histamine with amino acid residues in the 3rd and 5th transmembrane domains (TM3 and TM5) of the rat D_3- and H_2-receptors, respectively. The alleged interactions were deduced from site-directed mutagenesis studies of homologous amino acid residues in the β_2-adrenergic receptor.

family that bind biogenic amines and believed to form an ion pair with the protonated amine function of catecholamines. The two others are serine residues in TM5, responsible for hydrogen bonding the two vicinyl hydroxyl groups of the catechol ring: their substitution affects the affinity and efficacy of catechol agonists but does not affect the affinity of non-catechol antagonists. Interestingly, in the dog (Gantz *et al.*, 1991) and rat (Ruat *et al.*, 1991) histamine H_2-receptor, the two serine residues are replaced by aspartate and threonine residues which may play an equivalent role in binding the imidazole ring of histamine (Fig. 6.2). It seems likely that these agonist bindings at the level of TM5 are crucial for the triggering of receptor transconformations resulting in the interaction with G-proteins; indeed, this interaction appears mainly to involve the initial (N-terminal) segment of the third intracellular loop, i.e. a segment close to the segment of TM5 binding agonists. This would account for the absence of efficacy of catecholamine antagonists, i.e. compounds binding with high affinity to other areas but lacking the dihydroxyl groups necessary for interaction with the two serine residues in TM5.

These examples illustrate how molecular genetics are starting to throw some light on some fundamental problems of pharmacology such as the structural mechanisms underlying the differences between agonists and antagonists and the changes elicited by ligands.

From the viewpoint of drug design, it would be surprising if the unravelling of receptor residues binding the ligands should not ultimately result in a rational design of compounds in which these interactions are optimized. The next step towards this aim would be to study the binding of ligands in a crystallized receptor by X-ray diffraction analysis. However, no receptor has so far been crystallized in amounts adequate for X-ray diffraction studies. This challenge is more likely to be met now, however, with cloned receptors that can be obtained in large amounts from transfected mammalian cells. Furthermore, it has been shown that β-adrenoreceptors can be expressed in active form in bacteria, which offer the advantage of a rapid and abundant production.

It can safely be anticipated that cloned receptors will soon be available to various physical studies such as nuclear magnetic resonance (NMR). Progress in this analytical method will allow study of integral membrane proteins the size of receptors.

7 References

Arrang, J.M., Garbarg, M., Lancelot, J.C., Lecomte, J. M., Pollard, H., Robba, M., Schunack, W. & Schwartz, J.C. (1987). Highly potent and selective ligands for H_3-receptors. *Nature (London)*, **327**, 117–23.

Burgen, A.S.V. (1984). Muscarinic receptors. An overview. *Trends Pharmacol. Sci.*, **3** (Suppl.), 1–3.

Changeux, J.P. (1981). The acetylcholine receptor: an 'allosteric' membrane-protein. *Harvey Lect.*, **75**, 85–254.

Dohlman, H.G., Thorver, J., Caron, M.G. & Lefkowitz, R.J. (1991). Model systems for the study of seven-transmembrane-segment receptors. *Ann. Rev. Biochem.*, **60**, 653–88.

Galzi, J.L., Revah, F., Bessis, A. & Changeux, J.P. (1991). Functional architecture of the nicotinic acetylcholine receptor: from electric organ to brain. *Ann. Rev. Pharmacol.*, **31**, 37–72.

Gantz, I., Schäffer, M., Delvalle, J., Logsdon, C., Campbell, V., Uhler, M. & Yamada, T. (1991).

Molecular cloning of a gene encoding the histamine H_2-receptor. *Proc. Natl. Acad. Sci. USA*, **88**, 429–33.

Giros, B., Sokoloff, P., Martres, M.P., Riou, J.F., Emorine, L.J. & Schwartz, J.C. (1989). Alternative splicing directs the expression of two D_2 dopamine receptor isoforms. *Nature* (*London*), **342**, 923–6.

Giros, B., Martres, M.P., Pilon, C., Sokoloff, P. & Schwartz, J.C. (1991). Shorter variants of the D_3 dopamine receptor produced through various patterns of alternative splicing. *Biochem. Biophys. Res. Commun.*, **176**, 1584–92.

Johnson, J.W. & Ascher, P. (1987). Glycine potentiates the NMDA response in cultured mouse brain neurons. *Nature* (*London*), **325**, 529–31.

Lefkowitz, R.J., Kobicka, B.K. & Caron, M.G. (1989). The new biology of drug receptors. *Biochem. Pharmacol.*, **38**, 2941–8.

Libert, F., Parmentier, M., Lefort, A., Dinsart, C., Van Sande J., Maerhant, C., Simons, M.J., Dumont, J.E. & Vassart, G. (1989). Selective amplification and cloning of four new members of G-protein-coupled receptor family. *Science,* **244**, 569–72.

Libert, F., Schiffmann, S.N., Lefort, A., Parmentier, M., Gérard, C., Dumont, J.E., Vanderhaegen, J.J. & Vassart, G. (1991). The orphan receptor cDNA RDC7 encodes an A_1 adenosine receptor. *EMBO J.*, **10**, 1677–82.

Martres, M.P., Bouthenet, M.L., Salès, N., Sokoloff, P. & Schwartz, J.C. (1985). Widespread distribution of brain dopamine receptors evidenced with [^{125}I]iodosulpiride, a highly selective ligand. *Science*, **228**, 752–5.

Matsuda, L.A., Lolart, S.J., Brownstein, M.J., Young, A.C. & Bonner, T.I. (1990). Structure of a cannabinoid receptor and functional expression of the cloned cDNA. *Nature* (*London*), **346**, 561–4.

O'Malley, B. (1990). The steroid receptor superfamily: more excitement predicted for the future. *Mol. Endocrinol.*, **4**, 363–9.

Piomelli, D., Pilon, C., Giros, B., Sokoloff, P., Martres, M.P. & Schwartz, J.C. (1991). Dopamine activation of the arachidonic acid cascade as a basis for D_1/D_2 receptor synergism. *Nature* (*London*), **353**, 164–7.

Ruat, M., Traiffort, E., Arrang, J.M., Leurs, R. & Schwartz, J.C. (1991). Cloning and tissue expression of a rat histamine H_2-receptor gene. *Biochem. Biophys. Res. Commun.*, **179**, 1470–8.

Schwartz, J.C., Delandre, M., Martres, M.P., Sokoloff, P., Protons, P., Vasse, M., Costentin, J., Laibe, P., Wermuth, C.G., Gulat, C. & Lafitte, A. (1984). Biochemical and behavioral identification of discriminant benzamide derivatives: new tools to differentiate subclasses of dopamine receptors. In: E. Usdin, A. Carlsson, A. Dahlstrom and J. Engel (Eds.), *Catecholamines: Neuropharmacology and Central Neurons System*, pp. 59–72, Alan R. Liss, New York.

Simon, M.I., Strathmann, M.P. & Gantan, N. (1991). Diversity of G-proteins in signal transduction. *Science,* **252**, 802–8.

Snyder, S.H. (1984). Drugs and neurotransmitter receptors in the brain. *Science,* **224**, 22–31.

Sokoloff, P., Giros, B., Martres, M.P., Bouthenet, M.L. & Schwartz, J.C. (1990). Molecular cloning and characterization of a novel dopamine receptor (D_3) as a target for neuroleptics. *Nature* (*London*), **347**, 146–51.

Strader, D.C., Sigal, S.I. & Dixon, A.F.R. (1989). Mapping of functional domains of the β-adrenergic receptor. *Am. J. Respir. Cell Molec. Biol.,* **1**, 81–6.

7 A Structural Basis for Proteinase–Protein Inhibitor Interaction

W. BODE and R. HUBER
Max-Planck-Institut für Biochemie, D-8033 Martinsried, Germany

1 Introduction

Proteinase inhibitors are important tools of nature in regulating the proteolytic activity of their target proteinases, in blocking these in emergency cases, or in signalling receptor interactions or clearance. Endogenous inhibitors appear always to be proteins; only in micro-organisms are small non-proteinaceous inhibitors produced which impair the proteolytic activty of host proteinases.

The number of proteinaceous proteinase inhibitors isolated and identified so far is extremely large. In a now 'classical' review paper Laskowski & Kato (1980) for the first time introduced a rational nomenclature in that they grouped these diverse inhibitors into distinct protein families. Meanwhile, with the advent of many new inhibitor species, this list of families has considerably expanded and is still growing.

The majority of protein inhibitors known and characterized so far are directed towards serine proteinases. Within the last few years a large number of protein inhibitors of cysteine proteinases have also been discovered and characterized (Barrett *et al.*, 1986; Turk & Bode, 1991). In contrast, only a few protein inhibitors directed towards metallo-proteinases (TIMP and PCI; see Cawston, 1986; Woessner, 1991) or aspartyl proteinases (see Martzen *et al.*, 1990; Ritonja *et al.*, 1990; Baudys *et al.*, 1991) are known to date. The α_2-macroglobulin family presents an exception, because these proteins can inhibit each of these proteinases according to a 'molecular trap' mechanism by virtue of a promiscuous 'bait region' (see Sottrup-Jensen, 1990).

Until recently, X-ray crystal structures were available of only a few serine proteinase inhibitors, one carboxypeptidase inhibitor, and some of their complexes with cognate proteinases. The protein inhibitor X-ray crystal structures published up to 1985 have been reveiwed by Read & James (1986). Since 1986, several more proteinase inhibitor-related crystal structures have been determined, in particular some serine proteinase inhibitors of hitherto unknown folding (Tsunogae *et al.*, 1986; Grütter *et al.*, 1988, 1990; Bode *et al.*, 1989a; Greenblatt *et al.*, 1989; Rydel *et al.*, 1990, 1991), and the first two cysteine proteinase inhibitors (Bode *et al.*, 1988; Stubbs *et al.*, 1990). Recent structural studies of two pancreatic procarboxypeptidases (Coll *et al.*, 1991) revealed an inhibitor–proteinase complex-like stucture. A new aspect is provided by inhibitor structures elucidated by two-dimensional nuclear magnetic resonance (NMR) methods (see Clore & Gronenborn, 1989; Markley, 1989 for reviews). These data are often somewhat complementary to X-ray data, but are restricted to isolated inhibitors of relatively small molecular weight; no NMR structures of protein inhibitors have been reported until now for which there is no X-ray structure available.

In this review we shall attempt to illuminate the characteristic structural properties conferring inhibitory activity to proteins. Nature has used diverse approaches to achieve proteinase inhibition. This is particularly well illustrated by some more recently published structures. In a recent mini-review (Bode & Huber, 1991) we have surveyed and evaluated the most recent protein inhibitor structures. Here we will extend this survey to all proteinase protein inhibitors and their proteinase complexes for which the spatial atomic structure has been determined. These structures are summarized (following Read & James, 1986) in Table 7.1 together with some characteristic parameters. In addition a gallery of representative structure models will aid in demonstrating some characteristic features of these structures.

Table 7.1. Spatial atomic structures of protein proteinase inhibitors and their complexes with proteinases

Family/Structure	Abbreviation	Method	Resolution (Å)	*R* factor	Reference
1 **Serine proteinase inhibitors**					
1.1 BPTI (small Kunitz) family					
Bovine pancreatic trypsin inhibitor	BPTI (I)	X-ray	1.5	0.162	Deisenhofer & Steigemann (1975)
BPTI (crystal form II)†	BPTI (II)	X-ray	0.98	0.200	Wlodawer *et al.* (1987a)
BPTI (crystal form III)	BPTI (III)	X-ray	1.7	0.16	Wlodawer *et al.* (1987b)
BPTI (C30A/C51A)	BPTIC30A/C51A	X-ray	1.60	0.170	Eigenbrot *et al.* (1990)
Amyloid β-protein precursor inhibitor	APPI	X-ray	1.5	0.177	Hynes *et al.* (1990)
BPTI (crystal form IV)	BPTI (IV)	NMR			Wagner *et al.* (1987)
BPTI—bovine trypsin	BPTI:BT	X-ray	1.9	0.187	Huber *et al.* (1974); Marguart *et al.* (1983)
BPTI—anhydrotrypsin	BPTI-BTan	X-ray	1.9	0.175	Huber *et al.* (1975); Marquart *et al.* (1983)
BPTI—bovine trypsinogen	BPTI:BTgen	X-ray	1.9	0.200	Bode *et al.* (1978); Marquart *et al.* (1978)
BPTI:BTgen:Ile-Val	BPTI:BTgen:IV	X-ray	1.9	0.193	Bode *et al.* (1978); Marquart *et al.* (1978)
Arg-15–BPTI:BTgen: Val-Val	Arg-15-BPTI:BTgen: VV	X-ray	2.24	0.170	Bode *et al.* (1974)
BPTI:porcine glandular kallikrein	BPTI:PGK	X-ray	2.5	0.230	Chen & Bode (1983)
1.2 Kazal family					
Japanese quail ovomucoid third domain	OMJPQ3	X-ray	1.9	0.202	Papamokos *et al.* (1982)
Porcine pancreatic secretory trypsin inhibitor:BTgen	PSTI:BTgen	X-ray	1.8	0.195	Bolognesi *et al.* (1982)

Table 7.1. (Continued)

Family/Structure	Abbreviation	Method	Resolution (Å)	*R* factor	Reference
Turkey ovomucoid third domain:*Streptomyces griseus* proteinase B	OMTKY3:SGPB	X-ray	1.8	0.125	Read *et al.* (1983)
OMTKY3:Bovine α-chymotrypsin	OMTKY3:CHT	X-ray	1.8	0.168	Fujinaga *et al.* (1987)
Silver pheasant ovomucoid third domain	OMSVP3	X-ray	1.5	0.199	Bode *et al.* (1985)
OMTKY3:human leukocyte elastase	OMTKY3:HLE	X-ray	1.8	0.166	Bode *et al.* (1986b, 1992a)
Indian peafowl ovomucoid third domain:HLE	OMSPF3:HLE	X-ray	2.2	0.165	Bode *et al.* (1992a)
Human pancreatic secretory trypsin inhibitor (K18L,I19E,D21R, P32A):HLE	HSTI4a:HLE	X-ray	2.5	0.172	Epp *et al.* (1992)
Reactive site-cleaved OMJPQ3	OMJPQ3*	X-ray	1.55	0.192	Musil *et al.* (1991)
Reactive site-cleaved OMSVP3	OMSVP3*	X-ray	2.5	0.185	Musil *et al.* (1991)
HSTI(K18Y,I19E, D21R): Bovine chymotrypsinogen A	HSTI3:Chgen	X-ray	2.3	0.195	Hecht *et al.* (1991)
HSTI(K18L,I19E, D21R):Chgen	HSTI4:Chgen	X-ray	2.3	0.195	Hecht *et al.* (1991)
Bull seminal plasma inhibitor-IIa	BUSI-IIa	NMR			Williamson *et al.* (1985)
1.3 STI (large Kunitz family)					
Soybean trypsin inhibitor: porcine trypsin	STI:PT	X-ray	2.6	Not determined	Sweet *et al.* (1974)
Erythrina trypsin inhibitor	ETI	X-ray	2.5	0.208	Onesti *et al.* (1991)
Proteinase K/α-amylase inhibitor from wheat	PKI3	X-ray	2.5	0.21	Zemke *et al.* (1991)
1.4 SSI family					
Streptomyces subtilisin inhibitor	SSI	X-ray	2.3	0.27	Mitsui *et al.* (1979); Hirono *et al.* (1984)
SSI:subtilisin BPN′	SSI:SBPN	X-ray	2.2	0.34	Hirono *et al.* (1984)
Plasminostreptin	PS	X-ray	2.8	Not determined	Kamiya *et al.* (1984)
SSI(M73K):SBPN	SS173:SBPN	X-ray	1.8	0.178	Takeuchi *et al.* (1991)
SSI(M73K,M70G): SBPN	SSI:SBPN	X-ray	1.8	0.178	Takeuchi *et al.* (1991)
1.5 Potato inhibitor 1 (PI-1) family					
Barley chymotrypsin inhibitor 2: subtilisin novo	CI-2:SNOV	X-ray	2.1	0.154	McPhalen *et al.* (1985); McPhalen & James (1988)

Table 7.1. (Continued)

Family/Structure	Abbreviation	Method	Resolution (Å)	*R* factor	Reference
CI–2	CI–2	X-ray	2.0	0.198	McPhalen & James (1987)
Eglin c: subtilisin Carlsberg	Eglc:SCAR	X-ray	1.2	0.178	Bode *et al.* (1986a); Bode *et al.* (1987)
Eglc:SCAR	Eglc:SCAR	X-ray	1.8	0.136	McPhalen & James (1988)
CI–2	CI–2	NMR			Clore *et al.* (1987a)
Eglc:thermitase (I)	Eglc:THER(I)	X-ray	2.20	0.179	Gros *et al.* (1989a)
Eglc:thermitase (II)	Eglc:THER(II)	X-ray	1.98	0.165	Gros *et al.* (1989b)
Eglc:CHT	Eglc:CHT	X-ray	2.6	0.18	Bolognesi *et al.* (1990)
Eglc:SNOV	Eglc:SNOV	X-ray	2.4	0.169	Heinz *et al.* (1991)
Egle(L45R):SNOV	EglcL45R:SNOV	X-ray	2.1	0.186	Heinz *et al.* (1991)
Eglc(R53K):SNV	EglcR53K:SNOV	X-ray	2.4	0.159	Heinz *et al.* (1991)
1.6 Potato Inhibitor 2 (PI-2) family					
Chymotrypsin inhibitor-1:SGPB	CI-1:SGPB	X-ray	2.1	0.142	Greenblatt (1989)
1.7 Chelonianin family					
Mucous proteinase inhibitor:CHT	MPI-CHT	X-ray	2.5	0.19	Grütter *et al.* (1988)
1.8 Bowman–Birk family					
Azuki beans protease inhibitor:BT	AB-I:BT	X-ray	3.0	Not determined	Tsunogae *et al.* (1986)
Peanut inhibitor A-II	A-II	X-ray	3.3	Not determined	Suzuki *et al.* (1987)
Mung bean trypsin inhibitor:PT	MBTI:PT	X-ray	2.5	0.182	Lin *et al.* (1991)
Soybean trypsin/ Chymotrypsin, Bowman–Birk inhibitor	STCI	NMR			Werner & Wemmer (1991)
1.9 Squash seed inhibitors					
Cucurbia maxima trypsin inhibitor:BT	CMTI-I:BT	X-ray	2.0	0.152	Bode *et al.* (1989a)
CMTI-I	CMTI-I	NMR			Holak *et al.* (1989)
Echallium elaterium trypsin inhibitor-II	EETI-II	NMR/X-ray			Chiche *et al.* (1989)
1.10 Serpins					
Reactive-site modified α_1-proteinase inhibitor tetragonal form (I)	α_1-PI*T(I)	X-ray			Löbermann *et al.* (1984); Engh *et al.* (1989)
α_1-PI* hexagonal form	α_1-PIH*	X-ray	3.1	0.215	Engh *et al.* (1989)
α_1-PI*T(II)	α_1-PI*T(II)	X-ray	3.0	0.209	Engh *et al.* (1989)
α_1-PI (E264V)*H S-variant	α_1-PIS*H	X-ray	3.1	0.219	Engh *et al.* (1989)
Reactive-site modified α_1-antichymotrypsin	α_1-AChy*	X-ray	2.7	0.180	Baumann *et al.* (1991).
Reactive-site modified horse leukocyte elastase inhibitor	HLEI*	X-ray	2.3	0.177	Baumann *et al.* (1992)
Plakalbumin, cleaved chicken ovalbumin	PLA	X-ray	2.8	0.187	Wright *et al.* (1990)
Chicken ovalbumin	OVA	X-ray	1.95	0.169	Stein *et al.* (1990)

Table 7.1. (Continued)

Family/Structure	Abbreviation	Method	Resolution (Å)	*R* factor	Reference
1.11 Hirudin					
Desulphato-hirudin variant 2 K47:human α-thrombin	HIRV2:HUTHR	X-ray	2.3	0.173	Rydel *et al.* (1990); Rydel *et al.* (1991)
Desulphato-hirudin variant 1:HUTHR	HIRV1:HUTHR	X-ray	2.95	0.225	Grütter *et al.* (1990)
Desulphato-hirudin (I)	HIR-I	NMR			Folkers *et al.* (1989)
HIR-I (K47E)	HIRK47E-I	NMR			Folkers *et al.* (1989)
Desulphato-hirudin (II)	HIR-II	NMR			Haruyama & Wüthrich (1982)
2 Cysteine proteinase inhibitors					
2.1 Cystatins					
Chicken egg-white cystatin	CEWCYS	X-ray	2.0	0.19	Bode *et al.* (1988)
2.2 Stefins					
Stefin A (C3S):papain	STA:PAP	X-ray	2.4	0.19	Stubbs *et al.* (1990)
3 Metalloproteinase inhibitors					
3.1 PCI family					
Potato carboxypeptidase inhibitor:carboxy-peptidase A	PCI:CPA	X-ray	2.5	0.196	Rees & Lipscomb (1982)
PCI	PCI	NMR			Clore *et al.* (1987b)
Porcine carboxy-peptidase B	ProCPA	X-ray	2.3	0.169	Coll *et al.* (1991)
Pro-part	Pro	NMR			Vendrell *et al.* (1991)

* The crystallographic *R* factor defined as $\Sigma||F_{obs}|-|F_{calc}||/\Sigma|F_{obs}|$ is *one* measure for the quality of a crystal structure; lower values indicate agreement between the observed (F_{obs}) and the model calculated (F_{calc}) structure factor amplitudes. The quality of a structure depends, however, also on the restraint parameters; local errors are detected by other criteria, such as difference Fourier maps.

† Result of a joint X-ray neutron diffraction structure; neutron data were collected to 1.8 Å; the corresponding *R* factor was 0.197 (Wlodawer *et al.*, 1987a).

2 Protein inhibitors of serine proteinases

The protein inhibitors directed against serine proteinases can be grouped into at least 16 different families based on sequence homology, topological similarity, and mechanism of binding (Laskowski & Kato, 1980). For 11 of them at least one (often several) representative spatial atomic structure is known to date (see Table 7.1); for some others, such as an *Escherichia coli* trypsin inhibitor (McGrath *et al.*, 1991) and an *Ascaris* trypsin inhibitor (Gronenborn *et al.*, 1990) structural analyses are underway.

Most of the serine proteinase-directed inhibitors react with cognate enzymes according to a common, substrate-like 'standard mechanism' (Huber & Bode, 1978; Laskowski & Kato, 1980). This group of 'canonical' inhibitors comprises relatively 'small' proteins (or protein domains of multiheaded inhibitors) of between 29 and about 190 amino acid residues. They all possess an exposed binding loop of a characteristic canonical conformation, but are otherwise unrelated in structure.

The serpins (*ser*ine *p*roteinase *in*hibitors) (Carrell & Travis, 1985) form a family of homologous, large (glyco-) proteins comprising about 400 amino acid residues (Travis & Salvesen, 1983; Huber & Carrell, 1989). Like the canonical inhibitors, the serpin inhibitors seem to interact via an exposed binding loop with their cognate proteinases; the resulting complexes are, however, only transient and collapse under liberation of a cleaved form of different structure and stability. Most serpins exhibit inhibitory activity, exclusively towards serine proteinases; only a few (such as ovalbumin) apparently lack inhibitory properties and may have other functions.

Hirudin binds according to a very different mechanism (Grütter *et al.*, 1990; Rydel *et al.*, 1990, 1991). This non-canonical interaction may be used more frequently in serine proteinase inhibition, but has so far been defined only for hirudin.

2.1 *The 'canonical' serine proteinase protein inhibitors*

2.1.1 THE INHIBITOR STUCTURES

Figures 7.1–7.7 illustrate inhibitors belonging to seven different inhibitor families which have in common an exposed binding loop of a 'unique' canonical conformation, but are otherwise unrelated in structure. Previous suggestions that distinct structural elements (such as a β-hairpin loop following the binding loop (Bolognesi *et al.*, 1982), or a disulphide bridge in the vicinity of the scissile peptide bond (Laskowski & Kato, 1980)) might represent necessary structural elements for proper inhibitory function are not confirmed by recently established structures: the residual structure outside the binding loop can possess quite different folding motifs. In all cases, however, the inhibitors (or their single active domains) have a compact shape and contain a hydrophobic core which sometimes consists mainly of the cross-connecting disulphide bridges (see for example CMTI-I, Bode *et al.*, 1989a). The stability of the native inhibitor domains towards unfolding is generally high (the melting temperatures of BPTI (Moses & Hinz, 1983; Schwarz *et al*; 1987) and of OMTKY3 (Otlewski & Laskowski, 1992) are, for example, 95° and 85°C, respectively, and these proteins remain essentially native in 6 mol/l guanidinium chloride). The contribution of the disulphide bonds to overall stability has been intensively studied (Creighton & Goldenberg, 1983, 1984; Goldenberg, 1985). The three natural disulphide bridges in BPTI stabilize this inhibitor to different extents. The amount of stabilization is clearly related to 'effective concentrations' of thiol groups forming given disulphide bridges in the native-like state. The removal of a single disulphide bridge, although destabilizing, is not necessarily accompanied by large visible changes in the structure (Eigenbrot *et al.*, 1990).

The thermal unfolding of inhibitor domains seems to be highly cooperative, in case of BPTI occurring in a two-state transition (Schwarz *et al.*, 1987). Single cleavages in thecore backbone result in much lower melting temperatures (M. Laskowski, pers. commun.). All segments of this core domain interact as a

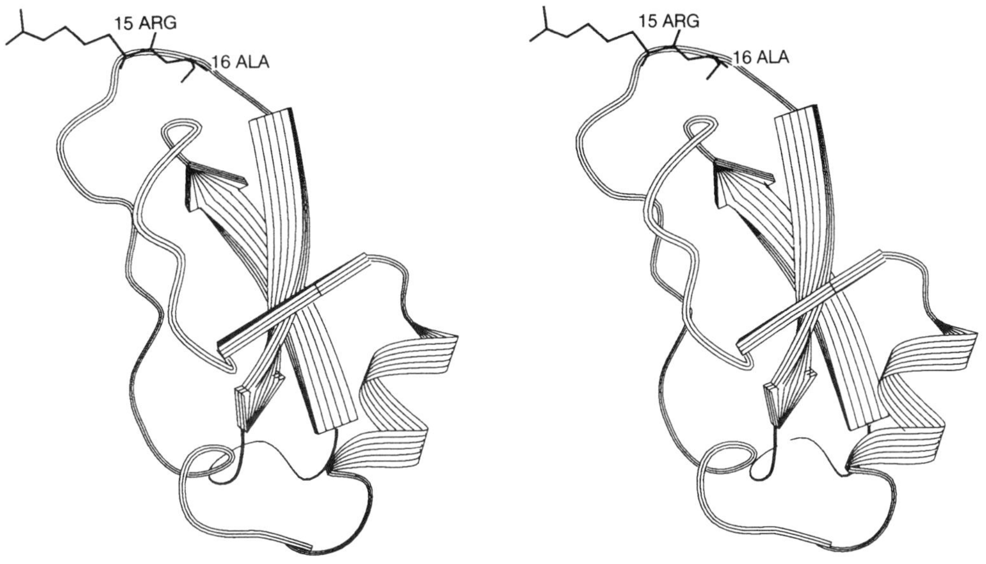

(a)

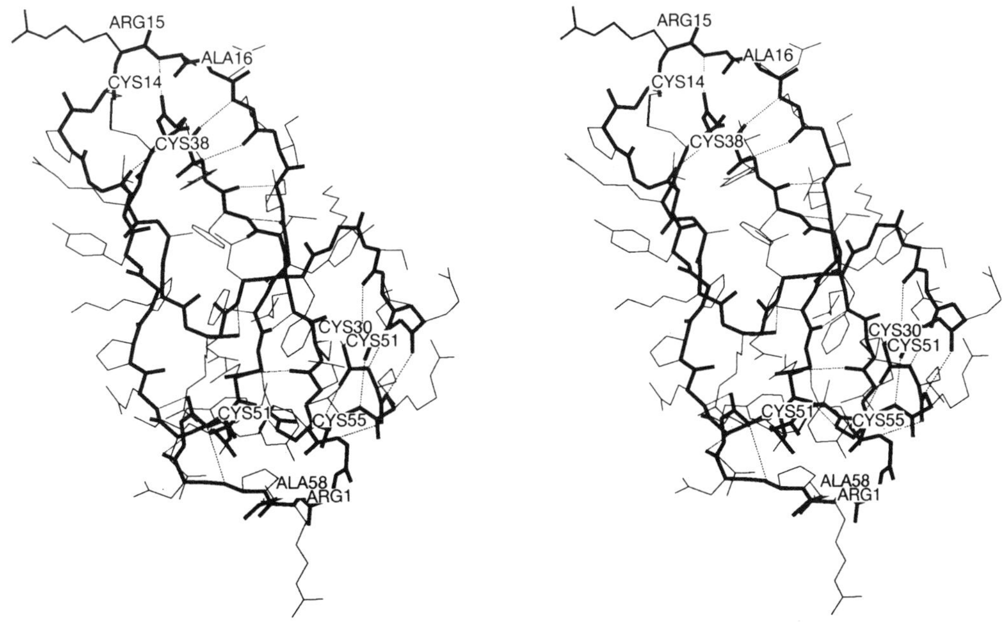

(b)

Figure 7.1. (a,b) Ribbon and skeletal drawing of basic pancreatic trypsin inhibitor (semi-synthetic Arg-15 form) as derived from the structure of its complex with trypsinogen and Val-Val (Bode *et al.*, 1984). A central antiparallel three-stranded β-sheet and a C-terminal α-helix form the core of the molecule to which the protease binding loop (P3–P4′ Pro-13–Ile-19) is attached. Arg-15–Ala-16 are the P1–P1′ sites. (Ribbon drawings have been made with a program by Priestle (1988)). In the skeletal drawings, main chains are indicated by bold lines, side-chains by thin lines and intermain-chain hydrogen bonds by broken lines. Disulphide linkages are also drawn. The views here and of Figures 7.2 to 7.7 have been chosen to present the scissile peptide group on top in similar orientations.

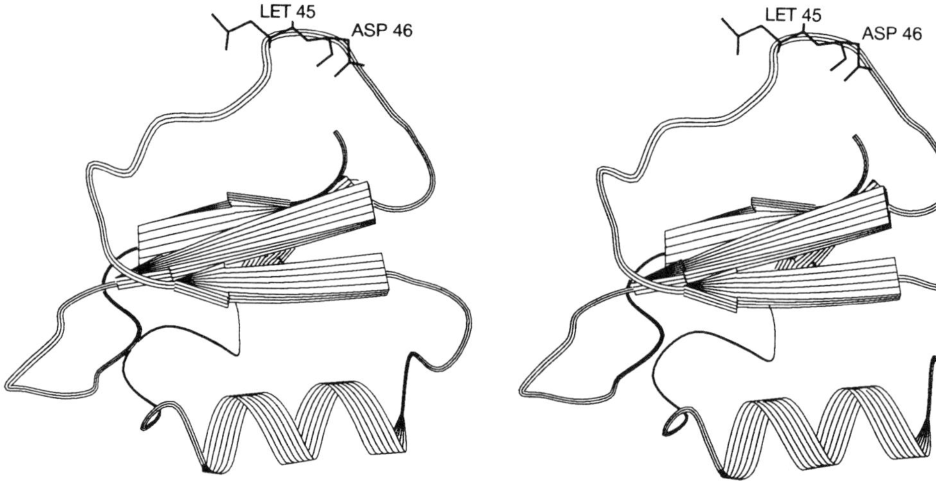

(a)

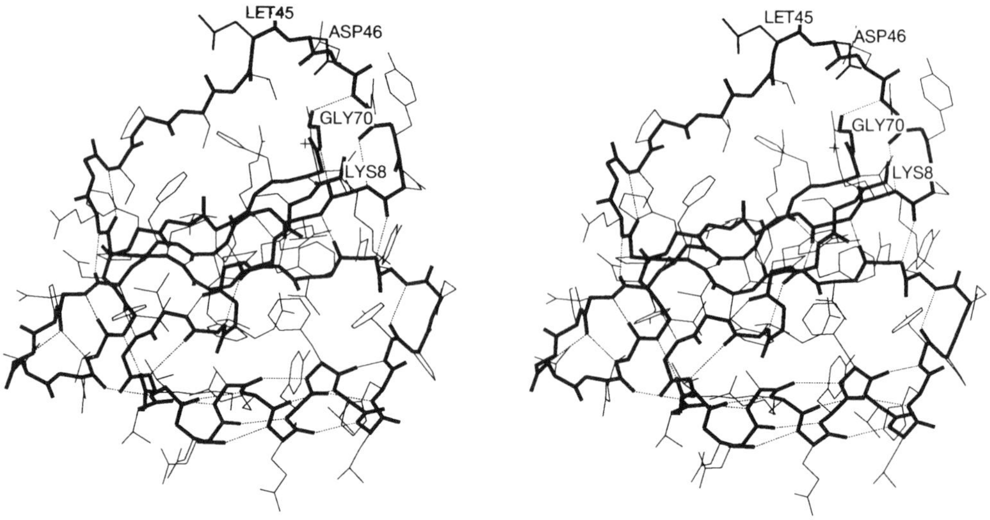

(b)

Figure 7.2. (a,b) Ribbon and skeletal drawing of eglin c as seen in the subtilisin–eglin complex (Bode *et al.*, 1986a, 1987). A four-stranded mixed antiparallel (β_1, β_4, β_3) parallel (β_3, β_2) β-sheet forms the molecule core. The α-helix connects strands β_1 and β_2. It is approximately parallel to the strand directions. The protinease binding loop (scissile peptide bond between P1 Leu-45 and P1′ Asp-46) is supported by arginine side-chains projecting from the β-sheet core.

'cooperative unit' which, as long as the protein remains folded (i.e. significantly below its melting temperature), forms the supporting scaffold for the exposed proteinase binding loop. This loop (the 'primary binding segment'), spanned between scaffold-based anchoring points, has a flat shape that fits into the active-site clefts of cognate serine proteinases (see Figs 7.1–7.7).

The loop strand has an extended conformation so that the side-chains flanking the scissile bond are (with the exception of P3*) exposed and project away from the

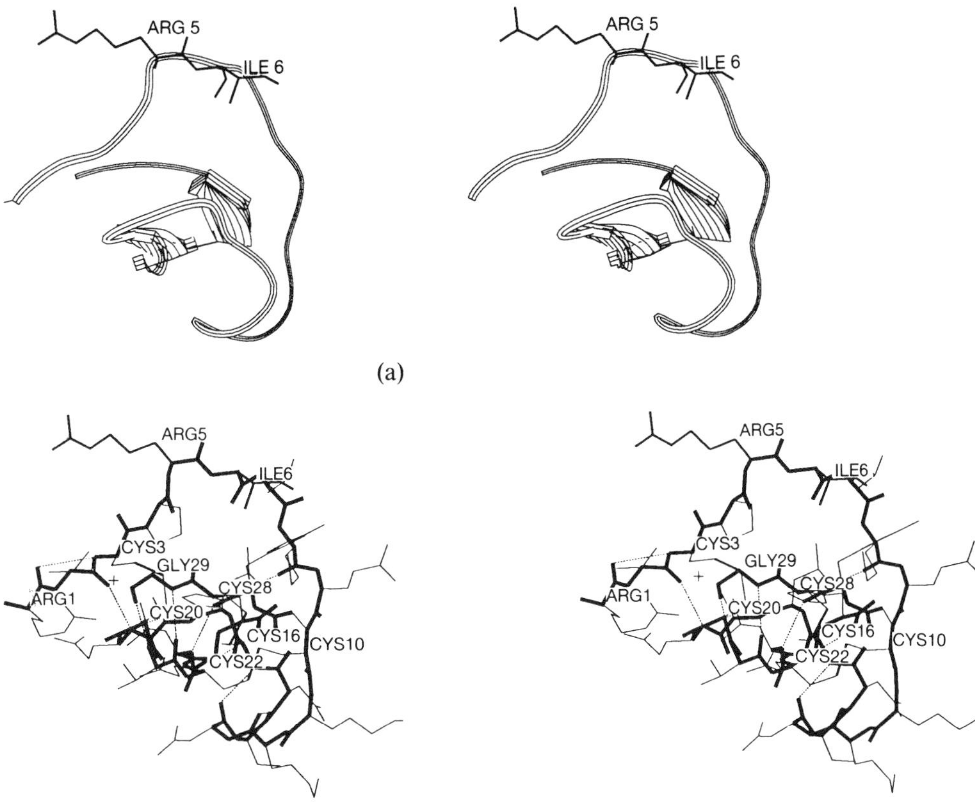

(b)

Figure 7.3. (a,b) Ribbon and skeletal drawing of squash seed inhibitor-I (CMTI-I) as derived from crystallographic studies of the complex with trypsin (Bode *et al.*, 1989a) and from NMR studies of the free inhibitor (Holak *et al.*, 1989). Apart from a very short central two-stranded β-ladder the molecule has no regular secondary structure. Its polypeptide chain is fixed by three disulphide links. The P1–P1′ site is Arg-5–Ile-6.

supporting scaffold. The conformation of the binding loop segment exhibits a quite characteristic conformation from P3 to P3′ (see, for example, Bode *et al.*, 1987): the main chain conformation and angles of antiparallel β-strands (φ: $-120°$ to $-140°$; ψ: 140° to 170°) at P3 (except BPTI), of polyproline II (φ: $-60°$ to $-100°$; ψ: 139° to 180°) at P2 and P1′, of an approximate 3_{10}-helix (φ: $-95°$ to $-120°$; ψ; 9° to 50°) at P1, and of parallel β-strands (φ: $-99°$ to $-140°$; ψ: 70° to 120°) at P2′ and P3′. The same inhibitors analysed in different crystal environments (Papamokos, 1982; Wlodawer *et al.*, 1987a; Eigenbrot *et al.*, 1990) or complexation states (compare Table 7.1) and/or by NMR techniques (Clore *et al.*, 1987a; Wagner *et al.*, 1987; Holak *et al.*, 1989) exhibit similar conformations. Thus this common canonical conformation, presumed also to be attained by a productively bound substrate, is an inherent property of the inhibitors themselves.

The exposed (often remarkably hydrophobic, see in particular Grütter *et al.*,

* P1, P2, P3, etc. and P1′, P2′ designate substrate/inhibitor residues at the amino- and carboxy-terminal of the scissile peptide bond, respectively, and S1, S2, S3, etc. and S1′, S2′, the corresponding subsites of the cognate proteinases (Schechter & Berger, 1967).

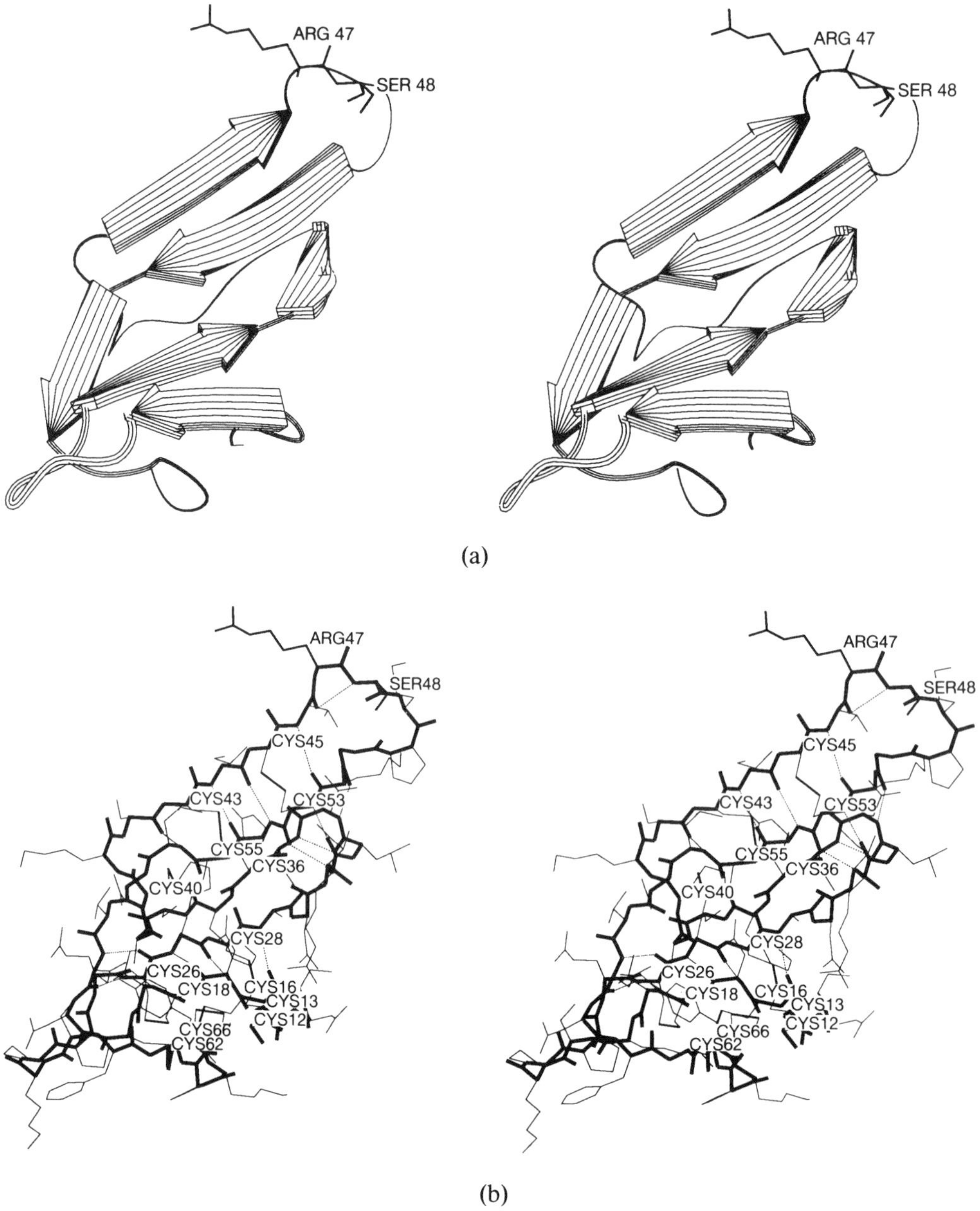

Figure 7.4. (a,b) Ribbon and skeletal drawing of mung bean trypsin inhibitor as seen in the ternary complex with porcine trypsin (Lin *et al.*, 1992). The molecule is double headed. Both domains are active against trypsin, where Arg-47-Ser-48 and Lys-20–Ser-21, respectively, occupy the S1–S1′ sites of the enzyme. The folding shows internal symmetry in the central β-sheet, which is divided into similar halves, and in the duplication of the protease binding segment. An NMR study of the free inhibitor from soybean is near completion (Werner & Wemmer, 1991).

1988) binding loop is further stabilized in all inhibitors by additional interactions between residues flanking the reactive site and the inhibitor core (see also Fig. 7.9a). In most canonical small inhibitors the P2-residue (BPTI and MPI (Grütter *et al.*, 1988); see Figs 7.1 & 7.5) or the P3-residue (in Kazal-, SSI-, Bowman–Birk-, squash

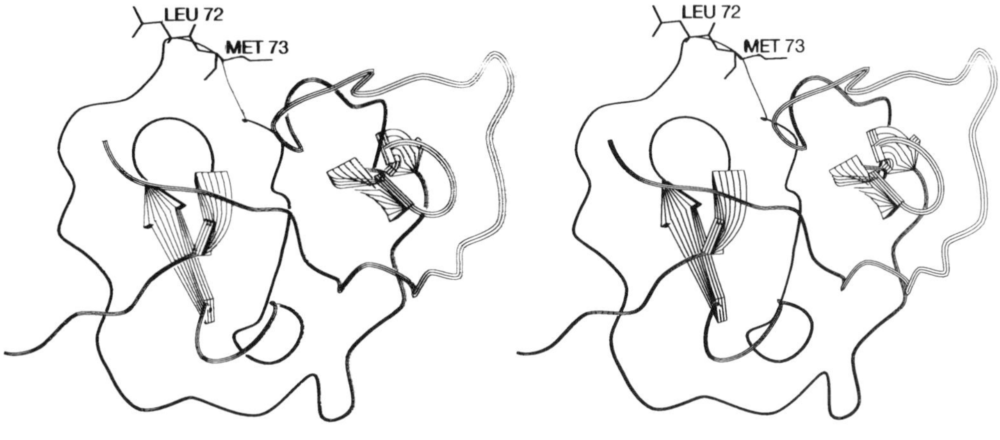

(a)

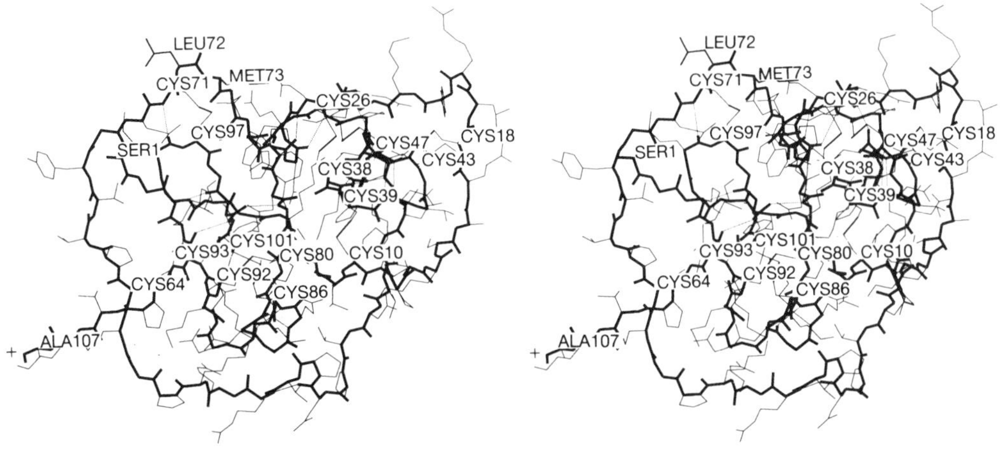

(b)

Figure 7.5. (a,b) Ribbon and skeletal drawings of MPI mucous proteinase inhibitor. The molecule consists of two similarly folded domains. Only the second domain has inhibitor activity. The P1–P1′ site (Leu-72–Met-73) is identified in the complex with α-chymotrypsin (Grütter *et al.*, 1988) and drawn. Apart from central double-stranded β-ladders in each domain there are no regular secondary structures. Four disulphide bonds tie the strands together in each domain.

seeds and potato inhibitors 2; see Figs 7.3, 7.4 & 7.7) is disulphide-connected with the hydrophobic core. In addition, side-chain 'spacers' and intermain-chain hydrogen bonds clamp the loop to the main body. In some inhibitors (potato inhibitors 1, STI; see Figs 7.2 & 7.6) stabilization of the binding loop is achieved instead by an elaborate electrostatic/hydrogen bond network through side-chains extending from the core to the binding loop (particularly well illustrated by the two parallel arginine side-chains of potato inhibitors 1 (Bode *et al.*, 1986a; McPhalen & James, 1988; see Figs 7.2a,b).

Scaffold and binding loop of the free inhibitors are not independent units. Replacements of amino acid residues in the binding loop in general have only small effects on the thermal stability (due to their mainly exposed nature). Replacement of conserved 'spacer' residues which are particularly engaged in (non)-covalent

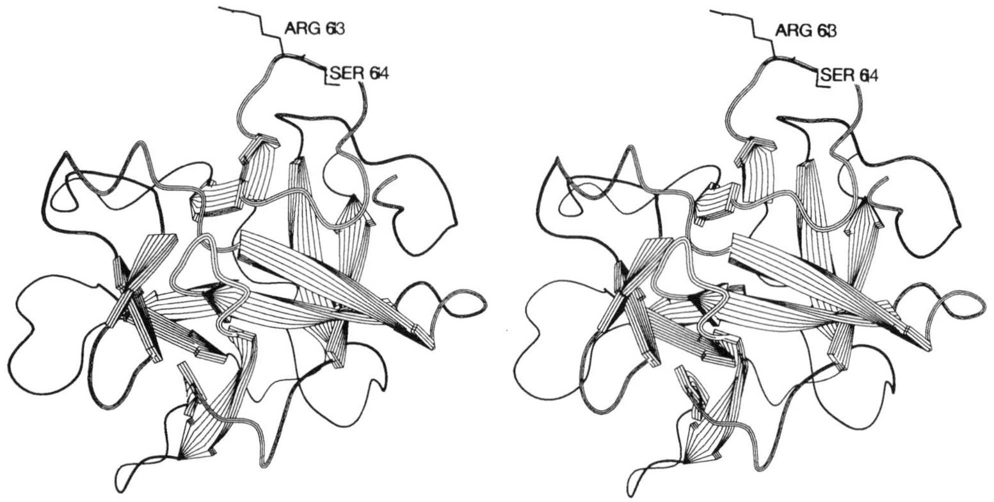

(a)

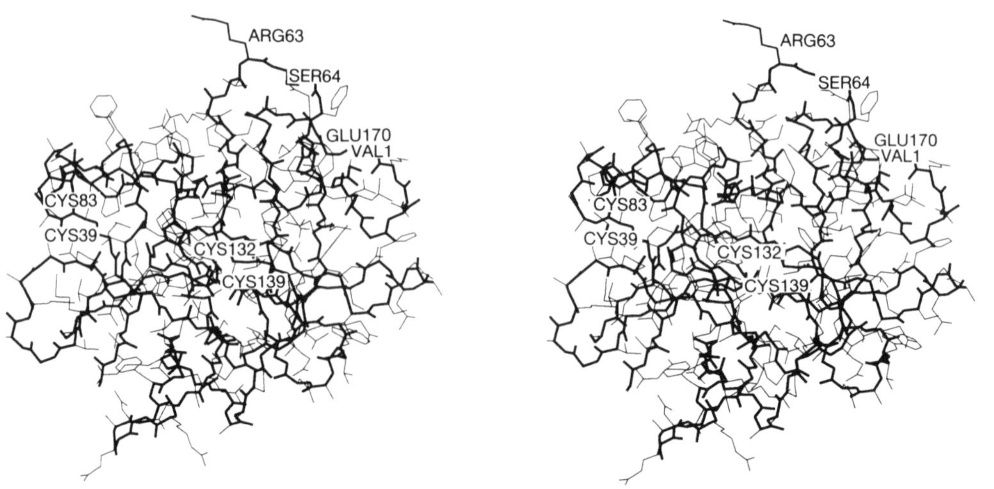

(b)

Figure 7.6. (a,b) Ribbon and skeletal drawings of erythrina trypsin inhibitor (ETI). ETI has 172 residues arranged in an all-β protein consisting of 12 antiparallel β-strands joined by long loops (Onesti *et al.*, 1991). Six of the strands form a β-barrel which is narrow at one end and open at the other end. Three loops join the adjacent strands at the narrow end, one of which contains the scissile peptide bond P1–P1′ Arg-63–Ser-64. The other end is capped by the other six strands. The barrel is approximately three fold symmetric. The barrel axis runs in a north-easterly direction.

interactions between the loop and the scaffold (see, for example, eglin and OMTKY3, Figs 7.2 and 7.7) can, however, result in considerable enhancement of the loop mobility (Hyberts & Wagner, 1990; Wagner *et al.*, 1990) and reduced rigidity of the scaffold (Goldberg *et al.*, 1989; Jandu *et al.*, 1990; Ardelt & Laskowski, 1991). Such an increase in mobility is also observed upon proteolytic cleavage at the reactive site 18–19 bond of ovomucoid (Kazal-type) inhibitors: the residues adjacent to the cleaved bond are partially disordered and some intraloop and loop–scaffold hydrogen bonds present in the uncleaved (virgin) inhibitors

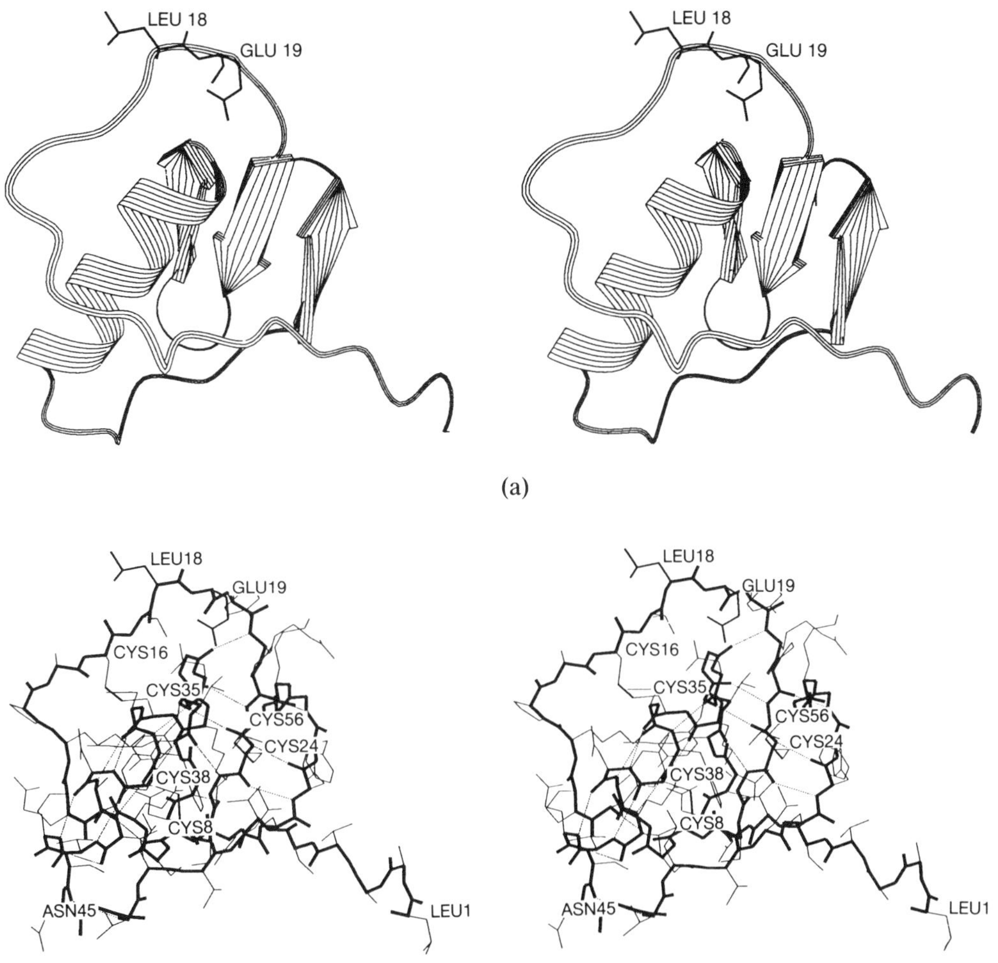

(a)

(b)

Figure 7.7. (a,b) Ribbon and skeletal drawing of the turkey ovomucoid third domain, a member of the large superfamily of ovomucoids and pancreatic secretory (Kazal-type) inhibitors (Bolognesi *et al.*, 1982; Papamokos *et al.*, 1982). A three-stranded antiparallel β-sheet and an α-helix approximately parallel to the β-strands conecting β_2 and β_3 form the nucleus of the molecule, onto which the amino-terminal segment, the protease binding loop (Leu-18–Glu-19 P1–P1′) and the α–β_3 connecting segment are attached.

disrupted, while the inhibitor scaffold essentially maintains its conformation (Rhyn & Markley, 1988; Musil *et al.*, 1991). The reactive-site-cleaved forms of the ovomucoid third domain and STI inhibitors denature at lower temperatures than the virgin forms and have a considerably lower free enthalpy of denaturation (Otlewski & Laskowski, 1992). On the other hand, the folded cleaved as well as the folded virgin (BPTI, Finkenstadt *et al.*, 1974; Quast *et al.*, 1978; and Kazal-type, Ardelt & Laskowski, 1985) inhibitors are of similar free-energy levels (the equilibrium constant of hydrolysis is close to 1). The difference in thermal stability observed experimentally for the virgin and the cleaved form seems therefore to be mainly due to the ring-opening entropy gained upon denaturation of the *cleaved* form (Ardelt & Laskowski, 1991).

2.1.2 THE 'CANONICAL' INTERACTION

Almost all of these 'small' virgin inhibitors (I) seem to interact (possibly with the exception of the cystine-free potato inhibitors 1) with cognate enzymes (E) according to the following minimal scheme (Finkenstadt *et al.*, 1974; Quast *et al.*, 1978)

$$E + I \underset{k_{off}}{\overset{k_{on}}{\rightleftarrows}} EI \underset{k_{on^*}}{\overset{k_{off^*}}{\rightleftarrows}} E + I^*$$

under rapid formation of stable complexes (EI), which dissociate usually very slowly into free enzymes and virgin or modified (I*) inhibitors, the latter being specifically cleaved at the scissile peptide bond P1–P1′ (Ardelt & Laskowski, 1983, 1985). Typically, the hydrolysis constant K_{hyd} = [I*]/[I] is close to unity at physiological pH values, but increases with increasing or decreasing pH (Ardelt & Laskowski, 1983, 1985).

In complexes, the inhibitor reactive-site loops bind across the catalytic residues of their cognate proteinases in a manner similar to that of productively bound substrates (see Fig. 7.9a). In the case of trypsin-like proteinases (see, for example, Fig. 7.8; Bode *et al.*, 1986b, 1992a), the segment amino-terminal to the scissile peptide bond adds as an antiparallel β-strand through main chain–main chain hydrogen bonds formed at P3 and P1 to the enzyme (a three-stranded antiparallel β-sheet is formed in subtilisin complexes; McPhalen *et al.*, 1985; Bode *et al.*, 1986a), while the carboxy-terminal flanking side interacts through another hydrogen bond at P2′. The reactive site of the inhibitor is close to the catalytic residues of the proteinase, with its P1 carbonyl carbon fixed in 'sub-van der Waals' contact with Ser-195 Oγ (typically around 2.7 Å; Huber *et al.*, 1974; Marquart *et al.*, 1983).

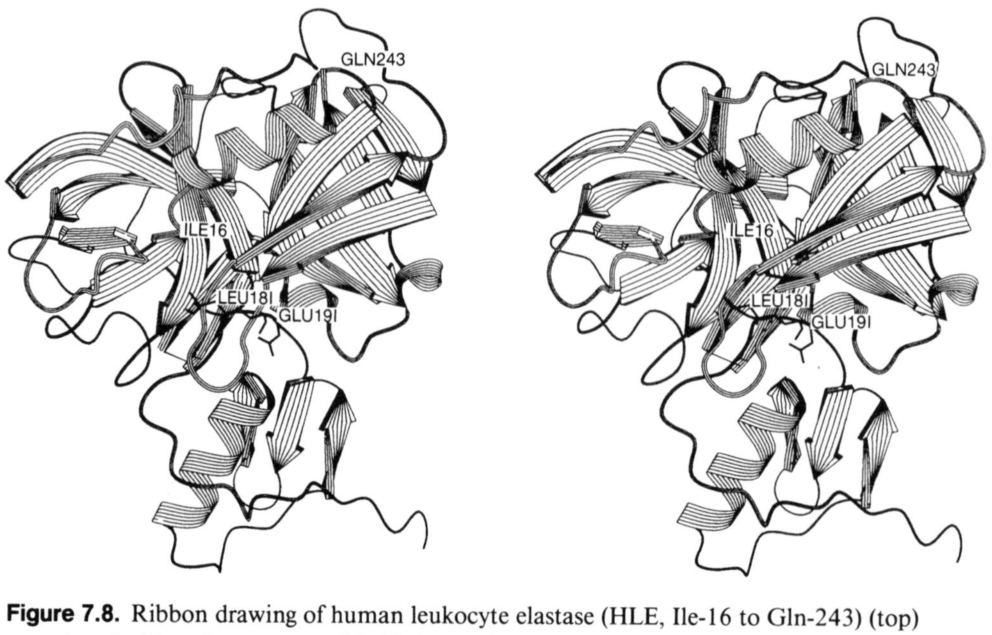

Figure 7.8. Ribbon drawing of human leukocyte elastase (HLE, Ile-16 to Gln-243) (top) complexed with turkey ovomucoid third domain (bottom) in a view similar to Figure 7.7. The amino- and the carboxy-termini of HLE and the P1 and P1′ residues of the inhibitor are marked (Bode *et al.*, 1986b).

The carbonyl group always projects into the 'oxyanion hole' (Robertus *et al.*, 1978) where it forms two hydrogen bonds with Gly-193 N and Ser-195 N presumably similar to transition-state complexes. The amide nitrogen of P1 points to Ser-195 Oγ (Bode *et al.*, 1987) rather than to Ser-214 O (this latter contact is believed to shorten in the catalytic cleavage process; James *et al.*, 1980; Bode & Huber, 1986). The scissile peptide bond remains intact with a slight 'out-of-plane' deformation of the carbonyl oxygen observed in some of the complexes (in particular in all BPTI complexes (Marquart *et al.*, 1983) and in the subtilisin–eglin complex (Bode *et al.*, 1987), but not in ovomucoid inhibitor complexes (Read & James, 1986; Bode

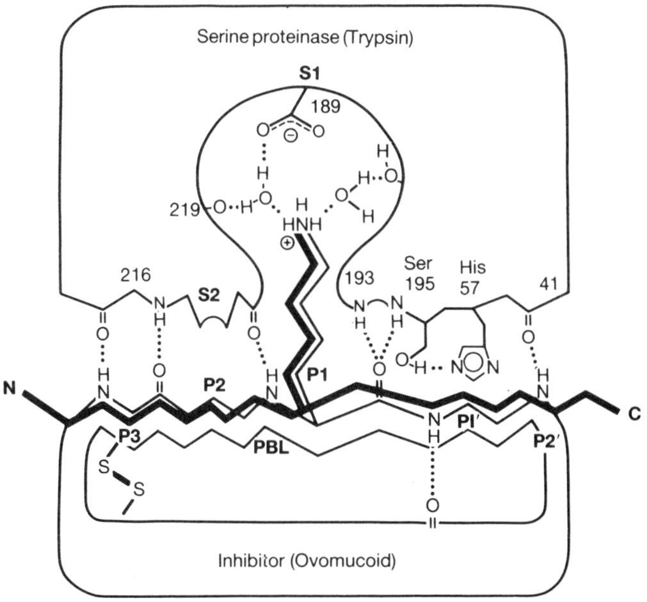

(a)

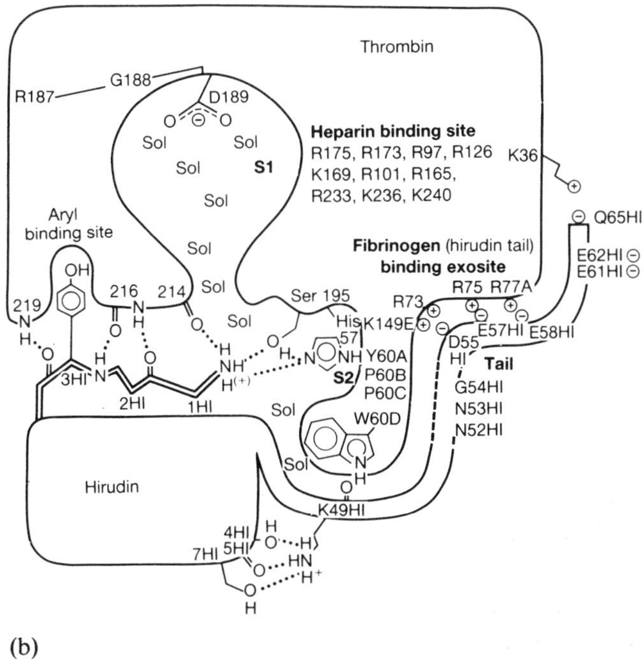

(b)

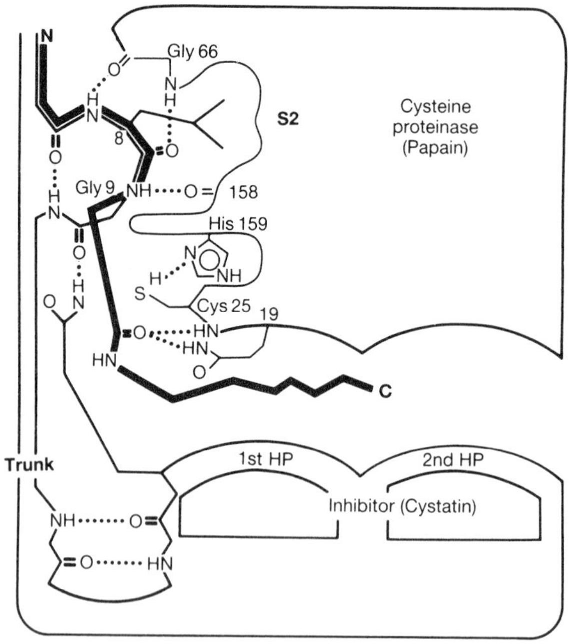

(c)

Figure 7.9. Comparison of the canonical substrate-like inhibition mode (a) with that of hirudin (b) and cystatin (c). (a) Schematic representation of a trypsin–ovomucoid inhibitor complex. The canonical binding loop (PBL) of the inhibitor is spanned between scaffold-anchored pillars and further connected to its molecular core through a disulphide bridge and other spacer elements. It is bound (similarly to a substrate, bold lines) to the proteinase binding site through several main chain–main chain hydrogen bonds, forming an antiparallel β-structure between P1–P3 of the inhibitor and 216–218 of the enzyme. The catalytic Ser-195 Oγ of trypsin is in 'sub-Van der Waals' contact to the P1 carbonyl carbon of the scissile peptide bond. (b) Schematic representation of the thrombin–hirudin complex (Rydel *et al.*, 1990, 1991). The three amino-terminal residues (1HI–3HI) align with thrombin segment Ser-214–Gly-219 through formation of a parallel β-sheet. The reactive Ser-195 is not blocked, and the specificity pocket is filled with water molecules. The carboxy-terminal tail loops around the characteristic thrombin insertion loop (Tyr-60A–Trp-60D) and binds to the fibrinogen binding exosite of thrombin via several salt bridges. (c) Schematic representation of the cystatin–papain complex (Bode *et al.*, 1988; Stubbs *et al.*, 1990). Cystatin binds largely via two hairpin loops (1st HP, 2nd HP) to papain surface areas adjacent to the catalytic residues. The amino-terminal 'trunk' lies over the reactive site Cys-25, but interacts via residues 8 and Gly-9 with subsite S2 in a manner similar to a productively bound substrate (Drenth *et al.*, 1976), which is shown with thick lines.

et al., 1992a); these inhibitor complexes therefore essentially resemble distorted Michaelis complexes (see Fig. 7.9a).

Several of the exposed side-chains of the (8–12) contacting loop residues around the reactive site (between P9 and P4′) make many-fold, mostly hydrophobic, interactions with proteinase subsites which comprise together 15–25 residues (see, for example, Fig. 7.8). In complexes of trypsin-like proteinases, the interactions of the P1 side-chain with the 'specificity pocket' (S1-subsite; see Fig. 7.9a) are energetically most important (compare Laskowski *et al.*, 1987): they thus determine primarily the specificity of a given inhibitor for a particular proteinase (arginine and lysine residues at P1, for example, confer 'trypsin-like' specificity); consequently, substitution of the P1-residue has a particularly large effect on the specificity towards cognate proteinases (Laskowski *et al.*, 1987, 1989; Beckmann *et al.*, 1988; Longstaff *et al.*, 1990). The side-chain contacts of loop residues P2, P1′ and P2′ with their opposing more shallow proteinase subsites rank second. In subtilisins, interactions with the S1-cleft play a less dominant role, while the fit of the P4-residue becomes of particular importance (see Hirono *et al.*, 1984; Bode *et al.*, 1987; Takeuchi *et al.*, 1991). The almost independent arrangement of the inhibitor loop side-chains in the interface explains the usually observed additivity of individual binding contributions to the overall binding. This in turn allows reliable affinity predictions for optional loop sequences from the knowledge of only a few parameters (see the 'sequence-function algorithm' of M. Laskowski; Laskowski, 1980; Laskowski *et al.*, 1990).

Besides these 'primary' interactions, most inhibitors contact their cognate proteinases additionally by a 'secondary' binding segment up to four residues long. These 'secondary' contacts (like those of more peripheral loop residues) are, however, in most cases not very specific and (except BPTI) apparently not important for tightness of binding. In the case of proteinases with quite narrow active-site clefts (such as thrombin; Bode *et al.*, 1989b) these 'secondary' elements can, however, cause severe sterical hindrance and thus prevent complex formation (Bode *et al.*, 1989b).

In the interfaces of the complexes, the amino acid residues of both components are as densely packed as in the interior of proteins or in amino acid crystals (Janin & Chothia, 1990). The contact surfaces between native inhibitors and their cognate proteinases are essentially complementary; complex formation is accompanied by only slight conformational rearrangements (with root mean square deviations in the order of 0.35 Å; Bode *et al.*, 1987; McPalen & James, 1988), but (except for BPTI; Wlodawer *et al.*, 1987a) result in considerable 'freezing' of the binding loop. The intermolecular contact area is restricted to a relatively small surface strip of 600–900 $Å^2$. The inhibitor loop does not occupy the binding cleft fully. The residual spaces left between inhibitor loop and cleft rims are often filled with a few localized solvent molecules allowing some freedom to adapt to external stress. Various eglin c complexes indeed show slightly different orientations of the inhibitor around its central binding loop 'axis' relative to the enzyme's binding site (Gros *et al.*, 1989b) although the main interactions are maintained.

This surface complementarity and the loop stabilization through the scaffold explain in part the tightness of these complexes: the inhibitor binding region does not lose as much conformational freedom upon binding as a flexible substrate; from a comparison of the binding constants of ovomucoid inhibitors or eglin c with those for octapeptides derived from their binding loops (Okada *et al.*, 1989) a scaffolding contribution of about −8 kcal/mol can be estimated (M. Laskowski, pers. commun.). Surface complementarity might also contribute to the fast association reaction with typical k_{on} values of 10^6 M^{-1} s^{-1} believed to be necessary for many physiologically important blocking reactions. However, modified 'small' inhibitors, specifically cleaved at their reactive sites, bind with similar affinity and sometimes even with an almost equal rate constants as their native compounds (Quast *et al.*,

1978; Ardelt & Laskowski, 1985; Read & James, 1986). These cleaved inhibitors associate with a proteinase to form the same stable complex as the virgin species, i.e. complexation involves peptide re-synthesis.

The binding loop of the inhibitor component in the complex is tightly packed and quite rigid. The elaborate interactions with the subsites of the proteinase and with the inhibitor's own core are obviously of quite favourable energetics to stabilize this complex. These two types of interactions mutually stabilize the binding loop in the complex and confer thermodynamic stability to it. Due to the cooperative behaviour of the inhibitor scaffold the binding constants of less-stable inhibitor variants ('weakened', for example, through main-chain cleavages), do not seem to be drastically changed (reduced) at temperatures significantly below their melting points compared with their intact counterparts (M. Laskowski, pers. commun.). Amino acid substitutions of inhibitor loop-core-spacers can seriously disturb the mutually stabilizing contacts and confer to the inhibitor more substrate-like properties (Wagner *et al.*, 1990).

These mutually stabilizing interactions apparently prevent the reactive-site deformation presumed to be important for peptide-bond cleavage (James *et al.*, 1980; Read & James, 1986), and thus slow down catalytic processing (a property which itself does not seem to be of great importance for inhibitor potency, but might rather be a side-effect). The stable complex is an energy sink so that the energy barriers slow dissociation to either product. (For more details see Longstaff *et al.*, 1990.) The height of these flanking barriers can vary considerably and is clearly not a function of the inhibitor structure alone, but of the proteinase and the proteinase–inhibitor complex (Quast *et al.*, 1978; Ardelt & Laskowski, 1985; Read & James, 1986). It has often been pointed out (Finkenstadt *et al.*, 1974; Ardelt & Laskowski, 1985) that most proteinase–protein inhibitor interactions are characterized by relatively large 'specificity constants' k_{cat}/K_M comparable with those measured for 'good' peptide substrates. However, the k_{cat}/K_M index characterizes enzyme–substrate reactions only at very low substrate concentrations; at higher concentrations ($[S] > K_M$) the reaction rate is governed by k_{cat}. This latter index is known to be extremely low in case of protein inhibitor–proteinase interactions. The inhibitor interaction with zymogens (in particular with trypsinogen; Bode *et al.*, 1976, 1978, 1984; Fehlhammer *et al.*, 1977; Huber & Bode, 1978; Bode, 1979) provides evidence that the mutual stabilization of inhibitor and proteinase allow complex formation even in the case of non-complementarity of the reacting components. The substrate binding site of these pro-enzymes is organized quite differently to that of the activated proteinase and often disordered. The re-organization of the 'activation domain' (Bode *et al.*, 1976; Fehlhammer *et al.*, 1977; Huber & Bode, 1978; Bode, 1979; Zbyrt & Otlewski, 1991) upon inhibitor binding is similar to that found upon activation cleavage and follows an induced-fit mechanism (Nolte & Neumann, 1978); the free energy needed to enforce ordering of the disordered trypsinogen segments of the activation domain is provided in part by the free energy of binding, with a concommittant reduction in affinity (Bode, 1979). This indicates again that not the structure of the single components, but rather that of the resulting complex confers stability (Read & James, 1986). This is further underlined by results showing that the same inhibitor (BPTI, Quast *et al.*,

1978; or turkey ovomucoid, Wlodawer *et al.*, 1987a) may interact with different proteinases as an inhibitor or as a substrate (see Estell & Laskowski, 1980).

2.2 *Serpins*

To date, atomic resolution crystal structures of four specifically modified serpin inhibitors—two α_1-proteinase inhibitor species (Löbermann *et al.*, 1984, Engh *et al.*, 1989), α_1-antichymotrypsin (Baumann *et al.*, 1991) and HLEI (Baumann *et al.*, 1992) and of two ovalbumin species (plakalbumin, Wright *et al.*, 1990; and

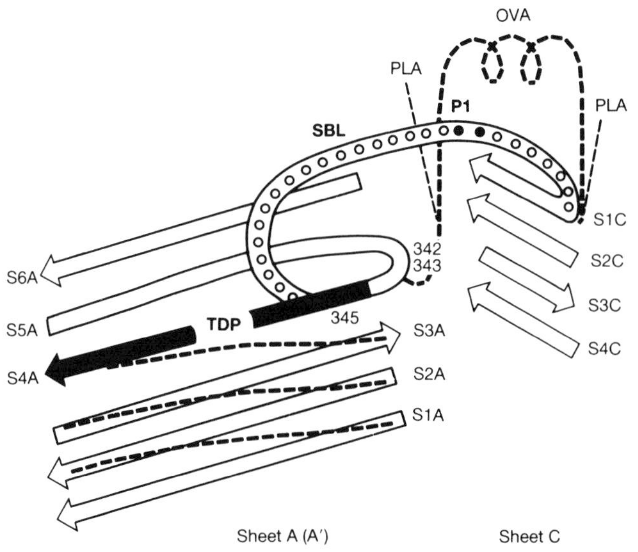

(a)

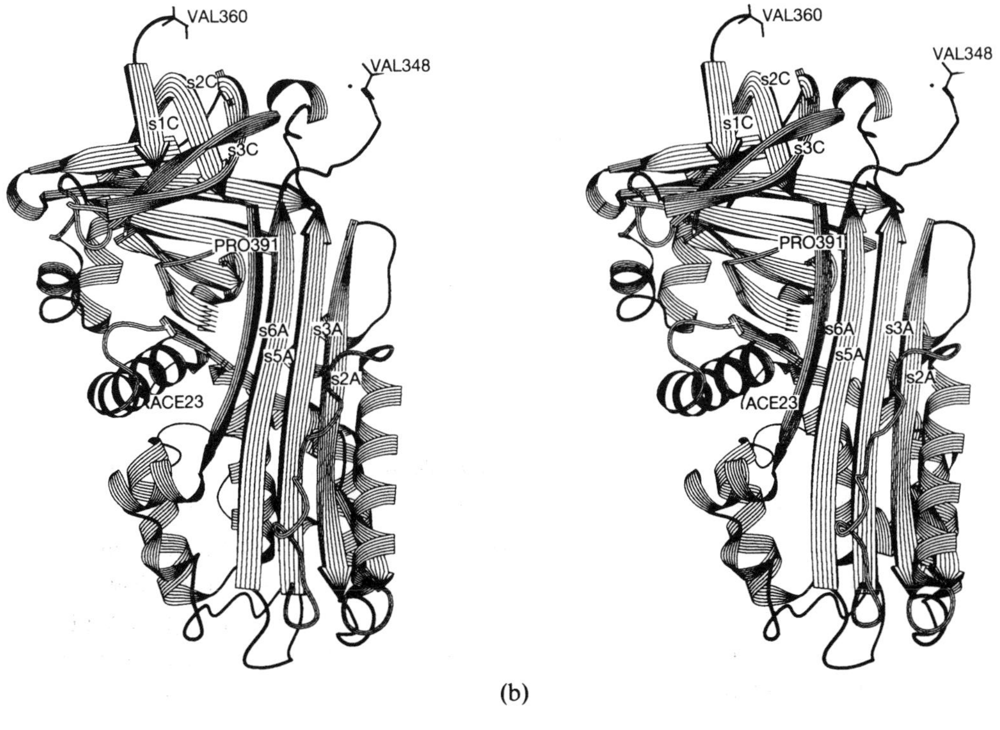

(b)

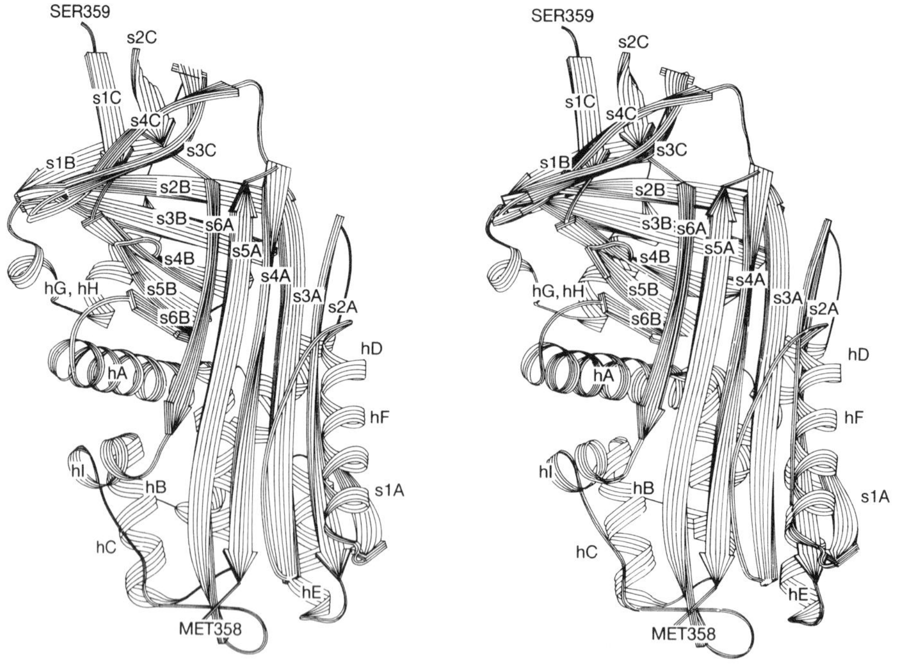

(c)

Figure 7.10. (a) Schematic representation of β-sheets A and C in serpins. Open arrows represent peptide strands as observed in cleaved inhibitors (see also (c)). In both ovalbumin structures, which are models for intact active serpins, strands s1A, s2A and s3A (dashed lines) are annealed with strand s5A to form a five-stranded β-sheet (see also (b)); in plakalbumin (PLA), the free ends project from the surface (Wright *et al.*, 1980), while in ovalbumin (OVA) (Stein *et al.*, 1990) the connecting peptide adopts a helical conformation. We presume that in functional native serpin inhibitors the serpin binding loop (SBL) is partially inserted in sheet A and spanned over the inhibitor surface favouring a canonical conformation of the protease binding segment. The functional native serpins would therefore appear to have structures between ovalbumin and cleaved serpins. Insertion of an exogeneous tetradecapeptide (Schulze *et al.*, 1991) (TDP, black arrow) leads to a binary complex with structural features very similar to cleaved inhibitors. The proteinase binding segment may attain a more coiled conformation leading to the observed non-inhibitory property. (b) Ribbon drawing of plakalbumin (Engh *et al.*, 1990; Wright *et al.*, 1990). It is acetylated at its N-terminus Thr-23 (ACE23) and has a C-terminal Pro-391. The five-stranded β-sheet A (s1A, s2A, s3A, s5A, s6A) lies in front and is vertically oriented. Cleavage of ovalbumin by subtilisin leads to plakalbumin by piecewise removal of a hexapeptide 353 to 358. Residues 348 and 360, respectively, are structurally undefined in the crystals. (c) Ribbon drawing of cleaved α_1-protease inhibitor in a similar orientation to plakalbumin (Löbermann *et al.*, 1984; Engh *et al.*, 1989, 1990; Baumann *et al.*, 1991). β-strands, s, and helices, h, are marked. The newly liberated chain ends are Met-358 and Ser-359. β-sheet A is six-stranded, s1A to s6A, by insertion of s4A in antiparallel fashion between s5A and s3A.

native ovalbumin, Stein *et al.*, 1990)—are available (see Figs 7.10a,b,c). In the first four cleaved serpin species, the newly formed terminal segments S4A and S1C (see Fig. 7.10a) are incorporated into two different β-pleated sheets A and C; the P1- and P1′-residues are separated by a distance of 70 Å (Fig. 7.10c). The more compact and better hydrogen-bonded structure is in agreement with the higher thermodynamic stability of the cleaved species compared with native inhibitors (Bruch *et al.*, 1988; Carrell & Owen, 1989; Engh *et al.*, 1990).

In the two ovalbumin species, the peptide segment equivalent to the carboxy-terminal strand S4A is not part of sheet A (Figure 7.10b); in the cleaved form (plakalbumin) it projects away from the molecule and is disordered (Wright *et al.*, 1990), whereas in the intact ovalbumin it is covalently connected to strand S1C through a helical segment with enhanced mobility (Stein *et al.*, 1990) (see Fig. 7.10a; we note, however, that binding loops of serpins with inhibitory function are very probably non-helical). In contrast to the modified serpin structures of α_1-PI*, α_1-Achy* and HLEI* (see Table 7.1), strands S1A, S2A and S3A in both ovalbumin structures are annealed with strands S5A and S6A forming a five-stranded sheet A′ (Figs 7.10a,b). An obvious reason for failure of strand S4A to incorporate into sheet A′ in plakalbumin is steric hindrance through the bulky side-chain of an arginine residue (Wright *et al.*, 1990); the equivalent Thr-345 (P14) of cleaved α_1-PI projects into the molecular centre. This residue is mostly threonine, rarely valine or serine, in all inhibitory serpins.

Neither the loop structure of isolated inhibitory serpins nor the geometry of their complexes is yet known. Several lines of evidence—the susceptibility of residues P10 to P2′ (residues 349 to 360 in α_1-PI) to proteolytic cleavage (Kress, 1986; Potempa *et al.*, 1986; Huber & Carrell, 1989); docking experiments with thrombin (Engh *et al.*, 1990), which places the tightest constraints on serpin models by its extraordinarily narrow binding cleft (Bode *et al.*, 1989b)—suggest that serpins might interact primarily with substrate binding sites of cognate proteinases through a particularly flat, exposed binding loop of canonical conformation. To confer stability to the formed complex without ready cleavage, the reactive site of this binding loop requires support through side-chains anchored in the inhibitor core, possibly in a similar manner to that observed for eglin c (see Fig.7.2, Bode *et al.*, 1986a).

In inhibitory intact serpins, about four (three to seven) residues of strand S4A (after passing the S5A–S4A tight turn, see Fig. 7.10a) may be located within sheet A, and the rest may loop in an extended conformation over the inhibitor surface to merge with strand S1C (Fig. 7.10a; Bode & Huber, 1991). Some evidence for this is provided by the observation that intact human α_1-proteinase inhibitor is rendered inactive upon insertion of a tetradecapeptide sequentially identical to Thr-345–Met-358 (Schulze *et al.*, 1991a) (Fig. 7.10a), i.e. its strand S4A. (Similar findings have recently been reported for cleaved antithrombin-III; Carrell *et al.*, 1991.) Obviously, in the binary peptide complex the integrated peptide prevents the partial re-entering of strand S4A. This may, however, be required to hold the inhibitor loop segment in a proper conformation close to the inhibitor surface.

Very recent results obtained for recombinant serpin inhibitor mutants with bulky residues at positions P14, P12, P10 and P8 (Bock, 1991; Carrell *et al.*, 1991; Schulze *et al.*, 1991b) seem to confirm earlier ideas (Bode & Huber, 1991; Schulze *et al.*, 1991a) according to which strand S4A must re-enter sheet A up to about its P10 residue to render an inhibitor functional. In PAI-I and antithrombin-III latent temporary inactive but non-cleaved serpin forms (Carrell *et al.*, 1991; Goldsmith *et al.*, 1991) are observed, in which this re-entering is obviously not occurring to a sufficient extent.

The amino acid sequences around the active sites of serpins, in particular their

P1 residues, match the requirements for canonical binding of their target proteinases. Thus, the geometry of the inhibitor loop–proteinase interaction might to a first approximation be similar to that observed for the 'small' protein inhibitors. In fact, the association kinetics of α_2-antiplasmin and plasmin do not differ drastically from that of small inhibitors (Longstaff & Gaffney, 1991). However, the serpin–proteinase interaction might differ in many respects such as:

• residues flanking the scissile peptide bond cannot be exchanged without loss of inhibitory potency (Rubin *et al.*, 1990) (this finding presumably points to the importance of the 'spacer fit' between residues of the binding loop and the scaffold for keeping the loop in an appropriate conformation);

• in a given serpin (e.g. α_2-antiplasmin; Potempa *et al.*, 1988) not only the 'natural' scissile peptide bond, but also an adjacent bond might serve as the reactive site (indicating some enhanced adaptability of the loop);

• very recent NMR studies on ^{13}C-labelled serpins (J. Travis, 1991, pers. commun.) have been interpreted as indicating a tetrahedral state of the P1-carbonyl carbon in serpin–proteinase complexes;

• the kinetics of some serpin–proteinase interactions are better understood if several semi-stable intermediates are assumed (Rubin *et al.*, 1990; Mast *et al.*, 1991).

2.3 *Interaction of hirudin with thrombin*

Hirudin, a 65- or 66-amino acid residue protein (Dodt *et al.*, 1985), is an extremely tight binding and selective inhibitor of the coagulation protease thrombin (Walsmann & Markwardt, 1981; Stone & Hofsteenge, 1986). The structural analyses of two recombinant hirudins by two-dimensional NMR (Folkers *et al.*, 1989; Haruyama & Wüthrich, 1989) revealed that hirudin segment 3I-47I (except 31I–36I) forms a globular, compact domain, whereas the first two residues and the carboxy-terminal 18 residues are flexible in solution (see the crystal structure of the hirudin component, Fig. 7.11a).

Recently, the X-ray crystal structures of two complexes formed between human α-thrombin and two slightly different recombinant hirudin variants have been solved at 2.3 Å (Rydel *et al.*, 1990, 1991) and 2.95 Å resolution (Grütter *et al.*, 1990). In these complexes hirudin binds in an extended manner (Fig. 7.11b) along the canyon-like active-site cleft of thrombin (Bode *et al.*, 1989b). Its globular domain contacts with characteristic thrombin surface patches adjacent to the thrombin active site (see Fig. 7.9b). The amino-terminal hirudin segment 1I–3I forms a parallel β-pleated sheet structure with thrombin segment Ser-214–Gly-219. This is in contrast to the antiparallel binding observed for the canonical proteinase inhibitors (compare Figs 7.9a and 7.9b). The catalytic residue Ser-195 of thrombin is not blocked, nor is its specificity pocket used by hirudin residues, but filled with several structured water molecules instead (Fig. 7.9b).

The extended carboxy-terminal 'tail' of hirudin (48I–65I; see Fig. 7.9b) runs in a long groove which extends from the active-site cleft of thrombin (Gütter *et al.*, 1990; Rydel *et al.*, 1990, 1991). Three non-polar and several acidic side-chains of this tail segment make numerous hydrophobic contacts and a few surface salt bridges with this positively charged putative 'fibrinogen secondary binding exosite'.

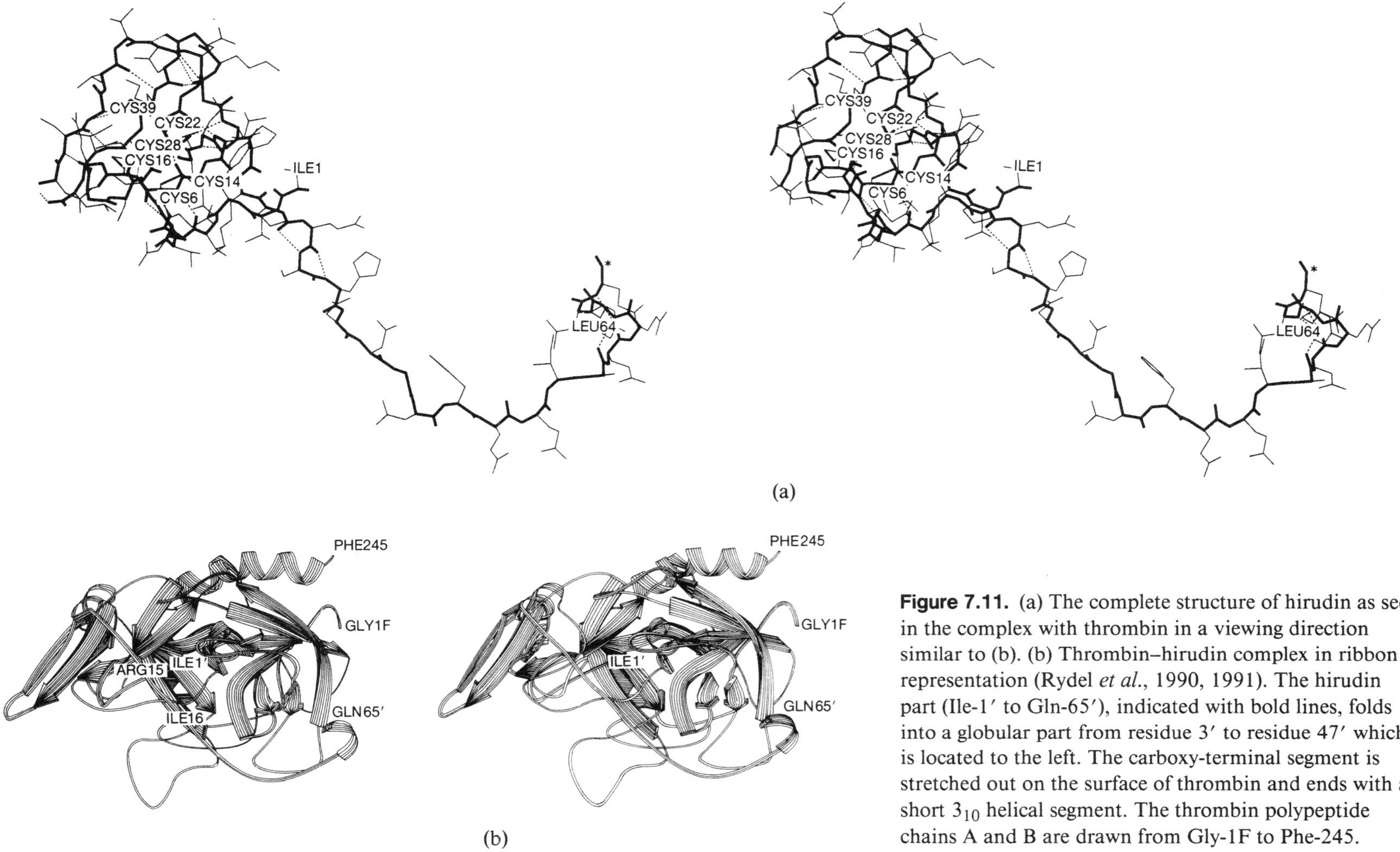

Figure 7.11. (a) The complete structure of hirudin as seen in the complex with thrombin in a viewing direction similar to (b). (b) Thrombin–hirudin complex in ribbon representation (Rydel *et al.*, 1990, 1991). The hirudin part (Ile-1′ to Gln-65′), indicated with bold lines, folds into a globular part from residue 3′ to residue 47′ which is located to the left. The carboxy-terminal segment is stretched out on the surface of thrombin and ends with a short 3_{10} helical segment. The thrombin polypeptide chains A and B are drawn from Gly-1F to Phe-245.

The major contribution to thrombin binding of these negatively charged hirudin tail residues has been demonstrated by the systematic removal of charges (Stone *et al.*, 1989) and by recent electrostatic calculations in our laboratory (Karshikov *et al.*, 1992). The contact surface area in the thrombin–hirudin complex (1800 Å^2) is considerably larger than that of complexes between small protein inhibitors and their cognate serine proteinases. The three distinct hirudin regions (the amino-terminal segment, the globular domain and the tail) participate in the intermolecular contacts to similar degrees (Rydel *et al.*, 1991). The globular domain (including the amino-terminal part) and the tail portion exhibit additive binding contributions, i.e. they bind independently of one another (Dennis *et al.*, 1990).

In approaching thrombin, hirudin might be preoriented by the complementary, strong electrostatic fields of the binding surfaces (Stone *et al.*, 1989; Bode *et al.*, 1992b). This would account for the almost diffusion-controlled association rate constant (Stone *et al.*, 1989). Kinetic data analysed at varying ionic strengths indicate that, in a first step, the negatively charged hirudin tail associates (and becomes rigid) with the positively charged exosite of thrombin, before the compact domain binds adjacent to the active site of thrombin through a favourable induced-fit of the amino-terminal segment (see Fig. 7.9b).

3 Cystatin–cysteine proteinase interaction

Cystatins and stefins are tight, reversibly binding, protein inhibitors of papain-like cysteine proteinases (Barrett *et al.*, 1986). Recently, the X-ray crystal structures of two representatives, of chicken egg-white cystatin (Bode *et al.*, 1988) and of stefin B in complex with papain (Stubbs *et al.*, 1990), have been elucidated. Cystatins/stefins consist of a long central α-helix, wrapped in a five-stranded antiparallel β-pleated sheet, with a subsidiary helix or strand, respectively (see Figs 7.12a,b). At one end of the sheet, an exposed 'first' β-hairpin loop (comprising a highly conserved 'QVVAG' or similar sequence) is flanked on both sides by the projecting amino-terminal segment and a second hairpin loop (see Figs 7.9c and 7.12c). The wedge-shaped hydrophobic edge is complementary in shape to the active-site cleft of papain. In the complex (Bode *et al.*, 1988; Stubbs *et al.*, 1990) both cystatin loops interact with conserved primed subsites adjacent to the papain catalytic residues; the (initially flexible; Bode *et al.*, 1990) amino-terminal segment (the 'trunk', see Fig. 7.9c) loops over the catalytic Cys-25 residue of papain, whose sulphur atom might be alkylated with only minor effects on the association constant (Björk & Ylinenjärvi, 1989), and interacts via two more amino-terminal residues with the putative subsites S2 and S3. In contrast to bound substrates (Drenth *et al.*, 1976) this inhibitor 'trunk' is removed from the catalytic residues in the complex and thus not cleavable (see Figs 7.9c and 7.12c).

The primarily hydrophobic side-chain interactions made by the rigid 'first' hairpin loop (1st HP in Fig. 7.9c) confer most of the stability to the complex; in cystatin complexes, the interactions in the S2-subsite would appear to strengthen complexes with papain considerably (Abrahamson *et al.*, 1987; Björk *et al.*, 1989; Machleidt *et al.*, 1989, 1991; Jerala *et al.*, 1990).

4 Interaction of carboxypeptidase-A with its potato inhibitor

No atomic spatial structure of any metallo-endoproteinase inhibitor is known to date. The X-ray crystal structure of a 39-amino acid residue protein inhibitor from potatoes has been determined in complex with carboxypeptidase-A (Rees & Lipscomb, 1982). The projecting carboxy-terminus of the potato inhibitor (the four residues S′1, S1, S2, S3) is inserted into the active-site cleft of the enzyme and forms the primary contact region; the carboxy-terminal residue Gly-39I is split off, but remains bound in the S1′-subsite, buried by the rest of the residual inhibitor moiety; a few additional 'secondary contacts' confer stabilization to the complex. This complex therefore represents an enzyme–product intermediate in the catalytic mechanism.

In the isolated carboxypeptidase-A inhibitor determined by NMR (Clore *et al.*, 1987b) the projecting carboxy-terminal residue exhibits considerable flexibility. Upon binding to carboxypeptidase-A it becomes ordered and binds, presumably in a substrate-like manner, with Gly-39I (which does not match the specificity requirements of carboxypeptidases) being slowly cleaved (Rees & Lipscomb, 1982). The intermolecular contacts made via its 'primary' contact residues P1, P2 and P3 and the 'secondary contact region' (Rees & Lipscomb, 1982) keep the truncated inhibitor in close position; they prevent fast dissociation as well as further intrusion into the active-site groove.

5 Procarboxypeptidases

Pancreatic procarboxypeptidases are activated by two tryptic cleavages of an amino-terminal segment which in pig procarboxypeptidase-B (PCP-B) is 95 amino acid residues long. This segment is folded into an open α,β-sandwich structure consisting of a four-stranded antiparallel β-sheet and two parallel α-helices (Coll *et*

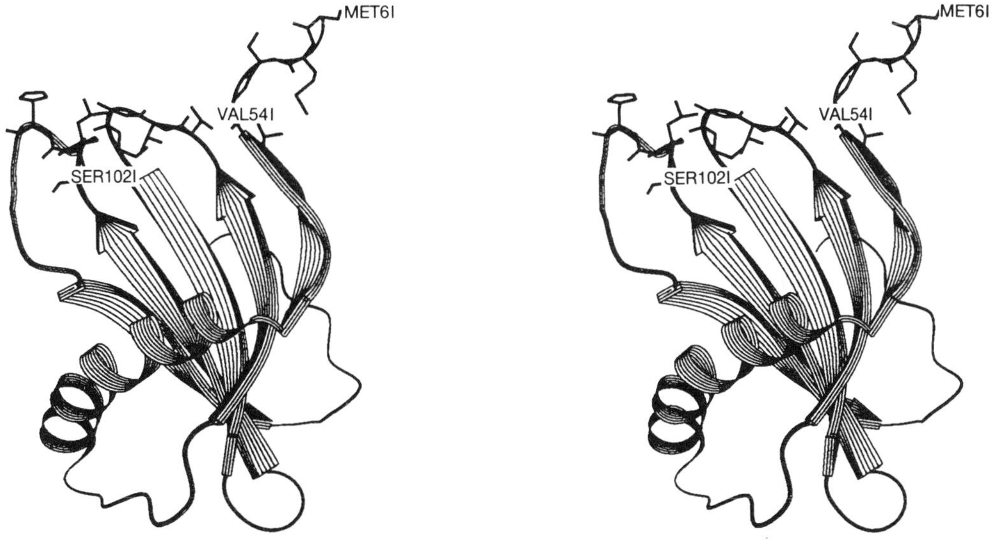

(a)

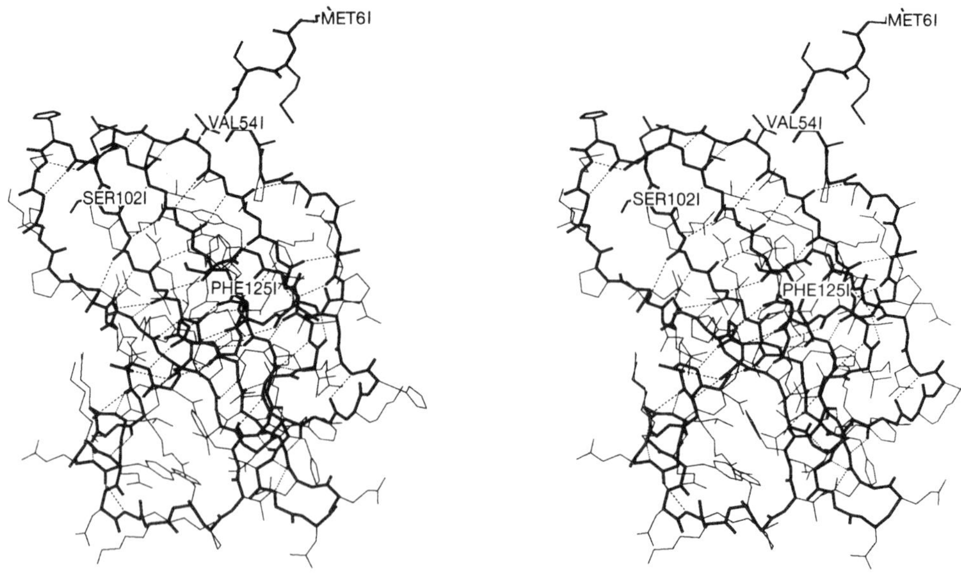

(b)

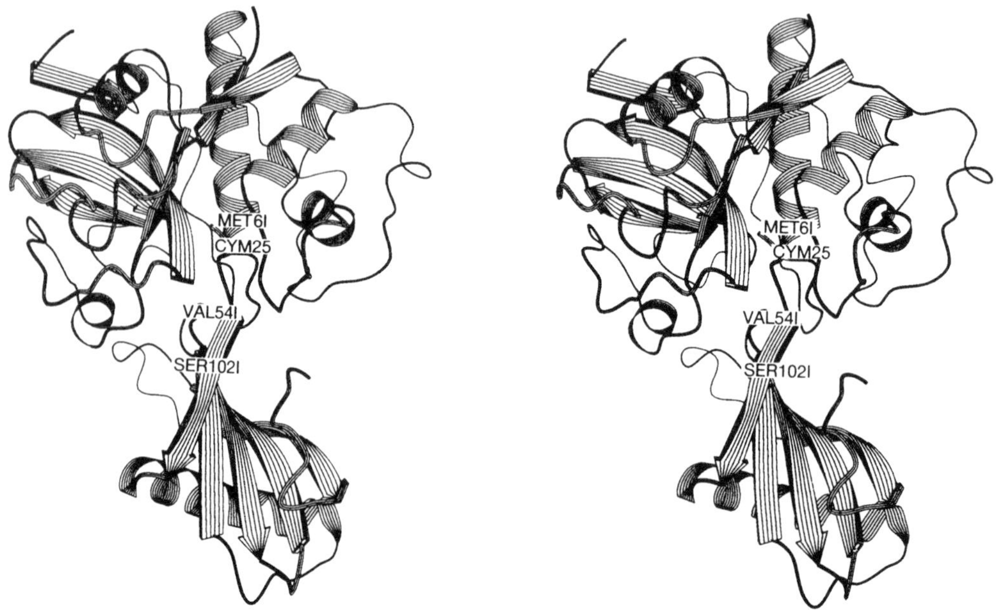

(c)

Figure 7.12. (a,b) Stefin in ribbon and skeletal representation (Bode *et al.*, 1988; Stubbs *et al.*, 1990). The viewing direction is approximately orthogonal to (c). The five-stranded antiparallel β-sheet is partly wrapped around the α-helix. The strand arrangement of the β-sheet is β_1, β_2, β_3, β_4, β_5. The α-helix connects β_1 with β_2. (c) Ribbon drawing of the papain stefin complex (Bode *et al.*, 1988; Stubbs *et al.*, 1990). The stefin component forms the lower stalk. Contacts of stefin with the enzyme are made by three non-contiguous loops at residues Gly-6I, Val-54I and Ser-102I. These form a wedge which fills the active-site cleft of papain. The active-site cysteine of papain is Cys-25 which was, in this complex, carboxymethylated (Cym-25).

al., 1991). It is an autonomous folding unit (Coll *et al.*, 1991) which retains its structural integrity when isolated (Vendrell *et al.*, 1991) (Fig. 7.13). The pro-segment of pig procarboxypeptidase-A (PCP-A) remains bound to the enyzme part after the first trypsinolytic cleavage thus resembling an inhibitor–proteinase interaction; after a second destabilizing cleavage it dissociates, giving rise to the free, active enzyme (Vendrell *et al.*, 1990). The main sites of interaction are between the loop connecting β_2 and β_3 strands of the pro-segment and the general substrate binding area of the enzyme (including residues specifically involved in the fixation of substrates) and between helix α_3 and enzyme. Several solvent molecules are integrated between the pro-structure and the enzyme; access to the active site of the enzyme is blocked for large substrates, but the small inhibitor benzamidine is bound in the S1 specificity pocket of crystals of PCP-B like an arginyl side-chain of a substrate (Coll *et al.*, 1991). PCP-B is enzymatically reactive while PCP-A displays activity against low-molecular-weight substrates (Uren & Neurath, 1974; Vendrell *et al.*, 1990; Burgos *et al.*, 1991).

6 Conclusion

All protein inhibitors of proteinases prevent access of substrates to the catalytic sites of proteinases through steric hindrance. One class of inhibitors (most serine proteinase inhibitors and the carboxypeptidase inhibitor) achieves this by binding with a peptide segment directly to the catalytic site in a substrate- or product-like

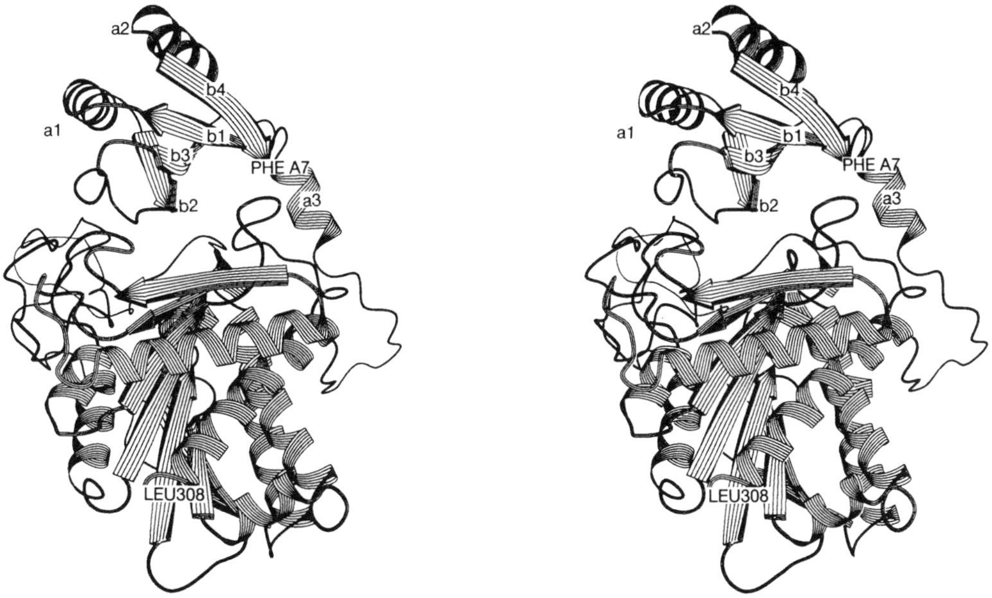

Figure 7.13. Ribbon drawing of pig procarboxypeptidase-B (Coll *et al.*, 1991). The pro-segment lies on top of the enzyme part. Its globular part has an open α, β sandwich topology with an antiparallel β-sheet with strand arrangement β_2, β_3, β_1, β_4. The α-helices connect β_1 and β_2 (α_1) and β_3 and β_4 (α_2) and are arranged antiparallel on the distal side with respect to the enzyme part. The wide loop connecting β_2 with β_3 lies over the active site of the enzyme part. Further extensive contacts with the enzyme part are made by the connecting α-helix α_3.

manner. In the case of product-like binding, the interactions are strong enough to prevent fast dissociation; in the case of substrate-like binding, the intra- and intermolecular interactions of the inhibitor's primary binding segment with the inhibitor's core (through spacer elements) and with the binding site of the enzyme mutually stabilize each other and are so tight that decomposition rarely occurs. Inhibition selectivity is achieved through utilization of the substrate recognition sites of the proteinase. The serpins differ from the small serine proteinase inhibitors by the formation of a postcomplex molecular form which is distinct in structure and activity from the active species.

The second class of inhibitors comprises the cysteine proteinase inhibitors, hirudin and related inhibitors and the activation segment of procarboxypeptidases. They bind mainly to surface sites adjacent to the catalytic residues of their cognate proteinases, utilizing substrate binding sites and other sites. If specific surface patches of the proteinases are involved, the interaction may be very selective. Hirudin is an excellent example of this case. Presumably, nature has used this concept more often to achieve selectivity (Waxman *et al.*, 1990). Procarboxypeptidase resembles complexes between proteinases and non-substrate inhibitors. Presumably, several other pro-proteinases (such as cysteine pro-proteinases, pro-subtilisins) achieve their inactivity by such a mechanism, i.e. through active-site blockage by the covalently bound amino-terminal chain extension.

7 Acknowledgements

We thank Monika Schneider and Karin Epp for help in preparing the figures, and Professor M. Laskowski and Dr J. Otlewski for many helpful discussions and access to some of their manuscripts prior to publication. The financial support of the Sonderforschungsbereich 207 der Universität München (projects H-1 and H-2) and of the Fonds der Chemischen Industrie are acknowledged.

8 References

Abrahamson, M., Ritonja, A., Brown, M.A., Grubb, A., Machleidt, W. & Barrett, A.J. (1987). *J. Biol. Chem.,* **262**, 9688–94.

Ardelt, W. & Laskowski, M. (1983). *Acta Biochim. Polon.*, **30**, 115–26.

Ardelt, W. & Laskowski, M. Jr. (1985). *Biochemistry,* **24**, 5313–20.

Ardelt, W. & Laskowski (1991). *J. Mol. Biol.*, **220**, 1041–53.

Barrett, A.J., Rawlings, N.D., Davies, M.E., Machleidt, W., Salvesen, G. & Turk, V. (1986). In: A.J. Barrett and G. Salvesen (Eds.), *Proteinase Inhibitors*, pp. 515–69, Elsevier, Amsterdam. .

Baudys, M., Gosh, M., Harlos, K., Mares, M., Fusek, M., Kostka, V. & Blake, C.C.F. (1991). *J. Mol. Biol.*, **218**, 21–2.

Baumann, U., Huber, R., Bode, W., Grosse, D., Lesjak, M. & Laurell, C.B. (1991). *J. Mol. Biol.*, **218**, 595–606.

Baumann, U., Bode, W., Huber, R., Travis, J. & Potempa, J. (1992). *Mol. Biol.* In press.

Beckmann, J., Mehlich, A., Schröder, W., Wenzel, H.R. & Tschesche, H. (1988). *Eur. J. Biochem.*, **176**, 675–82.

Björk, I. & Ylinenjärvi, K. (1989). *Biochem. J.*, **260**, 61–8.

Bock, S.C. (1991). In: L. Sottrup-Jensen, T.E. Petersen, B.F.C. Clark and H. Jörnvall (Eds.) *20th Linderstrøm-Lang Conference, Vingsted, Denmark*, pp. 149–51, Aarhus.

Bode, W. (1979). *J. Mol. Biol.*, **127**, 357–74.

Bode, W. & Huber, R. (1986). In: P. Desnuelle, H. Sjöström and O. Norén (Eds.), *Molecular and Cellular Basis of Digestion*, pp. 213–34, Elsevier, Amsterdam.
Bode, W. & Huber, R. (1991). *Curr. Op. Struct. Biol.*, **1**, 45–52.
Bode, W., Fehlhammer, H. & Huber, R. (1976). *J. Mol. Biol.*, **106**, 325–35.
Bode, W., Schwager, P. & Huber, R. (1978). *J. Mol. Biol.*, **118**, 99–112.
Bode, W., Papmokos, E. & Musil, D. (1987). *Eur. J. Biochem.*, **166**, 673–92.
Bode, W., Stubbs, M. & Laskowski, M. (1992a). In preparation.
Bode, W., Turk, D. & Karshikov, A. (1992b). *Protein Sci.*, **1**, 426–71.
Bode, W., Walter, J., Huber, R., Wenzel, H.R. & Tschesche, H. (1984). *Eur. J. Biochem.*, **144**, 185–90.
Bode, W., Epp, O., Huber, R., Laskowski, M. & Ardelt, W. (1985). *Eur. J. Biochem.*, **147**, 387–95.
Bode, W., Papamokos, E., Musil, D., Seemüller, V. & Fritz, H. (1986a). *EMBO J.*, **5**, 813–18.
Bode, W., Wei, A.-Z., Huber, R. Meyer, E., Travis, J. & Neumann, S. (1986b). *EMBO J.*, **5**, 2453–8.
Bode, W., Engh, R., Musil, D., Thiele, U., Huber, R., Karshikov, A., Brzin, J., Kos, J. & Turk, V. (1988). *EMBO J.*, **7**, 2593–9.
Bode, W., Greyling, H.J., Huber, R., Otlewski, J. & Wilusz, T. (1989a). *FEBS Lett.*, **242**, 285–92.
Bode, W., Mayr, I., Baumann, U., Huber, R., Stone, S.R. & Hofsteenge, J. (1989b). *EMBO J.*, **8**, 3467–75.
Bode, W., Engh, R., Musil, D., Laber, B., Stubbs, M., Huber, R. & Turk, V. (1990). *Biol. Chem. Hoppe-Seyler*, **371**, 111–18.
Bolognesi, M., Gatti, G., Menegatti, E., Guarneri, M., Marquart, M., Papamokos, E. & Huber, R. (1982). *J. Mol. Biol.*, **162**, 839–68.
Bolognesi, M., Pugliese, L., Gatti, G., Frigero, F., Coda, A., Antolini, L., Schnebli, H.P., Menegatti, E., Amiconi, G. & Aszenzi, P. (1990). *J. Mol. Recogn.*, **3**, 163–8.
Bruch, M., Weiss, V. & Engel, J. (1988). *J. Biol. Chem.*, **263**, 16626–30.
Burgos, F.J., Jalvà, M., Villegas, V., Soriano, F., Mendez, E. & Avilés, F.X. (1991). *Biochemistry*, **30**, 4082–9.
Carrell, R.W. & Owen, M.C. (1989). *Nature* (*London*), **317**, 730–2.
Carrell, R.W. & Travis, J. (1985). *Trends Biochem. Sci.*, **10**, 20–4.
Carrell, R.W., Evans, D.L., Marshall, C.J. & Stein, P.E. (1991). In: L. Sottrup-Jensen, T.E. Petersen, B.F.C. Clark and H. Jörnvall (Eds.) *20th Linderstrøm-Lang Conference, Vingsted, Denmark*, pp. 152–3, Aarhus.
Cawston, T.E. (1986). In: A.J. Barrett and G. Salvesen (Eds.), *Proteinase Inhibitors*, pp. 589–610, Elsevier, Amsterdam.
Chen, Z. & Bode, W. (1983). *J. Mol. Biol.*, **164**, 283–311.
Chiche, L., Gaboriand, C., Heitz, A., Mornou, J.-P., Castra, B. & Kollman, P.A. (1989). *Proteins: Struct., Funct., Genetics*, **6**, 405–17.
Clore, G.M. & Gronenborn, A.M. (1989). *Crit. Rev. Biochem. Mol. Biol.*, **24**, 479–564.
Clore, G.M., Gronenborn, A.M., James M.N.G., Kjaer M., McPhalen, C.A. & Poulsen, F.M. (1987a). *Prot. Eng.*, **1**, 313–18.
Clore, G.M., Gronenborn, A.M., Nilges, M. & Ryan, C.A. (1987b). *Biochemistry*, **26**, 8012–23.
Coll, M., Guasch, A., Avilés, F.X. & Huber, R. (1991). *EMBO J.*, **10**, 1–9.
Creighton, T.E. & Goldenberg, D.P. (1983). *Biopolymers*, **22**, 49–58.
Creighton, T.E. & Goldenberg, D.P. (1984). *J. Mol. Biol.*, **179**, 497.
Deisenhofer, J. & Steigemann, W. (1975). *Acta Crystallogr.*, **B31**, 238–50.
Dennis, S., Wallace, A., Hofsteenge, J. & Stone, S.R. (1990). *Eur. J. Biochem.*, **188**, 61–6.
Dodt, J., Seemüller, U., Maschler, R. & Fritz, H. (1985). *Biol. Chem. Hoppe-Seyler*, **366**, 379–85.
Drenth, J., Kalk, K.H. & Swen, H.M. (1976). *Biochemistry*, **19**, 3731–8.
Eigenbrot, C., Randal, M. & Kossiakoff, A.A. (1990). *Prot. Eng.*, **3**, 591–8.
Engh, R.A., Wright, H.T. & Huber, R. (1990). *Prot. Eng.*, **3**, 469–77.
Engh, R., Löbermann, H., Schneider, M., Wiegand, G., Huber, R. & Laurell, C.-B. (1989). *Prot. Eng.* **2**, 407–15.

Epp, O., Hörlein, H.D. & Bode, W. (1992). In preparation.
Estell, D.A. & Laskowski, M. Jr. (1980). *Biochemistry*, **19**, 124–31.
Fehlhammer, H., Bode, W. & Huber, R. (1977). *J. Mol. Biol.*, **111**, 415–38.
Finkenstadt, W.R., Hamid, M.A., Mattis, I.A., Schrode, I., Sealock, R.W., Wang, D. & Laskowski, M. Jr. (1974). In: H. Fritz, H. Tschesche, L.J. Greene and E. Truscheit (Eds.), *Proteinase Inhibitors. Proceedings of the Second International Research Conference—Bayer Symposium V*, pp. 389–411, Springer-Verlag, Berlin.
Folkers, P.J.M., Clore, G.M., Driscoll, P.C., Dodt, J., Köhler, S. & Gronenborn, A.M. (1989). *Biochemistry*, **28**, 2601–17.
Fujinaga, M., Sielecki, A.R., Read, R.J., Ardelt, W., Laskowski, M. & James, M.N.G. (1987). *J. Mol. Biol.*, **195**, 397–418.
Goldenberg, D.P. (1985). *J. Cell. Biochem.*, **29**, 321–35.
Goldenberg, D.P., Frieden, R.W., Haack, J.A. & Morrison, T.B. (1989). *Nature* (*London*), **338**, 127–32.
Goldsmith, E.J., Sheng-Cheng, C., Danley, D.E., Gerard, R.D., Geoghegan, K.F., Mottonen, J. & Strand, A. (1991). *Proteins*, **9**, 225–7.
Greenblatt, H.M., Ryan, C.A. & James, M.N.G. (1989). *J. Mol. Biol.*, **205**., 201–25.
Gronenborn, A.M., Nilges, M., Peanasky, R.J. & Clore, G.M. (1990). *Biochemistry*, **29**, 183–9.
Gros, P., Fujinaga, M., Dijkstra, B.W., Kalk, K.H. & Hol, W.G.J. (1989a). *Acta Crystallogr.*, **B45**, 488–99.
Gros, P., Betzel, C., Dauter, Z., Wilson, K.S. & Hol, W.G.J. (1989b). *J. Mol Biol.*, **210**, 347–67.
Grütter, M.G., Fendrich, G., Huber, R. & Bode, W. (1988). *EMBO J.*, **7**, 345–351.
Grütter, M.G., Priestle, J.P., Rahuel, J., Grossenbacher, H., Bode, W., Hofsteenge, J. & Stone, S.R. (1990). *EMBO J.*, **9**, 2361–5.
Haruyama, H. & Wüthrich, K.. (1989). *Biochemistry*, **28**, 4301–12.
Hecht, H.J., Szardenings, M., Collins, J. & Schomburg, D. (1991). *J. Mol. Biol.*, **220**, 711–22.
Heinz, D.W., Priestle, J.P., Rahuel, J., Wilson, K.S. & Grütter, M.G. (1991). *J. Mol. Biol.*, **217**, 353–71.
Hirono, S., Agawa, H., Iitaka, Y. & Mitsui, U. (1984). *J. Mol. Biol.*, **178**, 389–413.
Holak, T.A., Bode, W., Huber, R., Otlewski J. & Wilusz, T. (1989). *J. Mol. Biol.*, **210**, 649–54.
Huber, R. & Bode, W. (1978). *Acc. Chem. Res.*, **11**, 114–22.
Huber, R. & Carrell, R.W. (1989). *Biochemistry*, **28**, 8951–66.
Huber, R., Kukla, D., Bode, W., Schwager, P., Bartels, K., Deisenhofer, J. & Steigemann, W. (1974). *J. Mol. Biol.*, **89**, 73–101.
Huber, R., Bode, W., Kukla, D., Kohl, W. & Ryan, C.A. (1975). *Biophys. Struct. Mech.*, **1**, 189–201.
Hynes, T.R., Randal, M., Kennedy, L.A., Eigenbrot, C. & Kossiakoff, A.A. (1990). *Biochemistry*, **29**, 10018–22.
James, M.N.G., Sielecki, A.R., Brayer, G.D., Delbaere, L.T.J. & Bauer, C. -A (1980). *J. Mol. Biol.*, **144**, 43–88.
Jandu, S.K., Ray, S., Brooks, L. & Leatherbarrow, R.J. (1990). *Biochemistry*, **29**, 6264–9.
Janin, J. & Chothia, C. (1990). *J. Biol. Chem.*, **265**, 16027–30.
Jerala, R., Trstenjak-Aebanda, M., Kroon-Zitko, L., Lenarcic, B. & Turk, V. (1990). *Biol. Chem. Hoppe-Seyler*, **371**, 157–60.
Kamiya, N., Matsushima, M. & Sugino, H. (1984). *Bull. Chem. Soc. Jpn.*, **57**, 2075–81.
Karshikov, A., Bode, W., Tulinsky, A. & Stone, S.R. (1992). *Protein Sci.*, **1**, in press.
Kress, L.F. (1986). *J. Cell. Biochem.*, **32**, 51–8.
Laskowski, M. (1980). *Biochem. Pharm.*, **29**, 2089–94.
Laskowski, M. Jr. & Kato, I. (1980). *Annu. Rev. Biochem.*, **49**, 593–626.
Laskowski, M. Jr., Kato, I., Ardelt, W., Cook, J., Denton, A., Empie, M.W., Kohr, W.J., Park, S.J., Parks, K., Schatzley, B.L., Schoenberger, O.L., Tashiro, M., Vichot, G., Wheatley, H.E., Wieczorek, A., Wieczorek, M. (1987). *Biochemistry*, **26**, 202–21.
Laskowski, M.J., Park, S.J., Tashiro, M. & Wynn, R. (1989). In: *Protein Recognition of Immobilized Ligands*, pp. 149–68, Alan R. Liss, New York.

Laskowski, M., Apostol, I., Ardelt, W., Cook, J., Giletto, A., Kelly, C.A., Lu, W., Park, S.J., Qasim, M.A., Whatley, H.E., Wiezorek, A. & Wynn, R. (1990). *J. Prot. Chem.*, **9**, 715–25.
Lin, G., Engh, R., Bode, W., Huber, R. & Chi, C. (1992). In preparation.
Löbermann, H., Tokuoka, R., Deisenhofer, J. & Huber, R. (1984). *J. Mol. Biol.*, **177**, 531–56.
Longstaff, C. & Gaffney, P.J. (1991). *Biochemistry*, **30**, 979–86.
Longstaff, C., Campbell, A.F. & Fersht, A.R. (1990). *Biochemistry*, **29**, 7339–47.
McGrath, M.E., Erpel, T., Browner, M.F. & Fletterick, R.J. (1991). *J. Mol. Biol.*, **222**, 139–42.
Machleidt, W., Thiele, U., Laber, B., Assfalg-Machleidt, I., Esterl, A., Wiegand, G., Kos, J., Turk, V. & Bode, W. (1989). *FEBS Lett.*, **243**, 234–8.
Machleidt, W., Thiele, U., Assfalg-Machleidt, I., Förger, D. & Auerswald, E.A. (1991). *Biomed. Biochim. Acta*, **50**, 613–20.
McPhalen, C.A. & James, M.N.G. (1987). *Biochemistry*, **26**, 261–9.
McPhalen, C.A. & James, M.N.G. (1988). *Biochemistry*, **27**, 6582–98.
McPhalen, C.A., Svendsen, I., Jonassen, I. & James, M.N.G. (1985). *Proc. Natl. Acad. Sci. USA*, **82**, 7242–6.
Markley, J.L. (1989). *Methods Enzymol.*, **176**, 12–64.
Marquart, M., Walter, J., Deisenhofer, J., Bode, W. & Huber, R. (1983). *Acta Crystallogr.*, **B39**, 480–90.
Martzen, M.R., McMullen, B.A., Smith, N.E., Fujukawa, K. & Peanasky, R.J. (1990). *Biochemistry.*, **29**, 7366–72.
Mast, A.E., Enghild, J.J., Pizzo, S.V. & Salvesen, G. (1991). *Biochemistry*, **30**, 1723–30.
Mitsui, Y., Satow, Y., Watanabe, Y. & Iitaka, Y. (1979). *J. Mol. Biol.*, **131**, 697–724.
Moses, E. & Hinz, H.-J (1983). *J. Mol. Biol.*, **170**, 765–76.
Musil, D., Bode, W., Huber, R., Laskowski, M., Lin, T.-Y.L. & Ardelt, W. (1991). *J. Mol. Biol.*, **220**, 739–55.
Nolte, H.J. & Neumann, E. (1978). *Biophys. Chem.*, **10**, 253.
Okada, Y., Tsubri, S., Tsuda, Y., Nakaboyashi, K., Nagamatsu, Y. & Yamamoto, J. (1989). *Biochem. Biophys. Res. Commun.*, **161**, 272–5.
Onesti, S., Brick, P. & Blow, D.M. (1991). *J. Mol. Biol.*, **217**, 153–76.
Otlewski, J. & Laskowski, M. (1992). In preparation.
Papamokos, E., Weber, E., Bode, W., Huber, R., Empie, M.W., Kato, I. & Laskowski, M. (1982). *J. Mol. Biol.*, **158**, 515–37.
Potempa, J., Watorek, W. & Travis, J. (1986). *J. Biol. Chem.*, **261**, 14330–4.
Potempa, J., Shieh, B.-H. & Travis, J. (1988). *Science*, **241**, 699–700.
Priestle, J.P. (1988). *J. Appl. Cryst.*, **21**, 572–6.
Quast, V., Engel, J., Steffen, E., Tschesche, J. & Kupfer, S. (1978). *Biochemistry*, **17**, 1675–82.
Read, R. & James, M.N.G. (1986). In: A.J. Barrett and G. Salvesen (Eds.), *Proteinase Inhibitors*, pp. 301–36, Elsevier, Amsterdam.
Read, R.J., Fujinaga, M., Sielecki, A.R. & James, M.N.G. (1983). *Biochemistry*, **22**, 4420–33.
Rees, D.C. & Lipscomb, W.N. (1982). *J. Mol. Biol.*, **160**, 475–98.
Rhyn, G.I. & Markley, J.L. (1988). *Biochemistry*, **27**, 2529–39.
Ritonja, A., Krizaj, I., Mesko, P., Kopitar, M., Lucovnik, P., Strukelj, B., Pungercar, J., Buttle, D.J., Barrett, A.J. & Turk, V. (1990). *FEBS Lett.*, **267**, 13–15.
Robertus, J.D., Alden, R.A., Birktoft, J.J., Kraut, J., Powers, J.C. & Wilcox, P.E. (1978). *Biochemistry*, **11**, 2439–49.
Rubin, H., Wang, Z.-M., Nickbarg, E.B., McLarney, S., Naidoo, N., Schoenburger, O.L., Johnson, J.L. & Cooperman, B.S. (1990). *J. Biol. Chem.*, **265**, 1199–207.
Rydel, T.J., Ravichandran, K.G., Tulinsky, A., Bode, W., Huber, R., Roitsch, C. & Fenton, J.W. (1990). *Science*, **249**, 277–80.
Rydel, T.J., Tulinsky, A., Bode, W. & Huber, R. (1991). *J. Mol. Biol.*, **221**, 583–601.
Schechter, I. & Berger, A. (1967). *Biophys. Biochim. Res. Commun.*, **27**, 157.
Schulze, A.J., Baumann, U., Knof, S., Jaeger, E., Huber, R. & Laurell, C.B. (1991a). *Eur. J. Biochem.*, **194**, 51–6.

Schulze, A.J., Huber, R., Degryse, E., Speck, D. & Bischoff, R. (1991b). *Eur. J. Biochem.*, **202**, 1147–55.

Schwarz, H., Hinz, H.-J., Mehlich, A., Tschesche, H. & Wenzel, H.R. (1987). *Biochemistry*, **26**, 3544–51.

Sottrup-Jensen, L. (1990). *J. Biol. Chem.*, **264**, 11539–42.

Stein, P.E., Leslie, A.G.W., Finch, J.T., Turnell, W.G., McLaughlin, P.J. & Carrell, R.W. (1990) *Nature* (*London*), **347**, 99–102.

Stone, S.R. & Hofsteenge, J. (1986). *Biochemistry*, **25**, 622–8.

Stone, S.R., Dennis, S. & Hofsteenge, J. (1989). *Biochemistry*, **28**, 6857–63.

Stubbs, M.T., Laber, B., Bode, W., Huber, R., Jerala, R., Lenarcic, B. & Turk, V. (1990). *EMBO J.*, **9**, 1939–47.

Suzuki, A., Tsunogae, Y., Tanaka, I., Yamane, T., Ashida, T., Noriaka, S., Hara, S. & Ikenaka, T. (1987). *J. Biochem.*, **101**, 267–74.

Sweet, R.M., Wright, H.T., Janin, J., Chothia, C.H. & Blow, D.M. (1974). *Biochemistry*, **13**, 4212–28.

Takeuchi, Y. Noguchi, S., Satow, Y., Kojima, S., Kumagai, T., Miura, K-I., Nakamura, K.T. & Mitsui, Y. (1991) *Prot. Eng.*, **4**, 501–8.

Travis, J. & Salvesen, G.S. (1983). *Annu. Rev. Biochem.*, **52**, 655–709.

Tsunogae, Y., Tanaka, I., Yamane, T., Kikkawa, J., Achida, J.T., Ishikawa, C., Watanabe, K., Nakamura, S. & Takahashi, K. (1986). *J. Biochem.* (*Tokyo*), **100**, 1637–46.

Turk, V. & Bode, W. (1991). *FEBS Lett.*, **285**, 213–19.

Uren, J.R. & Neurath, H. (1974). *Biochemistry*, **13**, 3512–20.

Vendrell, J., Cuchillo, C.M. & Avilés, F.X. (1990). *J. Biol. Chem.*, **256**, 6949–53.

Vendrell, J., Billetter, M., Wider, G., Avilés, F.X. & Wüthrich, K. (1991). *EMBO J.*, **10**, 11–15.

Wagner, G., Braun, W., Havel, T.F., Schaumann, T., Go, N. & Wüthrich, K. (1987). *J. Mol. Biol.*, **196**, 611–39.

Wagner, G., Hyberts, S.G., Heinz, D.W. & Grütter, M.G. (1990). In: R.H. Sarma and M.N. Sarma (Eds.), *DNA Protein Complexes and Proteins*, Vol. 2, pp. 93–101, Adeniue Press, Guilderl and New York.

Walsmann, P. & Markwardt, F. (1981). *Pharmazie.*, **36**, 653–60.

Waxman, L., Smith, D.E., Arcuri, K.E. & Vlasuk, G.G. (1990). *Science*, **248**, 593–6.

Werner, M.H. & Wemmer, D.E. (1991). *Biochemistry*, **30**, 3356–64.

Williamson, M.P., Havel, T.F. & Wüthrich, K. (1985). *J. Mol. Biol.*, **182**, 295–315.

Wlodawer, A., Deisenhofer, J. & Huber, R. (1987a). *J. Mol. Biol.*, **193**, 145–56.

Wlodawer, A., Nachman, J., Gilliland, G.L., Gallagher, W. & Woodward, C. (1987b). *J. Mol. Biol.*, **198**, 469–80.

Woessner, J.F. (1991). *FASEB J.*, **5**, 2145–54.

Wright, H.T., Qian, H.X. & Huber, R. (1990). *J. Mol. Biol.*, **213**, 513–28.

Zbyrt, T. & Otlewski, J. (1991). *Biol. Chem. Hoppe-Seyler*, **372**, 255–62.

Zemke, K.J., Müller-Fahrnow, A., Jany, K.-D., Pal, G.P. & Saenger, W. (1991). *FEBS Lett.*, **279**, 240–2.

Detection of New Enzymes and Receptors by Hybridization Technique

8 New Approaches to Isolation, Expression and Molecular Modelling of Proteases

J.H. MCKERROW*,†,§, F.E. COHEN**,†, C.S. CRAIK**,‡ and J.A. SAKANARI*,§.
*Departments of *Pathology, **Pharmaceutical Chemistry, †Medicine and ‡Biochemistry, University of California, San Francisco, CA 94143, and §the San Francisco Veterans Administration Medical Center, CA 94121, USA*

1 Introduction

Enzyme inhibitors represent an important class of pharmaceuticals already in use for a variety of diseases. While many of these drugs were discovered by traditional drug-screening programmes, a number of technological advances in the last decade have opened up the possibility of more focused and cost-effective approaches to drug design. These new approaches of 'structure-based drug design' derive in large part from the avalanche of new information on enzyme gene sequences that has resulted from advances in molecular biology. Acquisition of information about the primary structure of an enzyme, which in the recent past could take years, now can take weeks. Concurrent with rapid advances in gene isolation and sequencing, there have been major advances in the ability to express active recombinant enzymes in a variety of host cells, and equally important advances in computer modelling of enzyme active sites based upon primary structure information. This chapter will review a potential strategy for drug design that utilizes each of these technologies in a new interactive fashion. The design of inhibitors for proteolytic enzymes will be presented as an example of the application of this strategy, but its more general applicability to other enzyme–inhibitor or receptor–inhibitor interactions should be apparent.

2 The place of structure-based drug design in the context of drug development for human or veterinary use

It is important to keep in mind the pragmatic aspects of structure-based drug design versus traditional drug screening in a drug development programme. Traditional drug screening can be carried out at the level of animal models of disease or in an *in vitro* system where a specific enzyme, cell or organism is being targeted. The advantage of an animal-model screen is that toxicity and pharmacokinetic information is acquired at the same time as that for the efficacy of the agent being screened. The disadvantages are cost, public concerns about animal welfare, and the lack of suitable animal models for many diseases.

In vitro screens can be more cost-effective and acceptable to the public. However, the major advantage of animal models, immediate access to toxicity and pharmacokinetic information, is lost. This means that the promising compounds from *in vitro* screens must still be evaluated in terms of drug delivery and toxicity. Furthermore, cells, organisms, or sufficiently purified enzyme or receptor targets may not be readily available. For example, in many of the important tropical parasitic diseases for which new drugs are needed, the infectious agent is not available in large quantities or cannot be readily maintained *in vitro*.

The strategy of structure-based drug design to be presented in this chapter provides a mechanism by which enzyme inhibitors are identified when only a minimum amount of material from the cell or organism of interest is available. The advantages of this strategy are:

1 A target gene can be isolated and sequenced using as little as 500 ng of genomic DNA.

2 Labour-intensive classical purification procedures are not necessary, and large amounts of starting material are not required.

3 While initial investment in trained personnel, computer graphics and molecular biological equipment is necessary, the procedures are very cost-effective compared to traditional screening techniques.

4 Computer-based screens of compounds are now very efficient and can often identify new compounds which would not be found in nature but could be synthesized by standard organic or inorganic techniques. Modifications of existing compounds can be predicted that will produce potential pharmaceuticals with more specificity and selectivity for the target protein.

Structure-based drug design, nevertheless, is not a panacea nor a replacement for traditional screening techniques. As with *in vitro* screening, toxicity and pharmacokinetic studies still need to be done, and these are often time-consuming and remain expensive. Nevertheless, this approach has already produced many promising compounds, some of which are now in clinical trials (Finke *et al.*, 1990; Maren *et al.*, 1990; Sugrue *et al.*, 1990).

3 An interactive strategy for developing new protease inhibitors as potential pharmaceutical agents

The design of new inhibitors for different classes of proteases will be used as an illustration. This has become an important field in medicinal chemistry because of the myriad roles proteases play in physiological and pathological processes, including fertilization, embryonic development, thrombosis and infectious diseases (Neurath, 1984, 1985, 1986; Barrett, 1986; Festoff, 1990). The proteases that are involved in this diverse array of activities are classified into four major families, based upon the key chemical groups at their active sites. These are the aspartyl (carboxy), metallo, serine and cysteine (thiol) proteases (McDonald, 1985; Barrett, 1986).

Inhibitors of all four protease classes are being tested as potential pharmaceuticals in a variety of diseases (Schnebli & Braun, 1986). For example, the aspartyl protease of the human immunodeficiency virus (HIV) is necessary for viral maturation and represents a potential target for new drugs for the treatment of AIDS (Kohl *et al.*, 1988). Protease inhibitors have also been used in the treatment of hypertension. Captopril inhibits angiotensin-converting enzyme, preventing the conversion of angiotensin-I to angiotensin-II. This results in a decrease in blood pressure. Finally, exogenous α_1-antitrypsin appears effective in halting the progressive emphysema seen in patients with a congenital deficiency of this protease inhibitor (Finke *et al.*, 1990; Hubbard & Crystal, 1990).

Although we are using proteases as an example, the strategy presented here could be applied to other enzymes or other protein–ligand interactions.

The key steps in the strategy can be outlined as follows:

1 *Identify target enzyme.* Usually this would be the result of previous work on the disease process of interest. For example, *de novo* purine biosynthesis is lacking in many human parasites (Wang & Simashkevich, 1981; Heyworth *et al.*, 1982; Wang *et al.*, 1983; Dovey *et al.*, 1984). Parasite-derived enzymes in key pathways for purine salvage have therefore been proposed as targets for structure-based design of new chemotherapy (Rainey *et al.*, 1983; Wang *et al.*, 1984; Dovey *et al.*, 1986).

2 *The gene coding for the enzyme of interest is directly identified and cloned using conserved consensus sequences and the polymerase chain reaction.* While this component of the strategy limits one to enzymes for which homologues are known in other organisms, it provides an extremely rapid and cost-effective access to structural information. Theoretically, cysteine, serine and aspartyl proteases from any eukaryotic and many prokaryotic organisms can be studied in this way.

3 *Computer modelling of enzyme active site based upon primary structure.* While crystallization and X-ray diffraction remains the gold standard for acquisition of structural information, this step provides a valuable shortcut for those proteins that are difficult to crystallize, or an important 'first look' at the structure that could help in analysing diffraction patterns obtained later.

4 *Expression of active recombinant enzymes.* In many instances, it is not possible to obtain sufficient pure natural product for large-scale screening of potential inhibitors. Linking polymerase chain reaction (PCR) acquisition of an enzyme gene to expression of recombinant enzyme provides a potentially unlimited source of enzyme to screen compounds identified by structural modelling.

5 *New synthetic inhibitors.* While synthesis strategies will not be reviewed in detail, it is important to recognize that a key element in the strategy is the availability of new peptide- and non-peptide-based inhibitors for proteases. New strategies for identifying the optimal peptide sequences for inhibitor design are forthcoming. For example, a complex mixture of peptides of various sequences can be synthesized and sequence analysis of the cleaved products of this mixture (after incubation with the target enzyme) can be achieved using robotics (Petithory *et al.*, 1991). Such advances will significantly enlarge the number of lead compounds that can serve as a foundation for computer graphics-derived modifications to increase specificity. Finally, in the case of many protease inhibitors, there is already substantial toxicity and pharmacokinetic data available from previous studies.

3.1 *Use of conserved structural motifs in design of oligonucleotide primers for protease gene amplification*

The PCR has made a revolutionary impact on forensic medicine, cloning techniques, site-directed mutagenesis and disease diagnosis (Guyer & Koshland, 1989). Since its development in 1985, there have been over 1200 publications using PCR (Perkin Elmer Cetus, 1990). This technique is now being used to isolate gene fragments from small amounts of DNA from parasitic organisms, a first step in a structure-based approach toward the design of drugs for chemotherapeutic use.

Prior to the development of PCR, genes were isolated from cells or organisms using conventional methods of screening libraries with antibodies, synthetic oligonucleotides and heterologous probes. These methods required milligram amounts of protein and micrograms of RNA and/or DNA, amounts often not easily obtained from many organisms. In addition, screening libraries with degenerate oligonucleotide probes was a time-consuming task and would often result in false-positive clones.

The PCR has now allowed us to isolate homologous gene fragments for use in screening libraries using nanogram amounts of DNA (even from single cells) to obtain full-length gene sequences which, in turn, can be used in expression systems and 3-D computer modelling (Fig. 8.1).

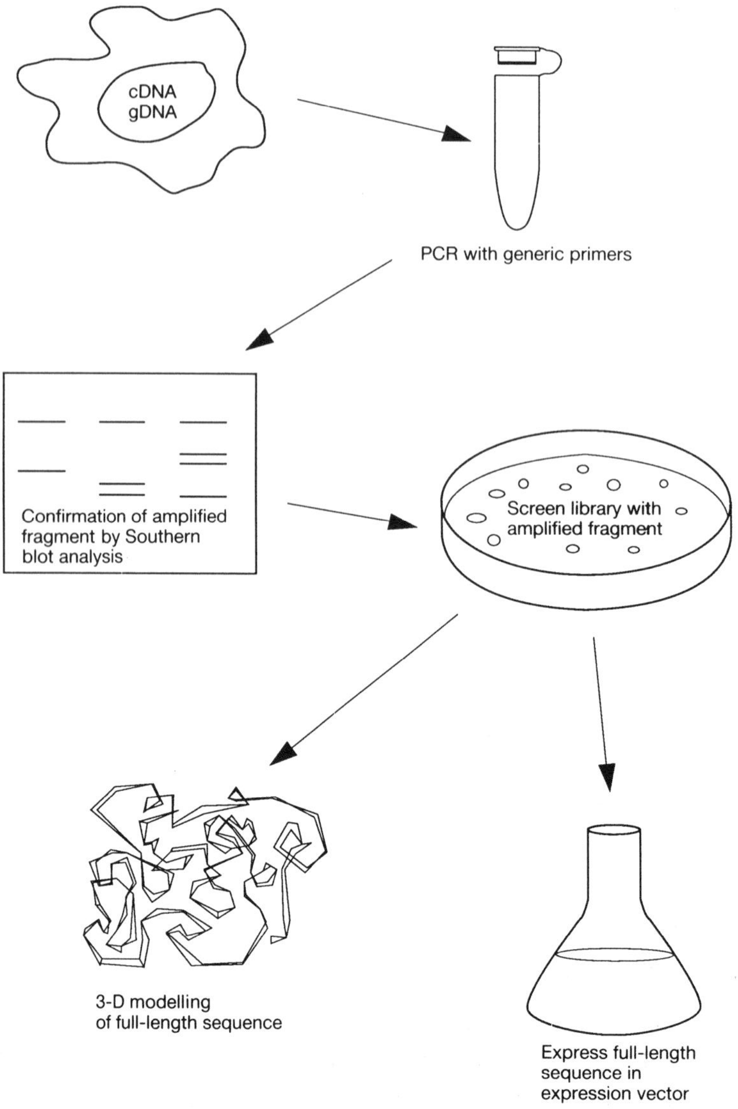

Figure 8.1. Schematic diagram of the approach used to isolate genes from parasites for use in expression systems and 3-D modelling.

Protease gene fragments can be amplified using 'generic' (highly degenerate) primers, the design of which is based on the consensus sequence of amino acids flanking the active sites of serine and cysteine proteases (Sakanari *et al.*, 1989; Eakin *et al.*, 1990). In previous successful applications of this strategy, serine protease primers were 2048- and 8192-fold degenerate, and cysteine protease primers were 1024- and 384-fold degenerate. Inosines were used to minimize degeneracy and maximize base-pairing promiscuity. We reasoned that under low stringency conditions (whereby primers are annealed at 25°C), the PCR might overcome the drawbacks of the degeneracy of the primers, since both the primers must hybridize for amplification to occur.

Amplified gene fragments were then used as probes for Southern blot analysis to confirm the authenticity of the PCR products. Once it is determined that the amplified gene fragment is legitimate, the fragment is sequenced. The authenticity of the fragment is also confirmed by comparing the sequence to homologues in the protein databanks.

The PCR gene fragments can then be used as homologous probes to screen cDNA or genomic libraries. A few examples of protease gene fragments that have been isolated by this method are from nematodes (Sakanari *et al.*, 1989), the parasitic protozoa *Trypanosoma cruzi* (Eakin *et al.*, 1990), *T. brucei* (Eakin *et al.*, 1990), *Entamoeba histolytica* (Eakin *et al.*, 1990), *Plasmodium falciparum* (Rosenthal & Nelson, 1992), fishes, gecko, newt (D. Banfield, Univerisity of British Columbia, pers. commun.), yeast (Smeekens & Steiner, 1990; Barr, 1991) and humans (Barr, 1991). This technique has also been used successfully to isolate genes such as the thymidylate synthase gene from *Pneumocystis carinii* (Edman *et al.*, 1989), ornithine decarboxylase gene from *Leishmania donovani* (B. Ullman, Oregon Health Sciences University, pers. commun.), and the transporter gene from *L. donovani* (S. Landfear, Oregon Health Sciences University, pers. commun.).

3.2 *Molecular modelling of proteases*

The amino acid sequence of a protein determines its precise 3-D structure. Although no general algorithm exists to translate sequence into structure, useful progress has been made in modelling the structure of an enzyme whose sequence is similar to a protein of known structure (Sali *et al.*, 1990). The serine proteases provide a useful example of this procedure (Greer, 1990).

X-ray crystallographers have determined the 3-D structure of more than 10 serine proteases to atomic resolution. These span a phylogenetic range from bacteria to humans. When a new protein sequence is obtained for a serine protease following the PCR primer strategy detailed above, this sequence can be aligned with the sequences of known structure. Emphasis is placed on aligning the structurally conserved regions (SCRs) and the catalytic residues (Fig. 8.2). A framework model of the protein of interest is then constructed for the SCRs by assuming that these residues can be superimposed on the structures of their aligned counterparts. The loops connecting the SCRs are modelled by analogy to loops from serine

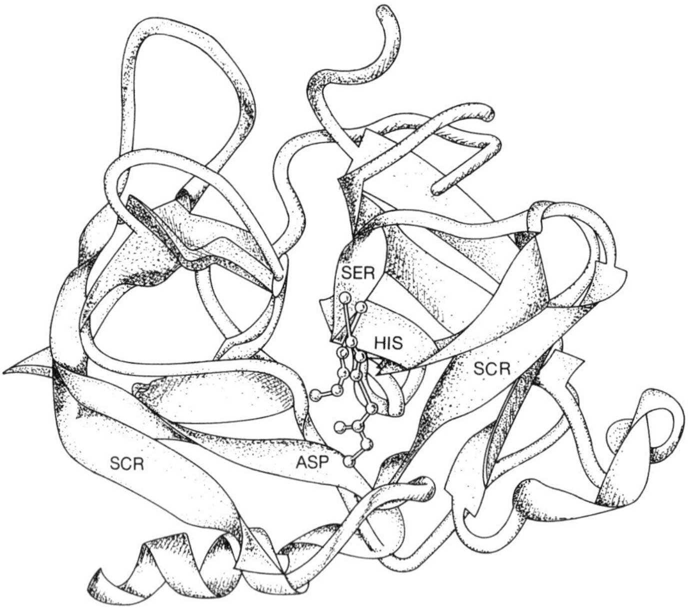

Figure 8.2. A computer model of the serine protease chymotrypsin. The catalytic triad residues are identified. The structurally conserved residues (SCRs) are the thickened segments of the chain.

proteases (Chothia *et al.*, 1989). If no convenient match is available, an analogous loop from an unrelated protein is sought which fits the geometric and sequential demands of the protein (Jones & Thirup, 1986). Amino acid side-chains are added using the known preferences for these groups (Ponder & Richards, 1987), and the preliminary all-atom model is refined to optimize the energy of the system. This process can be completed in 1–2 months and produces an approximate characterization of the enzyme active site and the substrate specificity pockets.

Having constructed a model of an enzyme 'receptor', attention must turn toward the identification of lead compounds for evaluation and development. Kuntz and colleagues (e.g. DesJarlais *et al.*, 1990) have devised a computer algorithm for examining a large database of small molecule structures and identifying compounds having a molecular shape which fits the target receptor. In the case of a protease, compounds would be sought which could bind at or near the active site of the enzyme, thereby blocking substrate entry. Of course, molecular shape is not the only determinant of binding affinity, and shape-based leads must be modified to satisfy the electronic constraints of the binding site. Analogues of these lead compounds must be synthesized to optimize the therapeutic index of the lead in order to create an effective pharmaceutical.

3.3 *Bacterial expression systems for proteases*

A fundamental requirement for developing a genetic approach to study structure–activity relationships in proteases is an efficient expression system. Rapid and convenient methods for generating 1–100 mg of enzyme are necessary for enzymological and structural studies. Even greater amounts of sample can be demanded by certain biophysical analyses such as calorimetry, spectroscopy or hydrodynamics. Therefore the method for producing reagent quantities and qualities of a protease must be inexpensive if such studies are to be considered. Finally, the protease must be produced in an authentic state, free from artifacts of the expression system for a accurate correlation between structure and function.

Escherichia coli is an ideal host for the expression of recombinant proteins for several reasons:

1 There are multiple well-designed bacterial strains and inducible vectors available.
2 Bacteria are very easy to manipulate genetically, including mutagenesis of the target protein and development of selection and screening capabilities.
3 The expressed protein can be localized in the cytoplasm or the periplasmic space of the bacteria or secreted into the extracellular medium.
4 A wide range of growth conditions including temperature and inducer concentrations are available for optimizing expression levels of the protein of interest.
5 The growth properties of bacteria permit rapid analysis of experiments.
6 It is relatively inexpensive to grow large quantities of recombinant bacteria in short time periods.

Bacterial expression systems can be problematic, however. Bacteria are often unable to produce eukaryotic proteins in an active state. The successful application of a genetic screen or selection based on the activity of the protein requires that any post-translational processing events that are required for activity are complete. *E. coli* cannot achieve *N*- and *O*-linked glycosylation, phosphorylation, formation of γ-glutamates, myristilation, sulphation, tyrosylation, palmitation, acetylation and proteolytic maturation. Often the initiator methionine is inefficiently removed from overexpressed, heterologous proteins in *E. coli.* If the protein contains intramolecular disulphide bonds, the reducing environment of *E. coli* can prevent correct formation of the disulphide linkage after the protein has been correctly folded. The proteins are usually found in an insoluble form with improper disulphide pairs. Similarly, for many other proteins that are hyperproduced in the cytoplasm of *E. coli*, the proteins generally precipitate as inclusion bodies (Harris, 1983; Marston, 1986). The separation of inclusion bodies from *E. coli* is straightforward and in some cases the protein can be dissolved with a denaturant and refolded (Marston *et al.*, 1984; Marston, 1986). However, refolding proteins generally result in low yields of properly folded material, restricts the ability to develop activity-based screens or selections and may cause artifacts due to chemical modifications of the protein by the denaturant. The formation of inclusion bodies can sometimes be avoided by modulating the level of expression (Browner *et al.*, 1991), but such results are unique to each protein and contrary to the goal of obtaining high-level expression.

Proteases are exceptionally difficult to overproduce in a heterologous expression system. Carboxypeptidase has been expressed in *Saccharomyces cerevisiae* from the α-factor promoter and directed to the extracellular media by the α-factor signal peptide (Gardell *et al.*, 1985). Subtilisin has been expressed in *Bacillus subtilis* from its own promoter and directed to the extracellular media by the native subtilisin signal peptide (Wells *et al.*, 1983). Papain has been expressed in baculovirus

(Vernet *et al.*, 1990) and α-lytic protease has been expressed in *E. coli* (Silen *et al.*, 1989). HIV protease has been expressed in an active form intracellularly in *E. coli* (Farmerie *et al.*, 1987) and secreted to the extracellular medium in yeast (Pichuantes *et al.*, 1989). Initial efforts to express trypsin entailed the use of a mammalian tissue culture system (Craik *et al.*, 1985, 1987). While mammalian cells provide a more native environment of rat anionic trypsin, tissue culture is slow and expensive compared to culturing micro-organisms. A more practical expression system was then developed that took advantage of the various attributes of *E. coli.*

The first step in establishing a system for the expression of trypsin was the construction of a complete coding sequence for rat anionic trypsin-II. A 'mini-gene' was obtained by fusing a genomic clone (Craik *et al.*, 1984) comprising the signal peptide and the first exon to a partial cDNA clone (MacDonald *et al.*, 1982). Deletion of the intervening sequence present in the genomic sequence yielded the trypsinogen mini-gene that comprised the entire pretrypsinogen coding sequence and 55 bp of the 3′-untranslated region (Craik *et al.*, 1985). A bacterial system was then created by splicing the trypsinogen-encoding fragment into a pBR322-based vector containing the alkaline phosphatase promoter and signal peptide.

The bacterial vector, pTRAP (plasmid *tr*ypsinogen, *a*lkaline *p*hosphatase signal peptide), directs the secretion of trypsinogen to the periplasmic space constitutively in phoR cells or in a regulated fashion by varying phosphate concentration (Graf *et al.*, 1987). It was presumed that a secretion system would be favourable for the expression of a mammalian protease containing disulphide bonds because the reducing environment of the cytoplasm would prevent correct disulphide bond formation and degradative intracellular proteases would have access to heterologous trypsin localized to the cytoplasm. In addition, active trypsin formed inside the cell would presumably be deleterious to the host. Furthermore, periplasmic localization of trypsin could aid in the eventual purification scheme.

The activation of trypsin from trypsinogen requires the addition of a separate protease that itself must be purified. To eliminate the need for this processing, a vector was subsequently created that could direct secretion of mature trypsin. In this vector pT3 (plasmid, *t*ac promoter, *t*rypsin, *t*erminator), the first codon of mature trypsin was abutted to the prepeptide sequence of alkaline phosphatase or the *his* J protein (Stern *et al.*, 1988; Higaki *et al.*, 1989; Vásquez *et al.*, 1989) (Fig. 8.3). Neither constitutive nor regulated secretion of the mature, native enzyme seems to affect the viability of the *E. coli* at the 1 mg/litre/optical density level of expression obtained in this system.

As with pTRAP, pT3 directs the expression of the enzyme plus a leader peptide. The protein is localized in the periplasmic space after processing of the signal peptide by signal peptidase. This processing is analogous to the activation of trypsin by its physiological activator because the loss of the signal peptide mimics the loss of the pro-peptide in that the remaining polypeptide undergoes a conformational change that results in active enzyme. The basis for the change is the formation of a salt bridge between the new amino terminus, isoleucine-16, and aspartic acid-194 (Stroud *et al.*, 1974; Bode & Schwager, 1975). Evidence for the

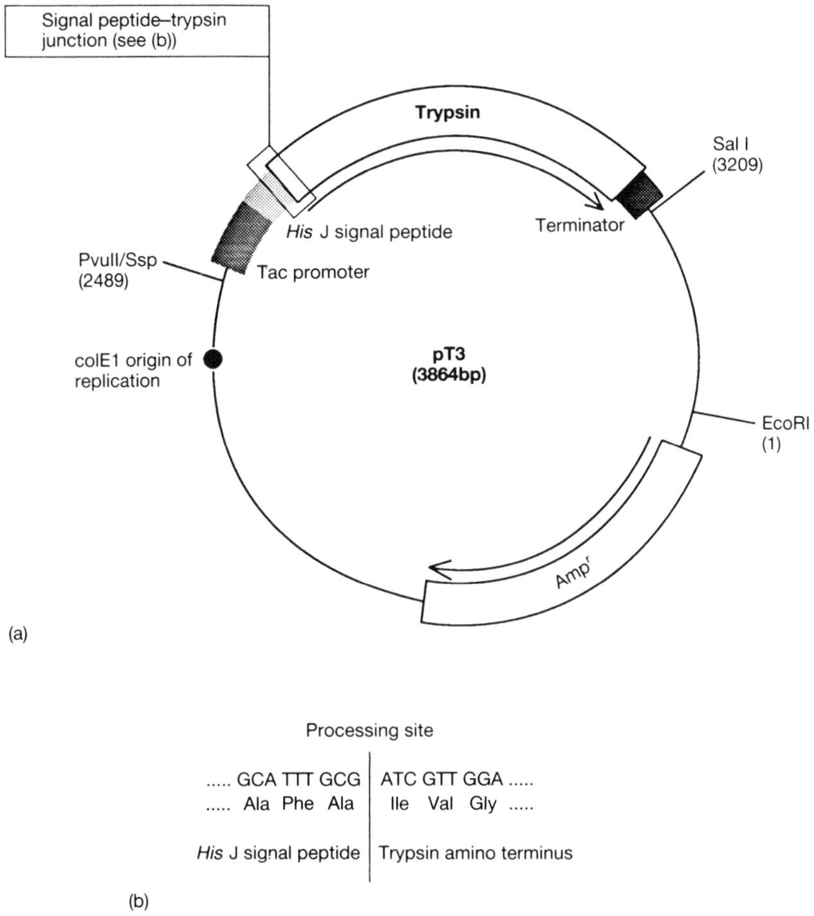

Figure 8.3. Schematic representation of the bacterial pT3 expression plasmid. (a) A schematic representation of the pT3 expression vector used to express trypsin. The area outlined in this figure shows the *his* J signal peptide–trypsin junction. The precise site of processing by endogenous signal peptidase at this junction is illustrated in more detail in (b). (b) The *his* J signal peptide–trypsin junction. Trypsin expressed in *E. coli* is directed through the inner membrane by the *his* J signal peptide and secreted into the periplasmic space. Endogenous signal peptidase cleaves the *his* J trypsin polypeptide on the C-terminal side of the *his* J Ala-Phe-Ala sequence, liberating mature trypsin containing the critical N-terminal Ile residue. (Reprinted with permission from Higaki *et al.* (1989).

creation of correctly folded trypsin is provided by the activity of the trypsin and by the ability of this trypsin to bind bovine pancreatic trypsin inhibitor in the purification procedure. The trypsin that is secreted to the periplasm can be released into the supernatant by generating spheroplasts from intact trypsin-producing cells. This expression system has been used to generate site-directed mutants of trypsin for the study of substrate binding (Craik *et al.*, 1985), catalysis (Higaki *et al.*, 1989) and metalloregulation (Higaki *et al.*, 1990). Recently, the system was used to develop a genetic selection to investigate the various arrangements of amino acids that could be used by trypsin for hydrolysis of arginine- and lysine-containing substrates (Evnin *et al.*, 1990).

4 Turning the system around: Identifying new enzymes for catalysing industrial and pharmaceutical reactions

While the strategy in this chapter was presented as a means of identifying new inhibitors of enzymes or other ligand–receptor interactions, the ability to amplify protease gene sequences (and, for that matter, any enzyme with conserved structural motifs) also opens up the possibility of obtaining and studying enzymes from unusual evolutionary niches. For example, enzymes which can function at high pH or salinity are of potential importance as catalysts for a variety of industrial reactions. Prokaryotic extremophiles could be screened for the presence of protease genes and enzyme expressed by recombinant systems. As noted above, time-consuming protein purification procedures are not necessary and a relatively small number of organisms are required. Enzymes identified and studied in this manner will be important not only for their use as potential catalysts, but also should provide new information relevant to mechanisms of catalysis and stability of protein structure.

Another potential use of this strategy would be in isolating and expressing genes for enzymes already in use as pharmaceutical reagents (for example, plasminogen activators) from new natural sources. Related enzymes from other organisms may have subtle differences in structure or biochemical properties that might make them even more attractive in terms of pharmaceutical use.

5 Acknowledgements

This work was supported by grants from the National Institutes of Health (No. AI 20452 to J.H.M., No. AI 29457 to J.A.S., No. GM 39900 to FEC); the World Health Organization TDR on Chagas' disease and filariasis No. T80/181/51 (J.H.M.); the American Heart Association No. 90–162 (J.H.M.); the Searle Scholars/Chicago Community Trust (F.E.C.); the Defense Advanced Research Projects Agency No. ONR N00014-86-K-0757 (F.E.C.); the National Science Foundation No. DMB 8904956 (C.S.C.); UNDP/World Bank/WHO Special Programme for Research and Training in Tropical Diseases No. L30/181/82 (J.A.S.); and the University of California, San Francisco, Academic Senate and School of Medicine (J.A.S.). We also thank Ramona J. Soto for her assistance in the preparation of this manuscript.

6 References

Barr, P.J. (1991). *Cell*, **66**, 1–3.

Barrett, A.J. (1986). In: A.J. Barrett and G. Salvesen (Eds.), *Proteinase Inhibitors. Research Monographs in Cell and Tissue Physiology*, Vol. 12, pp. 3–22, Elsevier Science Publishers, Cambridge.

Bode, W. & Schwager, P. (1975). *J. Mol. Biol.* **98**, 693–717.

Browner, M., Rasor, P., Tugendreich, S. & Fletterick, R. (1991). *Prot. Eng.*, **4**, 351–7.

Chothia, C., Lesk, A.M., Tramontano, A. *et al.* (1989). *Nature (London)*, **342**, 877–83.

Craik, C.S., Choo, Q.-L., Swift, G.H. Quinto, C., MacDonald, R.J. & Rutter, W.J. (1984). *J. Biol. Chem.*, **259**, 14 255–64.

Craik, C.S., Largman, C., Fletcher, T., Roczniak, S., Barr, P.J., Fletterick, R. & Rutter, W.S. (1985). *Science*, **228**, 291–7.
Craik, C.S., Roczniak, S., Largman, C. & Rutter, W.J. (1987). *Science*, **237**, 909–13.
DesJarlais, R.L., Seibel, G.L., Kuntz, I.D., Furth, P.S., Alvarez, J.C., Ortiz de Montellano, P.R., DeCamp, D.L., Babe, L.M. & Craik, C.S. (1990). *Proc. Natl. Acad. Sci. USA*, **87**, 6644–8.
Dovey, H.F., McKerrow, J.H. & Wang, C.C. (1984). *Mol. Biochem. Parasitol.*, **11**, 157–67.
Dovey, H.F., McKerrow, J.H., Aldritt, S.M. & Wang, C.C. (1986). *J. Biol. Chem.*, **261**, 944–8.
Eakin, A.E., Bouvier, J., Sakanari, J.A., Craik, C.S. & McKerrow, J.H. (1990). *Mol. Biochem. Parasitol.* **39**, 1–8.
Edman, U., Edman, J.C., Lundgren, B. & Santi, D.V. (1989). *Proc. Natl. Acad. Sci. USA*, **86**, 6503–7.
Evnin, L.B., Vásquez, J. & Craik, C.S. (1990). *Proc. Natl. Acad. Sci. USA*, **87**, 6659–63.
Farmerie, W., Loeb, D., Casavant, N., Hutchison, C. III, Edgell, M. & Swanstrom, R. (1987). *Science*, **236**, 305–8.
Festoff, B.W. (1990). In: B.W. Festoff (Ed.), *Serine Proteases and their Serpin Inhibitors in the Nervous System. Regulation in Development and in Degenerative and Malignant Disease*, pp. 1–359, Plenum Press, New York.
Finke, P.E., Ashe, B.M., Knight, W.B., Maycock, A.L., Navia, M.A., Shah, S.K., Thompson, K.R., Underwood, D.J., Weston, H. & Zimmerman, M. (1990). *J. Med. Chem.*, **33**, 2522–8.
Gardell, S., Craik, C.S., Hilvert, D., Urdea, M. & Rutter, W. (1985). *Nature (London)*, **317**, 551–5.
Graf, L., Craik, C.S., Patthy, A., Roczniak, S., Fletterick, R. & Rutter, W. (1987). *Biochemistry*, **26**, 2616–23.
Greer, J. (1990). *Proteins* **7**, 317–34.
Guyer, R.L. & Koshland, D.E. (1989). *Science*, **246**, 1543–6.
Harris, T.J.R. (1983). In: R. Williamson (Ed.), *Genetic Engineering 4*, pp. 127–85, Academic Press, London.
Heyworth, P.G., Gutteridge, W.E. & Ginger, C.D. (1982). *FEBS Lett.*, **141**, 106–10.
Higaki, J.N., Evnin, L.B. & Craik, C.S. (1989). *Biochemistry*, **28**, 9256–63.
Higaki, J.N., Haymore, B., Chen, S., Fletterick, R. & Craik, C.S. (1990). *Biochemistry*, **29**, 8582–6.
Hubbard, R.C. & Crystal, R.G. (1990). *Lung*, **168** (Suppl.), 565–78.
Jones, T.A. & Thirup, S. (1986). *EMBO J.*, **5**, 819–22.
Kohl, N.E., Emini, E.A., Schleif, W.A., Davis, L.J., Heimbach, J.C., Dixon, R.A., Scolnick, E.M. & Sigal, I.S. (1988). *Proc. Natl. Acad. Sci. USA*, **85**, 4685–90.
McDonald, J.K. (1985). *Histochem. J.*, **17**, 773–85.
MacDonald, R.J., Stary, S.J. & Swift, G.H. (1982). *J. Biol. Chem.* **257**, 9724–32.
Maren, T.H., Bar-Ilan, A., Conroy, C.W. & Brechue, W.F. (1990). *Exp. Eye Res.*, **50**, 27–36.
Marston, F.A.O. (1986). *Biochem. J.*, **240**, 1–12.
Marston, F.A.O., Lowe, P.A., Doel, M.T., Schoemaker, J.M., White, S. & Angal, S. (1984). *Bio/Technol.*, **2**, 800–4.
Neurath, H. (1984). *Science*, **224**, 350–7.
Neurath, H. (1985). *Fed. Proc.*, **44**, 2907–13.
Neurath, H. (1986). *J. Cell. Biochem.*, **32**, 35–49.
Perkin Elmer Cetus (1990). *PCR Bibliography,* Vol. 1, No. 5, pp. 128.
Petithory, J.R., Masiarz, F.R., Kirsch, J.F., Santi, D.V. & Malcolm, B.A. (1991). *Proc. Natl. Acad. Sci. USA*, **88**, 11510–14.
Pichuantes, S., Babé, L.M., Barr, P.J. & Craik, C.S. (1989). *Proteins: Structure, Function and Genetics,* **6**, 324–37.
Ponder, J.W. & Richards, F.M. (1987). *J. Mol. Biol.*, **193**, 775–91.
Rainey, P., Garrett, C.E. & Santi, D.V. (1983). *Biochem. Pharmacol.*, **32**, 749–56.
Rosenthal, R.S. & Nelson, R.G. (1992). *Mol. Biochem. Parasitol.*, **51**, 143–52.
Sakanari, J.A., Staunton, C.E., Eakin, A.E., Craik, C.S. & McKerrow, J.H. (1989). *Proc. Natl. Acad. Sci. USA*, **86**, 4863–7.

Sali, A., Overington, J.P., Johnson, M.S., Blundell, T.L. (1990). *Trends Biochem. Sci.* **15**, 235–40.

Schnebli, H.P. & Braun, N.J. (1986). In: A.J. Barrett and G. Salvesen (Eds.), *Proteinase Inhibitors. Research Monographs in Cell and Tissure Physiology,* Vol. 12, pp. 613–27, Elsevier Science Publishers, Cambridge.

Silen, J.L., Frank, D., Fujishige, A., Bone R. & Agard, D.A. (1989). *J. Bacteriol.*, **171**, 1320–5.

Smeekens, S.P. & Steiner, D.F. (1990). *J. Biol. Chem.*, **265**, 2997–3000.

Stern, M.J., Prossnitz, E. & Ames, G.F. (1988). *Mol. Microbiol.*, **2**, 141–52.

Stroud, R.M., Kay, L.M. & Dickerson, R.E. (1974). *J. Mol. Biol.*, **83**, 185–208.

Sugrue, M.F., Gautheron, P., Grove, J., Mallorga, P., Viader, M.P., Schwam, H., Baldwin, J.J., Christy, M.E. & Ponticello, G.S. (1990). *J. Ovular Pharmacol.* **6**, 9–22.

Vásquez, J.R., Evnin, L.B., Higaki, J.N. & Craik, C.S. (1989). *J Cell. Biochem.*, **39**, 67–78.

Vernet, T., Tessier, D.C., Richardson, C., Laliberte, F., Khouri, H.E., Bell, A.W., Storer, A.C. & Thomas, D.Y. (1990). *J. Biol. Chem.,* **265**, 16 661–6.

Wang, C.C. & Simashkevich, P.M. (1981). *Proc. Natl. Acad. Sci. USA,* **78**, 6618–22.

Wang, C.C., Verham, R., Rice, A. & Tzeng, S.-F. (1983). *Mol. Biochem. Parasitol.*, **8**, 325–37.

Wang, C.C., Verham, R., Cheng, H.-W., Rice, A. & Wang, A.L. (1984). *Biochem. Pharmacol.*, **33**, 1323–9.

Wells, J.A., Ferrari, E., Henner, D.J., Estell, D.A. & Chen, E.Y. (1983). *Nucl. Acids Res.*, **11**, 7911–25.

9 New Era of Biochemistry: Cytokine Signal Network and its Implication for Biomedical Research

K. ARAI[†*] and S. WATANABE[*]

[*]*Department of Molecular and Developmental Biology, Institute of Medical Science, University of Tokyo, 4-6-1 Shirokanedai, Minato-ku, Tokyo 108, Japan and* [†]*Department of Molecular Biology, DNAX Research Institute of Molecular and Cellular Biology, 901 California Avenue, Palo Alto, CA 94301, USA*

1 Biochemical discipline in medical science

Medical science, one of the oldest sciences in history, has provided fertile soil for various research fields including histology, pathology, microbiology, physiology, pharmacology and nutritional science. However, our understanding of the basic principles that regulate multicellular systems such as the human body has been elusive. Accordingly, medical research itself has remained rather descriptive. From medical science, modern biochemistry, molecular biology, cell biology and immunology have emerged. They evolved according to their own disciplines and developed relatively independently of each other. Among them, biochemistry played a pivotal role to provide the chemical basis for our understanding of living cells.

Biochemical research generally consists of two stages, i.e. resolution and reconstitution. The first stage is reductive and analytical. In general, biological phenomena observed at the cellular level are characterized biochemically at the subcellular level by employing functional assays. The overall reaction can be further resolved into elementary steps and the protein components for individual steps are identified, purified and their structures and functions are studied by using *in vitro* assays. The second stage is concerned with the reconstitution of biological processes initially observed at the subcellular level. The goal of this stage is to assemble biochemical reactions of much higher order of complexity from purified components in a test tube. One of the major motifs in biochemistry is to correlate biological activity to chemical structure. Biochemical research over decades has established that biological activities are mostly carried out by functional proteins. Protein, within its 3-D structure, clearly shows an important relationship between structure and function.

Living cells elicit physiological phenomena highly coordinated and vectorial in nature, such as the opening or closing of ion channels, signal transduction across the plasma membrane, cytoskeleton and cell motility, and condensation and segregation of chromosomes during mitosis. When biochemists disintegrate cells, many features of these highly integrated reactions observed in intact cells disappear and the components are distributed into various fractions. Important issues for biochemists are how to disintegrate living cells with minimum loss of function and how to assemble individual components into a functional unit in the test tube to reconstitute the phenomenon observed at a cellular level.

Some physiologists claim that cells or living organisms have to be studied in intact form and that life cannot adequately be studied once cells have been disrupted. However, without resolving overall reactions into elementary steps and learning the structure and function of each component involved, we do not have full understanding of the molecular basis of such phenomena. The genetic approach, employing mutant cells with altered components, offers the opportunity to study the structural and functional relationship with intact cells. Therefore, the genetic approach provides a link between physiological research at the cellular level and biochemical research at the subcellular level.

During its development, the basic disciplines of biochemistry have remained largely unchanged but the actual research targets have changed remarkably. To look into the 21st century, we would like to scrutinize the nature of such development by classifying biochemical reactions into several types.

2 Two types of biochemical reactions

Biochemical reactions can be divided into two classes: those employing the DNA template and those without it. Of course, overall free-energy change in both reactions should be thermodynamically favourable for the reaction to proceed. Thermodynamically unfavourable reactions can be driven by coupling to hydrolysis of adenosine triphosphate (ATP) (or guanosine triphosphate (GTP) in some cases).

2.1 *Type I: Intermediary metabolism*

Cells extract energy from environments and synthesize building blocks of their macromolecules through a highly integrated network of chemical reactions. The first type of biochemical research was concerned with an intermediary metabolism most actively studied from the 1930s to the 1950s (Fig. 9.1). Substrates which are generally organic compounds define the precursor–product relationship and accordingly the chemical structure of the substrate is the primary determinant in this type of reaction. A series of chemical transformations of the organic compounds is arranged linearly and distinct enzymes perform their functions at each step in the pathway. The concept of allosteric feedback regulation of enzyme activity has emerged from studies of metabolic pathways. Metabolic reactions often form a network defined by an array of organic compounds which are structurally related.

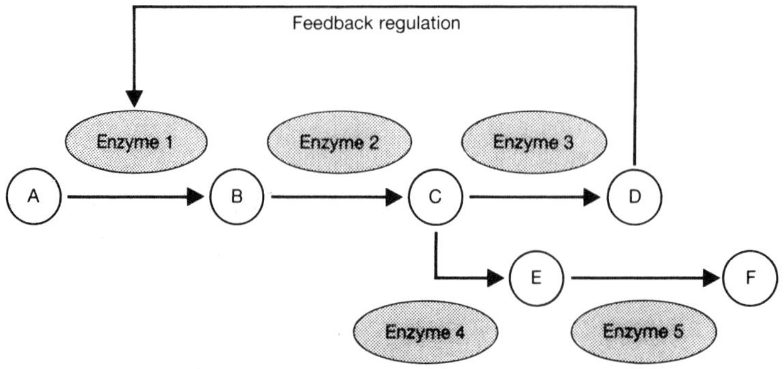

Figure 9.1. Type I reaction. Intermediary metabolism.

Enzymes play a role in determining the actual pathway among several possible alternatives. This reaction supplies energy and building blocks for biosynthetic reactions.

2.2 *Type II: Replication and expression of genetic information*

The second type of biochemical research is replication and phenotypic expression of genetic information stored in DNA templates (Fig. 9.2). Research on the type II reaction, which began in 1950s, involves three Rs in DNA research (i.e. *r*eplication, *r*ecombination and *r*epair) and expression of genetic information (transcription and translation). Genetic information is stored linearly on vectorial DNA templates. The essence of this type of research took shape in the 1960s and has been actively pursued for decades. Similar to other biosynthetic reactions, substrates (building blocks) for DNA, RNA or protein (i.e. nucleotides or amino acids) have to be activated prior to biosynthetic process. However, unlike the type I reaction which is governed by the chemical structure of organic compounds, master templates dictate the primary structure of the macromolecular products (DNA, RNA or protein). Biosynthesis of macromolecule is generally composed of three distinct steps, i.e. initiation, elongation and termination. Unlike the type I reaction where distinct enzymes perform reactions at each step, different templates employ a common enzyme complex which performs initiation, elongation and termination functions. The regulation is most likely at initiation and/or termination stages and the concept of allosteric regulation of enzyme level emerged from studies of bacterial operons. Besides activation of substrates, which is required for biosynthetic reaction *per se*, GTP hydrolysis (in protein synthesis) or ATP hydrolysis (in DNA replication, recombination and repair) are required for mechanochemical processes. These GTPase/ATPase reactions are the prototype of the 'molecular switch' employed in the signal transduction reaction (see the next section).

These two types of reaction represent two fundamental processes of life, i.e. generation and storage of metabolic energy and storage and expression of genetic information. Precise knowledge about these reactions should shed light on our understanding of genetic diseases and help to design drugs for intervention in

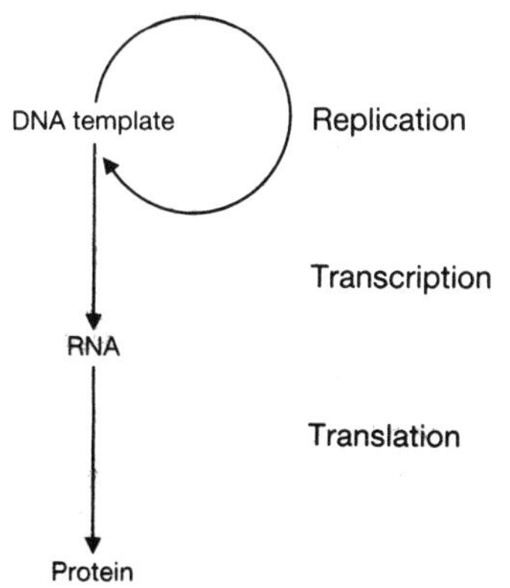

Figure 9.2. Type II reaction. Storage and expression of genetic information.

metabolic disorders. However, these reactions have to be coordinated in living cells which are in dynamic equilibrium with the surrounding environment. Such coordination is achieved by the third type of biochemical reaction, collectively termed 'signal transduction'.

3 Signal transduction: New frontier in biochemistry

3.1 *Components of the signal transduction pathway*

Coordination of a large number of reactions in biological processes is mediated by both extra- and intracellular signalling molecules. Intracellular targets of signal transduction pathways are the mechanics of transcription, replication and probably energy metabolism. Generally, an extracellular signal (ligand) interacts with plasma membrane receptors which are generally composed of a ligand-binding unit, transducer, effector and regulatory unit. Interaction of the ligand (growth factor, cytokine or hormone) with its receptor generates intracellular signals which are transmitted through the signal transduction network involving GTP-binding proteins and protein kinases located at the plasma membrane, cytoplasm and/or nucleus. A chain of reactions further downstream ultimately stimulates initiation of the DNA replication cycle or expression of a new genetic programme.

3.2 *Network formation among components of signalling pathways*

Studies on the mechanism of signal transduction which emerged in the 1970s will be one of the most active fields beyond this century. Several issues regarding the mechanism of signal transduction are (i) how are extra- or intracellular signals processed? (ii) how is the process regulated? and (iii) how does this process couple to energy metabolism? The complexity of signal transduction pathways may be comparable to that of metabolic pathways. In research on metabolic pathways, one can monitor the successive stages of chemical transformation of substrates by employing isotope tracer techniques. However, the situation is entirely different and the same strategy is not directly applicable in signal transduction research where the chemical structure of the input signal and that of the target are generally unrelated. Discovery of the basic rules of signal transduction mechanisms may be compared to finding a way out of the Labyrinth guided by Ariadne's spool.

In this signalling process, the signal triggers the response, but unlike the type I reaction, the chemical structure of the signal does not dictate the downstream process. Also, unlike the type II reaction, no master template exists to govern the signal transduction reaction. This process is arranged non-linearly, and probably forms a network mediated through a series of protein–protein interactions (Fig. 9.3). The protein, in a single molecule, has a unique ability to create multiple binding sites for a wide variety of substrates which are not structurally related. This provides the protein with a unique ability to establish a link between compound (or reaction) A at one site and compound (or reaction) X at another site. The nature of the link between A and X may depend on the nature of the allosteric interaction (either positive or negative) between A and X sites. In many signalling systems, GTP or ATP has been utilized to regulate the molecular switch

Figure 9.3. General scheme for signal transduction. L, ligand; E, effector(s); PK, protein kinase.

(i.e. signal transducing proteins). In these instances, free energy generated by GTP or ATP hydrolysis is utilized for mechanochemical reactions other than the formation of covalent bonds. This may be consumed in an entropic process to introduce 'order' by coordinating various biological reactions through a series of protein–protein interactions. Type III reactions, by employing several molecular switches, coordinate type I and type II reactions in terms of time, sequence and overall direction.

4 Molecular switch: Role of GTP/ATP hydrolysis in mechanochemical reactions

4.1 *GTPase/ATPase: Two-state protein comformation model*

In the 1970s, Kaziro (1978) and Arai *et al.* (1974), in the course of studying the role of GTP hydrolysis in protein synthesis, recognized a unique class of reactions which appear to have general implications to energy and signal transduction mechanisms. A two-state protein conformation model was developed in which the system acquires unidirectionality as a result of cyclic oscillation of a functional protein between active and inactive conformations induced by interconversion of nucleotide triphosphate/nucleotide diphosphate (NTP/NDP) ligands ('all or none' type transition) (Fig. 9.4). Essentially the same mechanism operates in many DNA-dependent ATPases (Arai & Kornberg, 1980a,b), helicases (Arai *et al.*, 1980) and topoisomerases involved in DNA replication and recombination. In this model, protein acts as a 'molecular switch' having 'on' and 'off' conformation. NTP binding is sufficient to complete the translocation step and the mechanochemical process re-cycles by inducing 'off' conformation through NTP hydrolysis. This step introduces 'unidirectionality' into the overall process. 'On' conformation is regenerated from guanosine diphosphate (GDP) bound and the 'off' conformation by the

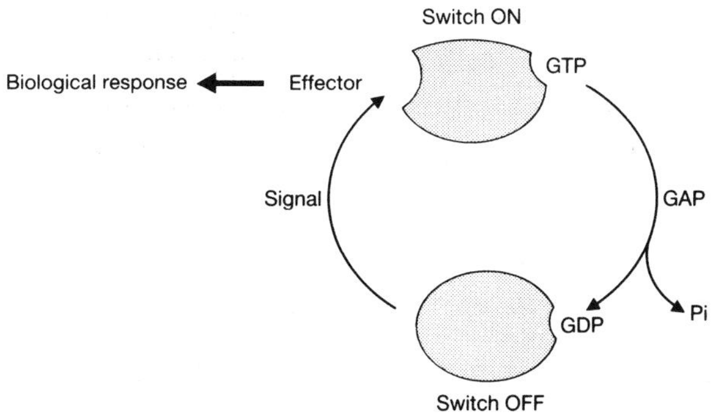

Figure 9.4. GTP-binding protein as a 'molecular switch'.

NDP–NTP exchange reaction. This step proceeds either in a signal-dependent (such as in polypeptide chain elongated factor tu (EF-Tu) or dnaA protein) or signal-independent (such as in EF-G or dnaB protein) manner. The essence of this system is the selection of one state out of two equally feasible states at the expense of free energy supplied by the catalytic hydrolysis of the NTP ligand. This model, which is applicable for many GTPase/ATPase reactions, extends the definition of the classic allosteric mechanism found in metabolic enzymes and regulators of gene expression such as repressors. Instead of a graded response induced by allosteric effectors, this system operates in an 'all or none' manner. This is achieved simply by converting the allosteric site into a catalytic site of hydrolysis that enables conversion of an active ligand into the inactive form. As predicted, a similar mechanism was found in many signalling reactions involving heterotrimeric G-proteins (such as Gs, Gi, Go) (Gilman, 1987) and small molecular weight GTP binding proteins (such as *ras* protein). GTPase-activating protein (GAP) or the effector molecules which generate intracellular signals may turn off the signal by promoting GTP hydrolysis.

4.2 *Protein kinase: Molecular switch or signal amplifier?*

The other important issue is the role of protein kinases in signal transduction (Hanks *et al.*, 1988). Generally, growth signals received by cell-surface receptors activate a battery of protein kinases. In many cases, growth factor receptor itself has tyrosine kinase activity which is directly activated by ligand (receptor-type tyrosine kinases). The activity of some serine/threonine kinases appears to be regulated by second messengers (cAMP, calcium ion or diacylglycerol) (Edelman *et al.*, 1987). Activation of protein kinase C or cAMP-dependent protein kinase turns on a set of genes through transacting transcription factors that bind to a specific DNA motif. Proliferation of mammalian cells proceeds through four distinct stages, i.e. G1, S, G2 and M phases of the cell cycle. The major control point in the cell cycle lies at the G1/S border. Many protein kinases play a role in cell-cycle regulation. For example, CDC2 kinase is essential for both G1/S transition as well as G2/M transition. How do these protein kinases generate the cascade of events which ultimately stimulates multiple origins of chromosome replication? Similar to the GTP/ATPase reaction, protein kinase employs a nucleotide ligand carrying a high-energy phosphate bond which in the end is split into adenosine diphosphate (ADP) and inorganic phosphate in collaboration with protein phosphatase (Fig. 9.5). It is generally assumed that phosphorylation of specific proteins at a specific site plays a critical role in mediating cellular response and phosphorylated proteins return to their original activity after removal of phosphate by protein phosphatase. However, the precise mechanism by which protein kinase transmits signals remains to be determined.

5 Intracellular cytokine network and receptor signal transduction

Progression of the cell cycle in mammalian cells from G0/G1 to S phase is controlled by growth factors. This process provides a typical model for the

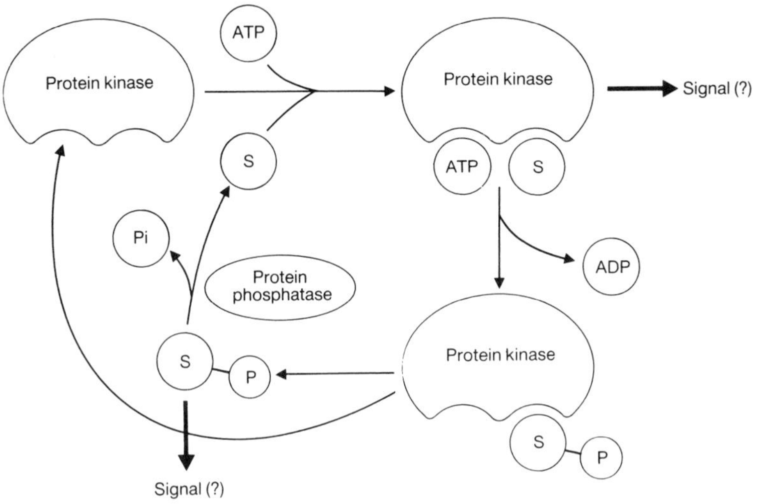

Figure 9.5. Possible reaction mechanism for protein kinase.

signal-transduction cascade. Along with epidermal growth factor (EGF), nerve growth factor (NGF), platelet derived growth factor (PDGF) and fibroblast growth factor (FGF), a battery of lymphokines and monokines, which are collectively termed cytokines produced by activated T-cells and macrophages has been added to the catalogue of growth factors (Fig. 9.6) (Miyatake *et al.*, 1988; Arai *et al.*, 1990; Watanabe *et al.*, 1991). Cytokines are mediators for cell-to-cell communication involved in viral infection, inflammation, immunity and haemopoiesis. In the immune response, T-cells help B-cells to differentiate into plasma cells by producing a set of lymphokines in response to antigen signals. Activation of helper T-cells by antigen has two distinct aspects: (i) recognition of the antigen by T-cell–antigen

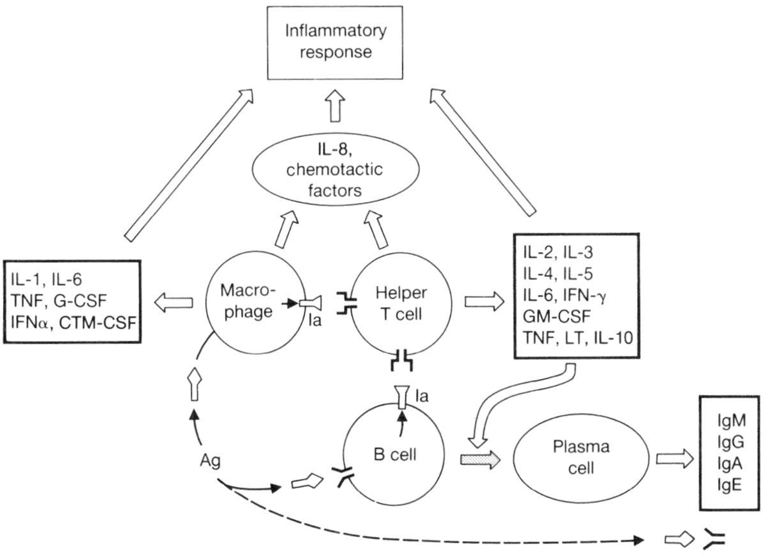

Figure 9.6. Interaction of T-cell, B-cell and macrophage in immune and inflammatory response.

receptor which generates intracellular signals for induction of a variety of lymphokines; and (ii) a T-cell effector function mediated by the produced lymphokines which regulate proliferation and differentiation of haemopoietic and lymphoid target cells.

Cytokines and/or haemopoietic growth factors were initially assumed to be cell-lineage specific and have been named based on their first described major target cells (M-CSF, G-CSF, GM-CSF, TCGF, BCGF, BCDF, etc.). This anticipated that cytokines acted in a co-linear manner through a series of cell-lineage-specific interactions and that no cross-communication exists among cells of different lineages (Fig. 9.7). However, cytokines are generally pleiotropic and we have learned that (i) a single cytokine can interact with more than one type of cell and has multiple biological activities, (ii) a single cell can interact with more than one cytokine, and (iii) many cytokines have overlapping activities. These results also indicate that a single target cell expresses receptors for multiple cytokines. Of

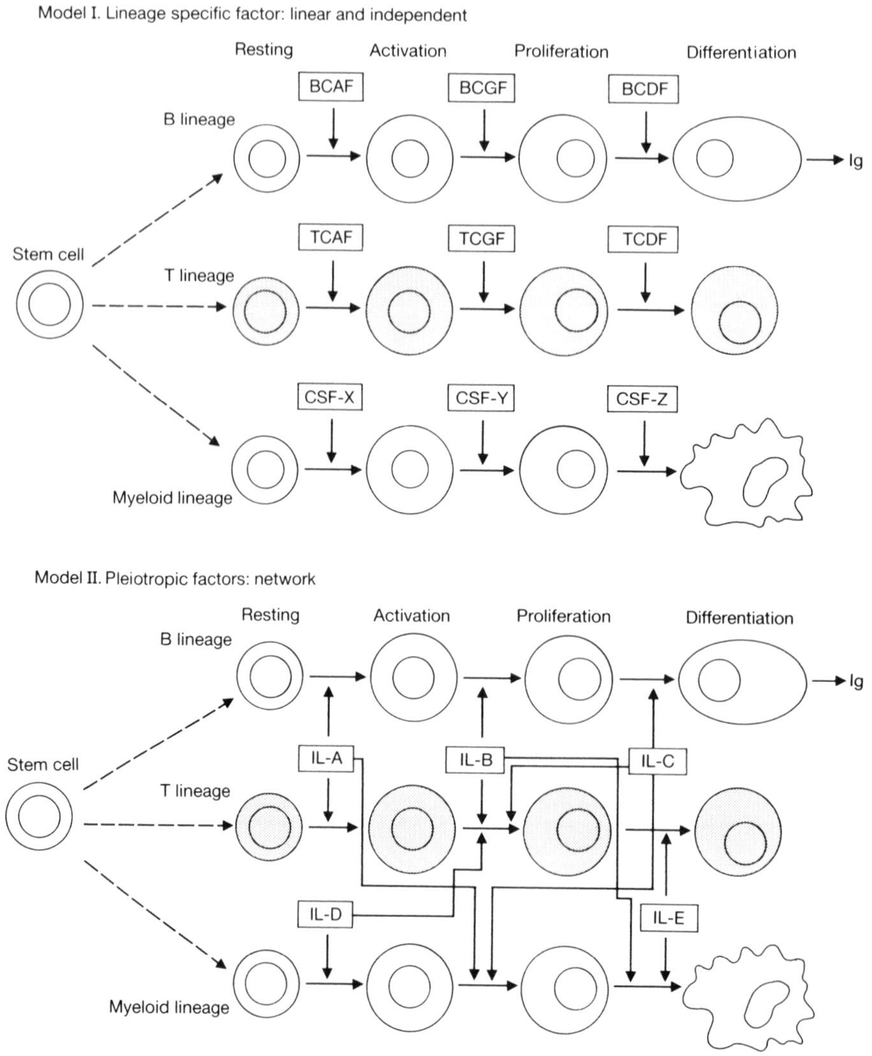

Figure 9.7 Models for relationship between cell lineages and activities of growth factors.

particular interest is the existence of myeloid cell lines whose survival and proliferation can be sustained by multiple cytokines, including colony-stimulating factors and interleukins (IL-3, GM-CSF, IL-4, IL-5, IL-6, G-CSF) and growth factors (steel factor) (Fig. 9.8). These results suggest that cytokine signalling pathways are non-linear and form a network with multiple cross-communication among different cytokines. In the course of lymphokine research, we recognized that the signal-transduction pathways of lymphokine receptors are organized into a complicated network distinct from that learnt from metabolic pathways or metabolism on DNA or RNA templates. Structural and functional studies on growing members of the cytokine receptor family and the interaction with their receptors may provide novel insights into the intracellular signal-transduction network. Characterization of intracellular signal-transduction pathways will also help our understanding of the complex intercellular cytokine network.

However, we do not know the basic rules that regulate this complex signal network and many questions remain unanswered:

- How many intracellular signal pathways exist?
- How are pathways organized: linear versus network?
- How do distinct multiple signals merge to common pathways?
- Which pathway will be used in response to given signals?
- Is the choice flexible or fixed?
- Is the signal network re-organized after repeated use?
- Is learning or memory involved in the signal network?

Many unanswered issues in developmental biology may be related to the organization of the signal-transduction network that determines the response of the progenitor cells to extracellular signals.

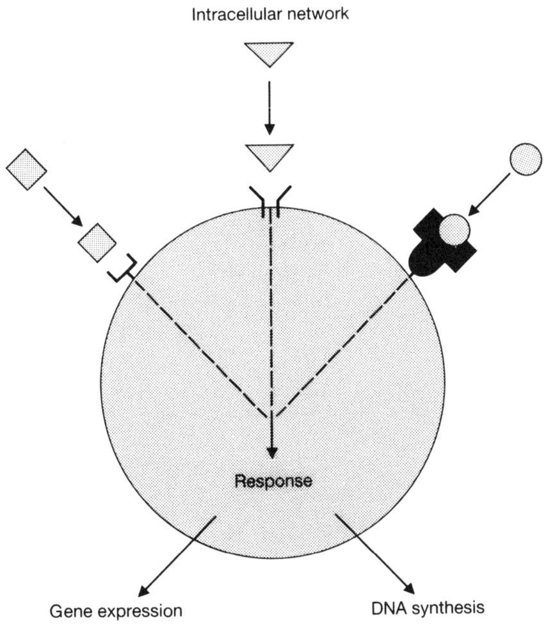

Figure 9.8. Multiple cytokines which act on single myeloid cell and intracellular signal-transduction network.

6 Cell-to-cell communications and intercellular networks

Growth of haemopoietic progenitor cells can be supported by soluble mediators such as cytokines or colony-stimulating factors. In their absence, these cells die quickly; however, in the presence of stromal cells, they survive. These observations suggest that cytokines or growth factors produced by stromal cells and/or cell-to-cell interaction of stromal cells with stem cells play a vital role in maintaining the viability of haemopoietic progenitors (Fig. 9.9). The microenvironment involving stromal cells may also be important for the development of a haemopoietic system. Requirement of growth factors or colony-stimulating factors undergoes a series of changes during development, for example FGF for mesodermal cells, IL-3 or steel factor for haemopoietic stem cells, IL-7 for pre-B-cells and IL-2 for T-cells. This strongly indicates that differential expression of growth factor receptors and re-organization of signalling pathways take place during development. This issue opens up a series of important questions in developmental biology. Many important issues regarding the role of stromal cells in a differentiation-inducing microenvironment remain to be determined:

1 Is there a fundamental difference in communication via soluble mediator versus cell-to-cell contact?

2 How are these signal-transduction pathways regulated in time- and position-dependent manner during early development?

3 Does re-organization of signalling pathways involve genome re-organization (re-arrangement or insertion of cassette)? If so, which component(s) of the signal-transduction pathway (ligand-binding unit, transducer of the receptor, or down-stream pathways) are re-organized?

4 Does re-organization of signalling pathways result from the cell's 'commitment' to differentiation or is this synonymous with the commitment itself?

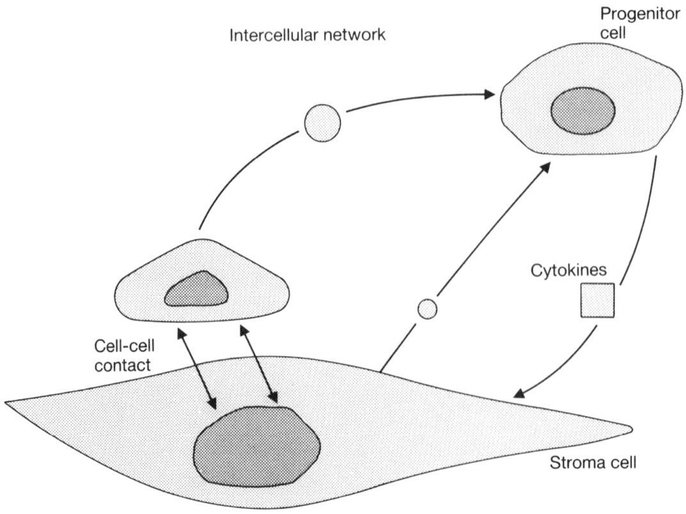

Figure 9.9. Intercellular network between haemopoietic stem cells and stromal cells.

7 Frontier of biomedical research and new challenge for biotechnology

With the advent of recombinant DNA technology, molecular biology, originally established in prokaryotic systems, emerged as a universal discipline and provided both basic concepts and tools to study biological phenomena in higher organisms. New molecular biology and biochemistry helped to amalgamate diverse research fields in medical science. Recent explosive developments in biology have eliminated the traditional boundaries separating the fields of cell biology, physiology, immunology, biochemistry and molecular biology, and have accelerated the mergers of several disciplines into a common molecular and cell biology research. New frontier research areas have been created such as receptor signal transduction, molecular oncology, molecular immunology, molecular haematology and neurobiology.

The early stage of recombinant DNA technology was mainly concerned with the structure of nucleic acids. More recently, the function of DNA elements has been studied by introducing normal and mutated DNA segments into living cells. The gene-transfer technique eliminated the wide gap between biochemical and physiological approaches and became an important tool to study the structure and function of mammalian cells. Needless to say, mammalian molecular biology (both somatic cells and germ cells) is still primitive. Important on the agenda is an improved gene targeting procedure based on homologous recombination, development of regulatable promoters, mini-chromosome vectors carrying centromeres and telomeres, and development of mammalian haploid cells. All are already available in the yeast system. These technologies should shed more light on the 'black box' of mammalian cells and will open up novel therapeutic approaches to medicine.

The other important but relatively unexplored area is designing the 3-D structure of functional proteins. Molecular genetics tells us that the primary structure of a protein is determined by the nucleotide sequence of a gene. In principle, we can chemically synthesize any kind of protein or gene. However, we know very little about which sequences among the myriad of possibilities are biologically significant. We have the techniques at hand but do not know how to employ them. Is it possible to create an artificial protein with specificity and plasticity comparable to that of a natural enzyme? The goal of protein engineering is to discover the basic rules that govern the folding of amino acid sequences into appropriate 3-D structure. In the near future, we wish to design a better molecular switch with properties even more sophisticated than GTP-binding proteins employed in signal transduction.

What will be the research environment in the 21st century? Traditional research required only small groups. However, today, with the growth of biotechnology, interaction between experts in many areas is required. This may alter traditional features of biomedical research within a decade. These developments are beginning to affect traditional departmental organizations of many universities in the world. Medical schools and affiliated hospitals are not immune to these revolutionary processes. Molecular biology and cell biology have already become common disciplines for life science. They will soon be integrated into medical science and the nature of the research environment in the next century will be significantly different. New frontiers in biochemistry such as signal transduction will grow and blossom in this rapidly developing new environment.

8 Acknowledgements

The authors would like to thank Drs Naoko Arai, Takashi Yokota, Atsushi Miyajima, Yoshito Kaziro and many colleagues at IMSUT and DNAX.

9 References

Arai, K. & Kornberg, A. (1980a). *Proc. Natl. Acad. Sci. USA*, **78**, 707–11.

Arai, K. & Kornberg, A. (1980b). *J. Biol. Chem.*, **256**, 5253–9.

Arai, N., Arai, K. & Kornberg, A. (1980). *J. Biol. Chem.*, **256**, 5287–93.

Arai, K., Kawakita, M., Kaziro, Y., Maeda, T. & Ohnishi, S. (1974). *J. Biol. Chem.*, **249**, 3311–13.

Arai, K., Lee, F., Miyajima, A., Miyatake, S., Arai, N. & Yokota, T. (1990). *Ann. Rev. Biochem.*, **59**, 783–836.

Edelman, A.M., Blumenthal, D.K. & Krebs, E.G. (1987). *Ann. Rev. Biochem.*, **56**, 567–613.

Gilman, A.G. (1987). *Ann. Rev. Biochem.*, **56**, 615–49.

Hanks, S.K., Quinn, A.M. & Hunter, T. (1988). *Science*, **241**, 42–52.

Kaziro, Y. (1978). *Biochim. Biophys. Acta*, **505**, 95–127.

Miyatake, S., Schreurs, J., de Vries, J., Arai, N., Yokota, T. & Arai, K. (1988). *FASEB J.*, **2**, 2462–73.

Watanabe, S., Yokota, T., Nakayama, N., Arai, K. & Miyajima, A. (1991). *Curr. Opin. Biotechnol.*, **2**, 227–37.

Signal Transduction Mechanisms

10 Molecular Mechanisms of Regulatory Signal Transduction

E.S. SEVERIN and M.V. NESTEROVA
Research Center of Molecular Diagnostics, Sympheropolsky Boulevard 8, Moscow 113149, Russia

Among the most important problems in the development of new drugs is the identification of a specific target to be effected in the case of some particular disease. To solve this problem, it does not suffice to examine individual aspects of the mechanism of pathogenesis of a certain disease. For correcting corresponding disorders, one should be able to control the main integrating systems of the organism, viz. nervous and endocrine. In this respect, studies on the functioning of regulatory systems that have very complex receptor–enzyme cascades are of prime significance. One of the basic strategic principles in the search for novel drugs is the identification of those stages of the regulatory system that are most efficient in influencing the disease process.

The regulatory signal pathway includes at least five steps. At the first step, a biologically active substance comes from the outside and is recognized by a receptor exposed on the cell surface. The second step is a complex process of transmembrane information transmittance from the outer to the inner cell surface. The third step of the cascade involves generation of an intracelluar effector (a second messenger) in response to the incoming information. An increase in the intracellular level of the second messenger triggers the fourth step in the external signal transduction, namely the functioning of cellular assemblies. The assemblies comprise enzymes whose activity is directly controlled by an intracellular effector. Often, these are protein kinases which regulate the functions of other cell proteins (Yeaman & Cohen, 1975; Nimmo *et al.*, 1976). So the enzyme regulator and the enzyme under control act as a single cellular assembly. In view of the key position of such regulatory enzymes in biological pathways, the cell has a rather sophisticated system for controlling their activity. The system determines qualitative and quantitative functional characteristics of these enzymes, and thereby the functioning of the whole system of second messengers (Rubin & Rosen, 1975; Potter & Taylor, 1979; Laks *et al.*, 1981; Abduragimov *et al.*, 1987).

For many regulatory systems the last step of functioning is nuclear translocation and intranuclear action of some of their components (Spielvogel *et al.*, 1977). This stage, complex in itself, can be further subdivided into several discrete steps. So directed regulation can be accomplished at the stage of membrane processes, the stage of an intracellular effector and the stage of cellular assemblies (Fig. 10.1). The results obtained in our study of the above levels of selective regulation of cell metabolism will be described below. We shall also consider the important task of modelling systems for directed protein delivery through the cell membrane and of creating, in this connection, drugs of a new generation.

The receipt of an external signal is the start of the long process of regulatory cascade stimulation which results in the final coordinated response of the cell. The

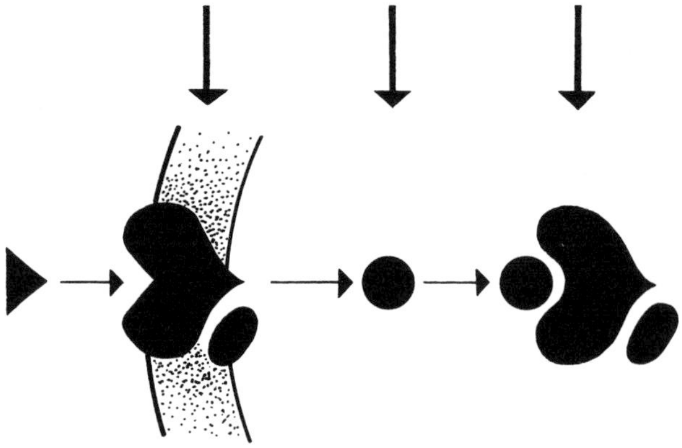

Figure 10.1. Approaches to regulation of cellular metabolism.

receptor network is subject to significant changes at various functional states of the healthy and diseased cell.

Nowadays, adrenergic and muscarinic acetylcholine receptors are proposed to play an important role in promoting lung diseases, in particular chronic obstructive respiratory diseases (Barnes *et al.*, 1988). These receptors are greatly affected in chronic bronchitis, bronchial asthma, chronic pneumonia, tuberculosis and lung cancer (Barnes *et al.*, 1980; Raaijmakers *et al.*, 1983, 1987; Kondratenko *et al.*, 1991). The same can be said of insulin, various growth factor receptors, where changes reflect the functional state of an organism (Ball *et al.*, 1986; Zick, 1989).

The above findings allow us to outline some new therapeutic approaches. In fact, the effect of many drugs used in various diseases is based on influencing the receptor system. However, practice shows that treatment with such preparations often fails to cause the expected results, and, in particular, the effect of desensitization develops very quickly. How can we account for this phenomenon in terms of molecular biology?

Cell receptors are a part of membrane complexes, such as adenylate cyclase which is an enzyme involved in the biosynthesis of cyclic adenosine monophosphate (cAMP). Adenylate cyclase activation and changes in the intracellular cAMP level mediate the action of a whole number of biologically active compounds, such as catecholamines, peptide and protein hormones. However, prolonged hormone influence leads to a reduced capacity of the cells to respond to the repeated hormone action. This process, called 'desensitization', appears to be one of the most important ways of regulating the cAMP level in the cell. It has been demonstrated (Popov *et al.*, 1984; Sibley *et al.*, 1984) that among the reasons for heterologous desensitization of pigeon erythrocyte adenylate cyclase are changes in β-adrenoreceptors which lead to disorders in the interaction of the latter with the guanosine triphosphate (GTP) protein. These disorders are caused by phosphorylation of β-adrenoreceptors.

Hence, the second messenger system is characterized by the phenomenon of autoregulation. The above example shows that cAMP can directly control its own synthesis through cAMP-dependent phosphorylation by the feedback principle. This means that an external stimulus does not in all cases induce the formation of intracellular mediators and trigger cell functioning. Inside the cell there exist certain mechanisms that interfere with the signal transduction. Among such mechanisms, cAMP-dependent phosphorylation of adenylate cyclase is particularly notable.

An external signal has to travel a long and involved way before a coordinated response of the organism is produced. In view of this, it is logical to suppose that if the effect is directed to earlier stages of the regulatory chain, the signal may become somewhat diffused and either reach the final link of the regulatory cascade in a strongly distorted manner or fail to reach it at all. Therefore, the nearer to the final stage of the cascade that the site of the exerted action is (drug application), the more likely is its adequate effect. The cell puts barriers in the way of external influence. In this connection, it is important to be able to penetrate into the cell and affect intracellular links of regulatory cascades.

Today, a whole series of second messengers, such as cyclic nucleotides, Ca^{2+}, phosphoinositides and oligoadenylate, have been reported. Nevertheless, despite the seeming diversity of intracellular effectors, all of them have the same object for their action, namely the phosphorylation system.

Among all protein kinases, cAMP-dependent protein kinases are the most studied. Activation of cAMP-dependent protein kinase causes induction of RNA synthesis (Miles *et al.*, 1981; Lamers *et al.*, 1982). So, nuclear translocation of this enzyme is an event of paramount importance for the external signal reception by the cell. It has been shown (Nesterova *et al.*, 1981; Aprikyan *et al.*, 1988) that compartmentalization of cAMP-dependent protein kinase plays a significant role in the regulation of cell activity. Disorders that occur as a result of neoplastic transformation directly involve nuclear translocation of protein kinase and its binding to structural elements of the genome. Nuclear translocation of this protein is a cAMP-dependent process; therefore, cAMP elevation can normalize protein kinase transport into the nucleus and, moreover, eliminate particular pathological states. Agents capable of increasing the cAMP level include site-selective analogues of cAMP. The normalizing effect of these compounds on a number of tumour markers has been reported (Tagliaferri *et al.*, 1988; Tortora *et al.*, 1988). In particular, the growth rate of many tumours has been shown to decrease *in vivo* (Cho-Chung *et al.*, 1989). Thus, specific drug action upon the final links of the regulatory cascade has the potential to cause a fairly efficient therapeutic effect.

Another bioregulator, analysed in our experiments, was 2′,5′-oligoadenylate (oligo(A)). Interest in this substance as a regulator of the biological activity of cells was evoked in connection with the detected increase in the oligonucleotide level in interferon-treated cells (Lengyel, 1981). We elucidated the role of 2′,5′-oligoadenylate in the regulatory mechanisms of the cell and its relationship with other second messengers. As a result, it became possible to select substances of quite a different nature but which fully imitiated the action of interferon, so permitting selective and directed modelling and control of the functioning system.

In a number of experiments, cAMP-dependent regulation of enzymes of oligo(A) metabolism was demonstrated. It was found that oligo(A) synthetase activation is

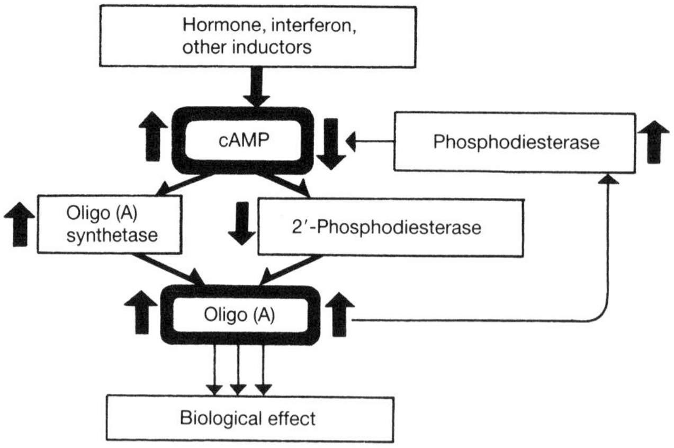

Figure 10.2. A scheme for interrelated regulation of cAMP and oligoadenylate levels.

probably connected with enzyme induction, while inhibition of 2′-phosphodiesterase is connected with cAMP-dependent phosphorylation (Kafiani *et al.*, 1983).

The above data allow us to propose a scheme of interrelated regulation of cAMP and oligo(A) levels (Fig. 10.2) that is based on the mechanism of negative feedback. A rise in cAMP concentration in the cell leads to the induction of oligo(A) synthetase, inhibition of 2′-phosphodiesterase and, hence, elevation of oligo(A). The latter activates phosphodiesterase cAMP, which results in decreased cAMP levels and the blockage of oligo(A) elevation.

According to the proposed scheme, a possible mechanism of cAMP involvement in the regulation of cell proliferation and induction of antiviral cell resistivity is as follows. The rise in cAMP level in human ovary carcinoma (CaOv) cells, induced by the action of theophyl-line or 1-methyl-3-isobutylxanthine, was shown (Itkes *et al.*, 1982) to cause essential inhibition of encephalomyocarditis virus growth; the maximal inhibition effect was time-dependent with maximal oligo(A) synthetase concentration in these cells. It can apparently be concluded that cAMP contributes to the cell resistivity to oligo(A)-sensitive viruses and that development of the antiviral effect requires time sufficient for maximal induction of oligo(A) synthetase in the cells.

Second messenger systems are not only employed by higher organisms. In the process of evolution, parasitic micro-organisms evolved so that their action became directed to second messenger systems of the host cell. Moreover, bacterial toxins can imitate the action of the real participants of the regulatory cascade, being characterized by properties identical with the latter. *Staphylococcus aureus* enterotoxin A (SEA) has proved to be one of them.

Limited proteolysis of SEA was found to result in the formation of the fragment bacteriomodulin (BacM), which is capable of activating, independently of Ca^{2+}, calmodulin-dependent phosphodiesterase and Ca^{2+}-Mg^{2+} ATPase (Dudkin *et al.*, 1984; Alakhov *et al.*, 1988b). BacM was shown to interact with the calmodulin binding site of these enzymes. The kinetic characteristics of calmodulin- and BacM-mediated activation of the enzymes are totally coincident (Dudkin *et al.*, 1984).

The suggested mechanism of SEA action is given in Fig. 10.3. The events that proceed under the effect of SEA on a target cell can be subdivided into two groups.

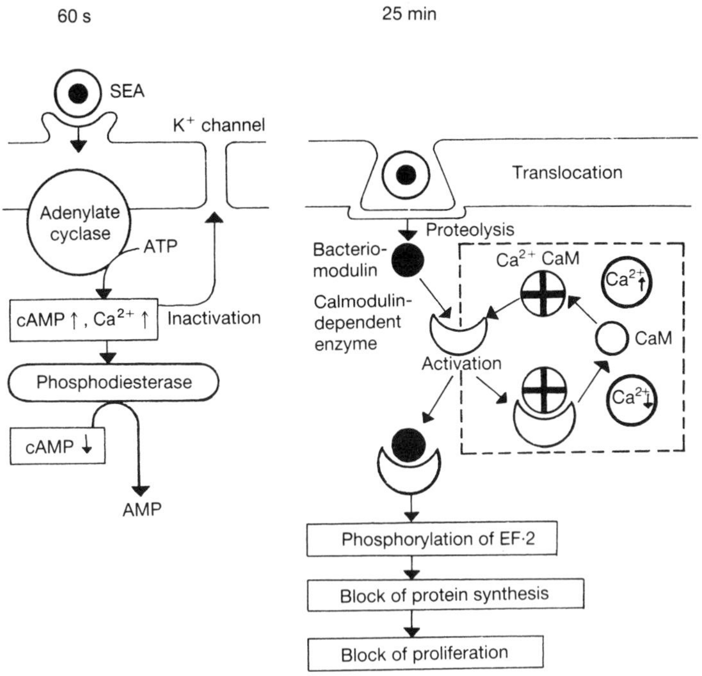

Figure 10.3. Mechanism of an antiproliferative action of SEA on SEA-sensitive cells.

The first group involves changes in second messenger systems caused by the formation of a complex of SEA with its specific receptor on the cell surface. During the first minutes after addition of SEA, adenylate cyclase becomes activated and cAMP levels increase, thus leading to the release of intracellular Ca^{2+}. The latter event induces translocation of protein kinase C into the membrane and blockage of potential-dependent K^+ channels. The result of these modifications is that SEA is translocated into the cell and undergoes proteolytic activation.

The second group of events is connected with BacM formation. Its appearance in the cells causes Ca^{2+}-independent activation of calmodulin-dependent enzymes. Hence feedback in the system's regulation is destroyed. The above enzymes retain their active state at decreased cellular Ca^{2+} concentrations. We believe that this results in the antiproliferative action of SEA. Staphylococcal enterotoxin, by imitating the calmodulin effect, actively interferes in the regulation of Ca^{2+}-dependent processes.

The processes described do not require penetration of the signal substance from the outside into the cell. However, in some cases a protein factor may penetrate the cell by non-specific interaction with the membrane. This interaction might be due, for example, to the presence in the protein molecule of a hydrophobic anchor. The literature is abundant with data documenting the very wide occurrence of the phenomenon of protein hydrophobization by phospholipids and fatty acids (Schmidt, 1983; Seeley *et al.*, 1984; Sefton & Buss, 1987; Hu *et al.*, 1988; Low & Saltiel, 1988).

An effective method of artificial conjugation of lipid molecules to proteins has been developed (Levashov *et al.*, 1984; Kabanov *et al.*, 1987). Systems of reversed micelles of surfactants in organic solvents have been used as a medium for protein modification (for instance, acylation with fatty acid chloranhydrides). These systems allow a high yield of artificially hydrophobic proteins containing one or two fatty acid residues per protein molecule.

Using model systems (liposomes, bilayer lipid membranes), it was demonstrated that the result of artificial hydrophobicity was that the proteins acquired the capacity for transmembrane penetration (Kabanov *et al.*, 1985).

The principle of imparting transmembrane properties to water-soluble proteins by fatty acid acylation can be utilized to create new drugs capable of penetrating target cells. This approach was used (Chekhonin *et al.*, 1988) for the directed delivery of F(ab) antibody fragments modified by stearic acid residues through the blood–brain barrier into the brain.

Today, immune biotechnology has at its disposal highly effective means for producing antibodies to various viral antigenic determinants. Rapid progress has been possible in the development of methods for diagnosis and investigation of viral diseases (Carter & Meulen, 1984; McCullough, 1986; Petrov, 1988). Why then are antiviral antibodies not employed in the therapy of such diseases, in particular for inhibiting viral reproduction? Antiviral antibodies are well known for being effective in protecting cells from viral infection (LeBarsy & Van Hoof, 1980). However, antibodies have no effect, as a rule, on disease development in already infected cells as these lack the potency to penetrate cells and block intracellular viral reproduction (Mandel, 1978). Our studies on influenza and respiratory syncitial viruses indicated that antiviral antibodies made hydrophobic with stearic acid residues are potent inhibitors of intracellular viral reproduction (Kabanov *et al.*, 1989).

It was found that incubation of influenza virus-infected MDCK cells for several hours with non-modified polyclonal antibodies against this virus does not affect the development of infection. Hydrophobic antiviral antibodies under the same conditions cause a two-fold decrease in virus replication. A similar result was found in the case of respiratory syncitial virus reproduction: stearoylated antiviral antibodies, unlike non-modified ones, block the viral reproduction in permissive HeLa cells.

It should be noted that although hydrophobic antibodies acquire the new property of inhibiting viral reproduction, their action still retains specificity. Indeed, incubation of infected cells with stearoylated normal immunoglobulin G (IgG) introduces no changes in the detected level of viral reproduction. Hydrophobic polyclonal antibodies against influenza A virus inhibit reproduction only of this virus, and not that of B viruses.

It seems very probable that hydrophobic antibodies penetrate infected cells where they block aggregation of viral particles and/or synthesis of virus components. This suggestion is supported by the significant antiviral action exhibited by stearoylated monoclonal antibodies against NP-protein, which is an internal antigen of the influenza virus and is accessible only to antibodies inside the cells (Wrigley, 1979).

Interesting data were recently obtained (Kabanov *et al.*, 1990) on the hydrophobicity of oligonucleotides. These authors modified an oligonucleotide at the 5′ end using undecanol, complementary to the loop-forming site of the RNA-encoding polymerase 3 of the influenza virus (type A). The modified oligonucleotide effectively suppressed influenza A/PR 8/34 (H1N1) virus reproduction and inhibited synthesis of virus-specific proteins in MDCK cells.

Thus, by modifying a biologically active compound, one can enhance its capacity for penetration of the cell membrane or even induce this ability if the molecule initially lacked this property. Such conjugate constructions of natural and synthetic macromolecules open up wide prospects for both studying fundamental aspects of biologically active compounds and developing systems for directed drug transport in organisms. This approach can also be used for producing artificial antigens and polymer-based drug preparations.

The concept of directed regulation of cell activity is highly attractive and has several aspects. First, it means specific delivery of a biologically active substance into a target cell. The second aspect concerns selectivity of action of the given substance on certain intracellular processes. Current progress makes it possible to design such substances with predetermined properties to serve a specific purpose. It seems especially expedient to use this principle for developing drugs. There is no doubt that the most effective drug action, with minimal side-effects, can be achieved when the preparation becomes active only at the point of contact with diseased organs or cells. This is especially important in tumour treatment requiring administration of cytotoxic and highly antigenic protein factors.

Interesting results in this respect have been shown by drugs comprising vector elements providing the non-specific factor with affinity for a definite cell type. Thus, for example, highly effective action was demonstrated by immunotoxins, i.e. hybrid molecules (Abelev *et al.*, 1963; Chazov *et al.*, 1987) of bacterial or plant toxins, or their fragments, conjugated with an antibody specific to a surface antigenic marker of diseased cells. These toxins ensure direct delivery of a drug preparation to the area of disease and subsequent removal of diseased cells.

Development and the possibilities of use of immunotoxins specific to various antigens have been reported by many authors; however, there exist certain obstacles to their widespread utilization in practice (Yamazumi, 1978; Pastan, 1986). First and foremost, these problems are associated with non-specific toxicity of immunotoxins. Another important disadvantage of these toxins is that not all surface antigens are conjugated with cellular systems allowing endocytosis of the hybrid molecule. The process can become significantly less effective after conjugation of antibodies with a toxin (Vittetta & Uhr, 1985). Hence the frequently observed low activity of immunotoxins. Finally, the concentration of antigen on the surface of a cell to be removed is, in some cases, too low, making the immunotoxin ineffective.

We have analysed a new family of compounds of directed transport function, called respecrins (receptor-specific screened toxins) (Alakhov *et al.*, 1988a). A respecrin comprises a physiologically active factor that is screened, thereby solving the problem of its non-specific activity. Interaction with a target antigen leads to activation of the factor and stimulates its specific action.

Figure 10.4 shows schematically the structure and mechanism of action of a respecrin, illustrating its property as a physiological factor reversibly screened by receptor-specific macromolecules. The factor is conjugated, by means of chemical modification, with the epitope-containing fragment of a marker antigen of the

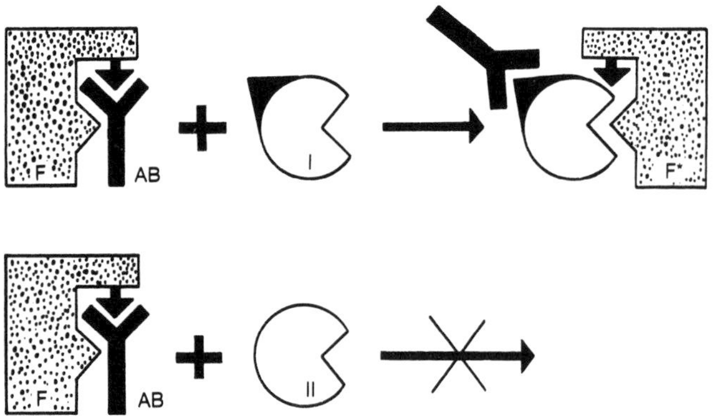

Figure 10.4. Structure and mechanism of action of respecrins: I, a target antigen-containing cell; II, a cell free from target antigen.

target cell. The result of this modification is that the factor should remain active within the conjugate but lose its activity on interaction with antibodies specific for the antigen. Antibodies thus combine the function of target recognition and that of a component screening the factor within the respecrin. It is also important to ensure that the affinity of antibodies for the antigenic determinant, introduced into the factor, should not exceed their affinity for the intact antibody. Once a respecrin comes into contact with the antigen, the immunocomplex dissociates and is accompanied by activation of the physiologically active factor.

A mathematical model of the respecrin activation, calculated by taking into account removal of screening antibodies due to their interaction with the target cell-surface antigen, shows that the proposed design can ensure highly specific activation of the factor even where there are small differences in the marker antigen concentration on the cell surface. There is therefore a significant advantage of respecrins over immunotoxins, since the high toxicity of immunotoxins makes them virtually unusable.

Respecrin functioning is independent of the antigen capacity for endocytosis, because the effect is through interaction of a biologically active factor with its own receptor. Moreover, it is possible to use as the complex activator not only a surface antigen but also a soluble one, for example a protein secreted by diseased cells.

It was found (Kabanov *et al.*, 1988) that conjugation of antibodies with some synthetic polymers, in particular poly-4-vinylpyrrolidone, leads to positive cooperation in the binding of the obtained derivative to the antigen. Utilization of a polymer component within a respecrin allows amplification of the respecrin opening. This, in turn, makes it possible to increase the selectivity of respecrin action still further.

As evidenced by experimental results (Alakhov *et al.*, 1990), the problem of screening the physiological activity of a factor proteinaceous in nature can be successfully solved. It was shown, in particular, that the cytotoxic properties of SEA, which were still retained after conjugation with mouse IgG, were effectively inhibited by rabbit antimouse antibodies. The SEA activity was restored, provided that the culture medium was supplemented with mouse IgG that acted as a target antigen.

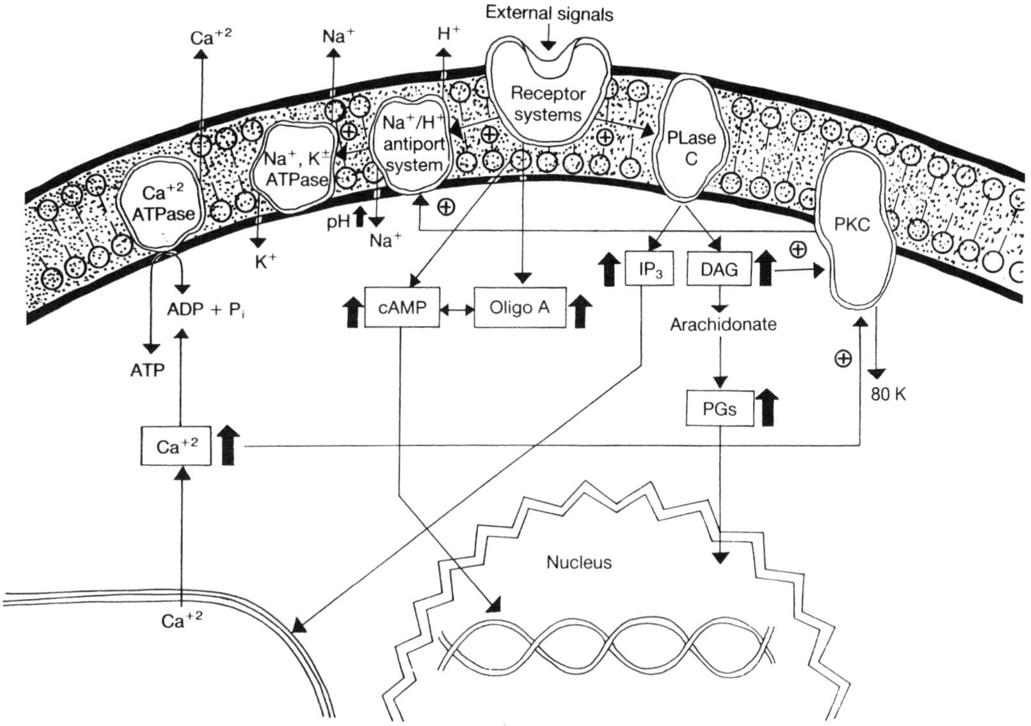

Figure 10.5. A scheme for interaction of various regulatory systems (IP_3, inositol triphosphate; DAG, diacylglycerol; PLase C, phospholipase C; PKC, protein kinase C; PGs, prostaglandins).

The idea of specific features common to any drug molecules was first formulated about a century ago by P. Ehrlich (see Himmelweit, 1960). Apart from the active component possessing a chemotherapeutic effect, a drug molecule must have groups providing for its specific organotropism and capacity for binding to the given cell. Scientific progress has now made construction of drug preparations with predetermined characteristics a reality.

While creating new drugs, it is necessary to analyse in detail their interactions with known receptor systems, effect on ion channels, systems of second messengers, the functioning of cellular components, the efficiency of membrane transport and gene expression (Fig. 10.5). In this respect, profound and comprehensive study of regulatory systems may be of great help in specifying substances that can produce their maximal effect at specific stages in the cascade.

1 References

Abduragimov, A.S., Itkes, A.V. & Kochetkov, S.N. (1987). *Biokhimiya,* **52**, 1798–807. (In Russian.)

Abelev, G.I., Petrova, S.D. & Krankova, N.I. (1963). *Biokhimiya,* **54**, 625–34. (In Russian.)

Alakhov, V.Yu., Arzhakov, S.A., Vasilenko, O.V., Voloshchuk, S.G., Glazkova-Stepanenko, I.S., Duvakin, I.A., Ishkov, A.G., Kabanov, A.V., Kabanov, V.A., Klinsky, E.Y., Kravtzova, T.N., Petrov, R.V., Sveshnikov, P.G. & Severin, E.S. (1988a). *Doklady Akademii Nauk SSSR,* **303**, 1494–7. (In Russian.)

Alakhov, V.Yu., Moskaleva, E.Y., Kravtzova, T.N., Smirnov, V.V., Duvakin, I.A., Loginov, B.V. & Severin, E.S. (1988b). *Biotechnol. Appl. Biochem.,* **10**, 563–7.

Alakhov, V.Yu., Arzhakov, S.A., Vasilenko, O.V., Voloshchuk, S.G., Glazkova-Stepanenko, I.S., Duvakin, I.A., Kabanov, A.V., Kabanov, V.A., Klinsky, E.Y., Kravtzova, T.N., Petrov, R.V., Sveshnikov, P.G. & Severin, E.S. (1990). *Biomed. Sci.*, **1**, 155–9.

Aprikyan, A.G., Nesterova, M.V. & Severin, E.S. (1988). *Biochem. Int.*, **16**, 601–4.

Ball, E.D., Sorensen, G.D. & Pettengill, O.S. (1986). *Cancer Res.* **46**, 2335–9.

Barnes, P.J., Karliner, J.S. & Dollery, C.T. (1980). *Clin. Sci.* **58**, 457–61.

Barnes, P.J., Minette, P. & Maclagan, J. (1988). *TIPS*, **9**, 412–16.

Carter, M.J. & Meulen, V. (1984). Adv. Virus Res., **29**, 95–130.

Chazov, E.I., Smirnov, V.N. & Torchillin, V.P. (1987). *Zhurnal Vsesojuznogo Khimicheskogo Obshchestva*, **32**, 485–7. (In Russian.)

Chekhonin, V.P., Morozov, T.V., Kashparov, I.A. & Ryabukhin, I.A. (1988). *Biotekhnologia*, **4**, 648–51. (In Russian.)

Cho-Chung, Y.S., Clair, T., Tagliaferri, P., Ally, S., Katsaros, D., Tortora, G., Neckers, L., Avery, T.L., Crabtree, G.W. & Robins, R.K. (1989). *Cancer Invest.*, **7**, 161–77.

Dudkin, S.M., Alakhov, V.Yu., Severin, E.S. & Shvets, V.I. (1984). *Doklady Akademii Nauk SSSR*, **276**, 1510–13. (In Russian.)

Himmelweit, F. (1960). In: *Collection of Papers of Paul Ehrlich*, Vol. 3, Pergamon Press, New York.

Hu, J.-S., James, G. & Olson, E.N. (1988). *BioFactors*, **1**, 219–26.

Itkes, A.V., Krispin, T.I., Shloma, D.V., Balandin, I.G., Tunitskaya, V. L. & Severin, E.S. (1982). *Biochem. Int.*, **5**, 388–98.

Kabanov, A.V., Nametkin, S.N., Levashov, A.V. & Martinek, K. (1985). *Biologicheskiye Membrany*, **2**, 985–95. (In Russian.)

Kabanov, A.V., Klebanov, A.L., Torchillin, V.P., Martinek, K. & Levashov, A.V. (1987). *Bioorganicheskaya Khimiya*, **13**, 1321–4. (In Russian.)

Kabanov, A.V., Alkahov, V.Yu., Klinsky, E.Y., Khrutskaya, M.M., Rahnyanskaya, A.S., Polinsky, A.S., Yaroslavov, A.A., Severin, E.S., Levashov, A.V. & Kabanov, V.A. (1988). *Doklady Akademii Nauk SSSR*, **302**, 735–8. (In Russian.)

Kabanov, A.V., Ovcharenko, A.V., Melik-Nubarov, N.S., Bannikov, A.I., Alakhov, V.Yu., Kiselev, V.I., Sveshnikov, P.G., Kiselev, O.I., Levashov, A.V. & Severin, E.S. (1989). *FEBS Lett.*, **250**, 238–40.

Kabanov, A.V., Vinogradov, S.V., Ovcharenko, A.V., Krivonos, A.V., Melik-Nubarov, N.S., Kiselev, V.I. & Severin, E.S. (1990). *FEBS Lett.*, **259**, 327–30.

Kafiani, C.A., Itkes, A.V., Kartasheva, O.N., Severin, E.S., Kochetkova, M.N. & Turpaev, K.T. (1983). *Adv. Enzyme Regul.* **21**, 353–65.

Kondratenko, T.Y., Kuzina, N.V., Severin E.S., Kornilova, Z.H., Tikhomirova, E.V. & Perelman, M.I. (1991). *Voprosy Meditsinskoi Khimii*, **37**, 20–1.

Laks, M.S., Harrison, J.J., Schwoch, G. & Jungmann, R.A. (1981). *J. Biol. Chem.*, **256**, 8775–85.

Lamers, W.H., Hansom, R.W. & Meisner, H.H. (1982). *Proc. Natl. Acad. Sci. USA*, **10**, 473–85.

LeBarsy, T. & Van Hoof, F. (1980). In: G. Gregoriadis and A.C. Allison (Eds.), *Liposomes in Biological Systems*, pp. 211–16, John Wiley & Sons, Chichester.

Lengyel, P. (1981). In: J. Gresser (Ed.), *Interferon 3*, pp. 78–99, Academic Press, New York.

Levashov, A.V., Kabanov, A.V., Khmelnitsky, Y.L., Berezin, I.V. & Martinek, K. (1984). *Doklady Akademii Nauk SSSR*, **278**, 246–8. (In Russian.)

Low, M.G. & Saltiel, A.R. (1988). *Science*, **239**, 268–75.

Mandel, B. (1978). In: H. Fraenkel-Conrat and R.R. Wagner (Eds.), *Comprehensive Virology*, pp. 37–121, Plenum Press, London.

McCullough, K.C. (1986). *Arch. Virol.*, **87**, 1–36.

Miles, M.F., Hung, H. & Jungmann, R.A. (1981). *J. Biol. Chem.*, **256**, 12 545–52.

Nesterova, M.V., Ulmasov, K.A., Abdukarimov, A., Aripdzanov, A.A. & Severin, E.S. (1981). *Exp. Cell Res.* **132**, 373–6.

Nimmo, H.G., Proud, C.G. & Cohen, P. (1976). *Eur. J. Biochem.*, **68**, 21–30.

Pastan, J., Willigham, M.C. & Fitzgerald, J.P. (1986). *Cell*, **47**, 641–8.

Petrov, R.V. (1988). *Zhurnal Vsesojuznogo Khimicheskogo Obshchestva* **33**, 484–93. (In Russian.)

Popov, K.M., Bulargina, T.V. & Severin, E.S. (1984). *Biokhimiya,* **49**, 1561–6. (In Russian.)

Potter, R.L. & Taylor, S.S. (1979). *Biol. Chem.,* **254**, 9000–5.

Raaijmakers, J.A.M., Terpstra, G.K., Van Rosen, A.J., Witter, A. & Kreukniet, J. (1983). *Clin. Sci.,* **66**, 215–20.

Raaijmakers, J.A.M., Beneker, C., Dol, R. & De Ruiter-Bootsma, A.L. (1987). *Cell. Mol. Biol.,* **33**, 515–18.

Rubin, C. S. & Rosen, O.M. (1975). *Ann. Rev. Biochem.,* **44**, 831–87.

Schmidt, M.F.G. (1983). *Curr. Top. Microb. Immunol.,* **102**, 101–29.

Seeley, P., Ruckenstein, A., Connolly, J. & Greens, L.A. (1984). *J. Cell Biol.* **98**, 417–26.

Sefton, B.M. & Buss, J.S. (1987). *J. Cell Biol.* **104**, 1449–53.

Sibley, D.R., Peters, J.R. Nambi, P., Caron, M.G. & Lefkowitz, R.J. (1984). *J. Biol. Chem.,* **259**, 9742–7.

Spielvogel, A.M., Mednieks, M.J., Eppenberger, V. & Jungmann, R.A. (1977). *Eur. J. Biochem.,* **73**, 199–212.

Tagliaferri, P., Katsaros, D., Clair, T., Ally, S., Tortora, G., Neckers, L., Rubalcava, B., Parandusch, Z., Chang, Y.-A., Revankar, G.R., Crabtree, G.W., Robins, R.K. & Cho-Chung, Y.S. (1988). *Cancer Res.,* **48**, 1642–50.

Tortora, G., Tagliaferri, P., Clair, T., Colamonici, O., Neckers, L.M., Robins, R.K. & Cho-Chung, Y.S. (1988). *Blood,* **71**, 230–3.

Vittetta, E.S. & Uhr. J.M. (1985). *Cell,* **41**, 653–4.

Wrigley, N.G. (1979). *Br. Med. Bull.,* **35**, 35–8.

Yamazumi, M. (1978). *Cell,* **15**, 245–50.

Yeaman, S.J. & Cohen, F. (1975). *Eur. J. Biochem.,* **51**, 93–104.

Zick, Y. (1989). *Crit. Rev. Biochem. Mol. Biol.,* **24**, 217–69.

11 Molecular Diversity of Signal Transduction Systems: A Puzzle for Pharmacologists of Today and Tomorrow

J. BOCKAERT

Centre CNRS-INSERM de Pharmacologie-Endocrinologie, Rue de la Cardonille, 34094 Montpellier Cedex 5, France

1 Introduction

One of the difficulties in the field of cellular intercommunication is to understand how a tremendous diversity of messages is generated by a limited number of signalling components. It was thought that this was achieved by the diversity of receptor molecules. This is probably true. Indeed, molecular cloning studies have extended this diversity to a point unexpected a few years ago. In addition, these studies have revealed that the intracellular components of signal-transduction pathways, including guanosine triphosphate (GTP) binding proteins (G-protein), effectors (phospholipases C and A_2), protein kinase C, ionic channels and other signalling components, have variety rivalling or exceeding that of membrane-bound receptors. It is no longer informative to speak about muscarinic receptors (at least five different molecules have muscarinic receptor activity) (Bonner *et al.*, 1987; Peralta *et al.*, 1987) or about protein kinase C (at least seven different molecules with different biochemical properties and localizations have been cloned) (Nishizuka, 1988). Within a given molecule, the possible subunit composition is variable. At least six α, three β, and two γ subunits can enter into the composition of γ-aminobutyric acid A ($GABA_A$) receptors to form dimeric, trimeric and possibly pentameric molecules (Olsen & Tobin, 1990). Therefore, within a single receptor, an enormous variety of variants could theoretically be obtained by changing its subunit composition. The same situation is found for many multimeric signalling molecules. When one takes into consideration that a given transduction pathway can involve the association of several of these molecules, it is easier to understand how a variety of signal patterns can be generated by a single regulatory molecule (hormone, neurotransmitter, growth factors, adhesive molecules).

In the 1980s, pharmacologists and molecular biologists were occupied in the identification of the diverse signalling molecules and their subunit composition. From now on, we will be faced with the following questions:

- Who is talking to whom in the puzzles constituted by the multiple transduction pathways?
- Which methods can be used to solve such puzzles?
- How we will be able to select specific drugs to act on these multiple molecules which are different, but generally very homologous?

This chapter is an attempt to give an idea of the diversity of signalling components and to answer to these specific questions.

2 Diversity of molecules involved in signal transduction

2.1 *Molecular diversity at the receptor level*

Over the past 20 years, the number of receptor subtypes discovered using specific radioligand techniques, molecular biology and cellular pharmacology techniques is considerable. The complete list is not given here, but a few examples of such receptors will be taken and the complexity of their transduction pathways discussed.

2.1.1 G-PROTEIN-COUPLED RECEPTORS

These receptors are composed of one subunit with seven transmembrane spanning domains, three intracellular and three extracellular loops, and are coupled to G-proteins (Birnbaumer *et al.*, 1990) (Fig. 11.1).

The first example of receptor diversity will concern the serotonin (5-HT) receptors. Twenty years ago, 5-HT receptors were not really characterized. Based on complex guinea-pig ileum contraction experiments, Gaddum & Picarelli (1957) defined the D (dibenziline-blocked) and M (morphine-blocked) receptors. We now know of four types of 5-HT receptor: the 5-HT_1, 5-HT_2, 5-HT_3 (Schmidt & Peroutka, 1989) and the recently discovered 5-HT_4 receptors (Dumuis *et al.*, 1988, 1989; Bockaert *et al.*, 1990; Craig & Clarke, 1990). The 5-HT_1 receptor type is divided into 5-HT_{1A}, 5-HT_{1B}, 5-HT_{1C} and 5-HT_{1D} subtypes having well-characterized pharmacological transduction mechanisms, cellular localization, interspecies variations and, sometimes, defined physiological functions (Schmidt & Peroutka, 1989). 5-HT_2 seems to be unique and corresponds to Gaddum & Picarelli's D receptors. 5-HT_3 receptors may be multiple but correspond to the M receptor described by Gaddum & Picarelli (Dumuis *et al.*, 1989). Therefore, at least seven different 5-HT receptor molecules have been described so far and there is no doubt that several others will be discovered. This situation is similar for other neurotransmitter receptors. In the early 1980s, it was of some concern to know whether or not these receptor subtypes,

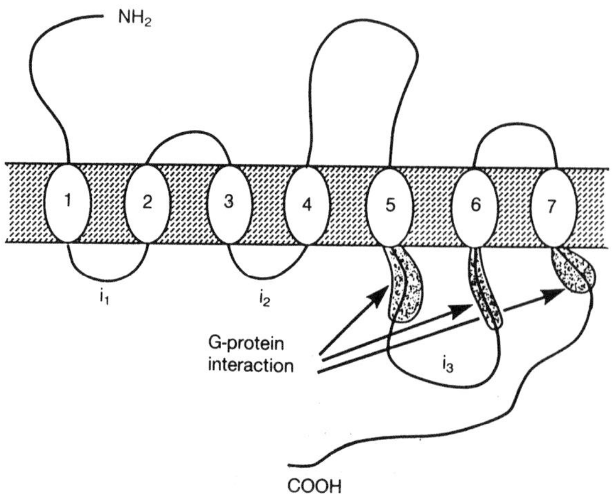

Figure 11.1. General structure of receptors, one to seven transmembrane domains, coupled to G-proteins. i_1, i_2, i_3 are intracellular loops.

described by binding or pharmacological experiments, were really different entity molecules. If one considers 5-HT, but also other neurotransmitter receptors, the response is encouraging; 5-HT_{1A}, 5-HT_{1C} and 5-HT_2 receptors have recently been cloned (Hartig, 1989) and, when expressed in eukaryotic cells or *Xenopus* oocytes, their pharmacology corresponds to what was expected from binding or pharmacological experiments. This great number of receptor subtypes revealed by pharmacological experiments has been further increased by the recent advances made in the cloning of receptor molecules. Two excellent examples can be given. Three acetylcholine muscarinic receptors were defined on the basis of pharmacological experiments (M_1, M_2, M_3). At the present time, the sequence of five muscarinic receptor species (m_1 to m_5) has been reported (Bonner *et al.*, 1987; Peralta *et al.*, 1987). Some species vary slightly in their length, depending on the mammalian species from which they were cloned. The sequence of a specific mammalian species is denoted by an upper-case prefix, e.g. Hm1 for the human m_1 sequence (R for rat, P for pig, M for mouse, D for *Drosophila*). It is likely, but not unambiguously proven, that the m_1 sequence corresponds to the M_1 receptor, m_2 to M_2 and m_3 to the M_3 receptor. The pharmacology of m_4 and m_5 receptor molecules, whilst being different from M_1, M_2 and M_3 receptors, is not distinct enough to provide a pharmacological characterization of what could be called M_4 and M_5 receptors. Here, we have genetically defined receptor molecules awaiting pharmacological tools enabling the recognition of their functions in cells and organisms.

A similar situation exists for dopaminergic receptors. Two well-characterized pharmacological dopaminergic receptors have been described: the D_1 and D_2 receptors. D_1 receptors increase adenylyl cyclase activity whereas D_2 receptors decrease it (Andersen *et al.*, 1990). The two receptors are differently distributed among cerebral neurones and endocrine cells. D_2 receptor transduction mechanisms are complex. It addition to the reduction of adenylyl cyclase, D_2 receptors decrease voltage-sensitive Ca^{2+} channel activity, increase K^+ channel activity and may decrease phospholipase C stimulated by thyrotrophin-releasing hormone (TRH) (Enjalbert *et al.*, 1990). It is not known whether the same D_2 molecular entity is able to trigger all these functions. Two isoforms of the receptor, termed D_2A and D_2B, produced by alternative messenger RNA splicing displaying the same pharmacology, have been cloned (Andersen *et al.*, 1990). D_2A and D_2B differ in their third intracellular loop (i_3) (Fig. 11.1). When compared to D_2B, the D_2A receptor contains an additional sequence encoding a 29-amino acid fragment. This is interesting because the i_3 domain of the G-protein-coupled receptor family has been recognized to be the main domain of interaction with G-proteins (Fig. 11.1) (Cheung *et al.*, 1989; Strader *et al.*, 1989; Wess *et al.*, 1989). Therefore, it is possible, but still not demonstrated, that D_2A and D_2B, although they have the same pharmacological recognition site, possess different transduction mechanisms. The diversity of DA receptors has recently been increased by the cloning of a new entity: the D_3 receptor (Sokoloff *et al.*, 1990). It differs from the D_1 and D_2 receptors in its pharmacology and brain localization. In particular, D_3 receptors seem to be specifically localized in limbic areas. Therefore, like m_4 and m_5 muscarinic receptors, D_3 receptors were discovered using molecular biology techniques. These receptors were not defined pharmacologically before their cloning.

This has led to the concept of 'reverse pharmacology', i.e. the cloning and expression of genes whose products remain to be matched with receptors for physiological response. Using such an approach, the cannabinoid receptor was recently discovered (Matsuda *et al.*, 1990).

2.1.2 RECEPTORS WHICH ARE ALSO IONIC CHANNELS (IONOTROPIC RECEPTORS)

We have described the increasing diversity of receptors composed of only one subunit molecule with seven transmembrane spanning domains. This example illustrates two mechanisms by which molecular diversity can be generated: (i) the multiplicity of genes encoding for a given receptor type (five genes for muscarinic receptors, for example) and (ii) the alternative splicing of some genes (D_2A and D_2B receptors, for example). A third type of diversity is observed in receptors from another family: the receptors which are ionic channels. In this family ($GABA_A$ receptors, nicotinic receptors, glycine receptors, glutamate receptors), each receptor molecule is composed of several subunits each having four transmembrane spanning domains (Changeux *et al.*, 1987; Hollmann *et al.*, 1989; Boulter *et al.*, 1990; Keinänen, 1990; Olsen & Tobin, 1990; Sommer *et al.*, 1990). Changing the number and nature of subunits which compose each receptor could lead to a considerable diversity of receptor entities. Cloning of $GABA_A$/benzodiazepine receptor subunits has provided several putative ligand-gated Cl^--conducting ion-channel receptor subunits, e.g. six α-subunits, two γ and at least one δ (Pritchett *et al.*, 1989; Lüddens *et al.*, 1990; Olsen & Tobin, 1990; Schofield, 1990). Co-expression of α_1, β_1 and γ_2-subunits in eukaryotic cells or *Xenopus* oocytes leads to Cl^- channels opened by GABA, potentiation by benzodiazepines and reduction by inverse benzodiazepine agonists (β-carbolines). This is a pharmacological situation similar to that described for naturally expressed GABA/benzodiazepine receptors. Homomeric γ_2 receptors, or even pairwise $\alpha_1\beta_1$ or $\beta_1\gamma_2$ combinations, form a GABA-activated channel, not potentiated by benzodiazepine (Pritchett *et al.*, 1989). The natural subunit content of GABA/benzodiazepine receptors are not known. It is possible that dimeric, trimeric and also pentameric GABA/benzodiazepine receptors are expressed *in vivo*. If this is the case, an extraordinary diversity is generated.

A similar situation exists for glutamate receptors. Recently, two groups cloned subunits of the α-amino-3-hydroxy-5-methyl-4-isoxazole propionic acid (AMPA) receptor which is one of the glutamate receptors (Hollmann *et al.*, 1989; Boulter *et al.*, 1990; Keinänen *et al.*, 1990; Sommer *et al.*, 1990). At least four subunits are expressed in the brain; each of them exists in two versions with different amino acid sequences (flip and flop sequences) originating from alternatively spliced messenger RNA (Sommer *et al.*, 1990). The electrophysiological responses obtained after transfection with each subunit are different from those obtained with pairwise subunits (Boulter *et al.*, 1990; Sommer *et al.*, 1990). The response also depends on the nature of the pair formed, the flip or flop nature of the subunit also giving a different conductance pattern (Sommer *et al.*, 1990). The same diversity is generated for nicotinic receptors (five α-subunits and four β), the receptor being pentameric (muscle), but possibly dimeric in the brain (Changeux *et al.*, 1987; Luetje *et al.*, 1990). It is likely that the pharmacology of receptors is dependent on

Table 11.1. Nature of receptor diversity

G-protein-coupled receptors (composed of one subunit)
Diversity of genes (five genes for muscarinic receptors, three for dopamine receptors)
Alternative RNA splicing (dopamine D_2A, and dopamine D_2B receptors)
Ionic channel receptors (composed of several subunits)
Diversity of genes coding for subunits
For $GABA_A$/benzodiazepine receptors: 10
For nicotinic receptors: 9
For AMPA receptors: 4
Multiplicity of subunit composition
Dimeric, trimeric or even pentameric for $GABA_A$/benzodiazepine receptors
Pentameric or dimeric for nicotinic receptors
Alternative RNA splicing, example 'flip and flop' sequences in AMPA receptor subunits

their subunit composition. The most extraordinary example has recently been published by Lüddens *et al.* (1990). These authors have cloned the α_6-subunit of $GABA_A$ receptors which is expressed only in cerebellar granule cells. When expressed in eukaryotic cells, recombinant receptors composed of $\alpha_6\beta_2\gamma_2$-subunits bound with high affinity to the GABA agonist [^{3}H]muscinol, but not to benzodiazepines or β-carbolines. However, it does bind RO 15-45-13, an imidazobenzodiazepine. This drug has been reported to be involved in antagonizing alcohol-induced impairment in motor performance. It is possible that the GABA receptor with this subunit composition ($\alpha_6\beta_2\gamma_2$) is a unique $GABA_A$ receptor among a whole series of other $GABA_A$ receptors having other subunit compositions to be involved in alcohol-induced impairment in motor performance. The possible sources of receptor diversity have been summarized in Table 11.1.

2.2 *Molecular diversity at the G-protein level*

At least 100 G-protein-coupled receptors have been discovered (Birnbaumer *et al.*, 1990). The number of G-proteins stimulated by these receptors is also high. G-proteins are heterotrimers with subunits designated as α, β, γ (Fig. 11.2). Differences in the α-subunits serve to distinguish the various G-protein oligomers. The α-subunits contain the binding site for GTP. They are adenosine disphosphate (ADP)-ribosylated by cholera (CT) or pertussis toxin (PT) except α_z (Fig. 11.2). Figure 11.2 gives the functions which can be assigned to each of them. Within each class of α-subunit, several forms can be distinguished. There are four forms of α (αsL_1, L_2, αsS_1, S_2) which can be produced by alternative splicing of a single mRNA (Northrup *et al.*, 1980). There are three αi (αi_1, αi_2, αi_3) which are products of different genes, whereas at least two α_o are produced by alternative splicing (Birnbaumer *et al.*, 1990a,b). Two forms of αt (transducin) products of different genes are synthesized (αt_1 in rods, and αt_2 in cones). β and γ are closely associated proteins. Whereas it was believed that the βγ complex was unique, recent data indicate that four distinct β-subunits may be associated with at least four different γ-subunits (two associated with αt—γt_1, γt_2—and at least two different γ-subunits, γ_5 and γ_6) (Birnbaumer *et al.*, 1990). However, the diversity may be even higher since

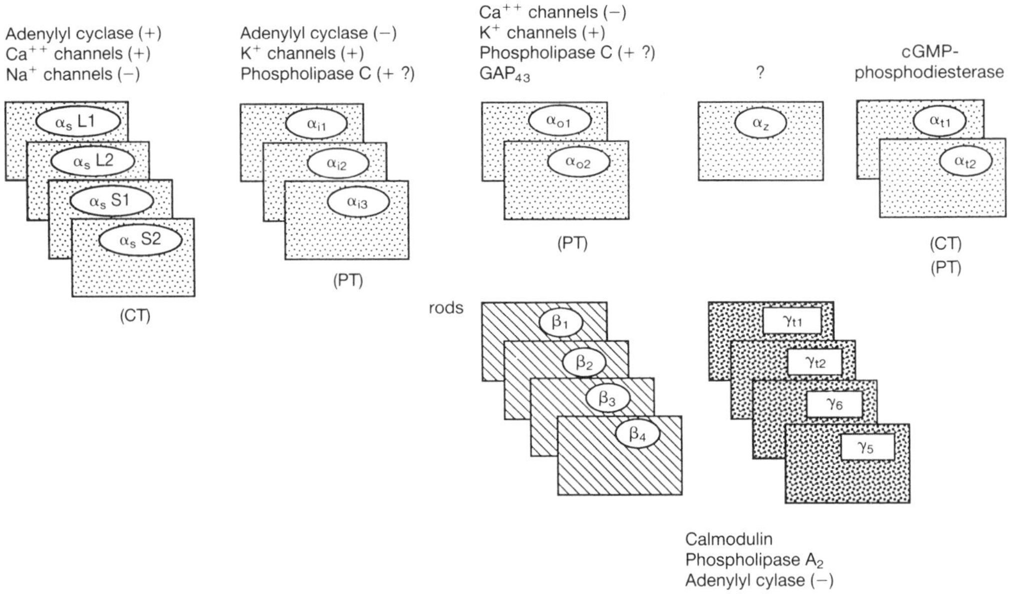

Figure 11.2. Identities and target of G-protein subunits. CT, cholera toxin; PT, pertussis toxin; GAP_{43}, growth cone-associated protein.

some γ-subunits may be phosphorylated, whereas others are not (Asano *et al.*, 1990). It is not known whether each α-subunit can be associated with several different βγ complexes or whether this results in different functional heterotrimeric molecules.

Alone, the βγ complexes may have some functions when dissociated from α-subunits (interaction with calmodulin, inhibition of adenylyl cyclase, stimulation of phospholipase A_2 (Fig. 11.2) (Birnbaumer *et al.*, 1990).

2.3 *Molecular diversity at the effector level*

The diversity of effectors stimulated by G-proteins is probably also very important. It is likely that each α-subunit stimulates several effectors. This is the case for all the α_s splice variants which stimulate both the adenylyl cyclase and voltage-sensitive Ca^{2+} channels (VSCC) (Mattera *et al.*, 1989). The α_o-subunit of Go certainly inhibits VSCC (Hescheler *et al.*, 1987; Harris-Warrick *et al.*, 1988), stimulating some K^+ channels in neurones (Von Dongen *et al.*, 1988) and in some cells stimulating phospholipase C (Moriaty *et al.*, 1990). 'Adenylyl cyclase', 'VSCC' or 'K^+ channels' are certainly not references to unique entities. Two adenylyl cyclases have been described biochemically. One which is Ca^{2+}–calmodulin dependent and one which is not (Minocherhomje *et al.*, 1987). Only one has been cloned (Krupinski *et al.*, 1989). However, in each category it is likely that many variants will be discovered.

The existence of multiple K^+ channels is well documented. Over 30 different K^+ channels have been characterized biophysically (Jan & Jan, 1990). They show different sensitivities to voltage and/or intracellular messengers, and have different kinetic or pharmacological properties. The wide range of K^+-channel properties reflects the multiple cellular functions that they serve including control of synaptic efficacy, heart beat, and endocrine and exocrine secretion. How might this tremendous diversity of K^+ channels be generated? Every source of diversity is found when considering K^+ channels:

1 alternative splicing in the Shaker gene of *Drosophila*, coding for a K^+ channel, called the A channel (Timpe *et al.*, 1988);

2 multiple genes encoding different K^+-channel polypeptides (Shaker subfamily, Shab subfamily, Shaw subfamily, Shal subfamily; each subfamily contains several genes) (Jan & Jan, 1990);

3 associations between channel polypeptides to form heteromultimeric channels (Christie *et al.*, 1990);

4 post-translational modifications may also contribute to channel diversity—K^+ channel expressed by transfection of a single Shaker cDNA in mammalian and in *Drosophila* muscle is charybdotoxin-sensitive and insentitive, respectively (Zagotta *et al.*, 1989).

Whether any of the observed differences are due to differences in posttranslational processing is not known, although modulation of K^+-channel activities by phosphorylation has been observed in a number of cases (Jan & Jan, 1990).

The situation is likely to be similar for other effectors. We have already pointed out that there are at least seven different protein kinase C molecules, having different localizations, activation mechanisms and likely different functions (Nishizuka, 1988). For example, one for them (protein kinase Cγ) is only localized in nervous tissues and activated by arachidonic acid (Nishizuka, 1988).

3 Who is talking to whom?

The diversity of transducing molecules implicated in the first steps following receptor stimulation gives an idea about how complex it is to understand 'who is talking to whom?', especially when considering the homologies between all these proteins. This cascade of complex interactions increases when one takes into account the downstream steps of receptor action (enzyme, channel, gene activations). A signalling molecule should no longer be considered as a trigger of a monodirectional transducing process but rather as a pleiotropic set of transducing events.

3.1 *A signalling molecule triggers a pleiotropic transduction pattern—example of the D_2-dopaminergic receptors in prolactin cells*

Figure 11.3 shows the pattern of events known to be triggered by D_2-dopaminergic receptors. These receptors (1) inhibit adenylyl cyclase (Giannattasio *et al.*, 1981; Enjalbert & Bockaert, 1983), (2) inhibit stimulated phospholipase C (Enjalbert *et al.*, 1990), (3) inhibit T- and L-VSCC (Lledo *et al.*, 1990), (4) activate I_A and I_K K^+ channels (Israël *et al.*, 1987; Lledo *et al.*, 1992) and (5) probably directly inhibit the final step of prolactin exocytosis in lactotroph cells. All these events are blocked by PT and are therefore mediated by either Gi_1, Gi_2, Gi_3, Go_1 or Go_2. Since we have already indicated that there are two D_2-dopaminergic receptors (D_2A and D_2B)

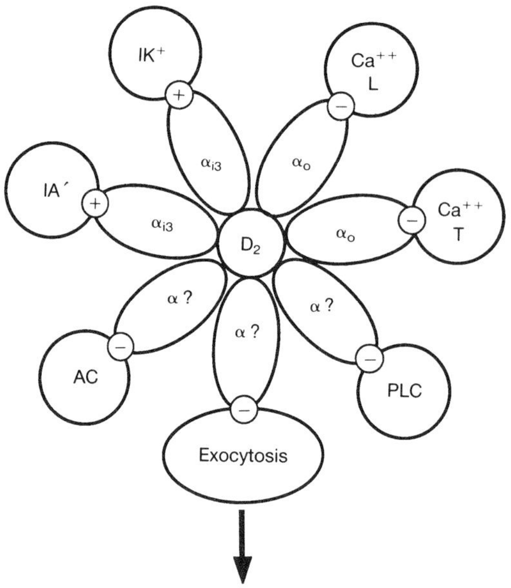

Figure 11.3. The D_2 receptors and its coupling to effectors by G_α-proteins. IA and IK, transient and sustained K^+ channels, respectively; Ca^{2+}L and Ca^{2+}T, voltage-sensitive channels of the L and T type, respectively; AC, adenylyl cyclase; PLC, phospholipase C.

(Andersen *et al.*, 1990) which differ in the domain of interaction with G-proteins (third intracellular loop), it is possible, but not demonstrated, that these receptors are specialized in triggering some of the events described in Fig. 11.3.

This example illustrates the problem of cross-talk between receptors, G-proteins and effectors. Are dopaminergic receptors able to activate only one G-protein which will modify the activity of several effectors or are dopaminergic receptors able to stimulate several G-proteins (all sensitive to PT), each of them modifying the activity of a specific effector? This problem will be discussed in Section 4.2.

3.2 *Advantages and disadvantages of heterologous transfections to solve the problem of cross-talking*

Obtaining genes for monomeric G-protein-coupled receptors or subunits of multimeric receptors is a big advantage in the understanding of receptor function. However, unexpected difficulties arise from the diversity of subunits which constitute ionotropic receptors or transducing components associated with monomeric G-protein-coupled receptors. Numerous ionotropic receptors can be theoretically generated into eukaryotic cells or *Xenopus* oocytes having different physiological and pharmacological properties depending on the association of subunits the researcher has decided to transfect (Boulter *et al.*, 1990; Luetje *et al.*, 1990; Olsen & Tobin, 1990; Sommer *et al.*, 1990). However, these types of experiments provide interesting information and the recent demonstration of the unique pharmacological property (interaction with RO 15-45-13, a drug antagonizing alcohol-induced impairment in motor performance) of the $GABA_A$ receptor obtained by transfection of $\alpha_6\beta_2\gamma_2$- subunits, as already discussed, is a perfect illustration of the interest of this method. However, one must be careful since these transfections do not always reproduce the compositions of natural receptors. A comparison between the physiological responses obtained in transfected cells with those obtained in a given natural preparation is required.

When considering the G-protein-coupled receptors, the difficulties are similar. Due to the great degree of homology between these G-protein-coupled receptors and G-proteins, transfections of a given receptor in a cell with its own set of G-proteins are likely to give non-natural coupling, especially if high concentrations of receptors are expressed. To give only one example, 5-HT_{1A} receptors have been shown to be negatively coupled to adenylyl cyclase and positively to phospholipase C when transfected in HeLa cells (Fargin *et al.*, 1989). We found only the negative coupling to adenylyl cyclase when the same cDNA clone was transfected into NIH-$3T_3$ fibroblasts (work in preparation). In contrast, in the brain, 5-HT_{1A} has been demonstrated to be negatively coupled to both adenylyl cyclase and phospholipase C (Claustre *et al.*, 1988; Dumuis *et al.*, 1988). These different results are probably due to the fact that the cells, in which 5-HT_{1A} receptors are expressed, contain different G-proteins or phospholipase C of a different nature.

3.3 *Methods which could be developed to analyse the problem of cross-talking*

The preceding analysis indicates that after the 'molecular taxonomy' consisting of the description of the puzzle's elements, we would have to analyse how they are associated in cells to trigger specific transducing functions. Transfection experiments will give a few clues but not all of them. Several other techniques have been developed to approach the problems which have arisen.

3.3.1 IMMUNOLOGICAL METHODS

The use of very specific antibodies directed toward specific epitopes of receptors or receptor subunits is certainly one of the more promising areas for solving the problems:

1 Immunolocalization of these receptors or receptor subunits, at the photonic or electron microscopic level (using immunogold techniques), can certainly be of great help in understanding the receptor and receptor subunit composition of a given cell.

2 Immunoprecipitation is indeed the method of choice to determine the subunit composition of a receptor and even the nature of the proteins associated with such receptors.

3 Immunoneutralization after injection of monospecific specific antibodies (monoclonals or polyclonals directed against synthesized peptide epitopes) into cells (microinjection or injection using the patch-clamp technique) can certainly be of great help to know 'who is talking to whom'.

We have recently shown (Lledo *et al.*, 1992) that the injection of antibodies against the α-subunit of Go, but not antibodies against the α-subunits of Gi_1, Gi_2 and Gi_3, block the coupling of D_2 receptors to T- and L-voltage-sensitive Ca^{2+} channels, but not the coupling of D_2 receptors to K^+ channels. On the contrary, antibodies against the α-subunits of Gi_3 but not antibodies against the α-subunit of

Gi_1 and Gi_2 or Go block the coupling of D_2 receptors to K^+ channels of I_A and I_K subtypes (Fig. 11.3). It is likely, but yet to be demonstrated, that the α-subunit which couples D_2 receptors to adenylyl cyclase is either αi_1 or/and αi_2. The nature of the D_2-receptor-activated α-subunit which is PT sensitive and controls the exocytosis at the final step is unknown (see Fig. 11.3).

3.4 *Antisense neutralization approach*

One of the methods which can be used selectively to suppress one protein from the complex network of transduction pathways in order to understand its function is the mRNA antisense approach. The principle is simple: the introduction of antisense RNA sequences, which hybridize with complementary mRNA sequences in a cell, result in hybridization arrest of translation (Green *et al.*, 1986; Marcus-Sekura, 1988). It has recently been suggested that RNAase H may play a role in this mechanism by cleaving the duplexes formed between RNA and antisense oligodeoxyribonucleotide (Minshull & Hung, 1986). Oligonucleotides can be introduced but some studies may be hampered by the relatively short half-lives of these compounds. In order to increase their metabolic stability, chemical modifications of nucleotides will certainly be used in the future. The length should be 12–15 bp and highest efficacy is obtained when sequences start at the initial methionine or overlap the initial methionine. Antisense oligonucleotides have been used predominantly in animal cell-culture systems. Blocking the synthesis would be efficient in suppressing the cellular content of a protein only if the turnover of this protein is rapid.

Several techniques could be used to trigger antisense oligonucleotides into cells:

1 The direct incubation of the cells with a high concentration of the nucleotide (5–200 μmol/l). This could be a relatively expensive experiment if many cells were required to test the activity of the protein one wanted to eliminate.

2 The direct microinjection of oligonucleotides by microinjection or by injection using the patch-clamp technique (Kluess *et al.*, 1991).

3 Introduction of plasmid constructs transiently or stably transferred to synthesize the antisense RNA (Marcus-Sekura *et al.*, 1988). It is likely that a strong promoter will be required. Since the mechanisms of hybridization-arrest are not well known, it has been reported that a large excess of a full-length antisense RNA is not necessarily sufficient to cause inhibition of gene expression.

The future will certainly provide us with chemically modified oligonucleotides. The poly(L-lysine) conjugates have demonstrable antiviral activity at concentrations as low as 100 nmol/l. This enhancement may be due to an increased cellular uptake of the oligonucleotides.

3.5 *Specific genetic deletion*

The introduction of additional genes coding for a given protein of the 'puzzle' into the hereditary genome in order to establish 'transgenic animals' will certainly be of interest in the future, especially to recognize the sequences of the genes which control its tissue-specific expression and its regulations. This can be done by introducing the genes into the egg by microinjection or by retrovirus-deficit recombinants. This will result in an untargeted introduction of genes into the genome.

In addition to this already classical method, one of the most promising new techniques in our understanding of the role of a given receptor or transducing molecule will be the use of targeted disruption of the gene coding for these proteins (Rossant, 1990). This method consists of the following:

1 Genetically modifying pluripotent embryonic stem cells using a double selection screening providing a way of selecting the cells in which homologous recombination has occurred (complete substitution of the natural gene by the introduced mutated gene) (see Thomas & Capecchi, 1990).

2 Introducing these cells with the specific mutation of one gene into mouse blastocytes.

3 Obtaining chimeric mice that transmit the mutant allele to its progeny. Then, using classical genetic methods, homozygous or heterozygous mice mutants can be selected.

The targeted disruption of *hprt*, *int2*, *int1*, *en-2*, *c abl*, *IGF-II* and β_2-microglobulin genes has been reported (Thomas & Capecchi, 1990). This is the method of choice in understanding the role of a given receptor, a G-protein or an effector, etc. at the cellular as well as the physiological level of the entire animal. This is also an excellent method for obtaining animal models of human genetic diseases.

4 Looking for new drugs that act specifically on different components of signal transduction

4.1 *Drugs at the receptor level*

We have seen that receptors coupled to G-proteins (those cloned so far) have a similar seven-transmembrane structure. They are likely to be derived from a common ancestor protein. Therefore, these homologies may explain why it has always been difficult to find antagonists which have an exclusive specificity for a given receptor. For example, it was a surprise to find that β-adrenergic blocking agents were also excellent 5-HT_1 antagonists and could also be partial agonists at 5-HT_{1D} receptors (Schoeffter *et al.*, 1988). Indeed, this cross-specificity has a molecular basis, since the 5-HT_{1A} receptor gene has been isolated by screening a human genomic library at low stringency with a full-length β_2-adrenergic receptor clone (Kobilka *et al.*, 1987). In several amino acid stretches, the 5-HT_{1A} receptor more closely resembles the β_2-adrenergic receptor than it does the other cloned 5-HT receptor (Hartig, 1989). For example, in the central part of the transmembrane helix VI, β_2-adrenergic receptors and 5-HT_{1A} receptors are identical in 18 out of 21 amino acids, while in the same region, the 5-HT_{1A} receptor only shares a maximum of 10 out of 27 amino acids with 5-HT_{1C} and 5-HT_2 receptors (Hartig, 1989). The homologies between the central parts of transmembrane helix VI of 5-HT_{1A} and β_2-adrenergic receptors are the following:

5-HT_{1A}: T L G I I M G T F I L C W L P F F I V A L
β_2: T L G I I M G T F T L C W L P F F I V N I

The problems facing the chemist in the future will be even more complex than in the past. Indeed, they will have to find specific drugs for subtypes of receptors which have been found on the basis of their homologies. For example, and only until very recently, the goal was to find specific neuroleptics able to discriminate between D_1- and D_2-dopaminergic receptors. Researchers have now to provide drugs which can act specifically on D_1, D_2 and D_3 receptors. It has been proposed that the D_3-dopaminergic receptor is implicated in schizophrenic symptoms and the D_2 receptor in the control of extrapyramidal locomotor activities (Sokoloff *et al.*, 1990). Therefore, a specific D_3 antagonist will be expected to have fewer undesirable parkinsonian-like effects than a specific D_2 antagonist. As discussed above, dozens of heteromultimeric GABA/benzodiazepines, glutamatergic or nicotinic receptors might exist, each of them having a different subunit composition and in many cases, a different pharmacological effect. It is certainly a fascinating challenge for chemists.

4.2 *Drugs at the intracellular level: A potential gold mine for future drugs*

Many of the drugs used in modern medicine act at the extracellular level on receptors or channels. It is likely that new drugs will be targeted to interact at the level of signal-transduction components localized at the intracellular level. Already potential loci can be proposed:

1 If one wants specifically to block the D_2A- or the D_2B-dopaminergic receptors, the only possible target is the 29-amino acid fragment present on the third intracellular loop (i_3) of D_2A and absent in D_2B (Andersen *et al.*, 1990). Indeed the recognition sites of the two proteins being identical, it is unlikely that a drug acting on these sites will specifically block one of them.

2 The third intracellular loop (i_3) of the G-protein-coupled receptors may be an interesting target for drugs. Mutagenesis studies indicate that deletion of i_3 in rhodopsin, β-adrenergic receptors and muscarinic receptors suppress their coupling to G-proteins (Fig. 11.1). Smaller deletions indicate that the eight residues at the N-terminus of the loop or the 12 residues at the C-terminus of this loop attenuate the coupling (Cheung *et al.*, 1989) (Fig. 11.1). In contrast, deletion of internal parts of the loop i_3 is without effect on the coupling. The region of the C-terminal tail adjacent to the transmembrane VII is also implicated (O'Dowd *et al.*, 1989). Similarly, a 16–17-amino acid sequence at the N-terminal portion of the i_3 is highly conserved in m_1, m_3 and m_5 muscarinic receptors (coupled to phosphoinositide (PI) hydrolysis) but not in m_2 and m_4 muscarinic receptors (negatively coupled to adenylyl cyclase). Replacement of the i_3 domain of m_2 by the corresponding m_3 sequence led to a chimera (m_2/m_3-i_3) which stimulates PI hydrolysis. Introduction of a short stretch of 16–17 residues from m_3 at the N-terminal portion of i_3 of m_2 (m_2/m_3-17aa) produces a chimera which is as potent as m_3 in stimulating PI hydrolysis (Wess *et al.*, 1989). The N- and C-termini of i_3 would form amphipathic α-helices (Strader *et al.*, 1989). This feature is interesting since the amphipathic peptide, mastoparan, is able to stimulate the Go protein directly (Higashijima *et al.*, 1988). Similarly, substance P (but not neurokinin A nor B) at concentrations exceeding those specifically interacting with their receptors (1–10 μmol/l), as well as the compound 48/80, are able directly to stimulate Go/Gi proteins (Mousli *et al.*, 1990). These compounds indeed have amphipathic structures. All these substances (mastoparan, substance P, compound 48/80, bradykinin) are able to stimulate histamine release through an exocytotic process and at concentrations similar to those used directly to stimulate G-proteins *in vitro*. Furthermore, these cellular effects are blocked by PT, suggesting that their actions are mediated by PT-sensitive G-proteins (Mousli *et al.*, 1989).

Other putative drugs acting on effectors are already known: forskolin, a hypotensive agent acting on adenylyl cyclase, phorbol esters or gangliosides interacting with protein kinase C. Many new drugs will be arriving, including specific activators or inhibitors of regulatory nucleotide sequences of genes.

5 Conclusion

By showing the complexity of the transduction mechanisms, this brief review should not discourage future researchers who want to understand the problem of cellular communication. Indeed, one can wonder if this complexity is not such that each individual might have its own network of transduction processes, in particular in the nervous system. This network is certainly an adaptive construction. Are not the great diversity of the transduction mechanism components, their functional associations and their cross-talk the molecular bases of individuality?

6 References

Andersen, P.H., Gingrich, J.A., Bates, M.D., Dearry, A., Falardeau, P., Senogles, S.E. & Caron, M. (1990). *TIPS*, **11**, 231–6.

Asano, T., Morishita, R., Kobayashi, T. & Kato, K. (1990). FEBS *Lett.*, **266**, 41–4.

Birnbaumer, L., Abramowitz, J. & Brown, A.M. (1990). *Biochem. Biophys. Acta*, **1031**, 163–224.

Bockaert, J., Sebben, M. & Dumuis, A. (1990). *Mol. Pharmacol.*, **37**, 408–11.

Bonner, T.I., Buckley, N.J., Young, A.C. & Brann, M.R. (1987). *Science*, **237**, 527–32, 1556–628.

Boulter, J., Hollmann, M., O'Shea-Greenfield, A., Hartley, M., Deneris, E., Maron, C. & Heinemann, S. (1990). *Science*, **249**, 1033–7.

Changeux, J.P., Giraudat, J. & Dennis, M. (1987). *TIPS*, **8**, 459–72.

Cheung, A.H., Sigal, I.S., Dixon, R.A.F. & Strader, C.D. (1989). *Mol. Pharmacol.*, **34**, 132–8.

Christie, M.J., North, R.A., Osborne, P.B., Douglas, J. & Adelman, J.P. (1990). *Neuron*, **4**, 405–11.

Claustre, Y., Benavides, J. & Scatton, B. (1988). *Eur. J. Pharmacol.*, **149**, 149–53.

Craig, D.A. & Clarke, D.E. (1990). *J. Pharmacol. Expl. Ther.*, **252**, 1378–86.

Dumuis, A., Sebben, M. & Bockaert, J. (1988). *Mol. Pharmacol.*, **33**, 178–86.

Dumuis, A., Sebben, M. & Bockaert, J. (1989). *Naunyn Schmiedeberg's Arch. Pharmacol.*, **340**, 403–10.

Dumuis, A., Bouhelal, R., Sebben, M., Cory, R. & Bockaert, J. (1988). *Mol. Pharmacol.*, **34**, 880–7.

Enjalbert, A. & Bockaert, J. (1983). *Mol. Pharmacol.*, **23**, 576–84.

Enjalbert, A., Guillon, G., Mouillac, B., Audinot, V., Rasolonjanahary, R., Kordon, C. & Bockaert, J. (1990). *J. Biol. Chem.*, **265**, 18 816–22.

Fargin, A., Raymond, J.R., Regan, J.W., Cotecchia, S., Lefkowitz, R.J. & Caron, M.G. (1989). *J. Biol. Chem.*, **264**, 14 848–52.
Gaddum, J.H. & Picarelli, Z.P. (1957). *Br. J. Pharmacol. Chemother.*, **12**, 323–9.
Giannattasio, G., De Ferrari, M.E. & Spada, A. (1981). *Life Sci.*, **28**, 1605–11.
Green, P.J. Pines, O. & Inouye, M. (1986). *Ann. Rev. Biochem.*, **55**, 569–97.
Harris-Warrick, R.M., Hammond, C., Paupardin-Tritsch, D., Homburger, V., Rouot, B., Bockaert, J. & Gerschenfeld, H.M. (1988). *Neuron*, **1**, 27–32.
Hartig, P.R. (1989). *TIPS*, **10**, 64–9.
Hescheler, J., Rosenthal, W., Trautwein, W. & Schultz, G. (1987). *Nature* (*London*), **325**, 445–7.
Higashijima, T., Uzu, S., Nakajima, Y. & Ross, E.M. (1988). *J. Biol. Chem.*, **263**, 6491–6.
Hollmann, M., O'Shea-Greenfield, A., Rogers, S.W. & Heinemann, S. (1989). *Nature* (*London*), **342**, 643–6.
Israël, J.M., Kirk, C. & Vincent, J.D. (1987). *J. Physiol.*, **390**, 1–22.
Jan, L.Y. & Jan, Y.N. (1990). *TINS*, **13**, 415–19.
Keinänen, K., Wisden, W., Sommer, B., Werner, P., Herb, A., Verdooln, T.A., Sakmann, B. & Seeburg, P.H. (1990). *Science*, **249**, 556–60.
Kleuss, C., Hescheler, J., Ewel, C., Rosenthal, W., Schultz, G. & Wittig, B. (1991). *Nature*, **353**, 43–8.
Kobilka, B.K., Frielle, T., Collins, S., Yang-Feng, T., Kobilka, T.S., Francke, U., Lefkowitz, R.J. & Caron, M.G. (1987). *Nature* (*London*), **329**, 75–9.
Krupinski, J., Coussen, F., Bakalyar, H.A. *et al.* (1989). *Science*, **244**, 1558–64.
Lledo, P.M., Israël, J.M. & Vincent, J.D. (1990). *Brain Res.*, **528**, 143–7.
Lledo, P.M., Hamburger, V., Bockaert, J. & Vincent, J.D. (1992). *Neuron*, **8**, 455–63.
Lüddens, H., Pritchett, D.B., Köhler, M., Killisch, I., Keinänen, K., Honyer, H., Spengel, R. & Seeburg, P.H. (1990). *Nature (London)*, **346**, 648–51.
Luetje, C.W., Patrick, J. & Séguéla, P. (1990). *FASEB*, **4**, 2753–60.
Marcus-Sekura, C.J. (1988). *Anal. Biochem.*, **172**, 289–95.
Matsuda, L.A., Lolait, S.J., Brownstein, M.J., Young, A.C. & Bonner, T.I. (1990). *Nature* (*London*), **346**, 561–4.
Mattera, R., Graziano, M.P., Yatani, A., Zhou, Z., Craf, R., Codina, J., Birmbaumer, L., Gilman, A. & Brown, A.M. (1989). *Science*, **243**, 804–7.
Minocherhomje, A.M., Selfe, S., Flowers, N.J. & Storm, D.R. (1987). *Biochemistry*, **26**, 4444–7.
Minshull, J. & Hunt, T. (1986). *Nucl. Acids Res.*, **14**, 6433–51.
Moriaty, T.M., Padrell, E., Carty, D.J., Omri, G.I., Landau, E.M. & Iyengar, R. (1990). *Nature* (*London*), **343**, 79–82.
Mousli, M., Bronner, C., Bueb, J.L., Tschirhart, E., Gies, J.P. & Landry, Y. (1989). *J. Pharmacol. Expl. Ther.*, **250**, 329–35.
Mousli, M., Bronner, C., Landry, Y., Bockaert, J. & Rouot, B. (1990). *FEBS Lett.*, **259**, 260–2.
Nishizuka, Y. (1988). *Nature* (*London*), **334**, 661–5.
Northrup, J.K., Sternweis, P.C., Smigel, M.D., Schleifer, L.S., Ross, E.M. & Gilman, A.G. (1980). *Proc. Natl. Acad. Sci. USA*, **77**, 6516–20.
O'Dowd, B.F., Hnatawich, M., Caron, M.G., Lefkowitz, R.J. & Bouvier, M. (1989). *J. Biol. Chem.*, **264**, 7564–9.
Olsen, R.W. & Tobin, A.J. (1990). *FASEB*, **4**, 1469–80.
Peralta, R.G. (1987). *EMBO J.*, **6**, 3923–9.
Pritchett, D.B., Lüddens, H. & Seeburg, P.H. (1989). *Science*, **245**, 1389–92.
Rossant, J. (1990). *Neuron*, **2**, 323–34.
Schmidt, A.W. & Peroutka, S.J. (1989). *FASEB*, **3**, 2242–9.
Schoeffter, P., Waeber, C., Palacios, J.M. & Hoyer, D. (1988). *Naunyn-Schmiedeberg's Arch. Pharmacol.*, **337**, 602–8.
Schofield, P.R. (1990). *TIPS*, **10**, 476–8.
Sokoloff, P., Giros, B., Martes, M.P., Bouthenet, M.L. & Schwartz, J.C. (1990). *Nature* (*London*), **347**, 146–51.

Sommer, B., Keinanen, K., Verdoorn, T.A., Wisden, W., Burnashev, N., Herb, A., Köhler, M., Takagi, T., Sakmann, B. & Seeburg, P.H. (1990). *Science*, **246**, 1580–5.

Strader, C.D., Sigal, I.S. & Dixon, R.A.F. (1989). *TIPS* (Suppl.), 26–30.

Thomas, K. & Capecchi, M.R. (1990). *Nature* (*London*), **346**, 847–50.

Timpe, L.C., Jan, Y.N. & Jan, L.Y. (1988). *Neuron*, **1**, 659–67.

Von Dongen, A., Codina, J., Olate, J., Mattera, R., Joho, R., Birnbaumer, L. & Brown, A.M. (1988). *Science*, **242**, 1988–92.

Wess, J., Brann, M.R. & Bonner, T.I. (1989). *FEBS Lett.*, **258**, 133–6.

Zagotta, W.N., Germaeraad, S., Garber, S.S., Hoshi, T. & Aldrich, R.W. (1989). *Neuron*, **3**, 773–82.

Part 3
Pharmacophore Studies
Pharmacophore Mimickry Modelling

12 The Future of Computer-aided Drug Design

G.R. MARSHALL
Center for Molecular Design, Washington University, Box 1099, 1 Brookings Drive, St Louis, MO 63130, USA

1 Introduction

Any attempt to forecast the future of computer-aided drug design can only be based on extrapolation of the trends perceived in recent changes in the field. Over the past decade, there has been an increasing acceptance of computational chemistry and molecular graphics as tools in structure–activity analyses and in the design of new therapeutics. While the number of success stories where a significant impact of this technology can be demonstrated is limited due to the lengthy time delay between research and disclosure in the pharmaceutical industry, the current impact can be clearly perceived by comparison of a current issue of the *Journal of Medicinal Chemistry* with one of a decade ago. With enhanced familiarity with these tools, however, comes a realization of their current limitations. Much has been promised, explicitly or otherwise, that cannot be delivered in general, except in those rare cases where the limiting assumptions of the computational model happen to coincide with the molecular reality.

In the course of this candid discussion, it will be necessary to comment, often unfavourably, about our current situation in order to contrast it with our hopes and expectations for the future.

2 Current status of computer-aided drug design

First, our current level of expertise is such that problem selection dominates the successful application of the techniques available. One is limited to extraction of information from experimental data, either pharmacological or spectroscopic (X-ray crystallography or nuclear magnetic resonance (NMR), to derive an intellectual model of the therapeutic target. If one has ready access to sufficient quantities of the receptor (the term is used generically to indicate that macromolecule within the body which forms a complex with the drug responsible for the desired therapeutic effect), one can attempt to determine a three-dimensional model experimentally, either by X-ray crystallography or by multidimensional NMR approaches (Fesik *et al.*, 1988). There are always interesting problems, however, where technical limitations preclude such direct experimental approaches. Angiotensin-converting enzyme (ACE) is a clear example where a significant effort, without success, was made to obtain crystals of adequate quality because of the enormous interest in development of inhibitors by the pharmaceutical industry. If one has a high-resolution crystal structure of the receptor under study, then sophisticated computational approaches, such as thermodynamic cycle perturbation (Kollman & Merz, 1990), become applicable, and one can hope to calculate the difference in affinity between a standard lead compound (whose affinity is known) and one postulated for synthesis with some degree of accuracy assuming that the difference in structure between the two compounds is not great. Currently, such an approach is emotionally very attractive, but only limited examples exist with the success stories coming primarily on model systems. Computationally, this is a herculean task if one is adequately to sample configurational space in order to derive good thermodynamic values.

If, however, one is willing to concentrate on design of novel chemical structures which are capable of interacting with the known receptor structure, then our methodology is much more robust as the geometric foundations of our science are much firmer than the thermodynamics ones. Techniques which exploit this aspect of the problem area, i.e. design of novel structures to interact with a known receptor site, are becoming more available and show promise. CAVEAT from the Bartlett group (Shea *et al.*, 1990) searches for chemical scaffolds in the Cambridge Crystallographic Database with the right geometrical properties for positioning pharmacophoric groups correctly for recognition. Kuntz and his colleagues (DesJarlais *et al.*, 1990b) have developed a cavity-matching algorithm, DOCK, to find molecules of the correct shape to interact with a receptor cavity. An interesting example is the prediction that haloperidol would show activity as an inhibitor of HIV protease. Visualization techniques (Ho & Marshall, 1990) have been developed to simplify the complex representations of ligand–receptor interactions in order to screen potential candidates for synthesis more effectively.

These approaches implicitly assume that the observed receptor cavity has some physical stability, i.e. a static view, and significant induced fit between ligand and receptor is not operative for each ligand molecule. While there is no guarantee that this is true for any particular case under study, the specificity seen in biological systems argues that a receptor site has some functional significance in imposing its unique steric and electrostatic characteristics in the molecular recognition and selection process. One must always be prepared, however, for binding to sites other than that anticipated, and possible exposure of cryptic sites which are not observed in the absence of the ligand. This problem has been clearly illustrated by Perutz *et al.* (1986), who showed that several compounds designed to bind to specific sites on haemoglobin actually bound within the hydrophobic core of the protein with re-organization of packing of amino acid side-chains. This clearly is due to the dynamic nature of the receptor. The current computational limits in molecular dynamics simulations restrict the chance of uncovering such alternative binding modes in our studies. In other words, if we can assume the binding mode of our candidate drug is nearly identical to that of a known compound, then we have a legitimate basis for thermodynamic perturbation calculations (Kollman & Merz, 1990). Multiple or alternate binding modes are a major problem. Naruto *et al.* (1985) have demonstrated a systematic approach to the determination of productive binding modes for mechanism-based inhibitors which could select starting structures for complexes for molecular dynamics simulations. Combinations of methods, such as Monte Carlo or systematic search, to generate multiple starting configurations for simulations to improve sampling and thermodynamic reliability will increase as adequate computational power to support these hybrid approaches becomes more readily available.

Three-dimensional structures from sequence data require a solution to the protein-folding problem. Incremental improvements in modelling by homology can be expected as the database of known structures increases and as our computational approaches improve their underlying heuristic assumptions. The goal, of course, is to come close enough with a predicted structure so that a minimization algorithm will locate the global minima and ligand–receptor modelling can begin with some confidence. Interpolation between known structures is becoming relatively routine with good software tools (Blundell *et al.*, 1987) being developed to assist this goal. Inherent uncertainties (Rooman & Wodak, 1988) in secondary-structure predictive schemes make it unlikely that sequence information will be routinely transformed into reliable structures, at least until a new conceptual breakthrough is made. Again, efforts to extrapolate novel structures from the database of known 3-D structures are inherently more reliable.

Many technical limitations remain to be overcome. Adequate modelling of electrostatics (Harvey, 1989; Davis & McCammon, 1990) remains elusive in many experimental systems of interest such as membranes. Newer derivations of force fields, such as MM3 (Allinger *et al.*, 1990), are attempting to represent the experimental data more accurately, while others include a broader spectrum of chemistry such as metals (Aqvist & Warshel, 1990; Vedani & Huhta, 1990; Allured *et al.*, 1991). Combinations of molecular mechanics with quantum chemistry (Arad *et al.*, 1990; Field *et al.*, 1990) are clearly necessary for problems in which chemical transformations are involved. Rather amazing agreement between calculation and experiment has been reported (Houk *et al.*, 1990) although there is some controversy (Menger & Sherrod, 1990). In any case, this is another area of rapid growth as adequate computational resources become available.

3 Interpolation versus extrapolation

Experience has shown that a conservative approach in science, which uses interpolation to predict activity of compounds within the boundaries of the parameter space explored by generation of an experimental data set, gives more reliable predictions than extrapolation. To forecast the future of molecular design, one must attempt to extrapolate from current trends and any predictions must, therefore, be more suspect. Extrapolation is also clearly more accurate the shorter the time frame over which the prediction is made, and most of my comments are reserved for the first decade of the 21st century.

3.1 *Experimental data*

A massive increase in experimental data available on which to perform analyses and generate hypotheses is certain. As enzymatic and bioassays are automated and synthetic capability increases to generate more compounds of diverse chemical structure, data management will play an increasingly important role. Integration of synthetic, analytical, conformational and biological data in an accessible and comprehensive database will provide the source of information for an increasingly more sophisticated approach based on statistical correlational analysis. Experimental determination of target structures will continue to dominate as NMR methodology (Fesk *et al.*, 1990) expands its capabilities, and robotics and array detectors increase the productivity of crystallographers. With NMR, each year sees the size limitation for solution structural determination increase as isotope-editing strategies (Fesk *et al.*, 1988) and pulse techniques advance. There are still many potential therapeutic targets which will elude solution NMR due to their physical state, i.e. membrane-bound or high-molecular-weight aggregates. In these cases, recent advances in solid-state NMR can provide (Marshall *et al.*, 1990) accurate determination of interatomic distances between nucleii of rare spins and crucial intra- and intermolecular distances determined under conditions where solution methodology cannot be applied. It should be possible, therefore, to check the validity of hypothetical conformations of drugs bound to integral membrane receptors which have been deduced from structure–activity data. Computational methods for interpretation of NMR data will become an adjunct of the molecular modelling tools of the near future, just as it currently exists for crystallography.

3.2 *Site models*

In many cases, if not most, pharmacology precedes structural data on the receptor and one must infer a model of the therapeutic target based on structure–activity data. Two classes of such site models exist: the pharmacophore model on which the initial formulation (Marshall *et al.*, 1979) of the Active Analog Approach was based, and the active-site model (Mayer *et al.*, 1987) which is the current preferred model. In simplest terms, the pharmacophore model assumes coincidence of functional groups in the set of compounds under analysis in order to minimize the degrees of freedom under consideration, while the active-site model assumes that the therapeutic target maintains functional groups in the same 3-D arrangement in order to conserve molecular recognition and specificity. The active-site model allows multiple binding models of the ligand to a rigidly determined receptor model. Because many problems do not have enough experimental data to distinguish between several self-consistent active-site models on a geometric basis, development of models which utilize binding data to distinguish between alternatives is of considerable value. In other words, one needs to utilize all the information in the experimental data. Three-dimensional, quantitative structure–activity relationships (3-D-QSAR) offer a solution to this problem. Comparative molecular field analysis (CoMFA) is the most focused and available example of such an approach (Clark *et al.*, 1990). In a recent analysis of 38 different chemical classes of ACE inhibitors, DePriest *et al.* (1990) found that a CoMFA analysis could distinguish between two alternatives for aligning the ACE inhibitors in an active-site model, based on significant differences in the predictive power of the resulting models derived from the two alternative geometries. Norinder (1990) has applied a similar statistical approach to corticosteroid binding with good predictive results. Alternative approaches to receptor-site modelling based on the distance geometry paradigm have been developed by Crippen and have been reviewed by Donne-Op den Kelder (1987).

An alternative to more analyses of a set of data to distinguish between hypotheses is to gather more experimental data on different compounds. Screening of diverse compounds in cellular and specific receptor-based assays will result in more structure–activity data, and new approaches which exploit the quantity of information available will be developed. Increasingly, novel structures as leads are being found by sophisticated screens of compound libraries and other sources, such as fermentation broths. The availability of a set of compounds of diverse chemical structures which interact with the same receptor is an ideal starting point for active-site analysis. As computational approaches to structure–activity analyses are primarily deductive in that they depend on transformation and extraction of information from an experimental set of observations, this increase in experimental data should complement the increase in screening by providing direction in setting priorities for compounds to be tested. This concept has already been demonstrated in several pharmaceutical compounds where identification of a hypothetical pharmacophore is coupled to selection of compounds for testing from the company's compound library.

3.3 *Knowledge-based approaches*

There is no doubt that heuristics has an important place in the development and improvement of computer-aided drug design tools and approaches. In other words, in order to simplify the computational complexity of a problem, one should incorporate into the software both chemical knowledge of the problem domain (Leach *et al.*, 1988) and the experience of an expert in terms of problem-solving strategies. A good example is found in the work of Perlman (1987) on generating 3-D models from molecular connectivities as embodied in the program CONCORD. Conformational analysis has also benefited from such an approach as exemplified in the work of Dolata *et al.* (1987), Leach *et al.* (1990b) and Leach & Prout (1990). Examples of areas which are likely to benefit from such developments are synthetic design strategies, alternative compound design based on bioisosteres, experimental design of compound series to be made and tested, as well as experimental protocols for optimal instrumentation use in structural determination by NMR and other analytical techniques. In other words, software can be developed which will help assist in branch-point decisions based on the particular information available, the desired information required, and the existing experience in similar problems.

The major challenge is the incorporation of current chemical knowledge within the software in a coherent and comprehensive manner so that the relevant experience of generations of chemists is available as one considers a particular problem. We are entirely too conditioned by our own particular set of experiences when we analyse a set of data and design an experiment. It is currently much easier simply to run a reaction than to carry out an exhaustive literature search to determine the reaction and its conditions which will minimize side-reactions, preserve chiral integrity, limit exposure to toxic reagents, or satisfy whatever arbitrary boundary conditions we establish. Access of relevant information is much too difficult at present, and one can anticipate that such access will have a major impact on the conduct of our research in the relatively near future.

3.4 *Computational constraints*

3.4.1 HARDWARE

The impact of computational chemistry has increased almost in direct proportion to the increase in available computational power. As capability improves, more and more of the limiting assumptions underlying the modelling can be removed and convergence with experimental observation anticipated. Distributed processing has become a reality and applications to molecular modelling are beginning to appear in the literature (Goodfellow *et al.*, 1990). As the inherent power of each microprocessor approaches that of our original supercomputers and arrays of such chips become readily available, we may be able to simulate molecular systems for sufficient time to sample adequately the ensemble of possible states, or alternatively, run enough simulations from different starting configurations to determine our sampling error. It would be naive, however, to assume that adequate computational power for realistic simulations of complicated molecular systems, such as membrane-bound receptors, will be available in the near future to model the binding of agonists and follow the subsequent transduction event.

Where the physical limitations of component density curtail the rapid increase in computational power is hard to predict accurately. Even if the rate of increase in computational power per component slows down, the cost per component is already sufficiently low that large arrays of such components designed optimally to solve a particular problem are feasible. Based on experience, one should focus on problems which are clearly beyond our computational capability so that their theoretical and algorithmic solution coincides with the availability of the necessary computational power. In other words, by the time one figures out the right approach to a complex and difficult problem, the increase in hardware capability has often transformed it from an impossible dream to a reality.

Enhancements in the user interface will continue although it is hard to predict how much increase in effective communication with users will occur, at least until we all have direct electronic interfaces to our CNS. There is little doubt that voice recognition and synthesis will become a commonplace mode of interaction. Three-dimensional displays based on holograms are technically feasible today. Having realistic images in realtime displaying molecular dynamics simulations does not offer much enhancement in insight, however. Simultaneous displays of multidimensional data in various transformations may be of more interest. The current problem is no longer one of the hardware being incapable of rapidly presenting adequately resolved images. One is now faced with analysis of huge quantities of computational data, and the problem is scientific; how does one distill and transform the information to present it in a manner whereby the user can quickly perceive its relevance to the question in hand?

3.4.2 SOFTWARE

Software is the real frontier. Most currently available software in computational chemistry is hardly more than a collection of functionalities, or research prototypes, testing new approaches. As increased applicability leads to a larger user community composed of the entire spectrum of the chemistry profession, then software will have to adapt to a more diverse set of users. Rather than the user learning to communicate with the rigidly formatted software, the programs will have to have a maleable interface which can interactively perceive the user's problem and expertise level and focus the responses on those aspects which are relevant in that problem's context. The comfort level of a user is related to experience, frequency of use, computer expertise, theoretical understanding of chemical and algorithmic foundation, etc. Both prompting and choice of functionality should be conditioned, not only by the user, but by the level of analysis of the particular problem under consideration. One does not need to be reminded of options which are inappropriate. By monitoring progress and classifying problem types, the software should be able to offer suggestions for proceeding, or overall strategies, based on an analysis of the data under consideration, the objectives, and other users' experience. HAL is not that far away, and should make an excellent research advisor. The power of the software should evolve as more problem types are solved and as the collective experience in applications increases. The software must become the repository of the codified experience of the medicinal chemistry community, and an active role of the software in collecting and codifying that experience will evolve. As an example of the type of approach which may prove useful, a self-organized knowledge base for organic chemistry has been designed (Wilcox & Levinson, 1986).

The rapid development of neural nets and the availability of enhanced computational hardware to run such programs on large data sets may alleviate much of the tedium of software development. An example is the ability of such a net to distinguish between complex carbohydrates based on their NMR spectra (Meyer *et al.*, 1991). Others are bringing this technology to bear on the protein-folding problem (Qian & Sejnowski *et al.*, 1988; Kneller *et al.*, 1990). By defining the desired functionality and supplying an adequately designed training set, one may be able to avoid the explicit analysis process which would be necessary before developing a conventional program to provide that function. This would mean that the chemist could concentrate on the logic and chemical basis of the approach, while leaving certain of the technical aspects of its implementation to neural nets. Of course, one sacrifices any important insights into the problem or related ones that might come from explicitly considering the problem, its theoretical foundations and alternative strategies for algorithm development. One can derive a functional module without gaining, or needing, insight into its internal algorithms.

A major need and, therefore, focus for the future is the area of scientific visualization. Almost all problems of interest involve multidimensional parameter space. Methods for presenting this information without losing the relevant correlations are essential in almost all areas of science. Use of statistical methodology to identify the principal components of such spaces is one way to reduce dimensionality (Hudson *et al.*, 1989). Other transformations of the data, such as Fourier analysis, need to be automatically invoked. A useful application (Dauber-Osguthorpe & Osguthorpe, 1990) of digital signal processing has been to extract important intramolecular anharmonic motions from molecular dynamics simulations analogous to normal mode analysis. Often we are looking for a signal in the midst of much noise. Only select areas of computational chemistry, such as spectroscopy, have much experience in this sort of approach. Ideally, the software will not only analyse the data, but find relevant correlations, and suggest self-consistent hypotheses. The role of the chemist will be to bring his or her expertise, experience and scientific intuition in establishing priorities for the testing of these hypotheses.

Currently, we do not have the capability to solve problems, in general, for the set of solutions consistent with the set of experimental constraints. Progress in using systematic search to identify all the conformations consistent with the experimental data in the case of cyclosporin A has been reported by Beusen *et al.* (1990) where an additional family of conformations was found which had not been discovered by either distance geometry or constrained molecular dynamics analyses of the experimental data. We are often blinded by the first self-consistent solution to a problem we find, while a real appreciation of the underdetermined nature of the problem would give us a more ego-free basis for thoughtful analysis of the next step.

4 Major development areas

In terms of drug design itself, there are several specific areas where future developments have the potential of significant impact in the next few years and certainly within a decade or two.

4.1 *Side-effects*

As our increased understanding of the underlying physiology and pharmacology in disease states leads to a more precise definition of the appropriate therapeutic target, one can design agents which are inherently more specific. One needs to define the pathology at the molecular level, and not simply at the functional level, in order to reap the full benefits of the molecular approach. It is not sufficient to determine that some anomaly exists in K^+-channel regulation; the question is which one of the myriad number of K^+-channel subtypes is manifesting the difference and in which tissues is this subtype expressed. What one often overlooks is the fact that increased complexity of the system offers more opportunity for selectivity and, ultimately, specificity. Medicinal chemistry is the history of serendipitous discovery of subtle differences in receptor subtypes based on their interactions with exogenous substances.

The use of biological screening in the pharmaceutical industry has a productive history and will continue to be a productive area in the foreseeable future. There has been one negative side-effect from this mode of operation resulting in enhanced side-effects of the compounds prepared. Medicinal chemists learned very early that molecules with several common pharmacological groups, i.e. basic amine, aromatic ring, etc., separated by flexible links were often active in one of the biological assays, even if it was not the one for which the compound was prepared. The compound files of many of our major pharmaceutical companies contain many molecules of this class. These compounds are multipotential and can interact with a variety of receptors depending on accessibility and induced fit. In fact, some of the beneficial therapeutic profiles seen in successful drugs may be due to interaction at a variety of sites. By determining the precise requirements for recognition at a particular therapeutic target, however, one can design relatively rigid molecules, frozen into the biologically relevant conformation, which have a significantly reduced likelihood of side-effects. Combination of such specific drugs could be used in optimal mixtures in order to produce a desired therapeutic profile rather than depend on a serendipitous ratio of affinities for different receptors inherent in one molecule.

As our knowledge of receptor subtypes and their specific requirements for recognition and activation increases, databases encoding this information will be established. This will enable compounds to be designed which are prescreened during the design process to be selective for one receptor subtype while possessing attributes which preclude their interaction with other receptors. The use of steric differences (Sufrin *et al.*, 1981) in receptor subtype selectivity has been demonstrated by Hibert *et al.* (1988) to be such an exploitable difference in work on the serotonin (5-HT) receptors.

It has become quite evident that much of a molecule acts simply as a scaffold to align the appropriate groups in the 3-D arrangement which is crucial for molecular recognition. By understanding the pattern for a particular receptor, one can transcend a given chemical series by replacing one scaffold with another of geometric equivalence. This offers a logical way to change dramatically the side-effect profile of the drug as well as its physical and metabolic attributes. Various software tools are already under development to assist the chemist in this design objective. Lewis & Dean (1989a,b) have described their approaches to molecular templates. An alternative approach, BRIDGE (Dammkoehler *et al.*, unpublished), is based on geometric generation of possible cyclic compounds as scaffolds given the constraints of chemistry which the chemist is willing to consider. CAVEAT is a program developed by Bartlett (Shea *et al.*, 1990) to find cyclic scaffolds by searching the Cambridge Crystallographic Database for the correct vectorial arrangement of appended groups. All of these approaches attempt to help the chemist discover novel compounds which will be recognized at a given receptor. Van Drie *et al.* (1989) have described a program ALADDIN for the design or recognition of compounds that meet geometric, steric or substructural criteria, and Bures *et al.* (1991) have described its successful application to the discovery of novel auxin transport inhibitors. As our knowledge base of receptors grows, such tools will prove increasingly useful. The ability to transcend the chemical structure of lead compounds while retaining the desired activity should dramatically improve the ability to design away undesirable side-effects.

4.2 *Metabolism*

Our knowledge base on metabolic pathways and enzymatic specificity of the relevant enzymes is expanding as well. Already, efforts to codify the known enzymatic transformations as a way of predicting metabolites have met with some success (Darvas, 1988; Seressiotis & Bailey, 1988). The use of the software developed for synthetic chemistry with the appropriate set of biological transforms (Gifford *et al.*, 1991) would appear to be a fruitful approach (Wipke & Hahn, 1989). One can envisage that a compound which is a candidate for synthesis would be submitted to such a program for a prediction of half-life and metabolites. Each metabolite would subsequently be screened for potential activity against the library of receptors whose recognition properties had been determined. Those compounds whose inherent properties at the target receptor were of sufficient potential would be synthesized, if none of their predicted metabolites was thought to possess undesirable side-effects.

4.3 *Potency*

The *in vitro* potency of a compound depends on a variety of properties:
1 its intrinsic affinity for the target receptor;
2 its physical properties which will effect distribution and alter excretion;
3 the possible role of alternative binding sites in the body which can alter distribution and half-life;
4 the rate metabolism and its effects on excretion.

As was mentioned above, computational chemistry has made great strides in calculating the affinity for a known site, but considerable improvements still need to be made before this becomes a routine and reliable predictive method. The ability to predict bioavailability, distribution and metabolism is certainly a long-term goal, but our fundamental understanding of the processes and characterization of the biological systems involved is very primitive. This is certainly an area for development in the 21st century.

4.4 *Synthesis*

The utility of databases and heuristic approaches to design of synthetic pathways has been amply demonstrated (Hendrickson & Toczko, 1989). There are two very different approaches to the synthesis design problem. One depends on codifying chemical transformation and the accumulation of such transformations in a library (Pensak & Corey, 1977; Wipke *et al.*, 1977); the other is based on a more theoretical definition of chemical transformations and can lead to prediction of novel chemical reactions (Brandt *et al.*, 1977; Jorgensen *et al.*, 1990). The power of the more pragmatic approach depends on the quantity of known chemical transforms which has been encoded. The power of the second approach depends on the quality of the theoretical underpinnings of the algorithms. As computational chemistry improves, one would expect the more theoretical approach to benefit, and ultimately surpass the pragmatic approach in power and applicability. Until that time, both methods offer a different spectrum of benefits and should be part of a comprehensive synthetic approach.

The one area for the future which seems obvious is an integration of this methodology in the design process. When alternative structures of similar predictive activity exist, the difficulty in synthetic preparation should be factored into the decision-setting priorities. As was emphasized above, the objective is to find that 'magic bullet' with the specificity and biological profile desired as quickly as possible and with minimum expenditure of effort.

5 Rational screening

It may seem strange that such a strong advocate of rational drug design should be such a strong advocate of screening. Two major considerations dominate this position:

1 the data for structure–activity studies, which represent the fundamental basis of rational drug design, are dependent upon biological assays;

2 most of the details of biological systems have yet to be elucidated.

Only by screening at various levels of organization of the biological systems can one be sure that all aspects of the relevant biology have been included. By this, one can envisage a hierarchical approach—isolated molecular target, organelles, cells, organs, and whole animal. The first three of these should be done in human tissue, if at all possible, or in a derivative cell line, to insure that the results have relevance to human disease and/or toxic profile. The role of computer-aided drug design is to increase the efficiency of the overall process by making maximum use of the information available at each decision process. It will be a job for the end of the 21st century to assess whether we have gathered sufficient details of the biological system *Homo sapiens* adequately to model the details and eliminate screening. At this time, this is just a dream, but a desirable goal for the future as we all would like to eliminate unnecessary animal testing as quickly as possible.

5.1 *Mining the tailings*

As the expense of chemical synthesis has increased, the company compound files assume a greater potential as a source of compounds to be screened. It is analogous to the effect of a dramatic rise in the price of gold on the behaviour of mining companies. As the price increases and new technology increases the efficiency of gold extraction, the tailings, or residual processed ore from goal mines, have become an economical source of gold. In a similar way, compounds which already exist and for which the cost of synthesis has already been invested should have priority for testing over those which have yet to be made. This depends, of course, on a rational basis for deciding which compounds are likely to have activity compared with analogues of known active compounds. Once a pharmacophoric pattern has been postulated, or an active-site hypothesis generated, then a criteria for selection and prioritization is established.

5.2 *Three-dimensional databases*

Several pharmaceutical companies have realized the logic expressed above and have developed 3-D databases for their compound files to help select candidates for testing (Martin *et al.*, 1988; Sheridan *et al.*, 1989). An essential component in such a system is a method for assessing similarity (Brint & Willett, 1988). As most compound databases were entered as 2-D structures, this has required conversion to a 3-D format. Programs have proven (Leach & Prout, 1990; Leach *et al.*, 1990a,b) useful in generating plausible 3-D structures from the connectivity data. Inherent to the use of such a database are methods for evaluating 3-D similarities. Because of the inherent flexibility in most compounds, the use of a single conformation to represent the 3-D potential of a molecule is a clear limitation. Development of real 3-D databases with a compact, coded representations of the conformational states available to each compound is a logical next step and under active investigation by several groups (Lewis, 1990), including our own. In addition to identification of compounds which are capable of presenting an appropriate 3-D pattern, compounds must also fit within the receptor cavity. Based on a shape-matching algorithm, Sheridan & Venkataraghavan (1987) screened candidate compounds to select those whose volumes would fit within the combined volumes of known active compounds. Previously, Sheridan and colleagues had used the same algorithm to help identify potential ligands for known active sites (DesJarlais *et al.*, 1990a), in this case papain and carbonic anhydrase, by screening compounds from the Cambridge Crystallographic Database. Screening of the active site of HIV protease (DesJarlais *et al.*, 1990b) identified haloperidol as an inhibitor of the enzyme and provided a novel chemical lead for further investigation. Burt and Richards (1990) have introduced flexible fitting of molecules to a target structure with assessment of molecular similarity as a means of dealing with the conformational problem.

5.3 *Receptor subtypes*

Molecular biology has provided a unique opportunity to exploit the biological diversity of receptors. Cloning and expressing a specific receptor subtype in a cell, such as the *Xenopus* oocyte which does not normally express that receptor, provides a unique biological assay system (Miledi *et al.*, 1989). Here one can hope to characterize the receptor and the functional consequences of receptor occupancy by a specific ligand. Each receptor subtype needs to be extensively studied in order to define its molecular requirements for recognition and activation as well as its steric constraints. Most of the available data on the myriad number of compounds which have been made and tested is of limited use as it was measured in non-homogeneous assays. The next century will certainly see the investment by the major pharmaceutical companies in development of databases of receptor subtypes and their pharmacological characterization. The use of company compound files and rapid screening paradigms will simplify this task.

The use of novel synthetic procedures to generate large number of compounds for screening is relatively new. Examples of such rapid screening paradigms and their rapid evolution can be seen in epitope mapping. Geysen (1985) introduced a procedure for the rapid synthesis and testing of peptides on a solid support which has been even further developed (Fodor *et al.*, 1991) and combined with photolithography resulting in the synthesis and testing of 1024 peptides on a surface of 1.6 cm^2. Peptide libraries generated on phage were used to survey for high-affinity peptide ligands for streptavidin (Devlin *et al.*, 1990) and antibodies (Scott & Smith, 1990). Tuerk and Gold (1990) have selected high-affinity nucleic acid ligands by selection from pools of variant sequences. Combination with the photolithographic synthetic scheme (Fodor *et al.*, 1991) would allow optimization of the lead nucleic acid sequence. The ability to generate thousands of compounds and measure their activity quickly offers an opportunity for coupling between structure–activity relations and the optimization of structure in an almost realtime interaction. This assumes that the analysis phase will be able to operate in a timely fashion so that it is not easier simply to make compounds in a random manner, because of a time lag in analysis.

6 Conclusions

The latter half of the 20th century has been dominated by the development of computational knowledge. Both the explosion of the microelectronics industry which supplies ever more capable hardware at less cost, and the software industry which develops more efficient and usable tools to exploit the hardware, are ample testimony to this fact. These developments have enabled chemists to bring theoretical chemistry to a level where practical utility can be demonstrated. This aspect of development is relatively new, however, and clearly far from mature. The impact of these developments will continue to spread, and our increased experience will lead us to develop more productive and insightful computational tools to assist the medicinal chemist in the design of novel compounds, and more importantly, the prioritization of the compounds which are prepared.

When one views the other half of the medicinal chemists' domain—biology—one is struck by a simultaneous revolution in technology. One routinely does today what was not conceivable one or two decades past, and it is likely that we cannot conceive what will be routine in two or more decades. The exciting challenge to the medicinal chemist is to maintain his or her balance while trying to ride the three charging stallions of theoretical chemistry, computational power and molecular biology. These are exciting times and the prospects for the future are extremely bright. One would not be surprised if most of the predictions made above had come true by the end of this century as most are relatively minor extrapolations of trends already apparent. Drugs by design with enhanced therapeutic profiles are already being developed, and it is only increases in the efficiency and reliability of the process which we have addressed.

7 Acknowledgements

Support from the NIH (GM24483) during the preparation of this review is gratefully acknowledged. The comments and suggestions of Dr Denise D. Beusen have also influenced my views and are thankfully credited.

8 References

Allinger, N.L., Li, F. & Yan, L. (1990). Molecular mechanics. The MM3 force field for alkenes. *J. Comput. Chem.*, **11**, 848–67.

Allured, V.S., Kelly, C.M. & Landis, C.R. (1991). Shapes empirical force field: New treatment of angular potentials and its application to square-planar transition-metal complexes. *J. Am. Chem. Soc.*, **113**, 1–12.

Aqvist, J. & Warshel, A. (1990). Free energy relationships in metalloenzyme-catalyzed reactions. Calculations of the effects of metal ion substitutions in staphylococcal nuclease. *J. Am. Chem. Soc.*, **112**, 2860–8.

Arad, D., Langridge, R. & Kollman, P.A. (1990). A simulation of the sulfur attack in the catalytic pathway of papain using molecular mechanics and semiempirical quantum mechanics. *J. Am. Chem. Soc.*, **112**, 491–502.

Beusen, D.D., Iijima, H. & Marshall, G.R. (1990). Structures from NMR distance constraints. *Biochem. Pharmacol.*, **40**, 173–5.

Blundell, T.L., Sibanda, B.L. Sternberg, M.J.E., & Thornton, J.M. (1987). Knowledge-based prediction of protein structures and the design of novel molecules. *Nature (London)*, **326**, 347–52.

Brandt, J., Friedrich, J., Gasteiger, J., Jochum, C., Schubert, W. & Ugi, I. (1977). Computer programs for the deductive solution of chemical problems on the basis of a mathematical model of chemistry. In: W.T. Wipke and W.J. Howe (Eds.), *Computer-Assisted Organic Synthesis*, American Chemical Society, Washington, D.C.

Brint, A.I. & Willett, P. (1988). Upperbound procedures for the identification of similar three-dimensional chemical structures. *J. Comput.-Aided Mol. Design*, **2**, 311–20.

Bures, M.G., Black-Schaefer, C. & Gardner, G. (1991). The discovery of novel auxin transport inhibitors by molecular modeling and three-dimensional pattern analysis. *J. Comput.-Aided Mol. Design*, **5**, 323–34.

Burt, C. & Richards, W.G. (1990). Molecular similarity: The introduction of flexible fitting. *J. Comput.-Aided Mol. Design*, **4**, 231–8.

Clark, M., Cramer, R.D.I., Jones, D.M., Patterson, D.E. & Simeroth, P.E. (1990). Comparative molecular field analysis (CoMFA). 2. Toward its use with 3D-structural databases. *Tetra. Com. Meth.*, **3**, 47–59.

Darvas, F.J. (1988). *Mol. Graphics*, **6**, 80–6.

Dauber-Osguthorpe, P. & Osguthorpe, D.J. (1990). Analysis of intramolecular motions by filtering molecular dynamics trajectories. *J. Am. Chem. Soc.*, **112**, 7921–35.

Davis, M.E. & McCammon, J.A. (1990). Electrostatics in biomolecular structure and dynamics. *Chem. Rev.*, **90**, 509–21.

DePriest, S.A., Shands, E.F.B., Dammkoehler, R.A. & Marshall, G.R. (1990). 3D-QSAR: Further studies on inhibitors of angiotensin-converting enzyme. *QSAR: Rational Approaches on the Design of Bioactive Compounds, Proceedings of the 8th European Symposium, Sorrento (Napoli), Italy, September 9–13, 1990*, Elsevier Science Publishers, Amsterdam.

DesJarlais, R.L., Seibel, G.L., Dixon, J.S. & Kuntz, I.D. (1990a). Using shape complimentarity as an initial screen in designing ligands for a receptor binding site of known three-dimensional structure. *J. Med. Chem.*, **31**, 722–9.

DesJarlais, R.L., Seibel, G.L., Kuntz, I.D., Furth, P.S., Alvarez J.C., Ortiz de Montellano, P.R., DeCamp, D.L., Babe, L.M. & Craik, C.S. (1990b). Structure-based design of nonpeptide inhibitors specific for the human immunodeficiency virus 1 protease. *Proc. Natl. Acad. Sci. USA*, **87**, 6644–8.

Devlin, J.J., Panganiban, L.C. & Devlin, P.E. (1990). Random peptide libraries: A source of specific protein binding molecules. *Science*, **249**(7), 249–406.

Dolata, D.P., Leach, A.R. & Prout, K. (1987). WIZARD: AI in conformational analysis. *J. Comput.-Aided Mol. Design*, **1**, 73–85.

Donne-Op den Kelder, G. (1987). Distance geometry analysis of ligand binding to drug receptor sites. *J. Comput.-Aided Mol. Design*, **1**, 257–64.

Fesik, S.W. (1989). Approaches to drug design using nuclear magnetic resonance. In: T.J. Perun and C.L. Propst (Eds.), *Computer-Aided Drug Design: Methods and Applications*, Marcel Dekker, New York.

Fesik, S.W., Luly, J.R., Erickson, J.W. & Abad-Zapatero, C. (1988). Isotope-edited proton NMR study on the structure of a pepsin/inhibitor complex. *Biochemistry*, **27**(22), 8297–301.

Fesik, S.W., Zuiderweg, E.R.P., Olejniczak, E.T. & Gampe, Jr, R.T. (1990). NMR methods for determining the structures of enzyme/inhibitor complexes as an aid in drug design. *Biochem. Pharmacol.*, **40**(1), 161–7.

Field, M.J., Bash, P.A. & Karplus, M. (1990). A combined quantum mechanical and molecular mechanical potential for molecular dynamics simulations. *J. Comput. Chem.*, **11**, 700–33.

Fodor, S.P.A. Read, J.L. & Pirrung, M.C. (1991). Light-directed, spatially addressable parallel chemical synthesis. *Science*, **251**(2), 767–73.

Geysen, H.M. (1985). Antigen–antibody interactions at the molecular level: adventures in peptide synthesis. *Immunol. Today*, **6**(12), 364–9.

Gifford, E., Johnson, M. & Tsai, C.-c (1991). A graph-theoretic approach to modeling metabolic pathways. *J. Comput.-Aided Mol. Design*, **5**, 303–22.

Goodfellow, J.M., Jones, D.M., Laskowski, R.A., Moss, D.S., Saqi, M., Thanki, N. & Westlake, R. (1990). Use of parallel processing in the study of protein ligand binding. *J. Comput. Chem.*, **11**, 314–25.

Harvey, S.C. (1989). Treatment of electrostatic effects in macromolecular modeling. *Proteins*, **5**, 78–92.

Hendrickson, J.B. & Toczko, A.G. (1989). SYNGEN program for synthesis design: basic computing techniques. *J. Chem. Inf. Comput. Sci.*, **29**, 137–45.

Hibert, M.F., Gittos, M.W., Middlemass, D.N., Mir, A.K. & Fozard, J.R. (1988). Graphics computer-aided receptor mapping as a predictive tool for drug design: development of potent, selective and stereospecific ligands for the 5-HT1a receptor. *J. Med. Chem.*, **31**, 1087–93.

Ho, C.M.W. & Marshall, G.R. (1990). Cavity search: An algorithm for the isolation and display of cavity-like binding regions. *J. Comput.-Aided Mol. Design*, **4**, 337–54.

Houk, K., Tucker, J.A. & Dorigo, A. (1990). Quantitative modeling of proximity effects on organic reactivity. *Acc. Chem. Res.*, **23**, 107–13.

Hudson, B., Livingston, D.J. & Rahr, E. (1989). Pattern recognition display methods for the analysis of computed molecular properties. *J. Comput.-Aided Mol. Design*, **3**, 55–65.

Jorgensen, W. L., Laird, E.R., Gushurst, A.J., Fleischer, J.M., Gothe, S.A., Helson, H.E., Paderes, G.D. & Sinclair, S. (1990). CAMEO: A program for the logical prediction of the products of organic reactions. *Pure Appl. Chem.*, **62**, 1921–32.

Kneller, D.G., Cohen, F.E. & Langridge, R. (1990). Improvements in protein secondary structure prediction by an enhanced neural network. *J. Mol. Biol.*, **214**, 171–82.

Kollman, P.A. & Merz, K.M. (1990). Computer modeling of the interactions of complex molecules. *Acc. Chem. Res.*, **23**, 246–52.

Leach, A.R. & Prout, K. (1990). Automated conformational analysis: directed conformational search using the A algorithm. *J. Comp. Chem.*, **11**, 1193–205.

Leach, A.R., Prout, K. & Dolata, D.P. (1988). An investigation into the construction of molecular models by the template joining method. *J. Comput.-Aided Drug Design*, **2**, 107–23.

Leach, A.R., Prout, K. & Dolata, D.P. (1990a). The application of artificial intelligence to the conformational analysis of strained molecules. *J. Comput. Chem.*, **11**, 680–93.

Leach, A.R., Prout, K. & Dolata, D.P. (1990b). Automated conformational analysis: Algorithms for the efficient construction of low-energy conformation. *J. Comput.-Aided Drug Design*, **4**, 271–82.

Lewis, R.A. (1990). Automated site-directed drug design: Approaches to the formation of 3D molecular graphs. *J. Comput.-Aided Mol. Design*, **4**, 205–10.

Lewis, R.A. & Dean, P.M. (1989a). Automated site-directed drug design: the concept of spacer skeletons for primary structure generation. *Proc. R. Soc. Lond. B.*, **236**, 125–40.

Lewis, R.A. & Dean, P.M. (1989b). Automated site-directed drug design: the formation of molecular templates in primary structure generation. *Proc. R. Soc. Lond. B.*, **236**, 141–62.

Marshall, G.R., Barry, C.D., Bosshard, H.E., Dammkoehler, R.A. & Dunn, D.A. (1979). The conformational parameter in drug design: The active analog approach. In: E.C. Olson and R.E. Christoffersen (Eds.), *Computer-Assisted Drug Design*, American Chemical Society, Washington, D.C.

Marshall, G.R., Beusen, D.D., Kociolek, K., Redlinski, A.S., Leplawy, M.T., Pan, Y. & Schaefer,

J. (1990). Determination of a precise interatomic distance in a helical peptide by REDOR NMR. *J. Am. Chem. Soc.*, **112**, 963–6.

Martin, Y., Danaher, E.B., May, C.S. & Weininger, D. (1988). MENTHOR, a database system for the storage and retrieval of three-dimensional molecular structures and associated data searchable by substructure, biologic, physical, or geometric properties. *J. Comput.-Aided Mol. Design*, **2**, 15–29.

Mayer, D., Naylor, C.B., Motoc, I. & Marshall, G.R. (1987). A unique geometry of the active site of angiotensin-converting enzyme consistent with structure–activity studies. *J. Comput.-Aided Mol. Design*, **1**, 3–16.

Menger, F.M. & Sherrod, M.J. (1990). Origin of high predictive capabilities in transition-state modeling. *J. Am. Chem. Soc.*, **112**, 8071–5.

Meyer, B., Hansen, T., Nute, D., Albersheim, P., Darvill, A., York, W. & Shellers, J. (1991). Identification of the H-NMR spectra of complex oligosaccharides with artificial neural networks. *Science*, **251**(2), 542–4.

Miledi, R., Parker, I. & Sumikawa, K. (1989). Transplanting receptors from brains into oocytes. *Fidia Research Foundation Neuro-Science Award Lectures*, **3**, 57–90.

Naruto, S., Motoc, I., Marshall, G.R., Daniels, S.B., Sofia, M.J. & Katzenellenbogen, J.A. (1985). Analysis of the interaction of haloenol lactone suicide substrates with α-chymotrypsin using computer graphics and molecular mechanics. *J. Am. Chem. Soc.*, **107**, 5262–70.

Norinder, U. (1990). Experimental design based 3-D QSAR analysis of steroid–protein interactions: Application to human CBG complexes. *J. Comput.-Aided Mol. Design*, **4**, 381–9.

Pensak, D.A. & Corey, E.J. (1977). LHASA—Logic and Heuristics Applied to Synthetic Analysis. In: W.T. Wipke and W.J. Howe (Eds.), *Computer-Assisted Organic Synthesis*, American Chemical Society, Washington, D.C.

Perlman, R.S. (1987). Rapid generation of high quality approximate 3D molecular structures. *Chem. Design Automation News*, **2**(1), 1–7.

Perutz, M.F., Fermi, G., Abraham, D.J., Poyart, C. & Bursaux, E. (1986). Hemoglobin as a receptor of drugs and peptides: X-ray studies of the stereochemistry of binding. *J. Am. Chem. Soc.*, **108**, 1064–78.

Qian, N. & Sejnowski, T.J. (1988). Predicting the secondary structure of globular proteins using neural network models. *J. Mol. Biol.*, **202**, 865–84.

Rooman, M.J. & Wodak, S.J. (1988). Identification of predictive sequence motifs limited by protein structure data base size. *Nature* (*London*), **335**, 45–9.

Scott, J.K. & Smith, G.P. (1990). Searching for peptide ligands with an epitope library. *Science*, **249**(7), 386–90.

Seressiotis, A. & Bailey, J.E. (1988). MPS: An artificially intelligent software system for the analysis and synthesis of metabolic pathways. *Biotech. Bioengineer*, **31**, 587–602.

Shea, G.T., Telfer, S.J., Waterman, S. & Bartlett, P.A. (1990). CAVEAT: A program to facilitate the design of organic molecules. In: S.M. Roberts (Ed.), *Molecular Recognition: Chemical and Biological Problems*, pp. 182–96, The Royal Society, London.

Sheridan, R. P. & Venkataraghavan, R. (1987). Designing novel nicotinic agonists by searching a database of molecular shapes. *J. Comput.-Aided Mol. Design*, **1**, 243–56.

Sheridan, R.P., Rusinko III, A., Nilakantan, R. & Venkataraghavan, R. (1989). Searching for pharmacophores in large coordinate data bases and its use in drug design. *Proc. Natl. Acad. Sci. USA*, **86**, 8165–9.

Sufrin, J.R., Dunn, D.A. & Marshall, G.R. (1981). Steric mapping of the L-methionine binding site of ATP: L-methionine S-adenosyltransferase. *Mol. Pharmacol.*, **19**, 307–13.

Tuerk, C. & Gold, L. (1990). Systematic evolution of ligands by exponential enrichment: RNA ligands to bacteriophage T4 DNA polymerase. *Science*, **249**(8), 505–10.

Van Drie, J.H., Weininger, D. & Martin, Y.C. (1989). ALADDIN: An integrated tool for computer-assisted molecular design and pharmacophore recognition from geometric, steric and substructure searching of three-dimensional molecular structures. *J. Comput.-Aided Mol. Design*, **3**, 225–51.

Vedani, A. & Huhta, D.W. (1990). A new force field for modeling metalloproteins. *J. Am. Chem. Soc.*, **112**, 4759–67.

Wilcox, C.S. & Levinson, R.A. (1986). A self-organized knowledge base for recall, design and discovery in organic chemistry. In: T.H. Pierce and B.A. Hohne (Eds.), *Artificial Intelligence Applications in Chemistry*, American Chemical Society, Washington, D.C.

Wipke, W.T. & Hahn, M.A. (1989). Analogy in computer-assisted design. In: J.L. Fauchere (Ed.), *QSAR: Quantitative Structure–Activity Relationships in Drug Design*, Alan R. Liss, New York.

Wipke, W.T., Braun, H., Smith, G., Choplin, F. & Sieber, W. (1977). SECS—Simulation and Evaluation of Chemical Synthesis: Strategy and Planning. In: W.D. Wipke and W.J. Howe (Eds.), *Computer-Assisted Organic Synthesis*, American Chemical Society, Washington, D.C.

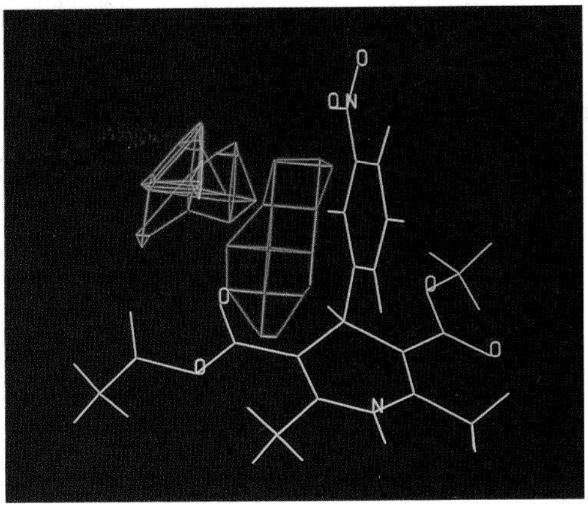

Plate 13.1. Nitrendipine molecule with sites of attraction for the probe (red). Contour lines indicate favourable locations of a protonated nitrogen as a model for a cationic receptor site. The probe acceptable regions are contoured at – 3.5 kcal/mole. (Display conditions are the same for all molecules. In the case of racemic compounds only one enantiomer was investigated.)

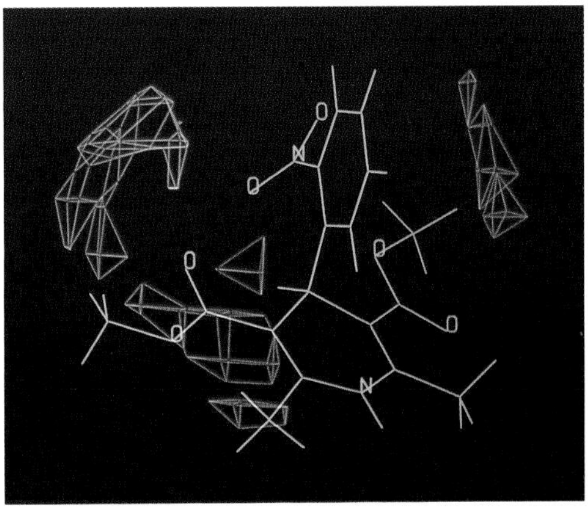

Plate 13.2. Shows the nifedipine molecule (atomic skeleton, cyan) and probe binding regions (magenta).

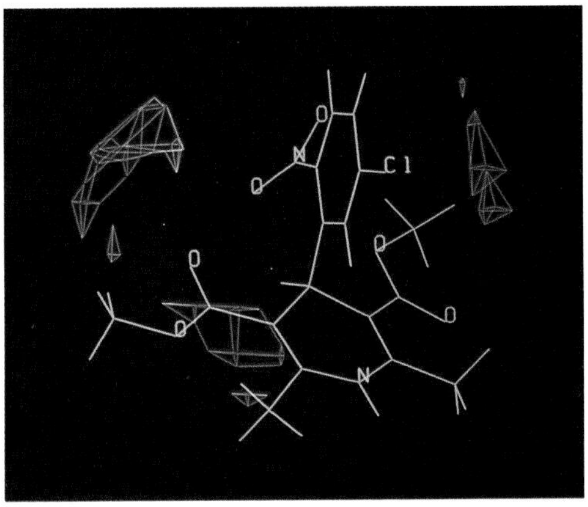

Plate 13.3. MR-1-12 is a nifedipine derivative with an additional chloro substituent on the aromatic ring. This substitution reduces the negative inotropic effects (see Table 13.3). Note that its contour volumes are correspondingly scaled down.

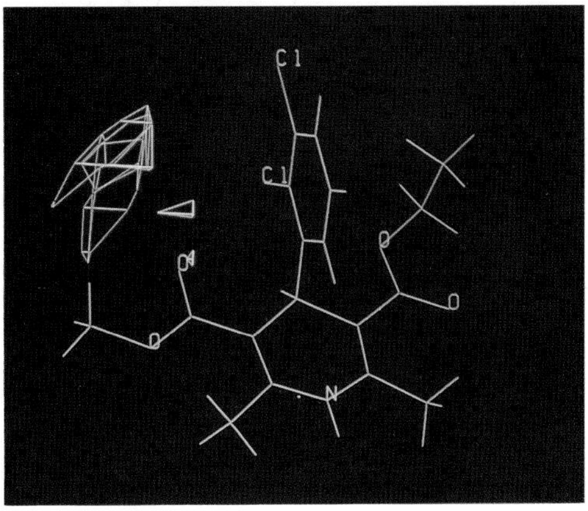

Plate 13.4. Felodipine, a DHP compound with chloro substituents, instead of a nitro group. From the contour map we can conclude that felodipine may not exhibit strong negative inotropic effects, a fact that is supported by pharmacological data.

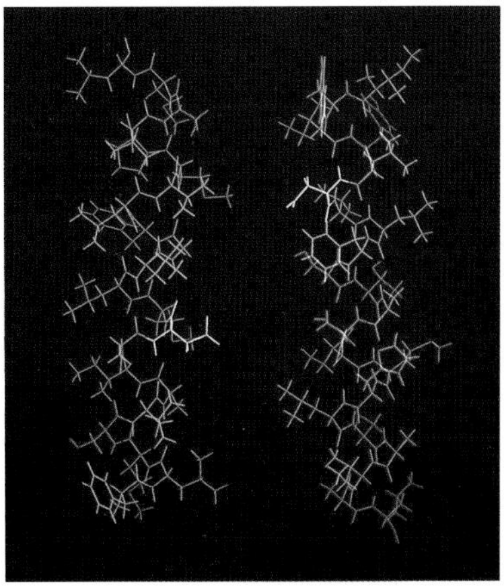

Plate 13.5. Model of α-helices II (left) and III of the 5-HT_2-receptor (hydrophobic amino acids, blue; polar amino acids, red). The two aspartic acids involved in binding of agonists or antagonists are shown (yellow).

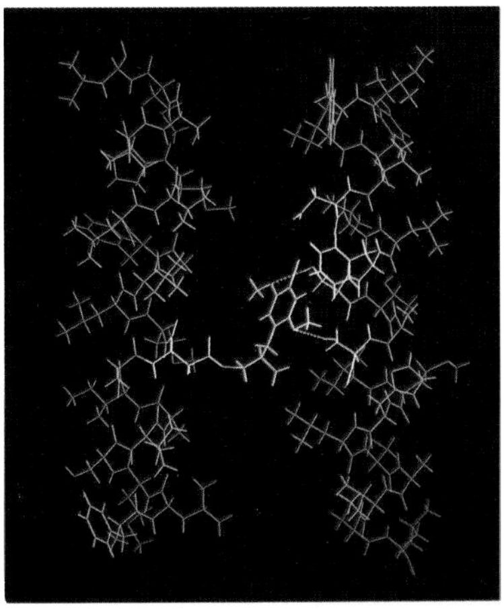

Plate 13.6. Model of agonistic binding site of the 5-HT_2-receptor. The agonist molecule is *R*-(–)-1-(2,5-dimethoxy-4-iodophenyl)-2-aminopropane (R-DOI).

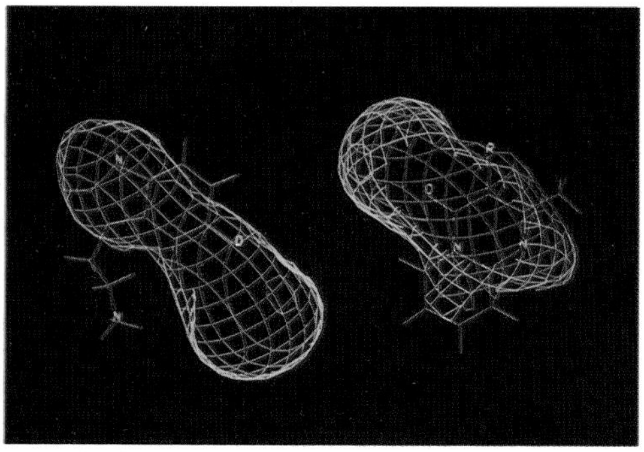

Plate 13.7. Comparison of the negative (1 kcal/mole) MEP volumes for the 5-HT_2-agonist serotonin (left) and the aromatic bicyclic system of pirenperone, which acts as a competitive antagonist at the same receptor.

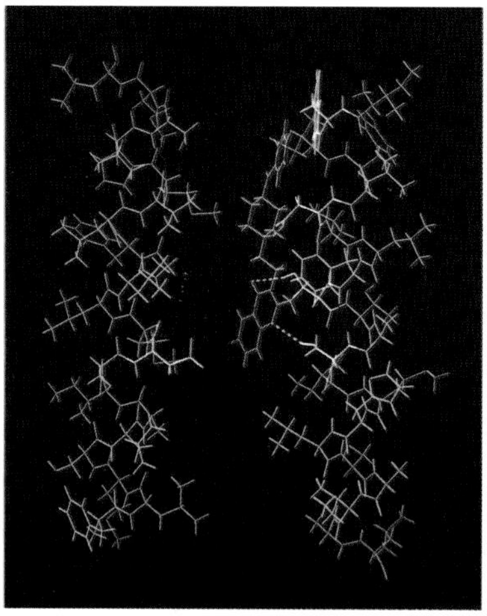

Plate 13.8. Model of the antagonistic binding site of the 5-HT_2-receptor. The antagonist pirenperone is shown (red).

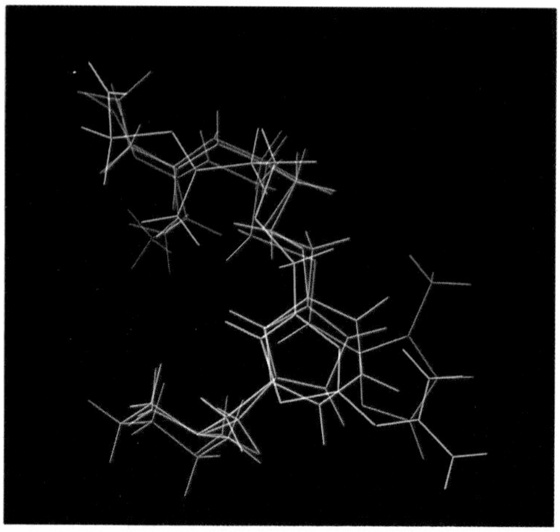

Plate 13.9. Superposition of all compounds: cimetidine (purple), ranitidine (red), roxatidine (magenta), famotidine (blue).

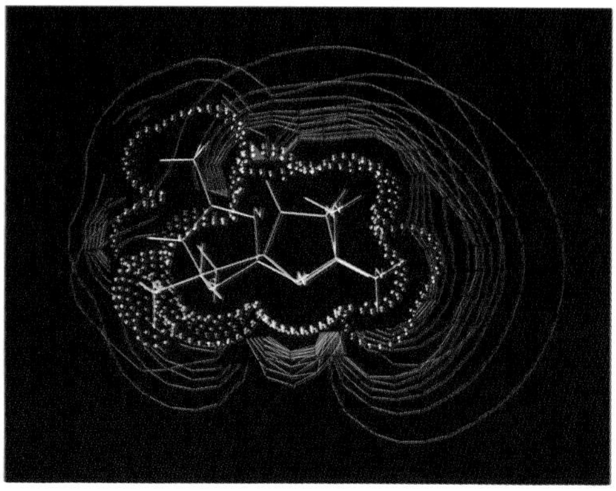

Plate 13.10. Superposition of histamine (green) and guanidiothiazole (white) in order to reach the best agreement of their electrostatic potentials. (The nitrogen of the histamine side-chain is kept unregarded.) The points correspond to the volumes of the molecules.

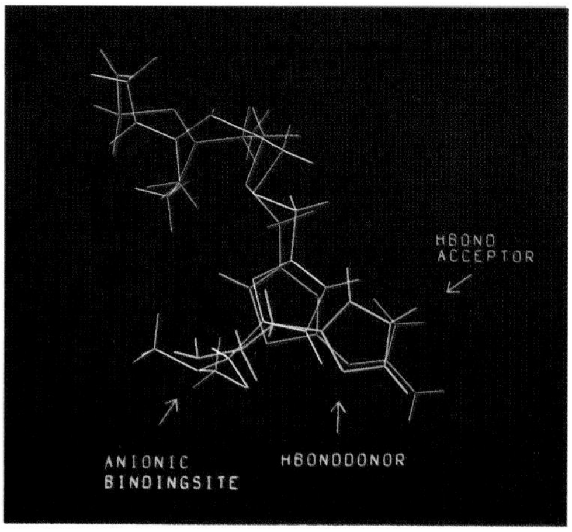

Plate 13.11. Pharmacophore model for H_2-histamine receptor antagonists: ranitidine (white), famotidine (blue), histamine (green).

13 Pharmacophore Identification Based on Molecular Electrostatic Potentials (MEPs)*

H.-D. HÖLTJE

Free University of Berlin, Institute of Pharmacy, Königin-Luise-Strasse 2 + 4, W-1000 Berlin 33, Germany

1 Introduction

The elucidation of the molecular mechanisms of drug action is a main goal of researchers in the field of molecular medicinal chemistry because knowledge of the molecular mechanisms is the basis of a rationale in drug design. However, up till now, information at the molecular level of receptors and drug-binding sites has been sparse. Therefore, the construction of hypothetical receptor models, capable of explaining at least known experimental data, provides a practicable way out of the dilemma. The astonishing progress in recent years in development of computer graphics hardware and software has been a powerful tool which can be used to design realistic drug–receptor model interaction complexes. Models of this kind possess the potential to yield information leading to more specific and more active drug molecules.

In our laboratory we have been engaged for years in the construction of drug–receptor models for various classes of drugs. Some recent findings shall be presented here. The list of methods used in our group includes several force-field and semi-empirical quantum chemical methods, the SYBYL/MENDYL molecular-modelling software package (Tripos), P. Goodford's GRIN/GRID (Molecular Discovery) as well as the INSIGHT/DISCOVER (Biosym) package. Our hardware equipment consists of Evans & Sutherland PS350 as well as PS390, Silicon Graphics IRIS workstations and Vax II or Convex C220 host computers.

The initial step in the formation of drug–receptor interaction complexes is a recognition event. The receptor has to recognize whether an approaching molecule possesses the properties necessary for specific and tight binding. This recognition process has to occur at rather large distances and precedes formation of the final interaction complex. The 3-D electrostatic field induced by molecules therefore plays a crucial role in recognition. As a corollary, studies on the determination of pharmacophores should, besides considering steric and conformational properties, include a detailed analysis of the molecular electrostatic potentials (MEPs) of drug molecules.

2 The 1,4-dihydropyridine pharmacophore

Recently we have proposed a hypothetical receptor model able to explain the contradictory effect of chiral 1,4-dihydropyridines on the potential-dependent Ca^{2+} channel (Höltje & Marrer, 1987):

* Dedicated to Professor Dr H. Oelschläger on the occasion of his 70th birthday.

In this series it was found that derivatives with a certain substitution pattern (Table 13.1), i.e. one large ester group and a small substituent in the meta-positions of the dihydropyridine ring, show dramatic differences in their pharmacological effects (Schramm *et al.*, 1983; Franckowiak *et al.*, 1985; Hof *et al.*, 1985; Gjoerstrup *et al.*, 1986). Only *R*-enantiomers behave as Ca^{2+} antagonists whereas their *S*-enantiomers own Ca^{2+}-channel activating properties. If both meta-substituents are different ester groups, then both enantiomers are Ca^{2+} antagonists but to varying extents. Our receptor model is based on striking differences in the molecular electrostatic potential which we discovered in a defined segment of space around the antagonistic and agonistic dihydropyridines (Fig. 13.1). All agonists possess a negative potential whereas all antagonists possess a positive potential in this area. In order to prove whether this difference can be transferred to a receptor structure, we constructed an interaction complex between dihydropyridines and a tryptophan molecule serving as a simplified receptor binding-site model. The geometry of the interaction followed the location of the discriminating potential area (Fig. 13.2). Using the SIMPOT program (Wise, 1985; Marrer, 1986) we were able to calculate effective changes in the MEP induced by the series of dihydropyridines (DHPs) on the receptor model tryptophan. These values are listed in Table 13.2 and show a clear-cut distinction between agonists and antagonists.

Table 13.1. Structure and pharmacological behaviour of 1,4-dihydropyridines

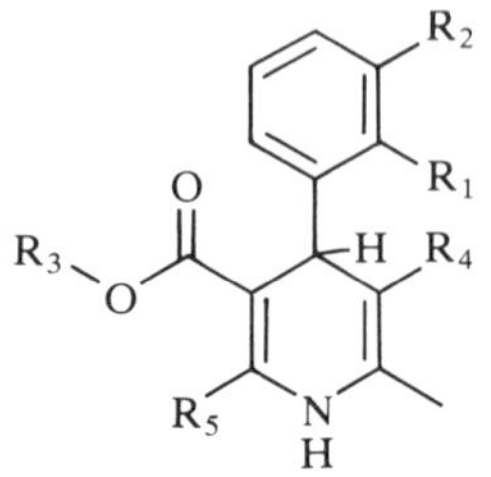

Compound	R_1	R_2	R_3	R_4	R_5	Ca^{2+} agonist	Ca^{2+} antagonist
Bay K 8644	$-CF_3$	-H	$-CH_3$	$-NO_2$	$-CH_3$	*S*	*R*
Sandoz 202 791	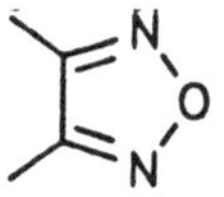		$-i-C_3H_7$	$-NO_2$	$-CH_3$	*S*	*R*
H 160/51	-Cl	-H	$-C_2H_5$	-H	$-NH_2$	*R**	*S**
PN 200 110	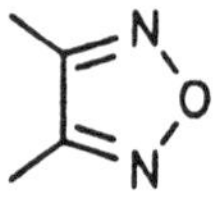		$-i-C_3H_7$	$-COOOCH_3$	$-CH_3$		*R/S*
Methyl-Bay E 6927	-H	$-NO_2$	$-i-C_3H_7$	$-COOOC_2H_5$	$-CH_3$		*R/S*
Fossheim	$-CF_3$	-H	$-CH_3$	$-COOCH_3$	$-CH_3$	(not chiral, Ca^{2+} antagonist)	

* Predicted configuration.

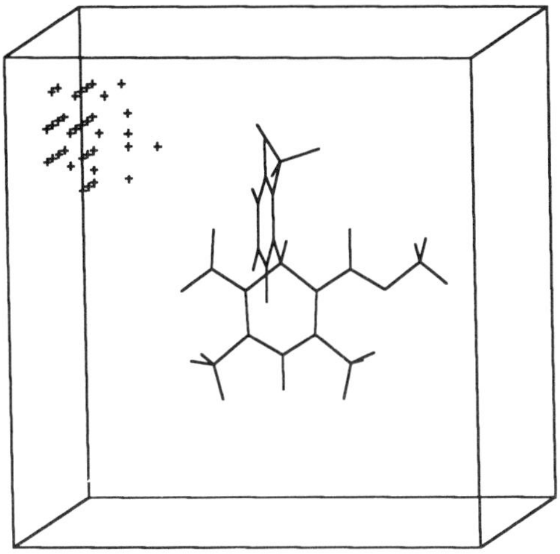

Figure 13.1. The box represents the volume of space for which the molecular electrostatic potential (MEP) has been calculated for all dihydropyridines under study. The molecule in the box is *S*-(−)-Bay K 8644. The crosses indicate the subspace where the potentials of agonists and antagonists differ.

The potential of tryptophan is decreased through agonists and increased through antagonists. We, therefore, came to the conclusion that the MEP of the DHP binding site can be influenced in opposite ways by binding of Ca^{2+}-channel modulating molecules with DHP structure.

Differences in phenyl-ring substitution of the 1,4-dihydropyridines may attribute quantitative and qualitative differences in activity including tissue selectiv-

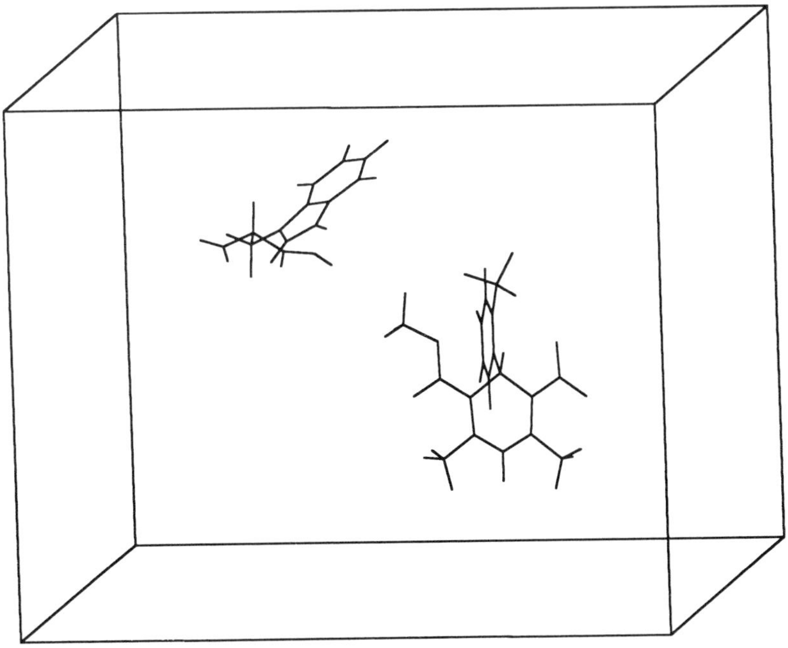

Figure 13.2. Interaction complex between the binding-site model of tryptophan and *R*-(+)-Bay K 8644. The tryptophan molecule occupies the subspace where potential differences exist.

Table 13.2. Calculated changes of the electrostatic potential (Δ-MEP) induced by 1,4-dihydropyridines (DHPs) on the receptor model tryptophan

DHPs	Δ-MEP (kJ)
Agonists	
(*S*)-(–)-Bay K 8644	– 316.1
(*R*)-H 160/51	– 34.8
(*S*)-(+)-Sandoz 202 791	– 320.3
Antagonists	
(*R*)-(+)-Bay K 8644	90.0
Fossheim	22.6
(*S*)-H 160/51	44.8
(*S*)-(+)-Pn 200 110	70.8
(*R*)-(–)-Sandoz 202 791	72.4
(*S*)-(–)-Methyl-Bay E 6927	38.9

ity (Triggle *et al.*, 1988). Variation of the chemical structure of DHPs also leads to differences in negative inotropic potencies. It is known that the negative inotropic effect of DHPs is dose-dependent (Lambert *et al.*, 1990), i.e. DHPs with high binding affinities show little or no myocardial depressant activity at doses needed for coronary dilatation. A further reason for differences in inotropic potencies may be related to structural features which control the efficacy or determine the selectivity for given channel types in different tissues as can be seen from QSAR studies (Goll *et al.*, 1986).

We have tried to identify and calculate molecular properties that can explain the modulation of negative inotropic activities using the GRIN/GRID program. It has been proposed that DHPs interact with the Ca^{2+}-channel via helical domains containing a regular pattern of positively charged amino acid residues and non-polar residues (Tanabe *et al.*, 1987). Thus to simulate these conditions, a methyl group was used

Table 13.3. Chemical structures of some DHP compounds

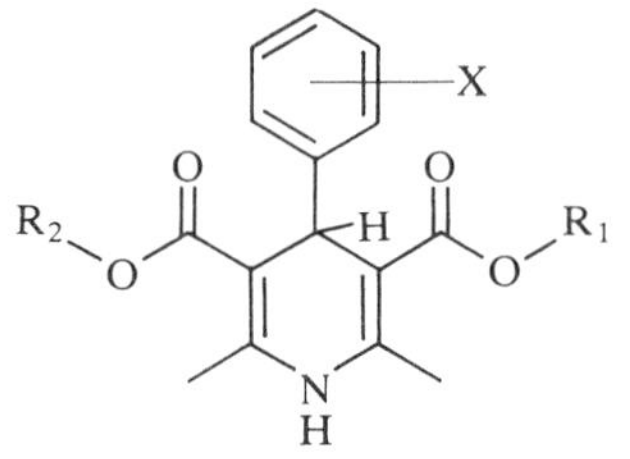

Compound	X	R_1	R_2
Nicardipine	*m*-NO_2	CH_3	$(CH_2)_2N(CH_3)CH_2Ph$
Nimodipine	*m*-NO_2	$CH(CH_3)_2$	$(CH_2)_2OCH_3$
Nitrendipine	*m*-NO_2	CH_3	C_2H_5
Nifedipine	*o*-NO_2	CH_3	CH_3
Nisoldipine	*o*-NO_2	CH_3	$CH_2CH(CH_3)_2$
MR-1-12	*o*-NO_2,*o*-Cl	CH_3	CH_3
Felodipine	*o*-Cl,*m*-Cl	C_2H_5	CH_3

as a model for non-polar receptor sites and a protonated nitrogen was used as a model for cationic receptor sites for the determination of corresponding interaction fields.

A similar interaction pattern was obtained for a series of DHPs with different negative inotropic effects (Table 13.3) with the methyl group, the only difference being the extension of the favourable binding sites in the case of more bulky ester side-chains. These results suggest that, although non-polar amino acids may be part of the receptor site for the DHPs, the interaction with these residues is not the discriminating factor determining inotropic activities. By contrast, remarkable differences occur among the DHPs using the second, positively charged, probe.

A common feature of all maps is a zone of attraction close to the two-ring system and the ester groups. Significant deviations in location and extension of the contour lines emerge between compounds having a nitro substituent in the meta-position like nitrendipine, nimodipine and nicardipine (Plate 13.1, facing page 180) in comparison to compounds where the nitro group is in the ortho position (nifedipine and nisoldipine, Plate 13.2).

Compound MR-1-12 is a nifedipine derivative with an additional chloro substituent. The chloro substitution reduces the negative inotropic response and in the contour map of MR-1-12 (Plate 13.3) the characteristic interaction region is diminished. Since felodipine, a DHP with chloro substituents in ortho- and meta-positions possesses low negative inotropic potency (Berntsson *et al.*, 1987) we have calculated its contour map to prove our results (Plate 13.4) and found that this derivative also shares the same characteristic features as all DHPs with low negative inotropic effects. These compounds lack distinct spheres of attraction for the positive nitrogen probe which, in contrast, can be found for nifedipine and nisoldipine, two compounds with comparable high binding affinities to the receptor but high negative inotropic effects.

3 The serotoninergic 5-HT$_2$ pharmacophore

On the basis of the low resolution structure of bacteriorhodopsin (Henderson *et al.*, 1990) and the detailed crystallographic analysis of two photosynthetic reaction centres (Diesenhofer *et al.*, 1985) together with the availability of many sequences derived from cloning, a general design principle of G-protein-coupled membrane receptors has been established. The predominant structural motif is characterized by a bundle of seven antiparallel membrane-spanning α-helices of about 23 residues in length. The α-helices are made up of a high proportion of apolar amino acids, which mostly are directed versus the lipid environment. The length of the linkage sequences which join consecutive α-helices varies considerably. Helices II and III of the 5-HT$_2$-receptor protein (Pritchett *et al.*, 1988) contain some highly conserved polar amino acids: II = Asp-98, III = Asp-133, Ser-137, Ser-140. From mutation experiments with β-adrenergic and muscarinic receptors, it is known that one of the aspartic acids (Asp-98) is essential for the agonistic effect (Chung *et al.*, 1988), whereas the other one (Asp-133) mainly determines binding of antagonists (Curtis *et al.*, 1989; Plate 13.5).

A conformation analysis of several 5-HT_2 agonists of different structural classes (serotonin derivatives, phenylethylamines and arylpiperazines) has led to the formulation of a stereospecific 5-HT_2 pharmacophore (Höltje & Briem, 1991). Using GRIN/GRID probes representing the most important structural characteristics of this pharmacophore with helices II and III of the 5-HT_2 protein, a binding-site model was constructed. In this model, the two electron-rich regions of the agonists are bound to the conserved Ser-137 and Ser-140 of helix III, whereas the amino tail of the neurotransmitter analogues is fixed at the Asp-98 on helix II (Plate 13.6).

Antagonists like ketanserine derivatives and analogues which block the serotoninergic 5-HT_2-receptor do not meet all structural prerequisites for binding to the agonistic binding site. They do, though, possess a heterocyclic ring system and a protonatable nitrogen at a suitable distance. An elaborate study of the MEPs revealed a close similarity in form and size of the electronegative −1 kcal isopotential volume for the heterocyclic parts of agonists and antagonists (Plate 13.7, facing page 181). This finding guides us to the assumption that antagonists can also be bound to the hydrogen-bond system of Ser-137 and Ser-140. Interestingly the MEP-directed positioning of the antagonist close to Ser-137 and Ser-140 automatically leads to a second contact of antagonists with the receptor protein via an electrostatic interaction of the protonated nitrogen and Asp-133. If the antagonists exist in the energetically favoured extended conformation, an additional binding contact can occur between the fluorinated aromatic system and Trp-129 of helix III. As a consequence, antagonists would be bound rather strongly only to one helix (helix III) of the receptor protein (Plate 13.8) and thereby impede binding of agonists. This might be the explanation for antagonists lacking efficacy. Although further experimental studies on the 3-D structure of the 5-HT_2-receptor may disclose that other helices take part in binding of agonists or antagonists, the characteristic features of the interaction certainly will remain unchanged.

4 The histaminergic H_2 pharmacophore

The H_2-receptor belongs to the same class of membrane receptors and is considered to operate in a similar way as the 5-HT_2-receptor mentioned above. H_2-receptor antagonists show a considerable structural diversity but they all contain at least one aromatic or heterocyclic system and a polar, planar π-electron system in common. These two substructures are connected by a lipophilic four-atom long linker. With respect to the cyclic system, H_2-antagonists can be divided into four main groups: the imidazole (cimetidine), the aminomethylfurane (ranitidine), the guanidinothiazole (famotidine) and the piperidinomethylphenoxy (roxatidine) series (see Fig. 13.3). Plate 13.9 shows that these H_2-antagonists are able to meet the same relative spatial arrangement of the polar planar end-group and the central hydrophobic part of the molecule. However, there is no agreement between the nitrogen substituents of the different heterocyclic systems. That means it is not possible to superimpose these atoms (Höltje & Batzenschlager, 1990).

This result is supported by consideration of the electronic properties in this region. The potentials of the guanidinothiazole part of famotidine and of the imidazole part of histamine are superimposed in Plate 13.10. The histamine

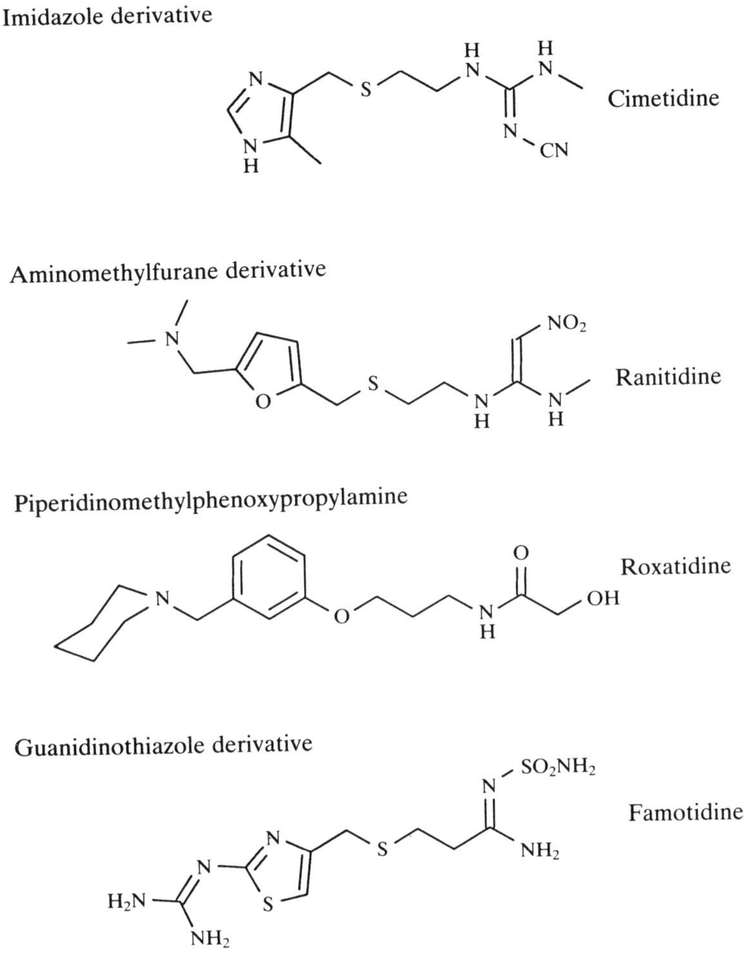

Figure 13.3. Structures of H_2-antagonists investigated in this study.

conformation used is the *trans,trans*-conformation which is postulated to be the bioactive one (Smeyers *et al.*, 1985). As can be seen, the MEP minimum of the guanidinothiazole is in the same position as the MEP minimum of histamine which is considered as essential for the receptor (Luque *et al.*, 1988). One nitrogen atom of the guanidino moiety of the guanidinothiazole corresponds to the τ-nitrogen of histamine which is supposed to interact with a hydrogen-bond acceptor site of the receptor; the other corresponds to the π-nitrogen of histamine which assumingly binds to a hydrogen-bond donor site (Weinstein *et al.*, 1985). It is evident that those H_2-antagonists which do not possess a guanidinothiazole system cannot meet this superimposition. But interestingly the protonable nitrogen of these substances can easily be superimposed with the nitrogen of the histamine side-chain. Weinstein *et al.* (1985) postulated an anionic binding site for this nitrogen atom.

Cimetidine is not able to meet the hydrogen-bond donor site, the hydrogen-bond acceptor site and the common position of the polar end-group simultaneously. This might be the reason for the lower activity of cimetidine. The resulting H_2-antagonistic pharmacophore is presented in Plate 13.11. Although there is not yet enough information about the H_2-receptor, several papers postulate that the H_2-receptor operates in a way comparable to the β-receptor or the 5-HT_2-receptor. The results presented here fit to the common ideas of G-protein-coupled membrane receptors and compare well to our binding-site model of the serotoninergic 5-HT_2-receptor.

5 Conclusion

Because molecules interact through their electron shells, it can easily be understood that MEPs determine to a large extent the chemical and pharmacological reactivities of all substances. As a corollary it should be acknowledged that pharmacophore models which have been established solely on the basis of conformational and stereochemical studies neglect an important piece of information about the molecular properties. MEPs and other molecular fields which can be generated by systematic interaction of drug molecules with chemically different probes represent data sets of high potential value especially in cases where experimental facts on the 3-D structure of a binding site do not exist. The utilization of this type of information may play a crucial role for drug development.

6 References

Berntsson, P., Johansson, E. & Westerlund, C. (1987). *J. Cardiovasc. Pharmacol.*, **10** (Suppl. 1), 60–5.

Biosym Technologies Inc., San Diego, California, USA.

Chung, F.-Z., Wang, C.-D., Potter, P.C., Venter, J.C. & Fraser, C.M. (1988). *J. Biol. Chem.*, **263**, 4052–5.

Curtis, C.A.M., Wheatley, M., Bansal, S., Birdsall, N.J.M., Eveleigh, P., Pedder, E.K., Poyner, D. & Hulme, E.C. (1989). *J. Biol. Chem.*, **264**, 489–95.

Diesenhofer, J., Epp, O., Miki, K., Huber, R. & Michel, H. (1985). *Nature (London)*, **318**, 618–23.

Franckowiak, G., Bechem, M., Schramm, M. & Thomas, G. (1985). *Eur. J. Pharmacol.*, **114**, 223–6.

Gjoerstrup, P., Harding, H., Isaksson, R. & Westerlund, C., (1986). *Eur. J. Pharmacol.*, **122**, 357–61.

Goll, A., Glossmann, H. & Mannhold, R. (1986). *Naunyn-Schmiedeberg's Arch. Pharmacol.*, **334**, 303–12.

Henderson, R., Baldwin, J.M., Ceska, T.A., Zemlin, F., Beckmann, E. & Downing, K.H. (1990). *J. Mol. Biol.*, **213**, 899–929.

Hof, R.P., Ruegg, U.T., Hof, A. & Vogel, A. (1985). *J. Cardiovasc. Pharmacol.*, **7**, 689–93.

Höltje, H.-D. & Marrer, S. (1987). *J. Comput.-Aided Mol. Design*, **1**, 23–30.

Höltje, H.-D. & Batzenschlager, A. (1990). *J. Comput.-Aided Mol. Design*, **4**, 391–402.

Höltje, H.-D. & Briem, H. (1991). *Quant. Struct.–Act. Relat. (QSAR)*, **10**, 193–7.

Lambert, C.R., Buss, D.D. & Pepine, C.J. (1990). *Circulation*, **81** (3), 139–47.

Luque, F.J., Sanz, F., Illas, F., Pouplana, R. & Smeyers, Y.G. (1988). *Eur. J. Med. Chem.*, **23**, 7–10.

Marrer, S. (1986). SIMPOT-Algorithm, Department of Pharmacy, Free University of Berlin, Germany.

Molecular Discovery Ltd., Oxford, UK.

Pritchett, D.B., Bach, A.W.J., Wozny, M., Taleb, O., Dal Toso, R., Shih, I.C. & Seeburg, P.H. (1988). *EMBO J.*, **7**, 4135–40.

Schramm, M., Thomas, G., Towart, R. & Franckowiak, G. (1983). *Nature* (*London*), **303**, 535–7.
Smeyers, Y.G., Romero-Sanchez, F.J. & Hernandez-Laguna, A. (1985). *J. Mol. Struct.*, **123**, 431–42.
Tanabe, T., Takeshima, H., Mikami, A., Flockerzi, V., Takahashi, H., Kangawa, K., Kojima, M., Matsuo, H., Hirose, T. & Numa, S. (1987). *Nature* (*London*), **328**, 313–18.
Triggle, D.J., Fossheim, R., Hawthorn, M., Joslyn, A., Triggle, A.M., Wei, X.-Y. & Zheng, W. (1988). In: *Recent Advances in Receptor Chemistry*, pp. 123–45. Elsevier Science Publishers, Amsterdam.
Tripos Associates Inc., St. Louis, Missouri, USA.
Weinstein, H., Mazurek, A.P., Osman, R. & Topiol, S. (1985). *Mol. Pharmacol.*, **29**, 28–33.
Wise, M. (1985). In: J.K. Seydel (Ed.), *QSAR and Strategies in the Design of Bioactive Compounds*, pp. 19–29, VCH, Weinheim.

14 New Rational Approaches for Structure–Activity Relationships and Drug Design

A. ITAI, N. TOMIOKA, Y. KATO, Y. NISHIBATA and S. SAITO
Faculty of Pharmaceutical Sciences, University of Tokyo, Japan

1 Introduction

Drug development has been achieved by repeated efforts of chemical syntheses and evaluations, improving the pharmacological activity of a lead compound by systematic modification of the structures. As the difficulties in synthesizing compounds have decreased by technical advances in organic chemistry, efficient design of bioactive molecules has become increasingly important. Although starting from an appropriate lead compound is most important for success, we cannot at present choose it as we would like. Most of the lead compounds developed so far have not been designed rationally; rather they have been discovered by screening natural and synthetic compounds with *in vivo* or *in vitro* assays. It has been desirable for a long time to establish a principle for designing lead compounds logically, without relying on chance, and trial and error.

In recent years, remarkable progress has been made in fields related to biological macromolecules. There is now a lot of knowledge on drug receptors that has brought a lot of wide possibilites for rational drug design. This knowledge can be used for analysing structure–activity relationships (SARs) and for designing new structures with improved activities. But even when the 3-D structure of the receptor is known in detail, great effort is still required to obtain new lead compounds because a general principle for lead generation has not yet been established.

Computers and computer graphics have become important tools in rational drug design. They can make up for human ability in manipulating and quantitizing 3-D structures, properties and interactions of molecules. However, the power of computers is still not fully utilized for the development of new bio-active compounds, because of the lack of efficient methodologies and software. As the molecular recognition between drug and receptor can be understood only through 3-D structures of molecules, methods not based on the 3-D structures, such as the conventional quantitative structure–activity relationships (QSAR) method, are insufficient. It is necessary to develop new concepts and new efficient computer methods based on 3-D structures and drug–receptor interactions in order to enable logical design of new bioactive compounds.

For the purpose of rational drug design, we have been developing several new computer methods and programs. Together with the details of our methods, we will describe the concepts and backgrounds of our strategies. Furthermore, we will discuss perspectives on the future of drug design.

2 Concepts and backgrounds

2.1 *Receptor*

Drug–receptor theory has become a very important basis for drug design (Dean,

1987). Drug molecules must satisfy two requirements, one for the process of reaching the receptor in the body, and the other for the specific interaction with the receptor site. Usually, the term 'receptor' has been used only for target macromolecules of hormones and neurotransmitters, which transduce signals between cells. In addition to these macromolecules, many protein molecules such as enzymes are known to play important biological functions in the body. Since it is widely accepted that all these macromolecules can be important targets of clinical drugs at present, here we use the term 'receptor' in a broad sense including all biological macromolecules such as receptors for hormones and neurotransmitters, enzymes, other proteins and nucleic acids. The molecular bases of recognition between these macromolecules and drugs or natural bioactive molecules are proposed to be very similar, although the events that ensue following the specific interactions are quite different.

Since the 1970s, great technological and conceptual advances have brought about a vast knowledge of structures and functions of receptors and drug–receptor interactions. The development of protein chromatography systems has enabled the isolation of purified proteins even from sources with very small content or even if the proteins are rather unstable. Recombinant gene technology has made it possible to isolate genes encoding receptor proteins from target organisms and to generate their products in the quantities necessary for experiments. Structure elucidation of these proteins has been greatly spurred by remarkable progress in techniques for X-ray crystallography (Blundell & Johnson, 1976). Nearly 1000 protein structures have been already elucidated to an atomic resolution, whose atomic coordinates are available world-wide from the Protein Data Bank (Brookhaven National Laboratory) (Bernstein *et al.*, 1977). It is expected that the number of structure-elucidated proteins will continue to increase and that receptor-based approaches to drug design will become more and more important.

2.2 *Drug–receptor interaction*

Some protein structures have been elucidated as complexes with ligand molecules. The structures provide us with a general structural basis of drug–receptor interaction, as well as details of the binding mode of individual ligand molecules. In protein–ligand complexes, the ligand molecule shows good complementarity in shape and properties with the binding-site cavity of the protein. Consequently, between the protein and ligand, strong intermolecular forces seem to be functioning, consisting of hydrogen bonds, electrostatic interactions, hydrophobic interactions and others. Among them, the hydrogen bond is considered to play an especially important role for recognizing ligand structures (Luecke & Quiocho, 1990) because of its strict geometrical requirements.

The fact that molecules with quite dissimilar chemical structures can bind to the same site of the same receptor is evidenced in many examples by binding assays to receptor or by crystallography. This strongly suggests that a receptor does not recognize the chemical structure itself, but recognizes the physical and chemical properties that are involved in the interaction with the receptor. In other words, it is the submolecular arrangement of physical and chemical properties that is important for the specific binding to a receptor. There are a vast number of receptor macromolecules in the body, and an individual receptor has a different requirement for ligand molecules according to the 3-D array of the properties at the ligand binding site. If the requirement of the individual receptor were known, more rational approaches could be undertaken for designing selective drugs.

2.3 *Active conformation*

Many of the natural and synthetic bio-active molecules have some degree of conformational freedom. Stable conformations of such molecules are known to change greatly according to environmental factors. In drug design, the receptor-bound structure, which is called the 'active structure' or 'active conformation', is especially important. Experimentally, the active conformation of a drug molecule can be estimated only from the syntheses of conformationally restricted analogues, except for structure determination of the complex with the target receptor by X-ray analyses. If the receptor structure is known, it can be estimated by docking study, as described in the next section. Even if the receptor structure is not known, if we had bioactive molecules with rigid structures or we could estimate the active conformations of molecules by any means, important clues to rational drug design would be provided.

Since it is very difficult to know the active conformation without a receptor structure, conformations are often assumed to be those found in crystal or in solution or of energetic global minimum of the ligand molecule alone. The assumption is very risky, because active conformation is not necessarily the same as any experimental states or computationally stable states of the ligand molecule itself. In order to investigate the relative stability of the active conformation, we have examined three ligand molecules (glycyltyrosine, citrate and methotrexate) by empirical energy calculation (M. Sugimori, N. Tomioka & A. Itai, unpublished results), whose structures were elucidated with and without their target enzymes (carboxypeptidase A, citrate synthase, dihydrofolate reductase) by X-ray crystallography. The active (enzyme-bound) conformation was found to be quite different from that in the ligand crystal and also from those of any energy-minimum structures of the ligand alone, including the global-minimum structure. For example, in the methotrexate–dihydrofolate reductase system, the active conformation is higher in energy than the global minimum by 7 kcal/mol even when the dielectric constant in water ($\varepsilon = 78.3$) is used. The difference becomes much larger when a lower dielectric constant ($\varepsilon = 1.0$) is used. The active conformation was near to (but not same as) the 87th stable conformer among all the possible local-minimum conformers ordered with their energy values.

This study suggested to us the difficulties in assuming the active conformation of a flexible molecule by energy calculation of the ligand molecule alone. That also goes for all conformation search techniques using energy calculation. Thus, active conformation should be searched widely within the stereochemically permissible ranges without severe steric hindrance, including halfway states between local minima, although the number of conformers to be considered might become enormous. They should be carefully assumed without being overly influenced by crystal structures or energetic stability.

2.4 *Three-dimensional structure–activity relationships*

Relationships between 3-D structures and biological activities (3-D-SAR) are made clear by various analyses using 3-D structures of active and inactive compounds, together with those of the receptor. Such analyses provide qualitatively different information from those obtained from statistical methods such as QSAR. If the receptor structure is available, more rational approaches can be undertaken for 3-D-SAR and more reliable information can be obtained. In this case, the favourable binding mode and active conformation for each ligand are investigated by docking simulation of the ligand molecule to the receptor cavity. Inter- and intra- molecular interaction energies are estimated using theoretical calculations. The total energies should correlate with the binding constant, inhibition constant or biological activity, whence the entropic contribution to the free energy of complex formation can be neglected. Functional groups in the ligand that are essential for the activity can be revealed from the binding mode. Finding unused cavity space or functional groups in the receptor would be useful for further modifications of structures to improve binding affinity or selectivity to the receptor. Furthermore, it might be useful, for obtaining improved pharmacological behaviour, to distinguish the submolecular region of the ligand which does not directly contact the receptor from the remaining region.

On the other hand, superposition of molecules and receptor mapping has been performed for 3-D-SAR in cases where the receptor structure is not available. Although this is the case for most of the present drug developments, 3-D-SAR results here are based on many assumptions and are rather risky because even a highly active ligand molecule cannot be a negative image of the target receptor cavity; also, all the functional groups or features of the ligand molecule are not necessarily essential for receptor binding. Nevertheless, the essential features for biological activity can be better presumed and the receptor environments can be better imagined from multiple molecules that are correctly superposed than from any single molecule. The first-order hypothesis can be constructed based on the assumption that structural features and properties common to all active molecules are essential for the activity. The active conformation of flexible molecules can also be estimated by superposition with rigid molecules or flexible molecules as the case may be. From superposition of active and inactive molecules, spatial volume around the key compound can be distinguished to determine whether the occupation is favourable or unfavourable for activity (Marshall *et al.*, 1979). A 3-D receptor model which is constructed based on the superposed structures of highly active molecules can be used for docking study in the same manner as a real receptor structure, in order to predict the binding mode and binding affinity. Presumption of the structural region that is not involved in the specific binding to the receptor may also be useful for designing new structures.

When the receptor structure is not available, elaborate effort is required to establish the correct SAR model which can explain all of the activity data without contradiction. A model constructed on the basis of a hypothesis should be improved or re-built through considerable iterations of chemical syntheses and evaluations. Compared with the QSAR approach, it seems to be difficult to deal with biological activities quantitatively in the 3-D-SAR approach in the absence of receptor structure. But important 3-D-SAR information is obtained not from a series of analogous compounds but from active compounds with dissimilar structures that can barely be analysed by QSAR.

2.5 *Lead generation*

Success stories in lead discovery have been quite rare. Therefore, it has long been believed that lead compounds cannot be generated artificially, and as consequence the words 'drug design' have meant 'lead optimization process'. However, we should be aware that the situation has changed. Nowadays, most drug developments are performed on the basis of biochemical processes. We are now able to use structures of the natural ligand or substrate of the target receptor even if the 3-D structure of the receptor is unknown. It can be said that artificial lead generation has become possible by making use of receptor structures or structures of known bioactive ligands. On the other hand, it is difficult to carry out the lead optimization process logically, such as for avoiding toxicity, since whole-body systems are complex.

We know that it is the complementarity in shape and properties that is required for drugs by the receptor, and many molecules with various structures can be accommodated in a common receptor. If we can discover or design molecules with such features, they might become lead compounds for receptor agonists, antagonists or inhibitors. But, it is not an easy task with only human power. On the other hand, it does not seem to be that difficult for computers to suggest possible structures, according to our requirements. Computers are competent for considering all possibilities for a given condition, unlike the human brain. So far, computers have been used to calculate various quantities for a given structure. For lead generation, we want inversely to generate structures that have the properties or functions we require by using computers. If computers can suggest a variety of structures that might bind to a receptor, a small number of excellent structures can be chosen from these as promising seeds for a lead compound.

3 Docking study

In the case where receptor structure is known, docking simulation is one of the most popular ways of making use of the receptor structure. Docking simulations are performed for elucidating modes and stabilities of binding of ligand molecules to the receptor site. It is useful not only for drug design but also for studies of biochemical mechanisms. Typically, stable locations of ligand molecules are looked for in the receptor cavity by interactive manipulation of the molecules using 3-D computer graphics, and the models of the complex are further refined by theoretical calculations. In the interactive docking process and in the first stage of the refinement, the protein structure is assumed to be rigid. But flexibility of the protein molecule should be considered in the further refinement process, especially in cases where significant conformational changes are known to occur upon the binding of ligand molecules.

As there are at least six degrees of freedom (translation, rotation and bond rotation) in the ligand molecule, there exist an enormous number of local-minimum states and energy barriers. Therefore, it is very difficult to reach the most stable state using only computational methods such as molecular mechanics. Even molecular dynamics calculations cannot go beyond a barrier which is too high. For obtaining the global-minimum solution in a docking study, it is necessary to prepare adequate starting structures by some means.

3.1 *Program* GREEN

The program GREEN was developed for performing docking simulation logically and efficiently (Tomioka *et al.*, 1986, 1987). The program provides realtime estimation of the receptor–ligand interaction energy and visualization of the spatial environment of the receptor site, by using the tabulated data on 3-D grid points inside the cavity of the receptor site. On each grid point, the program calculates van der Waals interaction (G_{vdw}), electrostatic potential (G_{elc}) and hydrogen-bonding flags (G_{hbd}), as shown in Fig. 14.1. The van der Waals term G_{vdw} is calculated between a probe atom placed on the grid point and the whole protein atoms by using an empirical potential function. (The program provides parameters of the AMBER program; Weiner *et al.*, 1984, 1986.) Energy values for several probe atoms such as carbon, nitrogen, hydrogen and oxygen are stored separately. The electrostatic potential G_{elc} is calculated by using atomic charges on the protein atoms. The value of G_{elc} is equivalent to the electrostatic interaction energy assuming that the probe atom bears a positive unit charge. If a grid point is at a favourable location where strong hydrogen-bonding interaction is expected from a functional group in the protein, the hydrogen-bonding flag (G_{hbd}) is turned on with the hydrogen-bonding character (donor, acceptor or both) of the functional group.

The G_{vdw} value is used for the definition of the 'atom acceptable region' which is represented by a bird-cage picture colour-coded by the G_{elc} values. Within the atom-acceptable region, centres of the ligand atoms can favourably exist energetically. Therefore, it is suitable for fitting a wire-frame model of the ligand molecule in the docking operation.

While a ligand molecule is manipulated in the gridded region, the protein–ligand interaction energy (E_{inter}) is calculated as shown in Fig. 14.2. The van der Waals energy is calculated simply by summing up the G_{vdw} value on the grid point nearest to the ligand atom (or the value interpolated from eight neighbouring grid

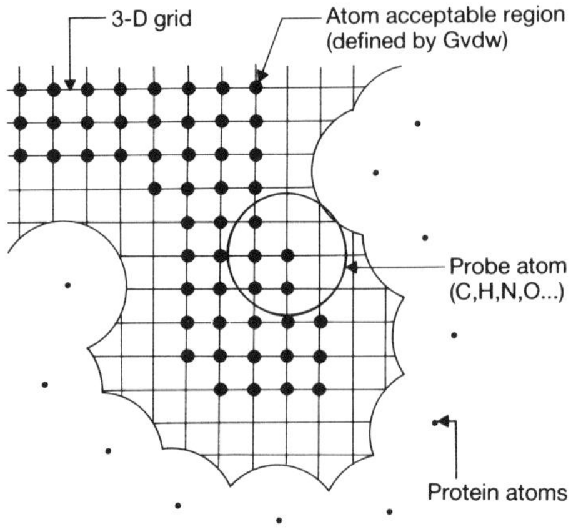

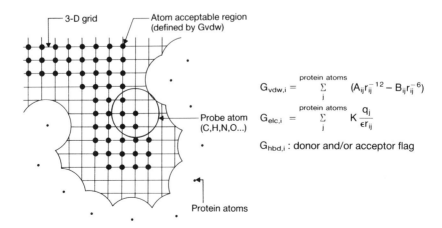

$$G_{vdw,i} = \sum_{j}^{\text{protein atoms}} (A_{ij} r_{ij}^{-12} - B_{ij} r_{ij}^{-6})$$

$$G_{elc,i} = \sum_{j}^{\text{protein atoms}} K \frac{q_j}{\epsilon r_{ij}}$$

$G_{hbd,i}$: donor and/or acceptor flag

Figure 14.1. Calculation of grid-point data in the GREEN program.

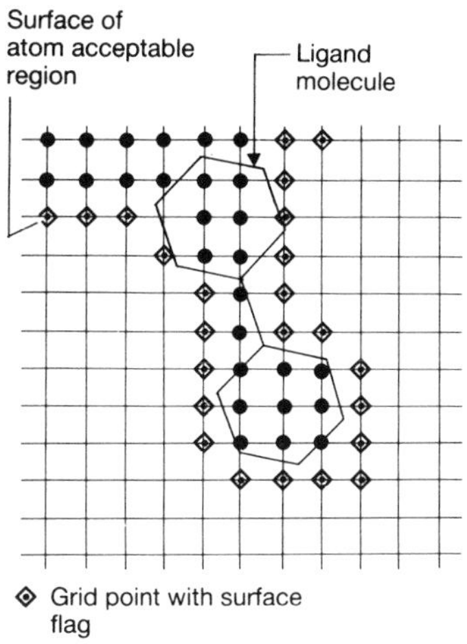

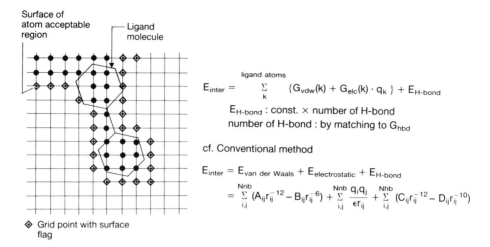

$$E_{inter} = \sum_{k}^{\text{ligand atoms}} \{G_{vdw}(k) + G_{elc}(k) \cdot q_k\} + E_{H\text{-}bond}$$

$E_{H\text{-}bond}$: const. × number of H-bond
number of H-bond : by matching to G_{hbd}

cf. Conventional method

$$E_{inter} = E_{van\ der\ Waals} + E_{electrostatic} + E_{H\text{-}bond}$$
$$= \sum_{i,j}^{Nnb} (A_{ij}r_{ij}^{-12} - B_{ij}r_{ij}^{-6}) + \sum_{i,j}^{Nnb} \frac{q_i q_j}{\epsilon r_{ij}} + \sum_{i,j}^{Nhb} (C_{ij}r_{ij}^{-12} - D_{ij}r_{ij}^{-10})$$

Figure 14.2. Calculation of interaction energy by using the grid-point data.

points), by using the tabulated values for the corresponding probe atom. Electrostatic interaction energy is calculated by summing the product of the atomic charge q_k of the ligand atom and the G_{elc} value of the nearest grid point. As for the hydrogen-bonding energy, the program does not use empirical energy functions, but simply counts the number of satisfied hydrogen bonds that are judged by the matching of ligand heteroatoms to the G_{hbd} flag. In order to evaluate conformational energy of the ligand molecule, intramolecular energy of the ligand is calculated by the empirical force field.

Calculation of the interaction energy is greatly accelerated by the grid-point approximation method compared to the conventional method such as the one shown in Fig. 14.2. With the GREEN program, the value of the interaction energy can be obtained in realtime while manipulating the ligand molecule on the graphics display. Furthermore, the program records the trajectory of the ligand manipulation with the energy values, from which the energy-minimum point can be located easily. The ligand coordinate determined by manual docking can be further optimized by the Monte Carlo method (Iga *et al.*, unpublished results), or by the Simplex method (Nedler & Mead, 1965), including the conformational freedom.

For designing new structures, the GREEN program has a function which shows the energetic contribution of individual atoms of the ligand molecule. This function is useful for modifying ligand structures to attain more favourable interactions by replacing, deleting or adding substituent groups. Interactive modelling functions of the program can be used for such modifications and also for creating new ligand structures. Realtime energy evaluation of the program is useful for designing more stable and better fitting structures throughout the design process. The grid-point method used in the GREEN program is also utilized in the program for automatic structure generation using the known receptor structure described below.

4 Molecular superposition and receptor mapping

It is supposed that there are some common structural features between molecules that can bind strongly to the same receptor. In order to extract these features, superposition of molecules is often performed. When receptor structure is not available, superposition of molecules is one of the most efficient ways of comparing active and inactive molecules three-dimensionally.

Most superposition studies have been done on the basis of the similarity of the chemical structures or of the correspondences of the atomic positions (Barakat & Dean, 1990). The least-squares method of atom positions, which is the most popular way to superpose molecules, can give us a numerical index which shows the degree of similarity, but the explicit specification of at least three atom-pairs between molecules is required. Consequently, the method cannot be applied to molecules between which correspondences of atoms are not obvious. On the other hand, visual superposition on 3-D graphic display by interactive manipulation of molecules does not require the atom-pair specification, but it is difficult to provide numerical indices.

As molecular superposition is meaningless unless it is based on the binding to the same receptor, it should be done in terms of the spatial arrangement of physical and chemical properties that are involved in the interactions with the receptor. Neither chemical structures themselves nor the positions of heteroatoms need to be superposed. Although it is easy to superpose molecules with similar structures, superposition of quite dissimilar structures would give us much more information. It thus is necessary to develop a new rational method which can superpose molecules with quite different structures.

4.1 *Program* RECEPS

Program RECEPS was developed for superposing molecules in terms of the spatial arrangement of physical and chemical properites (Itai *et al.*, 1986, 1992; Kato *et al.*, 1987). First a molecule with a rigid structure or with a limited number of conformational freedom should be chosen as a template molecule. (In the automatic superposition function described below, the program can treat a fairly flexible molecule as a template.) A 3-D grid is generated around the template molecule, so that the required part of the molecule is included with a sufficient margin. At each grid point, various physical and chemical properties are calculated and stored as grid-point data. Onto the graphic expression of these tabulated data as well as the structure of the template molecule, trial molecules are superposed by 3-D manipulation. The superposition is done so as to attain better values of 'goodness-of-fit indices' that are calculated by using the grid-point data at every step of the manipulation. Trial molecules are superposed one after another, and the atomic coordinates of the superposed molecules are successively stored into a file. The procedures are schematically shown in Fig. 14.3.

The grid-point data consist of occupancy flag, charge distribution inside the van der Waals volume of the molecule, electrostatic potential outside the van der Waals volume and flags for the expected hydrogen-bonding site in the receptor. Goodness-of-fit indices are calculated by using these tabulated data and atomic coordinates of the trial molecule as follows. The index for the molecular shape is calculated as a ratio of the number of commonly occupied grid points to the number of grid points occupied by the template molecule. The index for the electrostatic potential is calculated by Eqn. 14.1. It is a correlation coefficient of potential values from the

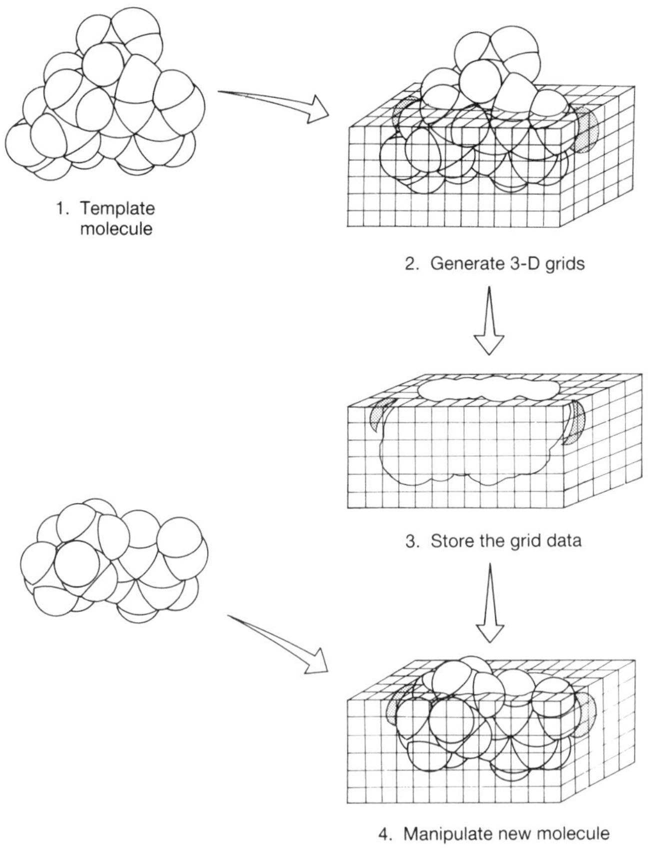

Figure 14.3. Procedure for superposing molecules by the RECEPS program in manual use.

template and trial molecules evaluated at surface grid points of the two molecules:

$$F_{elpo} = -\frac{\Sigma_i(V_{temp,i}V_{trial,i})}{\sqrt{\Sigma_i|V_{temp,i}|^2}\sqrt{\Sigma_i|V_{trial,i}|^2}} \tag{14.1}$$

where $V_{temp,i}$ is the electrostatic potential at the grid point i of the template molecule and $V_{trial,i}$ is the electrostatic potential at the grid point i of the trial molecule. The index for the hydrogen bond is calculated as a ratio of the number of common hydrogen-bonding grid points to those of the template. The common grid points are judged by the matching of hydrogen-bond characters from both molecules. These indices are rather tentatively defined, and should be modified or improved by further studies so that a correct superposing mode can be selected with high accuracy.

Based on the atomic coordinates of each superposed molecule, a set of grid point data is calculated. By unifying these sets of grid-point data, a receptor model can be constructed. The model provides information on the size and shape of the receptor cavity, electrostatic potentials on the cavity surface, expected locations of hydrogen-bonding heteroatoms, and so on. The receptor cavity is defined as the volume which is occupied by at least one molecule. The electrostatic potential on the cavity surface is defined as the average of potential values calculated from all the superposed molecules. An appropriate weighting scheme, such as with the potency of biological activity, is used in the averaging procedure. For the hydrogen-bonding sites and their character, the common ones are usually taken. But in some cases where there are ambiguities in donor or acceptor character, uncommon ones are also used by the judgement of the user. The receptor model can be modified or enlarged by further superposition of additional molecules.

4.2 *Superposition of tumour promoters*

We have applied the RECEPS program to TPA (tumour promoter)-type compounds (Itai *et al.*, 1988) shown in Fig. 14.4, namely teleocidin (Fujiki *et al.*, 1981), TPA (12-*O*-tetradecanoyl-13-*O*-acetylphorbol) (Hecker, 1978) and aplysiatoxin (Fujiki *et al.*, 1982). These compounds are supposed to bind to the same receptor, because they exhibit very similar biological activities in several assay systems (Fujika *et al.*, 1984). But, they are superficially so different that no structural similarity can be found among them. All these compounds have a large lipophilic group in common that is known to be essential for the potent tumour-promoting activity. The finding of remarkable activation of protein kinase C by TPA (Blumberg, 1988) has further stimulated interest in these compounds.

In the superposition study using the RECEPS program, we have assumed that these compounds bind to the same receptor site. In teleocidin, two stable conformers, twist and sofa form as shown in Fig. 14.5, are known to exist in an equilibrium ratio of 2:1 in solution (Endo *et al.*, 1986), although either of them is found in crystal structures of various teleocidins (Sakai *et al.*, 1984; Hitotsuyanagi *et al.*, 1984). The twist form has a *cis*-amide bond in the nine-membered lactam ring, whereas the sofa form has a *trans*-amide bond. Both conformers were tested as the template molecule onto which the TPA molecule was superposed independently, assuming either of them should be the active-ring conformation.

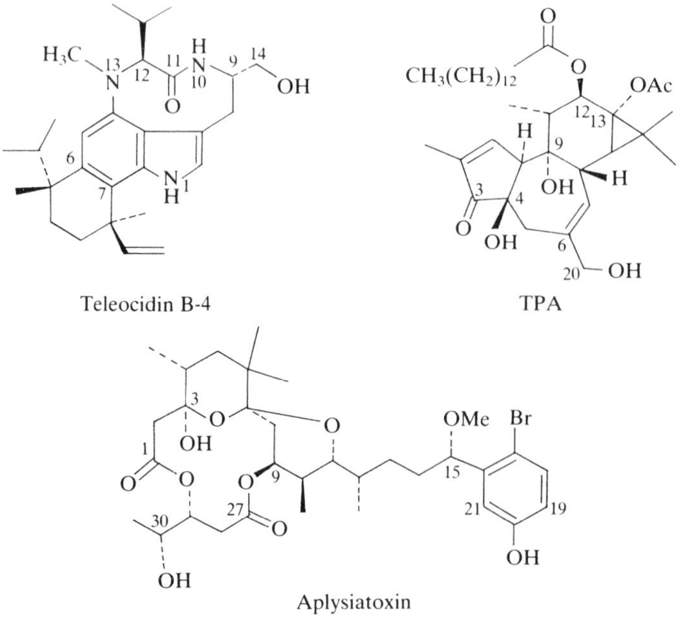

Figure 14.4. Chemical structures of tumour promoters.

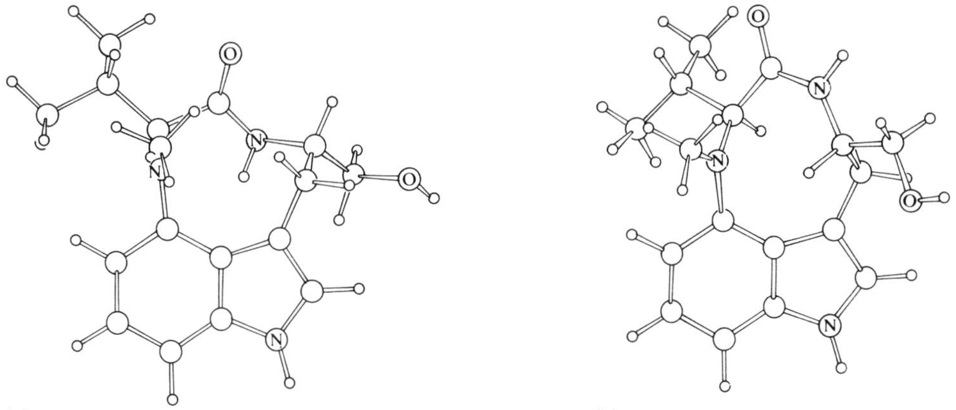

(a) (b)

Figure 14.5. Two stable conformers of teleocidin. Terpenoid moiety linked to the indol ring is omitted for clarity: (a) sofa form; (b) twist form.

As it seemed that hydrogen bondings play important roles in these compounds, best matching between the expected hydrogen bonding sites in the receptor from both molecules was looked for by rotation, translation and bond rotation of the trial molecule. The conformations of two substituent groups at C9 and C12 of teleocidin were adjusted interactively to attain the best superposition. At that time, the role of the lipophilic moiety had not been elucidated and there were two possibilities: the moiety is necessary for non-specific interaction with membrane, or it is required for specific binding to the receptor site. In this study, the latter case was assumed and macroscopic locations of the lipophilic groups in both molecules were determined after completion of the superposition of the hydrophilic region.

Figure 14.6(a) shows the best superposed structures of teleocidin sofa form and TPA, in which three pairs of hydrogen-bonding groups (a pair of CH_2OH groups, a pair of C=O groups, and a pair between amide NH in teleocidin and the OH group in TPA) are corresponded at the expected receptor sites. The similarity of the surfaces of two molecules was very good and the locations of the lipophilic groups agreed well. It was concluded that the superposition of TPA onto the sofa form is a little better than that onto the twist form, because only two hydrogen-bond pairs

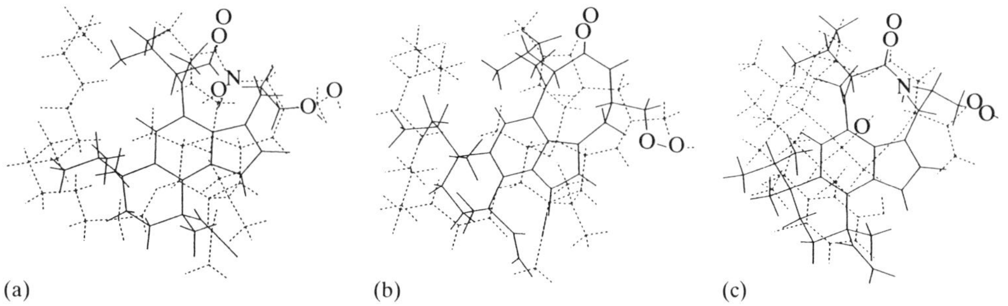

(a) (b) (c)

Figure 14.6. The best superposition models of tumour promoters. Corresponded hydrogen-bonding heteroatoms are marked with atomic symbols:
(a) TPA (------) onto teleocidin sofa form (——);
(b) TPA (------) onto teleocidin twist form (——);
(c) aplysiatoxin (------) onto teleocidin sofa form (——).

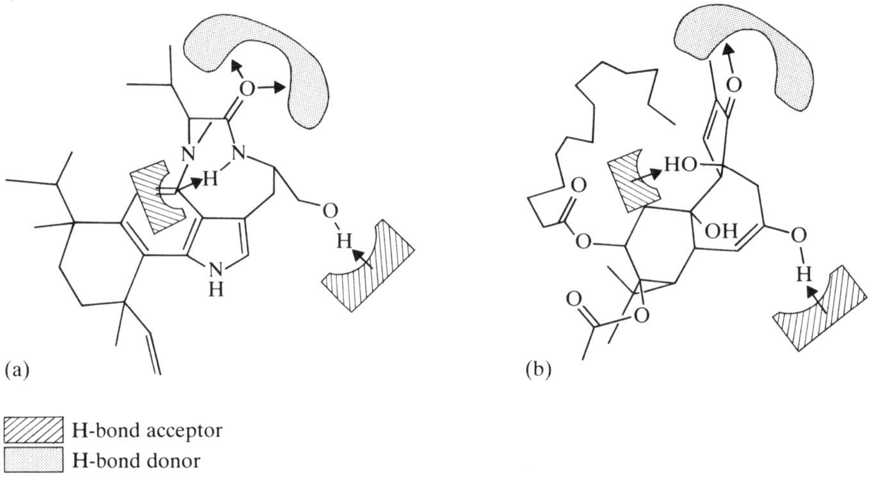

Figure 14.7. The 3-D–SAR model obtained from the superposition of teleocidin sofa form and TPA. Expected positions of hydrogen-bonding sites in the receptor are shown schematically as shaded moieties: (a) teleocidin sofa form; (b) TPA.

could be corresponded in the latter case as shown in Fig. 14.6(b). In the same way, superposition of aplysiatoxin onto the sofa-form teleocidin gave a good superposition model as shown in Fig. 14.6(c). The same conclusions were reached by the automatic superposition method described below.

Figure 14.7 schematically shows the results of the superposition of TPA onto sofa-form teleocidin. Three pairs of hydrogen-bonding groups of two molecules were correlated in terms of three expected sites in the receptor. The model is quite different from two superposition models that were previously proposed by Jeffrey & Liskamp (1986) and Wender *et al.* (1986): they did it with atomic positions of heteroatoms and without considering their roles and the chemical significance.

4.3 *Automatic superposing function* (AUTOFIT)

The program RECEPS was developed to establish a new rational concept for molecular superposition. But the manual superposition method is elaborate and requires care to obtain plausible results. The results are influenced by various factors, even if the goodness-of-fit indices are used as objective measures throughout the superposing process. For a truly rational superposition, the best superposing mode should be chosen without preconception from all the possible modes, taking into account the conformational freedoms in both molecules. Superposition itself is insignificant unless the molecules are superposed in their active conformations. But, it is not easy to elucidate the active conformation for flexible molecules and there are many systems where no rigid active molecule appropriate for use in superposition is available. In most molecular superposition studies, molecules have been superposed in the fixed conformation in crystal or that of energy minimum in spite of the lack of any physical meaning. When the active conformation is not known easily, the best superposing mode should be looked for by systematic rotations and bond rotations of the molecule, although this requires a vast amount of computation.

Although the manual use of the RECEPS program allows us arbitrarily to change both conformations and superposing modes, it is quite difficult to test all the possibilities. So we have developed a new function for the RECEPS program, called AUTOFIT, which can superpose molecules automatically, in order to cover all the possible modes of superposition (Kato *et al.*, 1992). We have adopted a method for testing all the possible combinations of hydrogen-bonding functional groups between molecules, instead of searching the whole space for the best superposing mode by systematic 3-D rotation of a molecule.

The procedures are as follows:

1 Generate all the combinations of hydrogen-bonding functional groups between two molecules.
2 For each combination, superpose the two molecules by a novel least-squares procedure, from which a superposed structure and a good-of-fit index are obtained.
3 Order the results for all the combinations by the indices in order to select promising superposition models.

This least-squares calculation superposes two molecules so that the expected hydrogen-bonding sites in the receptor that are deduced from corresponded hydrogen-bonding groups of two molecules match well in position and in direction as shown in Fig. 14.8. The quantity F to be minimized is shown in eqn. 14.2. The weight w_i , shown in Eqn. 14.3, is updated at every cycle of minimization by using the results of the previous cycle:

$$F = \sum_i w_i\{|\mathbf{Y}_{\mathrm{HB}_i} - \mathbf{X}_{\mathrm{HB}_i}|^2 + |\overrightarrow{\mathbf{DY}}_{\mathrm{HB}_i} - \overrightarrow{\mathbf{DX}}_{\mathrm{HB}_i}|^2\} - \sum_i w_i \mathbf{R}_{\mathrm{HB}}{}^2 \qquad (14.2)$$

$$w_i \propto 1/\{|\mathbf{Y}_{\mathrm{HB}_i} - \mathbf{X}_{\mathrm{HB}_i}|^2 + |\overrightarrow{\mathbf{DY}}_{\mathrm{HB}_i} - \overrightarrow{\mathbf{DX}}_{\mathrm{HB}_i}|^2\} \qquad (14.3)$$

where w_i = weight for the ith pair, $\mathbf{X}_{\mathrm{HB}_i}$ = expected site in the receptor deduced from the first molecule, $\mathbf{Y}_{\mathrm{HB}_i}$ = expected site in the receptor deduced from the second molecule, $\overrightarrow{\mathbf{DX}}_{\mathrm{HB}_i}$ = expected direction to $\mathbf{X}_{\mathrm{HB}_i}$, $\overrightarrow{\mathbf{DY}}_{\mathrm{HB}_i}$ = expected direction to $\mathbf{Y}_{\mathrm{HB}_i}$ and $\mathbf{R}_{\mathrm{HB}}$ = ideal hydrogen-bonding distance (assumed to be 3.1 Å). For selecting correct superposition models, hydrogen bonds are the most important but not enough. It is necessary to compare the models by using other properties such as the electrostatic potential and the molecular shape together with the hydrogen bonding. Then small numbers of the superposing models with favourable indices are refined by minimizing the united index which is the weighted sum of the goodness-of-fit indices from the 3-D grid point data. The AUTOFIT function has enabled us not only to choose the best superposing modes from all the possible ones without preconception but also to prepare promising starting structures for careful manual superposition. The function gives objectivity, reproducibility and reliability to the results of the superposition, with less labour and in a shorter time. Since this method is based solely on the hydrogen bonding, it would be effective for the molecular system where hydrogen bonds play important roles. For other systems, another efficient automatic method needs to be developed or the manual method should be used for the time being.

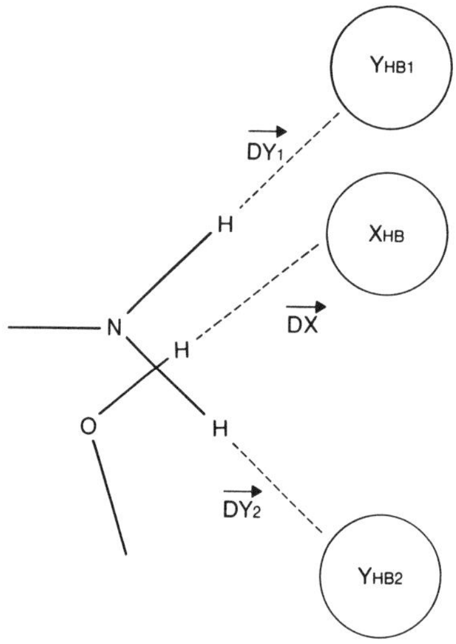

Figure 14.8. Scheme of the least-squares procedure of the automatic superposition (AUTOFIT), which makes matching of the position and direction of the expected hydrogen-bonding sites (○) in the receptor deduced from the two molecules.

4.4 *Superposition of dihydrofolate and methotrexate by the* AUTOFIT *function*

In order to show the usefulness of the AUTOFIT function, we have superposed dihydrofolate (DHF), which is the substrate for dihydrofolate reductase, onto the potent inhibitor methotrexate (MTX). Although the binding mode of MTX molecule to the enzyme has been elucidated by crystallographic study of the complex, no information from the crystal structure was used in the superposition. As shown in Fig. 14.9. the two molecules have quite similar structures where a carbonyl group in DHF is replaced by an amino group in MTX. In this study, only the pteridine part was superposed, neglecting the other parts.

In the superposition by the AUTOFIT function, each of the six heteroatoms (encircled in Fig. 14.9) in the two pteridine rings were combined between the two molecules, generating 720 (= 6!) combinations in total. For the 720 combinations, the result of the least-squares calculation of the AUTOFIT function was compared with that of the conventional least-squares matching of atomic positions. In the result of

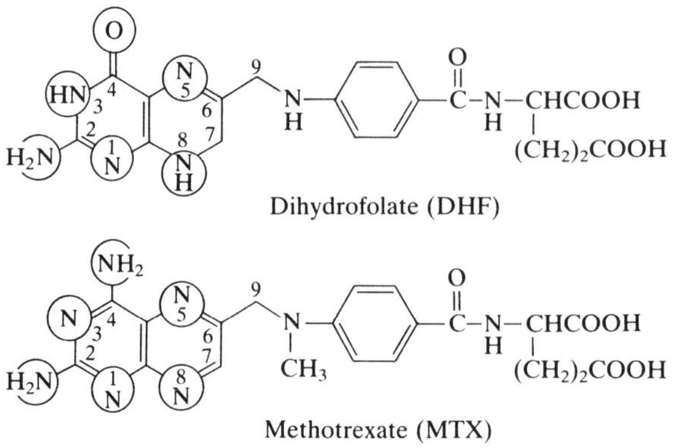

Figure 14.9. Chemical structures of dihydrofolate (DHF) and methotrexate (MTX).

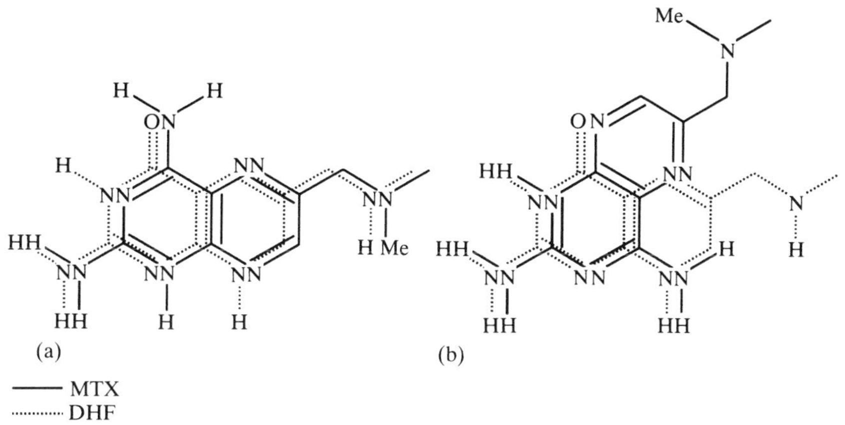

Figure 14.10. Two representative superposing models of DHF and MTX: (a) model A; (b) model B.

the conventional method, model A (Fig. 14.10(a)) and model B (Fig. 14.10(b)) were the best and the second best, respectively. In the result of the AUTOFIT function, model B was the best, and model A was ranked as 156th from the best among the 720 combinations. In model A, the two pteridine rings are just fitted, whereas in model B, the pteridine ring in DHF is reversed by rotation around the C6–C9 bond.

Which is the correct superposition, model A or model B? This question can be answered by considering the stereochemistry of the enzymic reaction. The hydrogen atom at C6 in tetrahydrofolate, which is the product of the reduction of DHF as shown in Fig. 14.11, was proved to originate from the cofactor NADPH (Charlton *et al.*, 1979). The required relative position between the pteridine ring of DHF and NADPH for the stereospecificity of this reaction can be presumed to be as follows. As the crystal structure of the ternary complex of enzyme–NADPH–MTX has been elucidated (Bolin *et al.*, 1982), we know from which side of the pteridine ring of MTX the attack of the hydride ion from NADPH occurs. If DHF binds to the enzyme with its pteridine ring oriented as in the superposing model A, the enzyme would produce anomeric tetrahydrofolate with the opposite stereochemistry at the C6 position compared to the real enzymatic product. The stereochemistry is satisfied if the pteridine ring of DHF is flipped over, as in the superposing model B. Thus, it can be concluded that model B is the correct superposition of DHF and MTX, when binding to a common receptor is assumed. Many experimental studies and docking simulation also support this conclusion. This example has shown that

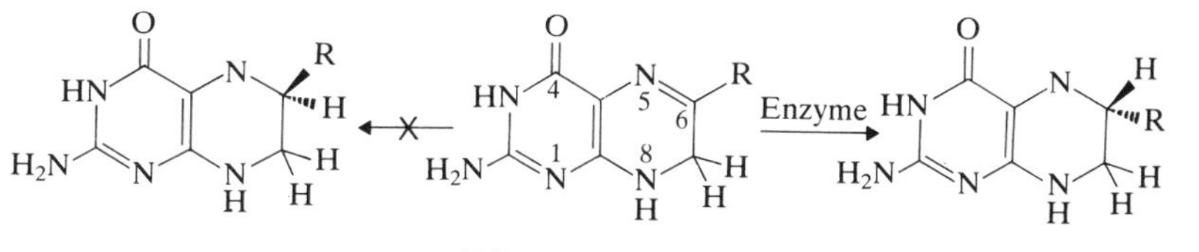

Figure 14.11. Stereochemistry of the reduction of dihydrofolate (DHF) to tetrahydrofolate (THF) by dihydrofolate reductase. The hydrogen atom at C6 in THF comes from the cofactor NADPH.

our method is very useful for choosing the correct superposition covering all the possibilities without any preconception.

5 Designing new structures

For design of new active molecules, it is necessary to start either from target receptor structure or from structures of known active molecules (drugs or natural active compounds in the body). Until now, quite a few drugs have resulted from attempts to design conformationally restricted analogues of known bioactive molecules with flexible structures (Patchett *et al.*, 1980; Thorsett *et al.*, 1986; Kuyper, 1989; Hangauer, 1989). In some cases, analogues whose conformations were restricted by various-sized ring structures were systematically synthesized and tested. In other cases, conformational analyses were performed using energy calculations in order to examine whether the designed analogues can stably adopt the active conformation of the known drug molecule.

Recently, as outcomes from protein crystallography have accumulated, receptor-based drug design has come to attract much attention. Based on the inspection of X-ray structures of protein–ligand complexes or docking simulation to the receptor cavity, many groups have reported successful results of modification of known ligand structures for better receptor binding (Beddell *et al.*, 1976; Kumar *et al.*, 1981; Kuyper & Roth, 1982; Abraham *et al.*, 1984). By elaborate iterations of design, computer simulation, chemical syntheses, biological assays and structure elucidation of the complex, new molecules with different skeletal structures have been successfully developed as promising lead compounds (Appelt *et al.*, 1991).

The GREEN program provides convenient functions for modifying structures and also for interactive *de novo* design of new structures. It provides functions for model building with fragmentary structural data, and functions for estimating energetic contributions for every change of structure. But, by gradual structure modification, it requires much effort and many iterations of the procedure to design new structures that deviate significantly from the known structures or have completely different skeletons. In order to design new lead compounds logically without relying on chance, we have come to the conclusion that we need new concepts and new methods which utilize computer power efficiently. The problem is how to discover structures which can fit well to the receptor cavity or those which maintain the structural features required for the biological activity. Fundamentally, two approaches are thought to be possible, one which makes use of structural databases such as the Cambridge Structural Database (Allen *et al.*, 1979) as a source of new structures and another which constructs possible structures under energetic considerations. The advantages of using computers in lead generation are serendipity, variety and coverage of the possibilities in designed molecules. One group has succeeded in searching the crystallographic structural database for skeletal structures whose molecular shapes fit well to the receptor cavity (DesJarlais *et al.*, 1988). In the application to the design of an HIV protease inhibitor, a molecule which is closely related to a known antipsychotic compound was selected by their method and was shown to have an inhibitory activity to the enzyme after adequate modification (DesJarlais *et al.*, 1990). Another group applied this method to the design of a ligand for the allosteric site of the enzyme phosphofructokinase (Sampson & Bartlett, 1991).

In order to make a starting point for artificial lead generation based on receptor structure, we have developed a program LEGEND (Nishibata & Itai, 1991). The program constructs new structures which can bind stably to the target receptor. We have also developed a program for constructing new molecules based on known bioactive molecules. It was shown that the program successfully generated molecules with various skeletal structures which maintain relative positions and orientations of chosen functional groups as in the active structure of the known molecule (Inoue *et al.*, unpublished results).

5.1 de novo *lead generation based on receptor structure*

The requirements for the structures generated by the program LEGEND are as follows:
1 Good fit to the receptor cavity or to the receptor model.
2 Specific interaction with the receptor with maximum hydrophilic interactions.
3 Stable geometry and conformation.
4 Wide variety of structures.

The LEGEND program generates new structures based on protein structure from the protein data bank by using random numbers and a force field as shown in Fig. 14.12. Starting from one of the hydrogen-bonding heteroatoms as an anchor, each molecule is generated by placing atoms one by one, determining an atom type, a bond type and atomic position (torsion angle) of the new atom by random numbers. As standard bond lengths and angles, those values in MM2 program (Allinger, 1976) are used. For the high-speed estimation of intermolecular van der Waals energy and electrostatic energy, the program makes use of tabulated data on 3-D grid points just like the GREEN program. If the new atomic position is not acceptable due to the severe violation of van der Waals radii of the previously

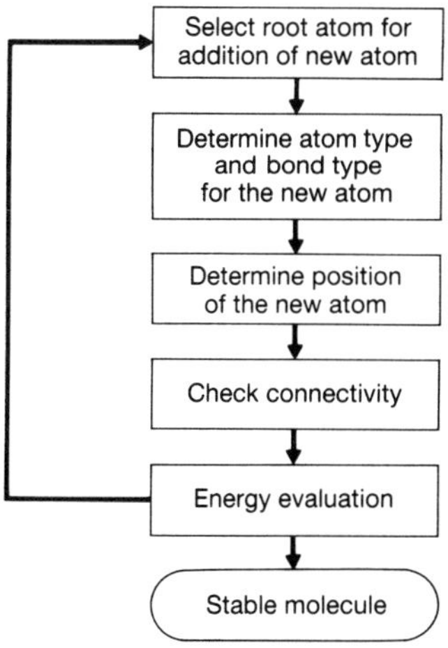

Figure 14.12. Flow chart of the program LEGEND.

generated atoms or unstable intermolecular interaction energy, the program reassigns the root atom and attempts to find an acceptable new atom. If the attempts fail after a given number of repeats, the program tracks back to the last step, i.e. it withdraws the last of the previously generated atoms and regenerates that atom. Carbon atoms at the grid point with very deep electrostatic potential are changed to heteroatoms. After generating atoms up to the specified molecular size, the program completes fragmentary aromatic rings by adding missing carbon atoms, and supplies hydrogen atoms for all the remaining valencies of non-hydrogen atoms. Finally, the structure is optimized by the Simplex method taking into account the intra- and intermolecular energy.

The LEGEND program goes on generating structures one after another up the maximum number of structures specified in the input data. From these generated structures, a small number of structures is selected by the program LORE, which selects structures on the basis of various energetic values, as well as some indices related to structural features.

5.2 *Application to* Escherichia coli *dihydrofolate reductase*

We have applied the LEGEND program to *E. coli* dihydrofolate reductase, whose structure was elucidated as a ternary complex of enzyme–NADPH–inhibitor folate (Bystroff *et al.*, 1990). Among 300 generated structures, Fig. 14.13 shows six structures that were selected by the program LORE using criteria on intra- and intermolecular energy, and number of hydrogen bonds (at least three in this case). In these selected structures, it was proved that heteroatoms were located fairly well without chemical inconsistency. Most of the heteroatoms can form hydrogen bonds to the enzyme. The hydrogen-bonding scheme of one of the generated structures in the binding-site cavity is shown in Fig. 14.14. The validity of the structures was tested by the stabilities of protein–ligand complexes after optimization by the AMBER program (Seibel *et al.*, 1989). The intra- and intermolecular energy values for the computer-generated ligands were similar, although they were rather unstable compared with those of the known ligands. Molecular orbital calculations were also applied to these compounds in order to check conformational stability.

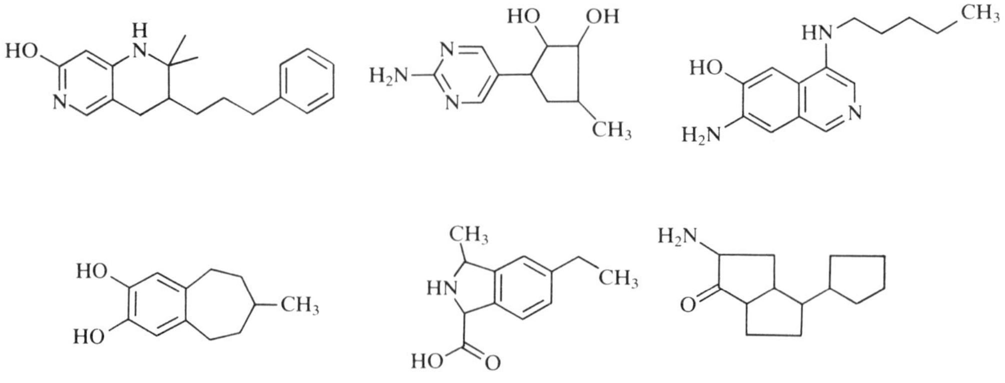

Figure 14.13. Lead-candidate structures, selected from 300 structures generated by the LEGEND program using *E. coli* dihydrofolate reductase as a receptor.

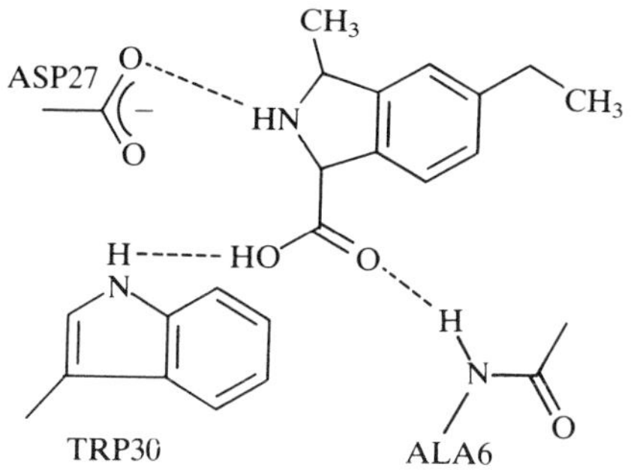

Figure 14.14. Hydrogen-bonding scheme of one of the compounds in Figure 14.13 in the binding-site cavity.

Remarkable structural changes were not caused by optimization using the PM3 method (Stewart, 1989) in the MOPAC program (Stewart, 1990).

As we must select a rather small number of promising structures from the vast number of generated structures where graphical selection is difficult, proper numerical selection is very important in our strategy. Efficient and reliable methods for selection should be established. Computer simulations on stabilities, physical properties and molecular interactions would be useful for further modification and selection, before synthesizing the most promising compounds. If a synthesized compound was proved to be active by receptor-binding assay, even if its potency was low, it might become a lead compound. After that, an elaborate optimization process for better biological activity and behaviour will be necessary just as in conventional drug development processes.

6 Perspective

Of the two requirements for drugs, reaching the receptor site and binding to the receptor, QSAR methods (Hansch & Fujita, 1964) are more competent for treating the former factors than 3-D approaches. QSAR analyses are helpful for increasing the efficiency of lead optimization processes. However, it would be more useful and clear if the information from different series of compounds could be merged into an analysis. This could be realized only by a 3-D approach of superposing those molecules. But for deeper understanding of drug delivery, it might be necessary to establish 3-D approaches to simulate various factors such as membrane permeability. Extensive basic research will be required to solve the problem.

As for binding to receptors, 3-D approaches give much more concrete information. Submolecular features and properties in 3-D space are more important for specific binding to the receptor than are whole-molecular ones. Protein crystallography has provided us not only individual receptor structures, but also the general concept of drug–receptor interaction. It seems to be most important and useful both for interpreting structure–activity relationships and for designing new bioactive structures. It is not the similarity of chemical structure but of physical and chemical properties involved in intermolecular interaction that is important for favourable drug–receptor interaction. Therefore, when we design new bioactive molecules, we should not be limited to skeletal structures of known active compounds too much, although molecules with similar structures are apt to show similar kinds of activities.

The most interesting subject in drug design in the 21st century is lead generation. Will we be able to discover or produce lead compounds as we would like, without relying on chance? To discover or produce minimum structures which satisfy the presupposition of drugs is the first step of artificial lead generation. To that purpose, several approaches have been attempted, searching for new drug structures based either on receptor structure or on known ligand structure. Although database approaches under the same presupposition would be useful for lead discovery, *de novo* construction is more challenging because of variety and serendipity in the generated structures. In a trial where computers suggest to us all the possible and promising structures, we have developed a program for automatically creating structures which can fit well to receptor cavity, providing sufficient intra- and intermolecular stability. We will be able to bring our ability into full play for selecting or modifying the structures for more favourable properties. This is also a trial to pioneer new ways of using computers, i.e. computers provide possible structures with certain features, as opposed to their use so far, in which we input structures into the computer for evaluating quantities. This might add a new page to the history of the relationship between humans and computers.

For the purposes of both structure–activity relationships and the design of new molecules, theoretical calculations are becoming more and more important. High reliability and accuracy is required for predicting drug–receptor interaction, physical properties and dynamic behaviour. Although the fundamental methodologies were established decades ago, there still remain many problems which should be solved or improved. As small energy differences determine relative affinity to the receptor (1.37 kcal/mol per order of binding constant), rigorous accuracy is required for the force-field functions and parameters, and atomic charges and dielectric constants used. For such accuracy, entropic effects such as free-energy contributions of hydrophobic interactions are not negligible. We realize the acute importance of further basic research in theoretical chemistry, especially for drug–receptor interactions, and for the necessity that we evaluate generated lead candidates in order to select promising structures.

In the 21st century, more detailed mechanisms both of normal biological systems and of disease will be made clear, and a considerable part of drug development will come about on the basis of receptor structure and computer-assisted drug design. It is expected that lead compounds will be prepared as we want, based on *de novo* lead-generation methods by using individual receptor structures. Selective drugs will be developed by making use of small structural differences between receptors, species, organs and receptor subtypes which control different biological activities.

7 References

Abraham, D.J., Kennedy, P.E., Mehanua, A.S., Potwa, D., & Williams, F.L. (1984). *J. Med. Chem.*, **27**, 967–78.

Allen, F.H., Bellard, S., Brice, M.D., Cartwright, B.A., Doubleday, A., Higgs, H., Hummelink, T.,

Hummelink-Peters, B.G., Kennard, O., Motherwell, W.D.S., Rodgers, J.R., & Watson, D.G. (1979). *Acta Cryst.*, **B35**, 2331–9.

Allinger, N.L. (1976). *Adv. Phys. Org. Chem.*, **13**, 1–76.

Appelt, K., Bacquet, R.J., Bartlett, C.A. *et al.* (1991). *J. Med. Chem.*, **34**, 1925–34.

Barakat, M.T. & Dean, P.M. (1990). *J. Comput.-Aided Mol. Design*, **4**, 295–316.

Beddell, C.R., Goodford, P.J., Norrington, F.E., Wilkinson, S. & Wootton, R. (1976). *Br. J. Pharmacol.*, **57**, 201–9.

Bernstein, F.C., Koetzle, T.F., Williams, G.J.B., Meyer, E.F., Brice, M.D., Rodgers, J.R., Kennard, O., Shimanouchi, T. & Tasumi, M. (1977). *J. Mol. Biol.*, **112**, 535–42.

Blumberg, P.M. (1988). *Cancer Res.*, **48**, 1–8.

Blundell, T.L. & Johnson, L.N. (1976). *Protein Crystallography*, Academic Press, London.

Bolin, J.T., Filman, D.J., Matthews, D.A., Hamlin, R.C. & Kraut, J. (1982). *J. Biol. Chem.*, **257**, 13 650–62.

Bystroff, C., Oatley, S.J. & Kraut, J. (1990). *Biochemistry*, **29**, 3263–77.

Charlton, P.A., Young, D.W., Birdsall, B., Feeney, J. & Roberts, G.C.K. (1979). *J. Chem. Soc. Chem. Comm.*, 922–4.

Dean, P.M. (1987). *Molecular Foundations of Drug–Receptor Interaction*, Cambridge University Press, Cambridge.

DesJarlais, R.L., Sheridan, R.P., Seibel, G.L., Dixon, J.S., Kuntz, I.D. & Venkataraghavan, R. (1988). *J. Med. Chem.*, **31**, 722–9.

DesJarlais, R.L., Seibel, G.L., Kuntz, I.D., Furth, P.S. & Alvarez, J.C. (1990). *Proc. Natl. Acad. Sci. USA*, **87**, 6644–8.

Endo, Y., Shudo, K., Itai, A., Hasegawa, M. & Sakai, S. (1986). *Tetrahedron*, **42**, 5905–24.

Fujiki, H., Mori, M., Nakayasu, M., Terada, M., Sugimura, T. & Moore, R.E. (1981). *Proc. Natl. Acad. Sci. USA*, **78**, 3872–6.

Fujiki, H., Suganuma, M., Nakayasu, M., Hoshino, H., Moore, R.E. & Sugimura, T. (1982). *Gann*, **73**, 495–7.

Fujiki, H., Suganuma, M., Tahira, T., Yoshioka, A., Nakayasu, M., Endo, Y., Shudo, K., Takayama, S., Moore, R.E. & Sugimura, T. (1984). In: H. Fujiki, E. Hecker, R.E. Moore, T. Sugimura and I.B. Weinstein (Eds.), *Cellular Interactions by Environmental Tumor Promoters*, pp. 37–45. Japan Sci. Soc., Tokyo; VNU Science, Utrecht.

Hangauer, D.G. (1989). In: T.J. Perun and C.L. Prost (Eds.). *Computer-Aided Drug Design. Methods and Applications*, pp. 253–95. Marcel Dekker, New York.

Hansch, C. & Fujita, T. (1964). *J. Am. Chem. Soc.*, **86**, 1616–26.

Hecker, E. (1978). In: T.J. Slaga, A. Sivak and R.K. Boutwell (Eds.), *Carcinogenesis: A Comprehensive Survey, Mechanisms of Tumor Promotion and Cocarcinogenesis*, pp.11–48. Raven Press, New York.

Hitotsuyanagi, Y., Yamaguchi, K., Ogata, K., Aimi, N., Sakai, S., Koyama, Y., Endo, Y., Shudo, K., Itai, A. & Iitaka, Y. (1984). *Chem. Pharm. Bull.*, **32**, 3774–8.

Itai, A., Kato, Y. & Iitaka, Y. (1986) In: Y. Iitaka, & A. Itai, (Eds.), *Proceedings of Symposium on Three-Dimensional Structures and Drug Action*, pp. 195–205. University Tokyo Press, Tokyo.

Itai, A., Tomioka, N. & Kato, Y. (1992). In: *QSAR: New Developments and Applications*, Elsevier Science Publishers, Amsterdam. In press.

Itai, A., Kato, Y., Tomioka, N., Iitaka, Y., Endo, Y., Hasegawa, M., Shudo, K., Fujiki, H. & Sakai, S. (1988). *Proc. Natl. Acad. Sci. USA*, **85**, 3688–92.

Jeffrey, A.M. & Liskamp, R.M.J. (1986). *Proc. Natl. Acad. Sci. USA*, **83**, 241–5.

Kato, Y., Itai, A. & Iitaka, Y. (1987). *Tetrahedron*, **43**, 5229–36.

Kato, Y., Inoue, A., Yamada, M., Tomioka, N. & Itai, A. *J. Comput.-Aided Mol. Design*. In press.

Kumar, A.A., Mangum, J.H., Blankenship, D.T. & Friesheim, J.H. (1981). *J. Biol. Chem.*, **256**, 8970–6.

Kuyper, L.F. (1989). In: T.J. Perun & C.L. Prost (Eds.), *Computer-Aided Drug Design, Methods and Applications*, pp. 327–69, Marcel Dekker, New York.

Kuyper, L.F. & Roth, B. (1982). *J. Med. Chem.*, **25**, 1120–2.

Luecke, H. & Quiocho, F.A. (1990). *Nature*, **347**, 402–6.

Marshall, G.R., Barry, C.D., Bosshard, H.E., Dammkoehler, R.A. & Dunn, D.A. (1979). In: E.C. Olson and R.E. Christoffersen (Eds.), *Computer-Assisted Drug Design*, ACS Sympasium series 112, pp. 205–26, American Chemical Society, Washington D.C.

Nedler, J.A. & Mead, R. (1965). *Computer J.*, **7**, 308.

Nishibata, Y. & Itai, A., (1991). *Tetrahedron*, **47**, 8986–90.

Patchett, A.A., Harris, E., Tristram, E.W. *et al.* (1980). *Nature*, **288**, 280–3.

Sakai, S., Aimi, N., Yamaguchi, K., Hitotsuyanagi, Y., Watanabe, C., Yokose, K., Koyama, Y., Shudo, K. & Itai, A. (1984). *Chem. Pharm. Bull.*, **32**, 354–7.

Sampson, N.S. & Bartlett, P.A. (1991). *J. Org. Chem.*, **56**, 7179–83.

Seibel, G., Singh, U.C., Weiner, P.K., Caldwell, J. & Kollman, P. (1989). *AMBER 3.0 Revision A*, University of California, San Francisco.

Stewart, J.J.P. (1989). *J. Comp. Chem.*, **10**, 209–20.

Stewart, J.J.P. (1990). *J. Comput.-Aided Mol. Design*, **4**, 1–105.

Thorsett, E.D., Harris, E.E., Aster, S.D. Peterson, E.R., Snyder, J.P., Springer, J.P., Hirshfield, J., Tristram, E.W., Patchett, A.A., Ulm, E. H. & Vassil, T. C. (1986). *J. Med. Chem.*, **29**, 251–60.

Tomioka, N., Itai, A. & Iitaka, Y. (1986). In: Y. Iitaka and A. Itai (Eds.), *Proceedings of Symposium on Three-Dimensional Structures and Drug Action*, pp. 186–96. University of Tokyo Press, Tokyo.

Tomioka, N., Itai, A. & Iitaka, Y. (1987). *J. Comput.-Aided Mol. Design*, **1**, 197–210.

Weiner, S.J., Kollman, P.A., Case, D.A., Singh, U.C., Ghio, C., Alagona, G., Profeta, S. Jr. & Weiner, P. (1984). *J. Am. Chem. Soc.*, **106**, 765–84.

Weiner, S.J., Kollman, P.A., Nguyen, D.T. & Case, D.A. (1986). *J. Comput. Chem.*, **7**, 230–52.

Wender, P.A., Koehler, K.F., Sharkey, N.A., Dell'Aquila, M.L. & Blumberg, P.M. (1986). *Proc. Natl. Acad. Sci. USA*, **83**, 4214–18.

Synthesis of Peptidomimetics

15 Rational Design of Peptidomimetics: Structural and Pharmacological Aspects

P.W. SCHILLER
Laboratory of Chemical Biology and Peptide Research, Clinical Research Institute of Montreal, 110 Pine Avenue West, Montreal, Quebec, Canada H2W 1R7

1 Introduction

In current efforts aimed at developing compounds containing non-peptide structural elements and capable of mimicking or antagonizing the biological action(s) of the natural parent peptide, two fundamentally different approaches are being taken. One strategy could be called the 'rational design' of peptidomimetics, since it uses the peptide of interest or its analogues as the starting point for further structural modification under consideration of conformational aspects. The other more empirical approach is based on screening of large numbers of compounds by means of a radioreceptor binding assay. Screening of synthetic compounds or of compounds from natural sources (e.g. fermentation broths) and subsequent structural modification recently led to the development of potent and receptor-selective non-peptide antagonists of cholecystokinin, angiotensin-II and substance P, as is discussed in Chapter 16. The major focus here is on the rational design of peptidomimetics.

Many of the smaller peptide hormones and neurotransmitters are structurally flexible molecules and, depending on the environment, may assume a number of different conformations of comparatively low energy. Due to the demonstrated fact that small linear peptides, such as the enkephalins, may exist in an equilibrium of different conformers in solution (for a review, see Schiller, 1984), conformational studies under these circumstances are of doubtful value because only average conformational parameters can be determined. Furthermore, the bioactive conformation of these peptides may be different from any of the conformers existing in solution and may only be assumed upon binding to the target site (receptor, enzyme, antibody, etc.) as the result of mutual conformational adjustments ('induced fit'). The inherent flexibility of many of the natural peptide hormones and neurotransmitters may also be the reason for their generally observed lack of selectivity towards different receptor types or subtypes, since conformational adaptation to different receptor topographies is possible. In view of this situation, the incorporation of conformational constraints into peptides without loss of biological activity appears attractive, not only because meaningful conformational studies aimed at determining bioactive conformations become possible, but also because compounds with improved receptor selectivity may result from such efforts. Conformational restriction of peptides represents a first step in the rational approach towards developing peptidomimetics because conformationally constrained analogues may serve as relatively rigid templates for further structural modification aimed at removing some of the less attractive peptide structural features (e.g. peptide backbone replacements).

Here we will illustrate these principles primarily with examples from the opioid peptide field which, of course, is of interest because opioid peptidomimetics (opiate alkaloids, etc.) were known long before the discovery of the opioid peptides. Furthermore, opioid pharmacology offers unique possibilities for testing new principles of peptidomimetic design because receptor-selective *in vitro* bioassays and receptor binding assays are available. The general significance of these principles of peptidomimetic design via constrained analogues will be documented with examples drawn from analogous developments with other peptide hormones and neurotransmitters. Fundamental pharmacological aspects related to the development of peptidomimetics will also be discussed. Issues to be addressed will include the mode of receptor binding of peptidomimetics, their 'efficacy' ('intrinsic activity') and their usefulness for the discovery of new receptor types or subtypes. Finally, a perspective on anticipated future developments expected to lead to a further rationalization and refinement of peptidomimetic design will be given.

2 The use of conformational restriction as a first step towards the development of peptidomimetics

Conformational restriction of a structurally flexible peptide hormone or neurotransmitter may affect various parameters related to its biological activity profile, including receptor affinity, receptor selectivity, efficacy and stability against enzymatic degradation. The effect of conformational restriction on receptor affinity can be discussed in terms of either the 'lock-and-key' model (Fischer, 1894) or the 'zipper model' (Burgen *et al.*, 1975). According to the lock-and-key model, one of the conformers present in the conformational equilibrium of the unrestricted peptide in solution would interact with the receptor without undergoing a significant conformational change. Freezing of this 'bioactive' conformation through introduction of conformational constraints would increase affinity, since all molecules rather than a fraction only would be able to bind to the receptor. In the 'zipper' model it is assumed that a flexible ligand undergoes a number of conformational changes in a stepwise binding process to assume finally the receptor-bound conformation. Thermodynamic considerations predict that, in comparison with the flexible parent peptide, an analogue conformationally constrained to assume the 'correct' bioactive conformation should again have higher receptor affinity because the loss of internal rotational entropy upon binding is smaller in the case of the more rigid ligand. The validity of such entropy considerations has been demonstrated with a series of differentially constrained clonidine-like imidazolidines binding to α-adrenergic receptors (Avbelj & Hadži, 1985). Thus, in theory, conformationally restricted peptide analogues should display higher receptor affinity than corresponding non-restricted peptides. The observation that in most cases conformational restriction produced an affinity decrease may be due to the fact that the interaction of certain pharmacophoric moieties with complementary receptor subsites may be less than optimal as a consequence of the introduced conformational constraints. Furthermore, it is also possible that conformational restriction may adversely affect the processes of binding to, and dissociation from, the receptor such that a net decrease in affinity results. In a situation of receptor heterogeneity, conformational restriction of peptide ligands has the demonstrated potential of leading to compounds with improved receptor selectivity, since the various receptor types may have different conformational requirements, as first shown in the case of the opioid receptors (Schiller & DiMaio, 1982). Ideally, the introduced conformational constraints would be compatible with the topography of only one preferred receptor type and detrimental to the interaction with all other receptor types. The conformational behaviour of a ligand (L) is not only implicated in its binding to, and dissociation from, the receptor (R) (initial binding equilibrium depicted in the left part of the scheme shown in Fig. 15.1), but also plays an important role in signal transduction, the extent of which is indicated by its efficacy (intrinsic activity). Signal transduction is thought to involve a ligand-induced conformational change leading to an activated form of the receptor (R^*), and the efficacy of a ligand is related to its ability to shift the equilibrium shown on the right-hand side of the scheme (Fig. 15.1) such as to increase the number of activated receptors. At present, little information on the relationship between structural flexibility of a ligand and its efficacy is available. Intuitively, one would expect that structural rigidification of a receptor ligand might reduce its efficacy, thereby resulting in partial agonists or antagonists; however, an efficacy enhancement leading to superagonists cannot be excluded *a priori*. Finally, it can also be expected that conformationally restricted peptide analogues might show improved stability against enzymatic degradation because the introduced conformational constraints might be incompatible with the substrate conformational requirements of peptidases.

Local conformational restriction of peptides can be achieved at particular amino acid residues either through appropriate modification of the peptide backbone or through incorporation of conformational constraints into side-chains, whereas more global conformational restriction is usually obtained through peptide cyclizations. Well-known modifications resulting in local conformational restriction of the peptide backbone (limitation of accessible ϕ, ψ angles) include N^α- and C^α-methylation (Marshall & Bosshard, 1972) and incorporation of a lactam bridge between the α-nitrogen of an amino acid residue and the α-carbon of the preceding residue (Freidinger *et al.*, 1982). A number of biologically active peptides have been modified in this manner and only a few examples are mentioned here. Insight into the bioactive conformation was obtained with C^α- and N^α-methylated analogues of angiotensin-II (Marshall & Bosshard, 1972). Incorporation of a γ-lactam ring structure into a peptide may stabilize a β-turn to some extent, even though other conformations are still possible. Lactam-bridged peptide analogues have been prepared in order to test β-turn models of the bioactive conformation. A γ-lactam analogue of luteinizing hormone-releasing hormone (LH-RH) was more potent than the native parent peptide in two bioassays and, thus, was in agreement with a proposed β-turn-containing bioactive conformation (Freidinger *et al.*, 1980), whereas various lactam-bridged analogues of cyclosporin showed low potency (Lee

$$R + L \underset{k_{-1}}{\overset{k_1}{\rightleftharpoons}} RL \underset{k_{-2}}{\overset{k_2}{\rightleftharpoons}} R^* L,$$

Figure 15.1. Scheme of receptor binding and activation equilibria.

et al., 1990). Sato & Nagai (1986) replaced the D-Phe-Pro dipeptide segments in gramicidin-S with a bicyclic dipeptide derivative in an attempt to stabilize the type II'β-turns centred on these residues. The retained antibacterial activity of the resulting compound demonstrated quite conclusively that there are indeed two β-turns in the bioactive conformation of gramicidin-S. Examples of peptide analogues with local backbone conformational restrictions showing improved receptor selectivity include N-methylated analogues of substance P (Wormser *et al.*, 1986) as well as a lactam-bridged substance P analogue (Cascieri *et al.*, 1986). Among various structural modifications aimed at restricting side-chain conformational flexibility, the substitution of dehydroamino acids or cyclopropylamino acids are best known. An illustrative example is a δ-opioid receptor-selective analogue of [D-Ala^2,Leu^5]enkephalin (Shimohigashi *et al.*, 1987). In a more recent development, substitution of a conformationally restricted methionine analogue in the C-terminal tetrapeptide of cholecystokinin resulted in a potent compound with B-receptor selectivity (Holladay *et al.*, 1991).

Cyclic peptide analogues can be obtained through end-group/end-group cyclization, side-chain/end-group cyclization or cyclization between two side-chains. Examples of the last type include cystine-bridged cyclic peptides such as exist also in nature (oxytocin, somatostatin, calcitonin, etc.) and cyclic lactam analogues resulting from amide bond formation between the side-chain amino group of an α,ω-diamino acid residue (Lys, Orn, etc.) and the side-chain carboxyl function of an Asp or Glu residue. The design of such cyclic analogues should be based on careful consideration of structure–activity data obtained with linear analogues and, possibly, also of conformational features of the unrestricted parent peptide (with the caveat that the crystal structure or the predominant solution conformation may be different from the receptor-bound conformation).

3 Biologically active cyclic analogues of peptide hormones and neurotransmitters

3.1 *Prototypes of receptor-selective cyclic opioid peptide analogues*

It is now generally accepted that at least three major opioid receptor types (μ, δ, κ) exist (Paterson *et al.*, 1984). The development of stable opioid peptide analogues and opioid peptidomimetics displaying high receptor selectivity continues to be an important challenge for peptide chemists and medicinal chemists. Whereas the preparation of linear analogues of opioid peptides resulted in a number of receptor-selective compounds (for a review, see Schiller, 1991), the major focus here will be on cyclic opioid peptide analogues because they represent a promising starting point for the design of novel peptidomimetics.

In an early effort cyclic analogues of [Leu^5]enkephalin were obtained through substitution of an α,ω-diamino acid (Daa) with D-configuration in the 2-position of the peptide sequence and subsequent amide bond formation between the side-chain amino function and the C-terminal carboxyl group (Fig. 15.2) (DiMaio & Schiller, 1980; DiMaio *et al.*, 1982; Schiller & DiMaio, 1982). Subtle variation in the degree of conformational restriction was achieved by altering the side-chain length in the 2-position (Daa = A_2pr (α,β-diaminopropionic acid), A_2bu (α,

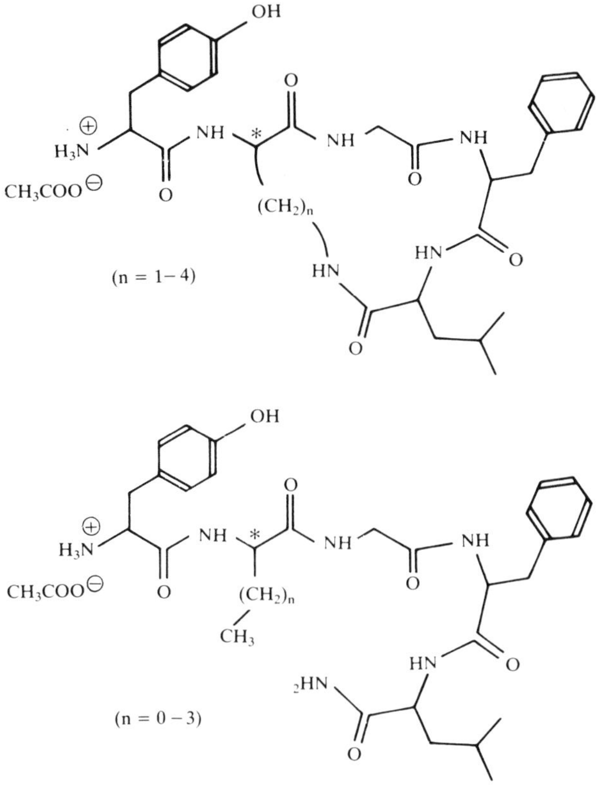

Figure 15.2. Structural formulas of cyclic enkephalin analogues (top) and their linear correlates (bottom).

γ-diaminobutyric acid), Orn or Lys). The resulting compounds, H-Tyr-cyclo[-D-Daa-Gly-Phe-Leu-], all had good affinity for μ-opioid receptors but relatively low affinity for δ-receptors and, therefore, showed considerable μ-selectivity, whereas corresponding open-chain analogues (Fig. 15.2) were non-selective. Since the cyclic analogues are structurally nearly identical to their linear counterparts—the only distinction being the carbon–nitrogen ring-closure bond—these results indicate unambiguously that their μ-receptor selectivity is the direct consequence of the conformational restriction resulting from cyclization. Furthermore, these findings demonstrate that μ- and δ-opioid receptors have indeed different conformational requirements. It is also of interest to point out that all four cyclic analogues had lower μ- and δ-receptor affinity than the corresponding open-chain analogues. These compounds were also tested in the *in vitro* guinea-pig ileum (GPI) and mouse vas deferens (MVD) bioassays, which are representative for μ- and δ-opioid receptor interactions, respectively. Determination of the IC_{50}(MVD)/IC_{50}(GPI) ratios of the cyclic analogues confirmed their μ-receptor selectivity. Finally, one of these compounds, H-Tyr-cyclo[-D-A_2bu-Gly-Phe-Leu-], was shown to be highly resistant to enzymatic degradation (DiMaio & Schiller, 1980).

The cyclic tetrapeptide analogue H-Tyr-D-Orn-Phe-Asp-NH_2 represents the prototype of a side-chain/side–chain cyclized opioid peptide analogue of the lactam type (Schiller *et al.*, 1985a) (Fig. 15.3). In contrast to the cyclic enkephalin analogues, this compound contains the Phe residue in the 3-position of the peptide

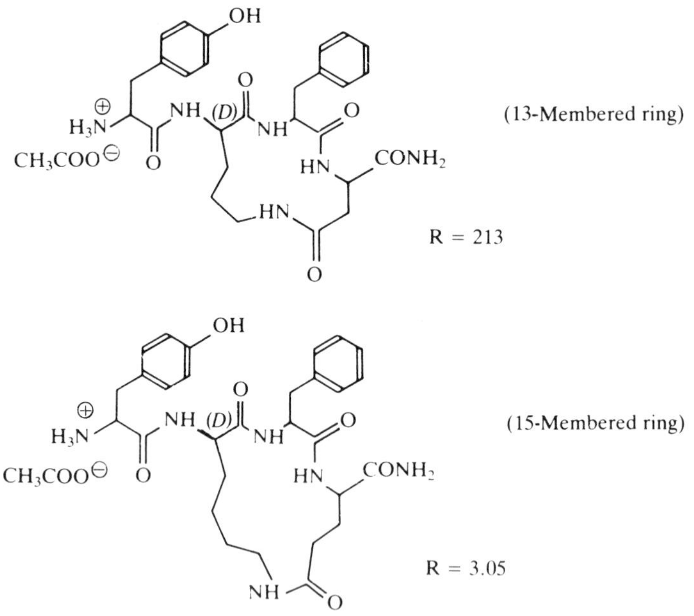

Figure 15.3. Structural formulas and K_i^{δ}/K_i^{μ} ratios (R) of cyclic dermorphin analogues.

sequence and, therefore, is structurally related to the dermorphins. Performance of receptor-binding assays revealed that this compound is highly μ-selective as a result of its very weak binding to δ-receptors. In comparison with the cyclic enkephalin analogues described above, H-Tyr-D-Orn-Phe-Asp-NH$_2$ shows at least tenfold higher μ-selectivity and it represents the most selective cyclic opioid peptide analogue with μ-agonist properties reported to date. Expansion of the 13-membered peptide ring structure contained in the latter analogue to a 15-membered one, as achieved by preparation of the analogue H-Tyr-D-Lys-Phe-Glu-NH$_2$ (Fig. 15.3), resulted in an almost complete loss of μ-receptor selectivity due to a drastic increase in δ-receptor affinity. These results demonstrate that variation in the degree of conformational restriction can have a profound effect on receptor affinity and selectivity. Transposition of the Orn and Asp residues in the parent tetrapeptide analogue led to a compound, H-Tyr-D-Asp-Phe-Orn-NH$_2$, which retained high μ-receptor affinity and selectivity (Schiller *et al.*, 1985b). Interestingly, the latter analogue showed about fivefold higher μ-affinity than its linear correlate, H-Tyr-D-Asn-Phe-Nva-NH$_2$, thus representing one of the rare examples where it could be demonstrated unambiguously that conformational restriction *per se* did produce an affinity enhancement.

Enkephalin and dermorphin analogues containing a cystine bridge represent a third prototype of cyclic opioid peptide analogues. Cyclic enkephalin analogues having the structure H-Tyr-D-Cys-Gly-Phe-D(or L)-Cys-X were synthesized independently by Sarantakis (1979) and by Schiller *et al.* (1981, 1985c). Analogues of this type with a C-terminal carboxamide function (X = NH$_2$) were essentially non-selective, whereas the corresponding free acids (X = OH) showed moderate preference for δ-receptors over μ-receptors. The compound H-Tyr-D-Cys-Phe-Cys-NH$_2$ represents an example of a cystine-bridged cyclic dermorphin analogue, containing a very small (11-membered) ring structure and displaying considerable preference for μ-receptors over δ-receptors (Schiller *et al.*, 1987).

In comparison with their linear correlates, several of the cyclic opioid peptide analogues discussed above showed relatively higher potency in the GPI bioassay than was expected on the basis of their μ-receptor affinities determined in the receptor binding assay. Whether or not this interesting finding is due to an efficacy enhancement remains to be further investigated.

3.2. *Further structural modification of cyclic opioid peptide analogues*

The receptor selectivity of several of the cyclic opioid peptide analogues described in the previous section could be improved or altered through further structural modification. For example, the reversal of two peptide bonds in the ring structure of H-Tyr-cyclo[-D-A_2bu-Gly-Phe-Leu-] led to a compound, H-Tyr-cyclo[-Glu-Gly-gPhe-D-Leu-], which was three times more μ-selective than the cyclic parent peptide (gPhe denotes the *gem* diamino equivalent of Phe) (Berman *et al.*, 1983). Similarly, replacement of the peptide bond in the 3–4 position of H-Tyr-cyclo[-D-Lys-Gly-Phe-Leu-] with a thioamide moiety produced a compound, H-Tyr-cyclo[-D-Lys-Glyψ[CSNH]Phe-Leu-] showing about a tenfold improvement in μ-selectivity (Sherman *et al.*, 1989).

Replacement of one or both Cys residues in H-Tyr-D-Cys-Gly-Phe-D(or L)-Cys-OH with a penicillamine (Pen) residue resulted in analogues with markedly improved δ-selectivity. The most δ-selective analogues of this type turned out to be H-Tyr-D-Pen-Gly-Phe-D-Pen-OH (DPDPE) and H-Tyr-D-Pen-Gly-Phe-L-Pen-OH (DPLPE) (Mosberg *et al.*, 1983). In a recent study, Mosberg *et al.* (1987) conclusively showed that the δ-receptor preference of these two analogues is due to the presence of the bulky *gem* dimethyl groups in the 2-position side-chain which causes more severe steric interference at the μ-receptor than at the δ-receptor. Replacement of Phe^4 in DPDPE with *p*-chlorophenylalanine (*p*Cl-Phe) resulted in a compound, H-Tyr-D-Pen-Gly-*p*Cl-Phe-D-Pen-OH showing a further fivefold improvement in δ-selectivity due to an increase in δ-affinity (Thót *et al.*, 1990). Finally, structural modification of the μ-selective cyclic dermorphin analogue H-Tyr-D-Cys-Phe-Cys-NH_2 (Schiller *et al.*, 1987) through substitution of D-Pen for Cys in the 4-position and replacement of the C-terminal carboxamide function with a free carboxyl group led to a compound, H-Tyr-D-Cys-Phe-D-Pen-OH, which ranks among the most potent and most selective δ-agonists (Mosberg *et al.*, 1988).

3.3. *Examples of cyclic analogues of other peptide hormones and neurotransmitters*

During the past decade, cyclizations of a number of peptide hormones and neurotransmitters have been successfully performed. Only a few examples of cyclic analogues showing increased receptor selectivity, high potency or antagonism will be discussed here briefly. In a most elegant development based on conformational studies and computer modelling, Veber *et al.* (1981, 1984) developed highly potent somatostatin analogues with greatly reduced ring size. Among various prepared cyclic hexapeptide analogues, the most active one was cyclo[-NMe-Ala-Tyr-D-Trp-

Lys-Val-Phe-], being 50–100-fold more potent than somatostatin for the inhibition of insulin, glucagon and growth hormone release. These analogues also showed improved resistance to degradation by trypsin, presumably as a result of their enhanced structural rigidity. Another interesting example of a reduced-size somatostatin analogue is the potent, enzyme-resistant octapeptide H-D-Phe-Cys-Phe-D-Trp-Lys-Thr-Cys-Thr-ol (SMS 201-995) which selectively inhibits growth hormone release (Bauer *et al.*, 1982) and has potential for the treatment of acromegaly. Interestingly, SMS 201-995 also turned out to be a potent opioid antagonist with selectivity for the μ-receptor (Maurer *et al.*, 1982). Subsequent structural modification of SMS 201-995 resulted in several highly μ-selective opioid antagonists with negligible affinity for somatostatin receptors, such as the octapeptide H-D-Phe-Cys-Tyr-D-Trp-Orn-Thr-Pen-Thr-NH_2 (CTOP) (Pelton *et al.*, 1986). Cyclic analogues of the C-terminal octapeptide of cholecystokinin (CCK) were prepared either by substituting D-lysine in position 29 and cyclizing its side-chain to the side-chain of the Asp (or Glu) residue in position 26 (Charpentier *et al.*, 1988), or by bridging the side-chains of two lysine residues substituted in positions 28 and 31 with a succinic moiety (Rodriguez *et al.*, 1990). Both types of analogues showed high preference for CCK-B receptors over CCK-A receptors. Among several cystine-bridged analogues of substance P and neurokinin B prepared by Ploux *et al.* (1987), [$Cys^{3,6}$, Tyr^8, Ala^9]substance P turned out to be about as potent and as selective for the NK1 receptor as the linear parent peptide, whereas [$Cys^{2,5}$]neurokinin-B showed high preference for the NK3 receptor. Various analogues of α-melanotropin (α-melanocyte-stimulating hormone, α-MSH) cyclized between the side-chains of two half-cystine residues substituted in positions 4 and 10 were 10–30-fold more potent than the native hormone in stimulating frog skin darkening (Cody *et al.*, 1984). Substitution of homocysteine residues in positions 3 and 5 of angiotensin-II and of its antagonist [Sar^1, Ile^8]angiotensin-II followed by oxidative disulphide bridge formation resulted in cyclic analogues which retained high agonist and antagonist potency, respectively, and had receptor affinities comparable to those of their linear parent peptides (Spear *et al.*, 1990). Interestingly, a corresponding analogue with reduced ring size, [$Cys^{3,5}$]angiotensin-II, appeared to be a partial agonist, indicating that efficacy may be reduced as a result of further structural rigidification. A cyclic lactam analogue of gonadotropin-releasing hormone, Ac-D-2-Nal-D-*p*F-Phe-D-Trp-Asp-Tyr-D-Arg-Leu-Arg-Pro-A_2pr-NH_2 (2-Nal = 2-naphthylalanine; *p*F-Phe = *p*-fluorophenylalanine), had about half the antagonist potency *in vitro* and one-third the receptor affinity of the linear parent peptide (Rivier *et al.*, 1988). Finally, the development of a conformationally restricted bicyclic analogue of vasopressin with antidiuretic antagonistic properties represents an excellent example of peptide analogue design based on conformational considerations (Skala *et al.*, 1984).

3.4. *Conformational studies of cyclic opioid peptide analogues*

Both theoretical and spectroscopic approaches have been used to study the conformational behaviour of the various cyclic opioid peptide analogues described above (for a review, see Schiller & Wilkes, 1988). These investigations clearly demon-

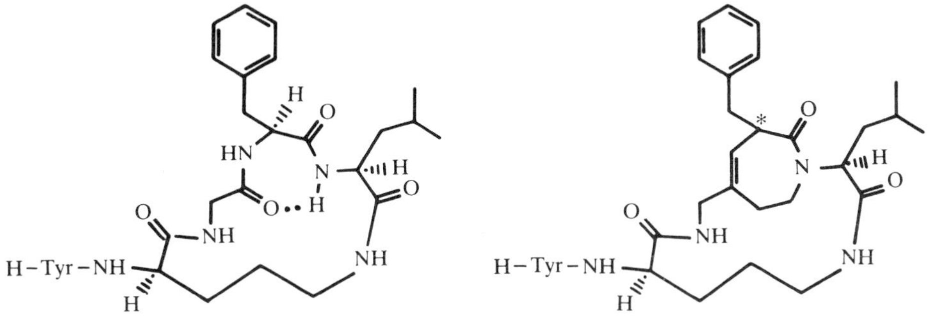

Figure 15.4. Structural formulas of the cyclic enkephalin analogue H-Tyr-cyclo[-D-Orn-Gly-Phe-Leu-] (left) and of a bicyclic analogue containing a γ-turn mimic (right).

strated that the ring structures in many of these compounds still retain some structural flexibility and that the various intramolecular hydrogen bonds observed are constantly formed, broken and re-formed again, as shown most convincingly in molecular dynamics studies (Mammi *et al.*, 1985). In the case of both H-Tyr-cyclo[-D-A_2bu-Gly-Phe-Leu-] and H-Tyr-cyclo[-D-Orn-Gly-Phe-Leu-] the results of NMR studies and computer simulations indicated the existence of a hydrogen bond which defines a γ-turn centered on Phe^4 (Kessler *et al.*, 1985; Mammi *et al.*, 1985) (Fig. 15.4). A molecular mechanics study of the μ-selective cyclic dermorphin analogue H-Tyr-D-Orn-Phe-Asp-NH_2 showed that its 13-membered ring structure is quite rigid and that its lowest energy conformation is characterized by a tilted stacking interaction between the Tyr^1 and Phe^3 aromatic rings (Wilkes & Schiller, 1987). Extension of the latter study to a series of eight cyclic tetrapeptide analogues structurally related to H-Tyr-D-Orn-Phe-Asp-NH_2 and showing considerable diversity in μ-receptor affinity led to the suggestion that the tilted stacking arrangement of the two aromatic rings may represent an important structural requirement for high μ-receptor affinity of the examined cyclic dermorphin analogues (Wilkes & Schiller, 1990). However, this theoretical analysis also revealed that the exocyclic Tyr^1 residue and the Phe^3 side-chain still enjoy considerable orientational freedom and that these moieties also need to be conformationally restricted in order to obtain more definitive insight into the receptor-bound conformation. The same holds true for the Tyr^1 and Phe^4 residues in the δ-selective cyclic enkephalin analogue H-Tyr-D-Pen-Gly-Phe-D-Pen-OH (Wilkes & Schiller, 1991). The N-terminal Tyr^1 residue and the Phe side-chain in position 3 or 4 of the amino acid sequence of opioid peptides represent crucial pharmacophoric moieties for opioid activity. The demonstrated fact that these residues have still considerable structural flexibility in all the cyclic analogues described above may be the reason why no consensus regarding the distinct bioactive conformation at the μ- and the δ-opioid receptor has been reached so far.

4 Further structural rigidification of cyclic opioid peptide analogues

In efforts to develop peptidomimetics with potential for therapeutic applications, it would often be desirable to replace parts of the peptide backbone with other appropriate structural elements in order to eliminate susceptibility to enzymatic degradation and to reduce the relatively polar character of peptide molecules which prevents them from being well absorbed and from crossing certain barriers, such as, for example, the blood–brain barrier. It is imperative that the spatial disposition of side-chains crucial for the interaction of the native peptide with the receptor be retained in the peptidomimetic. Whereas the various modifications of individual peptide bonds described above represent a first step in the direction of developing peptidomimetics, the substitution of structural elements mimicking larger portions of the peptide backbone has also been attempted. The latter approach is particularly applicable to peptides with a relatively well-defined conformation, such as the semi-rigid cyclic analogues described above. An example is the peptidomimetic derived from the cyclic enkephalin analogue H-Tyr-cyclo[-D-Orn-Gly-Phe-Leu-], shown in Fig. 15.4 (Huffman *et al.*, 1989). In this compound the γ-turn structure detected in the cyclic parent peptide (see above) was replaced with a *trans*-olefin and the oxygen and hydrogen atoms engaged in the hydrogen bond (C=O H–N) are replaced with an ethylene moiety. The two diastereomers of the bicyclic compound with either *R* or *S* configuration at the chiral centre of the γ-turn mimic could be obtained separately and were both found to have very low activity. Possible explanations for the lack of significant activity of this peptidomimetic may be: (i) the γ-turn structure detected in the conformational studies of the monocyclic parent peptide may not be a structural feature of the bioactive conformation; (ii) the Gly^3 carbonyl group which is no longer present in the γ-turn mimic may be important for the interaction with the receptor; and (iii) the ethylene bridge introduced in the mimic may interfere with the receptor binding process either due to unfavourable steric interactions or due to the fact that it produces too much rigidity in the molecule. Despite the lack of activity, the preparation of this compound represents a conceptually interesting first step towards the goal of developing rationally designed peptidomimetics of enkephalin and further efforts in this direction will certainly be worthwhile.

Promising efforts towards the development of novel β- and γ-turn mimics are currently being undertaken in several laboratories. Kahn *et al.* (1988) synthesized an 11-membered ring *bis*-lactam structure designed to mimic type I and type II β-turns. A further excellent example of β-turn mimicry is the design of a nine-membered ring lactam structure and its incorporation into the cyclic depsipeptide jaspamide, resulting in a depsipeptidomimetic with insecticidal activity comparable to that of the native parent compound (Kahn *et al.*, 1991). Another interesting β-turn mimic based on a nine-membered ring lactam has been developed by Olson *et al.* (1990).

In an effort to reduce the structural flexibility of the exocyclic Tyr^1 residue and of the Phe^3 side-chain in the slightly μ-receptor-selective cyclic dermorphin analogue H-Tyr-D-Orn-Phe-Glu-NH_2, conformationally restricted aromatic amino acid residues were substituted in positions 1 and 3 of the peptide sequence (Schiller *et al.*, 1991a,b) (Fig. 15.5). The analogue H-Tyr-D-Orn-Aic-Glu-NH_2 (Aic = 2-aminoindan-2-carboxylic acid) turned out to be a potent agonist with high preference for μ-receptors over δ-receptors. Opening of the five-membered ring of the indan moiety of Aic in the latter analogue, as achieved through substitution of C^{α}-methylphenylalanine (C^{α}MePhe) or *o*-methylphenylalanine (*o*-MePhe) (Fig. 15.5),

H-Tyr-D-Orn-X-Glu-NH_2

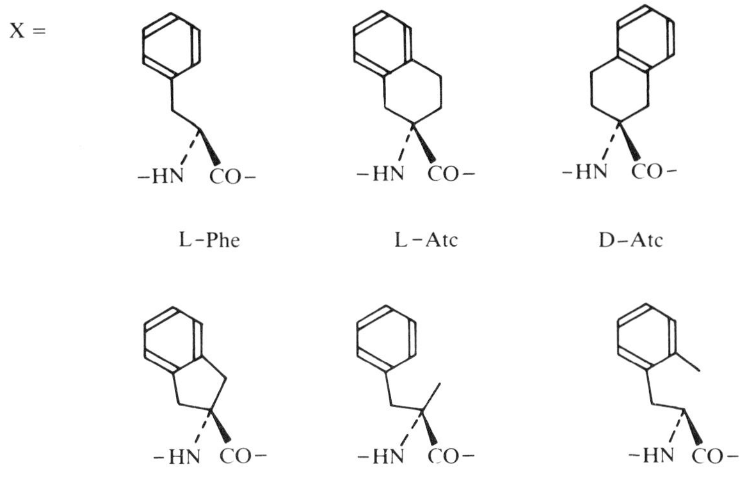

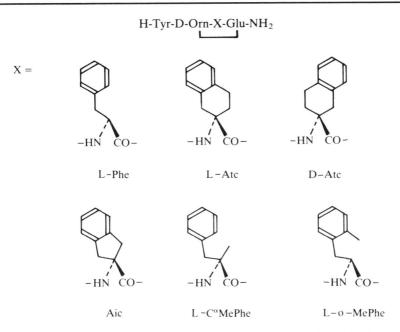

Figure 15.5. Substitution of conformationally restricted phenylalanine analogues in position-3 of a cyclic dermorphin analogue.

resulted in only slightly μ-selective compounds, indicating that the high μ-selectivity of the Aic analogue is exclusively the consequence of the imposed side-chain conformational restriction. Both diastereomers of H-Tyr-D-Orn-L(or D)-Atc-Glu-NH_2 (Atc = 2-aminotetralin-2-carboxylic acid) were highly μ-selective and, in contrast to the weak affinity observed with the D-Phe^3-analogue as compared to the L-Phe^3-analogue, both had similar potency. Thus, stereospecificity was lost as a consequence of side-chain conformational restriction. Substitution of a corresponding conformationally restricted Tyr analogue, 6-hydroxy-2-aminotetralin-2-carboxylic acid (Hat), in position 1 of the Aic^3-analogue resulted in a compound, H-(D,L)-Hat-D-Orn-Aic-Glu-NH_2, which still retained quite high μ-receptor affinity and μ-selectivity. The latter analogue represents the structurally most rigid, rationally designed opioid peptidomimetic reported to date, since essentially it contains only two freely rotatable bonds (Fig. 15.6).

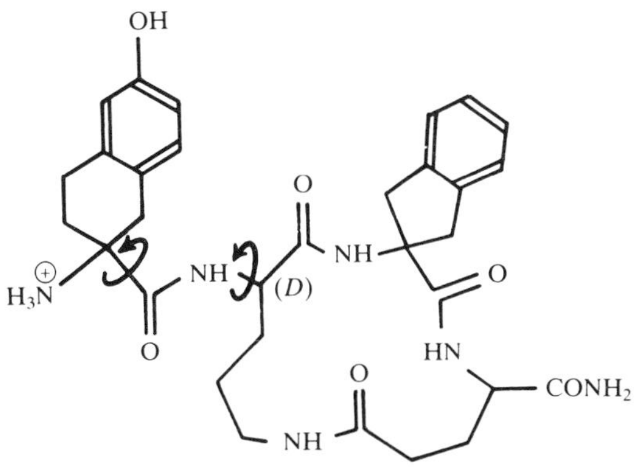

Figure 15.6. Structural formula of the opioid peptidomimetic H-Hat-D-Orn-Aic-Glu-NH_2.

5 Pharmacological aspects related to the design of peptidomimetics

The development of peptidomimetics raises some fundamental issues of ligand–receptor interactions which will be addressed below. Points of particular interest concern the mode of receptor binding of peptidomimetics, their efficacy (intrinsic activity) in comparison with their native parent peptides and their role in the discovery of new receptor types or subtypes.

5.1. *Mode of receptor binding*

Over 25 years ago, comparison of the effect of configurational inversion and analogous substitutions on analgesic activity led to the proposal that different classes of opiates may have different modes of binding to the receptor (Portoghese, 1965). According to this concept, binding of various families of ligands to the same receptor may involve the interaction with both common and different receptor subsites. In particular, it was suggested that the positively charged nitrogen atom present in all opioid compounds would be engaged in an electrostatic interaction with a common anionic receptor site, whereas other structural moieties contained in the various types of opioid analgesics might not all interact with the same region of the receptor topography.

The question of whether conformational restriction of peptide ligands might affect their mode of receptor binding represents an important issue. Structure–activity data obtained with analogues of cyclic enkephalins of the type H-Tyr-cyclo[-D-Daa-Gly-Phe-Leu-] (Fig. 15.2) and H-Tyr-D-Cys-Gly-Phe-D(or L)-Cys-NH_2 revealed that the cyclic peptides not only had the same configurational requirements in positions 1, 2, 4 and 5 of the peptide sequence as linear enkephalins, but also showed parallel behaviour with regard to the effect of introducing various substituents in *para*-position of the Phe^4 residue (Schiller & DiMaio, 1983). These data represent strong evidence to indicate that the mode of receptor binding of cyclic and linear enkephalin analogues is indeed the same.

As indicated above, the μ-receptor affinity of the cyclic dermorphin analogue H-Tyr-D-Orn-Phe-Glu-NH_2 is more than a 1000-fold higher than that of the corresponding D-Phe^3-analogue, whereas both diastereomers of H-Tyr-D-Orn-D(or L)-Atc-Glu-NH_2 show high affinity for μ-receptors (Schiller *et al.*, 1991a,b). The loss of stereospecificity as a consequence of the side-chain conformational constraint may be due to the fact that the D-Atc^3 analogue binds to the receptor in a manner different from that of the D-Phe^3 analogue. In the case of the D-Phe^3 analogue, a stepwise process of binding according to the 'zipper'-type model may occur such that the D-Phe^3 side-chain never has a chance to bind to the hydrophobic receptor site with which the L-Phe^3 aromatic ring interacts. The observation that in the case of a corresponding cyclic analogue with homophenylalanine (Hfe) substituted in position 3, the L-Hfe^3 analogue is again over a 1000-fold more potent than the D-Hfe^3 analogue (Schiller *et al.*, 1991b) is of interest in view of the fact that Atc represents a conformationally restricted analogue of both Phe and Hfe (with reversed configurational relationships!). The receptor binding site with which the 3-position aromatic rings of the L-Phe^3 and L-Hfe^3 analogues interact appears to be fairly large, since it is also able to accommodate the naphthyl ring of the cyclic analogue H-Tyr-D-Orn-1-Nal-Glu-NH_2 (1-Nal = 1-naphthylalanine) which retains high μ-receptor affinity (Schiller *et al.*, 1991b). Configurational inversion of Atc produces a change in the way the aromatic ring is fused to the cyclohexane structure by shifting it from one position to an adjacent one (see Fig. 15.5). This shift in the positioning of the aromatic ring of the 3-position residue is tolerated by the 'large' hydrophobic binding site on the receptor and, due to the side-chain conformational restriction, the process of binding of not only the L- but also the D-Atc^3 analogue is such that the Atc aromatic ring is forced to interact with this binding site. Thus, the low activity of the D-Phe^3 analogue as compared to the D-Atc^3 analogue may be explained with a different process and mode of receptor binding.

The receptor binding mode of many peptidomimetics lacking any peptide structural elements may be different from that of the native peptide ligands. The comparison of opioid peptides with morphine and morphine-derived compounds represents a classical example. While the functional correspondence between the tyramine segment in the morphine molecule and the tyrosine in the 1-position of opioid peptides is generally accepted, no consensus has been reached to date with regard to a possible correspondence between other structural elements present in the two classes of compounds. It may well be possible that only the tyramine moieties share a common binding site and that other structural elements of the morphine framework and of opioid peptides may interact with different receptor subsites. The recently discovered non-peptide antagonists of cholecystokinin, angiotensin-II and substance P may also have receptor binding modes that differ from those of the agonist parent peptides and that may be characterized by interactions with auxiliary binding sites not accessible to the native peptide ligands. Recently performed crystal-structure determinations of endothiapepsin complexed with various inhibitors represent an excellent experimental demonstration of the concept of different binding modes, insofar as it was evident that certain identical moieties of the various inhibitors examined were oriented differently such as to interact with different moieties in the active site of the enzyme (Foundling *et al.*, 1987).

5.2 *Efficacy of peptidomimetics*

According to a plausible model (Franklin, 1980), part of the receptor binding energy of a ligand with agonist properties is used to lower the energy requirement for the conformational transition of the receptor in the ground state (R) to an excited state (R^*) which leads to the biological response (Fig. 15.1). The efficacy of a receptor ligand (L) is thought to be related to the rate of formation of excited receptors (k_2) as well as to the final equilibrium between excited receptors and receptors in the ground state. In comparison with full agonists, partial agonists have a diminished ability to activate the receptor and antagonists would not be able to produce a productive conformational change in the receptor at all. Therefore, the entire receptor interaction energy of antagonists is expressed as tightness of binding and, for this reason, the binding equilibrium dissociation constants of antagonists are generally lower than those of agonists.

The extent of structural flexibility of a peptide ligand or peptidomimetic may be of importance for its ability to induce a productive conformational change in the receptor molecule such as required for agonist activity. However, to date it has not been possible to establish clear-cut relationships between the relative molecular rigidity of receptor ligands and their efficacy. Nevertheless, it is of interest to point out that most of the structurally more rigid peptidomimetics that have been discovered to date either by chance or by the 'screening' approach are either antagonists or partial agonists. Examples are morphine, which has been shown to be a partial agonist (Smith & Rance, 1983), as well as the relatively rigid non-peptide antagonists of cholecystokinin, angiotensin-II and substance P. However, the efficacy loss observed with the latter peptidomimetics may not merely be a consequence of their diminished structural flexibility but possibly may also be due to an altered mode of binding which may only partially, or not at all, produce the conformational change in the receptor required for signal transduction. The fact that 'screening' projects based on the use of a radioreceptor assay so far have almost exclusively led to the discovery of antagonists is not surprising, since this process identifies compounds that are all able to bind to the receptor in one mode or the other, but only in rare cases may also fulfil the structural requirements for receptor activation.

5.3 *Role of peptidomimetics in the discovery of new receptor types or subtypes*

Natural peptide ligands and their conventional analogues may be able to bind to two different receptor types with equal affinity because their inherent structural flexibility permits them to fulfil equally well the structural (conformational) requirements of both receptor classes. However, drastic structural rigidification, such as achieved with the development of various types of peptidomimetics, may result in differential binding and thus permit the demonstration of receptor heterogeneity. This has, for example, been the case with somatostatin, where the use of conformationally restricted analogues with reduced ring size in receptor binding assays resulted in the identification of two distinct receptor types (Tran *et al.*, 1985). The debate on whether or not more than one angiotensin receptor existed went on for decades and only binding experiments performed with the recently developed non-peptide antagonists led to the discovery of two different angiotensin-II binding sites (AII-1 and AII-2) which have equal affinity for the structurally flexible native peptide and its analogues (Chiu *et al.*, 1989). Finally, substitution of the conformationally restricted phenylalanine analogue (2*S*,3*R*)-*E*-cyclopropylphenylalanine in position-4 of [D-Ala^2,Leu^5]enkephalin and testing of the resulting analogue in various *in vitro* assay systems provided the first indication for the existence of two δ- opioid receptor subtypes (Shimohigashi *et al.*, 1987). The development of conformationally restricted peptide analogues and of peptidomimetics will undoubtedly play an important role in future identifications of new receptor types or subtypes and will complement the use of cloning techniques in the receptor discovery process.

6 Conclusion and future prospects

This chapter has reviewed current strategies related to the rational design of peptidomimetics. Obviously, the development of peptidomimetics through progressive conformational restriction of the native peptide and subsequent major peptide bond replacements is still at an early stage and the design process needs to be further refined. It must be realized that the conformational behaviour of peptide ligands and of peptidomimetics is of importance not only for the initial event of binding to the receptor but also for signal transduction and for the final dissociation from the receptor. The comparison of enkephalins with morphine-derived opiates represents an illustrative example, since these two classes of opioid compounds not only have different efficacies but also differ from one another in their binding kinetics. Thus, it has been shown that [Met^5]enkephalin associates with, and dissociates from, opioid receptors more slowly than rigid opiates (Simantov *et al.*, 1978), presumably as the consequence of a time-consuming conformational change which may occur in a binding process according to the 'zipper' model.

Future design strategies can be expected to be based on a more extensive evaluation of the conformational behaviour of planned peptidomimetics by computer simulations (e.g. molecular dynamics studies) prior to synthesis. Ideally, the conformational space available to the peptidomimetic and to the peptide to be mimicked should be identical at least as far as the essential pharmacophoric moieties are concerned. For example, various types of β- or γ-turn mimics displaying different degrees of structural flexibility might be formally incorporated into a peptide. Computer simulations would then first be performed with the resulting peptidomimetic structures in comparison with the original peptide structure in order to determine which type of turn mimic would be most suitable for mimicking the conformational behaviour of the parent peptide. Obviously, such an approach not only requires the use of the most advanced theoretical and nuclear magnetic resonance (NMR) spectroscopic techniques for conformational analysis but also the development of novel synthetic methodology.

Currently, the pharmacological characterization of peptidomimetics *in vitro* is in most cases limited to the determination of receptor-binding affinities and selectivities and to the measurement of potencies in bioassays. In future developments of peptidomimetics a more detailed analysis of the receptor interaction of these compounds might turn out to be advantageous for the design process. In particular, the binding kinetics and the efficacy of newly developed peptidomimetics should be carefully determined and interpreted in relation to their conformational behaviour. Such correlations might contribute to a more rational design of peptidomimetics with the desired activity profile.

The study of lipid interactions of peptide hormones and neurotransmitters is likely to play a more important role in the development of peptidomimetics in the future. In the case of acetylcholine, it has recently been shown that the predominant conformation in solution is different from that in the lipid-bound state, which in turn is identical with the receptor-bound conformation (Behling & Jelinski, 1990). It is quite probable that the conformation of a peptide ligand in a lipid environment might be different from its average conformation in solution and also might more closely resemble the bioactive conformation assumed in the receptor complex. Therefore, conformational studies of native peptides and of conformationally restricted analogues in the lipid-bound state might provide information likely to be useful for the design of peptidomimetics. Furthermore, the innovative idea that peptide–lipid interactions may be a major factor in determining receptor selectivity (Schwyzer, 1986) should be further pursued and taken into consideration in future developments of receptor-specific peptidomimetics.

Finally, it can be expected that cloning, sequencing and functional expression in host cells of an ever-increasing number of receptors will eventually provide exciting new information on conformational aspects related to the binding process and signal transduction of peptide hormones and neurotransmitters. In particular, some information on the nature of the receptor residues that constitute the binding site should become available and hopefully will lead to plausible models of the bioactive conformation of the peptide ligands which might turn out to be useful for the design of peptidomimetics. Whether or not it will become possible to obtain a detailed 3-D picture of the peptide–receptor complex by X-ray diffraction or NMR spectroscopic techniques remains to be seen. However, other interesting experiments will certainly also become feasible. For example, reconstitution of receptor–effector systems in membranes and the use of appropriate fluorescence techniques might provide fundamental new insight into the nature and dynamics of ligand-induced conformational changes in the receptor. The feasibility of such experiments has been demonstrated in a pioneering study on the galactose receptor, which showed that binding of galactose causes a propagated conformational change in the receptor protein extending to a site at least 30 Å apart from the binding site (Zukin *et al.*, 1977). Performance of such experiments with a series of agonists, partial agonists and antagonists might lead to interesting correlations between ligand efficacy, ligand structural flexibility and ligand-induced conformational changes in the receptor. Information of this kind might also turn out to be relevant to the design of peptidomimetics.

7 Acknowledgements

The author's work described in this chapter was supported by operating grants from the Medical Research Council of Canada (MT-5655 and MA-10131), the Heart and Stroke Foundation of Quebec and the U.S. National Institute on Drug Abuse (DA-04443).

8 References

Avbelj, F. & Hadži, D. (1985). *Mol. Pharmacol.*, **27**, 466–70.

Bauer, W., Briner, U., Doepfner, W., Haller, R., Huguenin, R., Marbach, P., Petcher, T.J. & Pless, J. (1982). *Life Sci.*, **31**, 1133–40.

Behling, R.W., & Jelinski, L.W. (1990). *Biochem Pharmacol.*, **40**, 49–54.

Berman, J.M., Goodman, M., Nguyen, T.M.-D. & Schiller, P.W. (1983). *Biochem. Biophys. Res. Commun.*, **115**, 864–70.

Burgen, A.S.V., Roberts, G.C.K., & Feeney, J. (1975). *Nature (London)*, **253**, 753–5.

Cascieri, M.A., Chicchi, G.R., Freidinger, R.M., Colton, C.D., Perlow, D.S., Williams, B., Curtis, N.R., McKnight, A.T., Maguire, J.J., Veber, D.F., & Liang, T. (1986). *Mol. Pharmacol.*, **29**, 34–8.

Charpentier, B., Pelaprat, D., Durieux, C., Dor, A., Reibaud, M., Blanchard, J.-C. & Roques, B.P. (1988). *Proc. Natl. Acad. Sci. USA*, **85**, 1968–72.

Chiu, A.T., Herblin, W.F., McCall, D.E., Ardecky, R.J., Carini, D.J., Duncia, J.V., Pease, L.J., Wong, P.C., Wexler, R.R., Johnson, A.L. & Timmermans, P.B.M.W.M. (1989). *Biochem. Biophys. Res. Commun.*, **165**, 196–203.

Cody, W.L., Wilkes, B.C., Muska, B.J., Hruby, V.J., de L. Castrucci, A.M. & Hadley, M.E. (1984). *J. Med. Chem.* **27**, 1186–90.

DiMaio, J. & Schiller, P.W. (1980). *Proc Natl. Acad. Sci. USA*, **77**, 7162–6.

DiMaio, J., Nguyen, T.M.-D., Lemieux, C. & Schiller, P.W. (1982). *J. Med. Chem.*, **25**, 1432–8.

Fischer, E. (1894). *Ber. Dtsch. Chem. Ges.*, **27**, 2985–93.

Foundling, S.I., Cooper, J., Watson, F.E., Cleasby, A., Pearl, L.H., Sibanda, B.L., Hemmings, A., Wood, S.P., Blundell, T.L., Valler, M.J., Norey, C.G., Kay, J., Boger, J., Dunn, B.M., Leckie, B.J., Jones, D.M., Atrash, B., Hallett, A. & Szelke, M. (1987). *Nature* (*London*), **327**, 349–52.

Franklin, T.J. (1980). *Biochem. Pharmacol.*, **29**, 853–6.

Freidinger, R.M., Perlow, D.S. & Veber, D.F. (1982). *J. Org. Chem.*, **47**, 104–9.

Freidinger, R.M., Veber, D.F., Perlow, D.S., Brooks, J.R. & Saperstein, R. (1980). *Science*, **210**, 656–8.

Holladay, M.W., Lin, C.W., May, C.S., Garvey, D.S., Witte, D.G., Miller, T.R., Wolfram, C.A.W. & Nadzan, A.M. (1991). *J. Med. Chem.*, **34**, 455–7.

Huffman, W.F., Callahan, J.F., Codd, E.E., Eggleston, D.S., Lemieux, C., Newlander, K.A., Schiller, P.W., Takata, D.T. & Walker, R.F. (1989). In: J. Tam and E.T. Kaiser (Eds.), *Synthetic Peptides: Approaches to Biological Problems* (*UCLA Symposia on Molecular and Cellular Biology*, New Series, Vol. 86), pp. 257–66, A.R. Liss, New York, NY.

Kahn, M., Wilke, S., Chen, B., Fujita, K., Lee, J.Y.-H. & Johnson, M.E. (1988). *J. Mol. Recognition*, **1**, 75–9.

Kahn, M., Nakanishi, H., Su, T., Lee, J.Y.-H. & Johnson, M.E. (1991). *Int. J. Peptide Protein Res.*, **38**, 324–34.

Kessler, H., Hölzemann, G. & Zechel, C. (1985). *Int. J. Peptide Protein Res.*, **25**, 267–79.

Lee, J.P., Dunlap, B. & Rich D.H. (1990). *Int. J. Peptide Protein Res.*, **35**, 481–94.

Mammi, N.J., Hassan, M., & Goodman, M. (1985). *J. Am. Chem. Soc.*, **107**, 4008–13.

Marshall, G.R. & Bosshard, H.E. (1972). *Circ. Res.*, **30** (Suppl. II), 143–50.

Maurer, R., Gaehwiler, B.H., Buescher, H.H., Hill, R.C. & Roemer, D. (1982). *Proc. Natl. Acad. Sci. USA*, **79**, 4815–17.

Mosberg, H.I., Omnaas, J.R. & Goldstein, A. (1987). *Mol. Pharmacol.*, **31**, 599–602.

Mosberg, H.I., Hurst, R., Hruby, V.J., Gee, K., Yamamura, H.I., Galligan, J.J. & Burks, T.F. (1983). *Proc. Natl. Acad. Sci. USA*, **80**, 5871–4.

Mosberg, H.I., Omnaas, J.R., Medzihradsky, F. & Smith, G.B. (1988). *Life Sci.*, **43**, 1013–20.

Olson, G.L., Voss, M.E., Hill, D.E., Kahn, M., Madison, V.S. & Cook, C.M. (1990). *J. Am. Chem. Soc.*, **112**, 323–33.

Paterson, S.J., Robson, L.E. & Kosterlitz, H.W. (1984). In: S. Udenfriend and J. Meienhofer (Eds.), *The Peptides: Analysis, Synthesis, Biology,* Vol. 6, pp. 147–89, Academic Press, Orlando, FL.

Pelton, J.T., Kazmierski, W., Gulya, K., Yamamura, H.I. & Hruby, V.J. (1986). *J. Med. Chem.*, **29**, 2370–5.

Ploux, O., Lavielle, S., Chassaing, G., Julien, S., Marquet, A., D'Orléans-Juste, P., Dion, S., Regoli, D., Beaujouan, J.-C., Bergström, L., Torrens, Y. & Glowinski, J. (1987). *Proc. Natl. Acad. Sci. USA*, **84**, 8995–9.

Portoghese, P.S. (1965). *J. Med. Chem.*, **8**, 609–16.

Rivier, J., Kupryszewski, G., Varga, J., Porter, J., Rivier, C., Perrin, M., Hagler, A., Struthers, S., Corrigan, A. & Vale, W. (1988). *J. Med. Chem.*, **31**, 677–82.

Rodriguez, M., Amblard, M., Galas, M.-C., Lignon, M.-F., Aumelas, A. & Martinez, J. (1990). *Int. J. Peptide Protein Res.*, **35**, 441–51.

Sarantakis, D. (1979). U.S. Patent 4148786.

Sato, K. & Nagai, U. (1986). *J. Chem. Soc. Perkin Trans. I,* 1231–4.

Schiller, P.W. (1984). In: S. Udenfriend and J. Meienhofer (Eds.), *The Peptides: Analysis, Synthesis, Biology*, Vol. 6, pp. 219–68, Academic Press, Orlando, FL.

Schiller, P.W. (1991). In: G.P. Ellis and G.B. West (Eds.), *Progress in Medicinal Chemistry,* Vol. 28, pp. 301–40, Elsevier, Amsterdam.

Schiller, P.W. & DiMaio J. (1982). *Nature* (*London*), **297**, 74–6.

Schiller, P.W. & DiMaio J. (1983). In: V.J. Hruby and D.H. Rich (Eds.), *Peptides: Structure and Function* (*Proc. 8th Amer. Peptide Symp.*), pp. 269–78, Pierce Chemical Company, Rockford, IL.

Schiller, P.W. & Wilkes B.C. (1988). In: R.S. Rapaka and B.N. Dhawan (Eds.), *Recent Progress in the Chemistry and Biology of Opioid Peptides* (*NIDA Research Monograph 87*), pp. 60–73, U.S. Government Printing Office, Washington D.C.

Schiller, P.W., Eggimann, B., DiMaio, J., Lemieux, C. & Nyuyen, T.M.-D. (1981). *Biochem. Biophys. Res. Commun.*, **101**, 337–43.

Schiller, P.W., Nguyen, T.M.-D., Maziak, L.A. & Lemieux, C. (1985a). *Biochem. Biophys. Res. Commun.*, **127**, 558–64.

Schiller, P.W. Nguyen, T.M.-D. Lemieux, C. & Maziak, L.A. (1985b). *J. Med. Chem.*, **28**, 1766–71.

Schiller, P.W., DiMaio, J. & Nguyen, T.M.-D. (1985c). In: Y.A. Ovchinnikov (Ed.), *Proc. 16th FEBS Congress, Part B*, pp. 457–62, VNU Science Press, Utrecht.

Schiller, P.W., Nguyen, T.M.-D. Maziak, L.A., Wilkes, B.C. & Lemieux, C. (1987). *J. Med. Chem.*, **30**, 2094–9.

Schiller, P.W., Weltrowska, G., Nguyen, T.M.-D., Lemieux, C., Chung, N.N. & Wilkes, B.C. (1991a). In: E. Giralt and D. Andreu (Eds.), *Peptides 1990* (*Proc. 21st Eur. Peptide Symp.*), pp. 661–2, ESCOM, Leiden.

Schiller, P.W., Weltrowska, G., Nguyen, T.M.-D., Lemieux, C., Chung, N.N., Marsden, B.J. & Wilkes, B.C. (1991b). *J. Med. Chem.*, **34**, 3125–32.

Schwyzer, R. (1986). *Biochemistry*, **25**, 6335–42.

Sherman, D.B., Spatola, A.F., Wire, W.S., Burks, T.F., Nyugen, T.M.-D. & Schiller, P.W. (1989). *Biochem. Biophys. Res. Commun.*, **162**, 1126–32.

Shimohigashi, Y., Costa, T., Pfeiffer, A., Herz, A. Kimura, H. & Stammer, C.H. (1987). *FEBS Lett.*, **222**, 71–4.

Simantov, R., Childers, S.R. & Snyder, S.H. (1978). *Eur. J. Pharmacol.*, **47**, 319–31.

Skala, G., Smith, C.W., Taylor, C.J. & Ludens, J.H. (1984). *Science*, **226**, 443–4.

Smith, C.F.C. & Rance M.J. (1983). *Life Sci.*, **33** (Suppl. I), 327–30.

Spear, K.L., Brown, M.S., Reinhard, E.J., McMahon, E.G., Olins, G.M., Palomo, M.A. & Patton, D.R. (1990). *J. Med. Chem.*, **33**, 1935–40.

Thót, G., Kramer, T.H., Knapp, R., Lui, G., Davis, P., Burks, T.F., Yamamura, H.I. & Hruby, V.J. (1990). *J. Med. Chem.*, **33**, 249–53.

Tran, V.T., Beal, M.F. & Martin, J.B. (1985). *Science*, **228**, 492–5.

Veber, D.F., Freidinger, R.M., Perlow, D.S., Paleveda, W.J., Holly, F.W., Strachan, K.G., Nutt, R.F., Arison, B.H., Homnick, C., Randall, W.C., Glitzer, M.S., Saperstein, R. & Hirschmann, R. (1981). *Nature* (*London*), **292**, 55–8.

Veber, D.F., Saperstein, R., Nutt, R.F., Freidinger, R.M., Brady, S.F., Curley, P., Perlow, D.S., Paleveda, W.J., Colton, C.D., Zacchei, A.G., Tocco, D.J., Hoff, D.R., Vandlen, R.L., Gerich, J.E., Hall, L., Mandarino, L., Cordes, E.H., Anderson, P.S. & Hirschmann, R. (1984). *Life Sci.*, **34**, 1371–8.

Wilkes, B.C. & Schiller, P.W. (1987). *Biopolymers*, **26**, 1431–44.

Wilkes, B.C. & Schiller, P.W. (1990). *Biopolymers*, **29**, 89–95.

Wilkes, B.C. & Schiller, P.W. (1991). *J. Comput.-Aided Mol. Design.*, **5**, 293–302.

Wormser, U., Laufer, R., Hart, Y., Chorev, M., Gilan, C., & Selinger, Z. (1986). *EMBO J.*, **5**, 2805–8.

Zukin, R.S., Hartig, P.R. & Khoshland, D.E. Jr. (1977). *Proc. Natl. Acad. Sci. USA,* **74**, 1932–6.

16 Non-peptide Ligands for Peptide Receptors

R.M. FREIDINGER
Department of Medicinal Chemistry, Merck Research Laboratories, West Point, PA 19486, USA

1 Introduction

During the past 40 years, knowledge of the structures, properties and functions of biologically active peptides has blossomed. These compounds play important roles in normal and disease physiology as hormones and neurotransmitters in mammalian systems, and modulation of their cell-surface receptors with agonists or antagonists or by inhibiting enzymes involved in their biosynthesis can have important therapeutic implications (Samanen, 1985). For example, angiotensin-II is a pressor substance, and limitation of its formation by inhibiting angiotensin-converting enzyme has proved to be an effective method of treating hypertension. Such agents are extremely valuable for gaining a better understanding of the biochemistry and pharmacology of these peptides as well.

In some cases, peptides or analogues have proved to be suitable therapeutic agents. A classic case is insulin. In addition, synthetic analogues of luteinizing hormone-releasing hormone (LH-RH) and somatostatin have become marketed drugs for treating certain tumours, and the selective vasopressin agonist desmopressin acetate is used for treatment of diabetes insipidus. Special formulations have also been important for development of the once a month dosage of the LH-RH superagonist goserelin and orally administered cyclosporin.

Unfortunately, in many instances, peptide-based structures are not suitable for use in therapy where an orally bioavailable, selective, long-duration agent would be desirable (Plattner & Norbeck, 1990). One problem is a short duration of action due to cleavage by proteases. Furthermore, a peptide can often interact with more than one receptor or subtype resulting in an unsatisfactory lack of selectivity for the desired effect. Poor oral bioavailability is a third major limitation. Significant progress in dealing with the first two issues has been made in recent years. In the case of peptide agonists, superpotent analogues that are very resistant to proteolytic degradation have been synthesized by sequence truncation, incorporation of unnatural amino acids and cyclic amino acid derivatives, and preparation of cyclic peptides. Consideration of possible receptor-bound 'bioactive conformations' has played an important role in design of some of these analogues. For example, cyclic hexapeptide analogues of somatostatin which are resistant to proteolysis and have increased duration of action and greater specificity for inhibition of insulin, glucagon and growth hormone release over gastric acid secretion were developed by such an approach. These analogues still have low oral bioavailability, however, demonstrating that other factors such as poor absorption or rapid first-pass clearance by the liver can be limiting.

Potent peptide receptor antagonists (Regoli, 1985) have been obtained by employing many of these same strategies, and the same problems have been encountered. In addition, undesirable partial agonist activity which can be species-specific has been encountered. Significant structural similarity between antagonist analogues and the native peptide agonist often remains, and this resemblance may result in this difficulty which in the case of vasopressin antagonist analogues has resulted in the termination of human clinical trials.

In recent years, medicinal chemists have begun trying to overcome these limitations of peptide structures through the development of 'pseudopeptide' or 'peptide mimetic' compounds (Plattner & Norbeck, 1990). The key premise of these approaches is that replacement of one or more amide bonds in the structure will result in improved properties. Here, the development of completely non-peptidal ligands for peptide receptors will be considered (Freidinger, 1989a). Such compounds offer a much broader range of structures and many examples of non-peptide drugs already exist. That non-peptide ligands for peptide receptors can be found is illustrated by morphine and the many related agonists and antagonists that are potent non-peptide ligands for endogenous opioid receptors.

The rational design of these non-peptides based on knowledge of the native peptide structure/conformation/function relationships and/or receptor structure is attractive, but its routine practice will likely not occur until the 21st century. To date, only angiotensin-converting enzyme inhibitors, which originated from a nonapeptide lead, were developed in this way. The alternative, which is now seeing increasing success, is to identify suitable lead structures from natural or synthetic sources utilizing a biological screen. Such leads can then be developed by systematic application of established medicinal chemistry principles into agents with suitable properties for testing in human therapy. The first successful application of this approach from studies at Merck will be described.

2 Cholecystokinin (CCK) and gastrin antagonists

CCK and gastrin are closely related in terms of structure and biology. CCK was originally isolated from porcine intestine as a 33-amino acid peptide (CCK-33), but the core active sequence has been found to be the carboxy-terminal octapeptide (CCK-8). The classical actions of CCK include the stimulation of gallbladder contraction and pancreatic enzyme secretions. Many studies have also focused on its role as a satiety agent. CCK has also been shown to be one of the most abundant peptides in the brain, and evidence supports its function as a neurotransmitter. Recent research has focused on the involvement of CCK in analgesia and anxiety disorders. Gastrin was first isolated as a 17-amino acid peptide from porcine antral mucosa, but the active core is the carboxy-terminal tetrapeptide known as tetragastrin. The last five amino acids in the CCK and gastrin sequences are identical, and CCK-4 and tetragastrin are the same peptide. The best known action of gastrin is stimulation of gastric acid secretion.

Regarding CCK receptors, present evidence indicates two or three subtypes. The CCK-A receptor was originally characterized in peripheral tissues such as pancreas and gallbladder. More recently, it has been found in discrete regions of the CNS. For peptide agonists, CCK-8 sulphate has high affinity for this receptor subtype, but the desulphated form and gastrin or CCK-4 bind only weakly. The CCK-B receptor is ubiquitious in the CNS and has properties similar to the gastrin receptor found in the stomach. All of the indicated peptides have high affinity for CCK-B and gastrin receptors. It is not clear from current evidence whether or not CCK-B and gastrin receptors are actually different. For the non-peptide antagonists to be discussed, affinity for these receptors is very similar, so only data from CCK-A and CCK-B binding studies will be presented.

Selective CCK antagonists (Freidinger, 1989b) are of considerable interest for potential new therapy of gastrointestinal disorders such as pancreatitis and irritable bowel syndrome and for CNS utilities such as treatment of pain and anxiety. At the time that the Merck development of CCK antagonists began, peptide antagonists and weakly potent amino acid-derived antagonists had been discovered. There was a clear need, however, for more potent, selective, and orally active antagonists.

The key modern technology which led to the discovery of the first moderately potent non-peptide ligand for the CCK receptor is receptor-based screening. Assays utilizing displacement of ^{125}I-CCK from rat pancreas (CCK-A receptor) (Innis & Snyder, 1980) and guinea-pig brain tissue (CCK-B receptor) (Satio *et al.*, 1981; Chang *et al.*, 1983) were established. Screening with these assays resulted in the discovery from *Aspergillus alliaceus* of the novel structure asperlicin (1) (Chang *et al.*, 1985; Goetz *et al.*, 1985), which has an IC_{50} of about 1 μmol/l for binding to the pancreas receptor, but has much weaker affinity for the brain receptor. This compound helped firmly to establish the existence of the two different CCK receptor subtypes. Structurally, asperlicin may be viewed as a 1,4-benzodiazepine with a large 3-substituent derived from tryptophan and leucine and a quinazolone fused to the 1,2-positions (Liesch *et al.*, 1985). Asperlicin was shown to be competitive and selective for CCK receptors with no agonist activity (Chang *et al.*, 1985). In spite of its non-peptide nature, however, asperlicin is not orally active and greater potency was desirable.

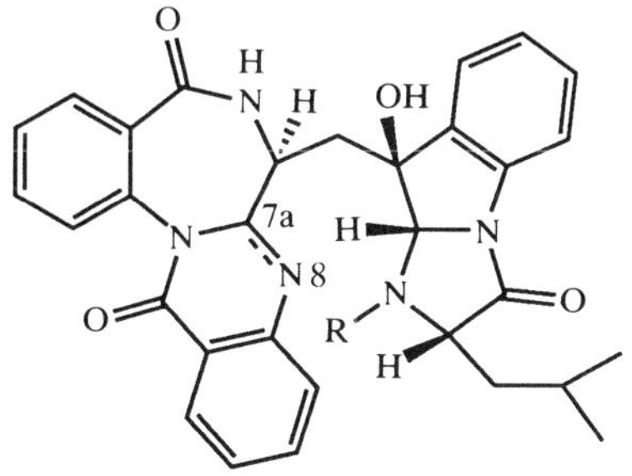

(1) R = H, Bond 7a-8 unsaturated

(2) R = H, Bond 7a-8 saturated

(3) R = C(=O)-CH_2-CH_2-CO_2Na, Bond 7a-8 saturated

Many analogues of asperlicin were prepared by direct chemical modification (Bock *et al.*, 1986) and biosynthesis (Houck *et al.*, 1988), and some of these had improved properties. For example, reduction of the 7a-8 imine linkage gave a sevenfold potency gain (2), and acylation of this antagonist with succinic anhydride produced (3) which has much improved water solubility and retains the potency and selectivity of asperlicin (Bock *et al.*, 1986). The desired oral activity was not achieved with these compounds, however.

2.1. *Benzodiazepine CCK-A selective antagonists*

An alternative to the asperlicin derivatization approach to improved antagonists involved attempts to design simpler, totally synthetic CCK antagonists based on the asperlicin lead. Evans *et al.* (1986) reasoned that a structure combining the elements of diazepam (4) with D-tryptophan might mimic asperlicin and have CCK receptor affinity. Such a structure and its relationship to asperlicin is outlined in Fig. 16.1. Compounds such as (5) are readily synthesized, for example from an aminobenzophenone and a D-trytophan ester, and (5) proved to have CCK-A receptor affinity identical to that of asperlicin with good selectivity. In accord with the design hypothesis, the binding of structures like (5) to the CCK receptor was found to be stereospecific, with the *S*-enantiomers having lower affinity. Furthermore, this new class of antagonists displays oral activity (Evans *et al.*, 1987).

Efforts to optimize the CCK-A antagonist activity of these benzodiazepines also proved very successful. Early structure–activity studies showed that potency could be increased severalfold with *N*-1-methyl and 2′-fluoro substituents, and *N*-1-carboxymethyl increased water solubility (Table 16.1) (Evans *et al.*, 1987). Subsequently, it was found that the nature of the group linking the benzodiazepine with the 3-position aromatic group as well as the point of attachment to the aromatic group were critical factors (Table 16.2) (Evans *et al.*, 1986, 1988). When the 3-position methylene linking group in compound (5) is replaced with a carboxamide and the indole is attached at its 2-position rather than the 3-position, the new analogue (10) is obtained. The increase in CCK-A receptor-binding affinity

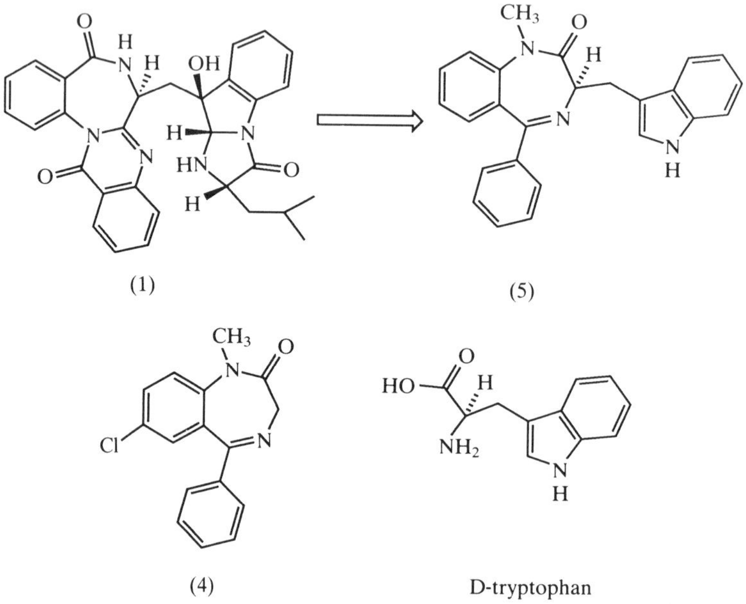

Figure 16.1. Design of benzodiazepine CCK-A selective antagonists.

Table 16.1. Receptor binding data for 3-indolymethyl-5-phenyl-1,4-benzodiazepines

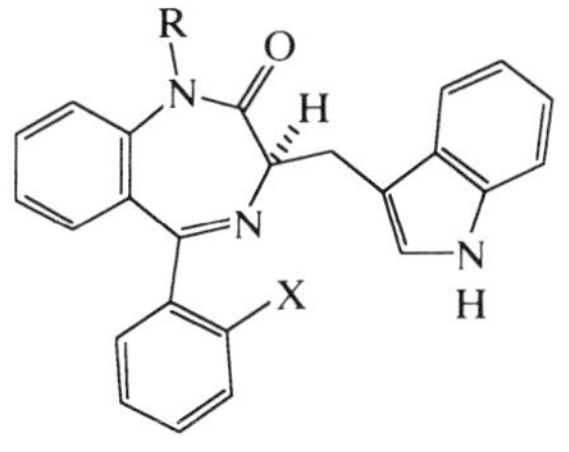

Compound	R	X	IC_{50} (μmol/l)* Rat pancreas	Guinea-pig brain
1	—	—	1.4	>100
5	H	H	1.2	50
6	H	F	0.5	80
7	CH_3	F	0.3	10
8	CH_2COOH	F	0.3	23

* Inhibition of ^{125}I-CCK binding.

with these changes is substantial, about 500-fold. The 3-indolyl derivative (9), on the other hand, shows almost no increase in binding. Reduction of the amide carbonyl to methylene as in the case of (11) also gives an analogue less potent than (10) thus illustrating the key role of the amide. *N*-1 methylation of (10) and resolution provided the enantiomers (12) and (13). This comparison points out that methylation further enhances potency and that the 3-position stereochemistry is important just as for the original lead (5).

Compound (13) is the optimal CCK antagonist in this series with a binding constant for the CCK-A receptor comparable to that of the native peptide CCK

Table 16.2. Receptor binding data for 3-amino-5-phenyl-1,4-benzodiazepine derivatives

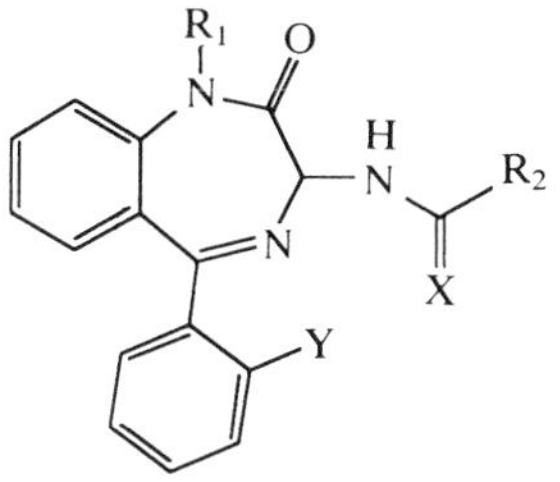

Compound	R_1	X	Y	R_2	Stereo	IC_{50}(μmol/l)* Rat pancreas	Guinea-pig brain
9	H	O	H	3-indolyl	*R,S*	1.1	8.4
10	H	O	H	2-indolyl	*R,S*	0.0047	8
11	H	H,H	F	2-indolyl	*R,S*	0.087	100
12	CH_3	O	H	2-indolyl	*R*	0.0083	3.7
13	CH_3	O	H	2-indolyl	*S*	0.00008	0.27
17	CH_3	O	F	*p*-Cl-phenyl	*S*	0.002	2.9
18	CH_3	O	F	*m*-Br-phenyl	*S*	0.0035	3.5
19	CH_2COOH	O	H	2-indolyl	*R,S*	0.0014	6.0

*Inhibition of ^{125}I-CCK binding.

(Chang & Lotti, 1986; Evans *et al.*, 1986, 1988). This compound, which is known as devazepide (also L-364,718 or MK-329), is a competitive antagonist with high selectivity versus CCK-B, gastrin, and other peptide and neurotransmitter receptors (Chang & Lotti, 1986). Devazepide has also been shown to be highly potent by several routes of administration, including oral, in a variety of functional assays such as gastric emptying and gallbladder contraction in a number of different species, and no agonist activity has been observed (Lotti *et al.*, 1987; Pendleton *et al.*, 1987). It has been demonstrated that devazepide crosses the blood–brain barrier efficiently (Pullen & Hodgson, 1987). This antagonist is the most potent known CCK-A antagonist of any structural type being at least 100-fold more potent than loxiglumide, a glutamic acid derivative developed by Rotta (Setnikar *et al.*, 1987a,b). Devazepide was chosen for clinical trials to evaluate its therapeutic potential in humans (Berlin & Freidinger, 1991).

The 3-aminobenzodiazepine derivatives such as devazepide are readily prepared by total synthesis (Fig. 16.2) (Bock *et al.*, 1987). The unsubstituted 1-methyl-5-phenyl-1,4-benzodiazepine (14) is first prepared in four steps from aminobenzophenone (15). This intermediate is then nitrosated followed by reduction to give the parent racemic 3-aminobenzodiazepine (16). An elegant resolution–racemization procedure converts most of the racemic amine to the desired *S*-enantiomer (Reider *et al.*, 1987). Acylation with an indole-2-carboxylic acid derivative provides devazepide in high yield.

A number of structurally related CCK-A antagonists with nanomolar level potency have been found (Table 16.2). For example, indole can be replaced with

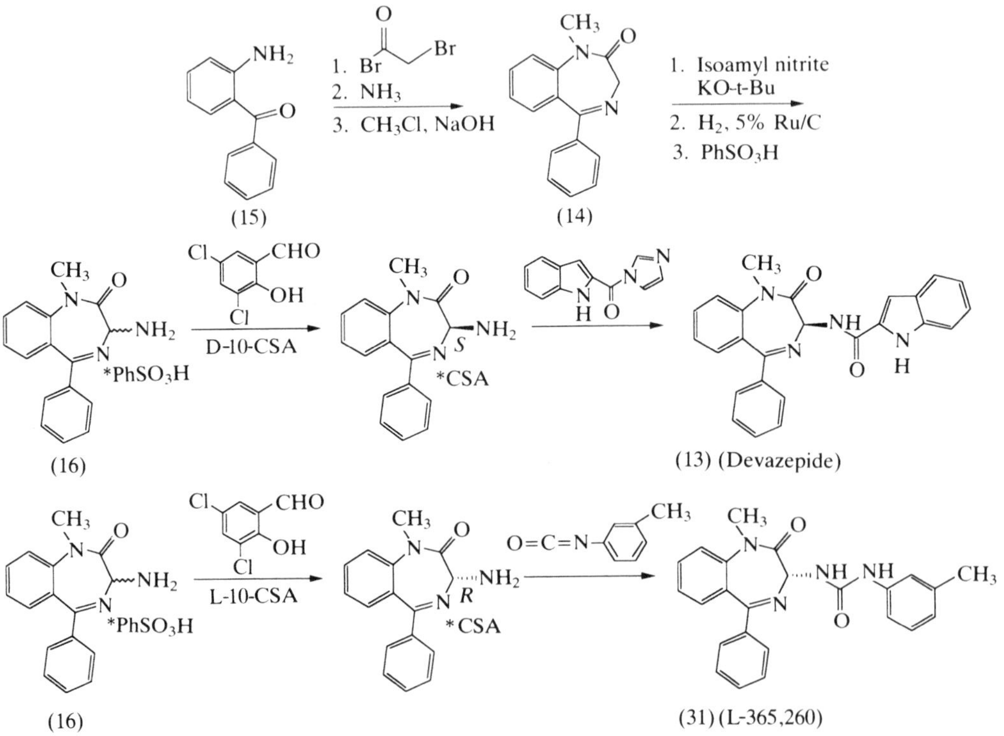

Figure 16.2. Synthesis of devazepide and L-365,260.

p- or *m*-substituted phenyl (e.g. 17 and 18). Groups providing increased water solubility such as an *N*-1-carboxymethyl can be incorporated as in (19) (Evans *et al.*, 1988). Other excellent antagonists within the 1,4-benzodiazepine class include triazolo compounds such as (20), which is very potent *in vitro* (IC_{50} = 0.2 nmol/l, pancreas binding) and *in vivo* (mouse gastric emptying, p.o.), but has somewhat lower selectivity than devazepide (Bock *et al.*, 1988).

In addition to the 3-substituted benzodiazepines, certain 2-substituted structures were also found to be CCK-A selective antagonists. One of these compounds is tifluadom (21), which had previously been shown to be a potent opioid agonist (Romer *et al.*, 1982). Tifluadom has moderate and selective affinity for the CCK-A versus the CCK-B receptor (IC_{50} = 47 nmol/l and >100 μmol/l, respectively), but it has about 30-fold greater affinity for the opioid receptor (Chang *et al.*, 1986). The *S*-(−)-tifluadom enantiomer has higher affinity than the *R*-(+)-enantiomer (sign of rotation refers to toluene) for both CCK and opioid receptors (Petrillo *et al.*, 1985; Chang *et al.*, 1986). A number of tifluadom analogues were synthesized in an effort to separate the CCK and opioid activities (Bock *et al.*, 1990). While some potent analogues such as the indole-2-carbonyl derivative (22) were obtained, the desired change in receptor selectivity was not achieved. Importantly, devazepide and related structures have low affinity for opioid receptors (Chang & Lotti, 1986).

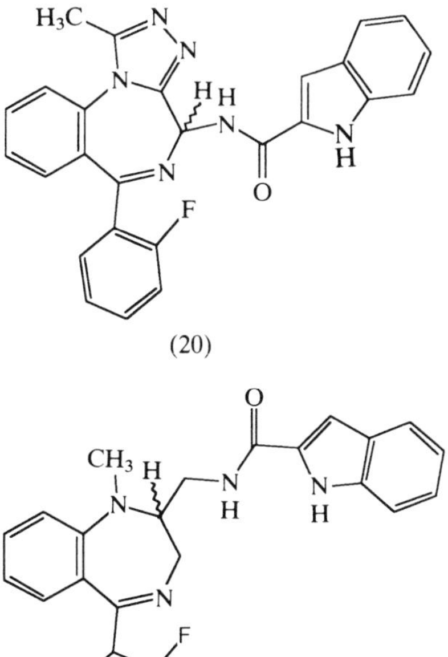

(20)

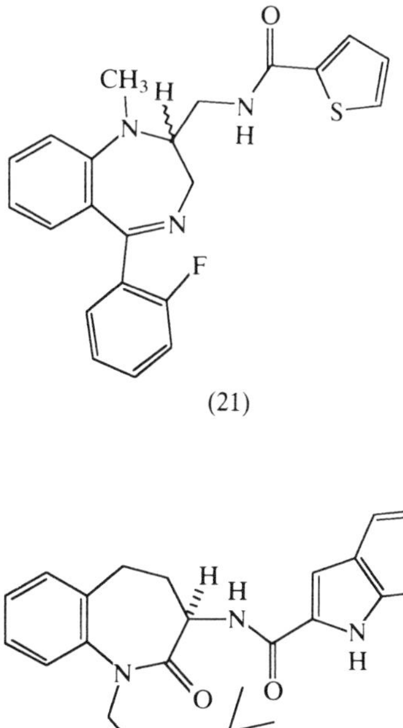

(21)

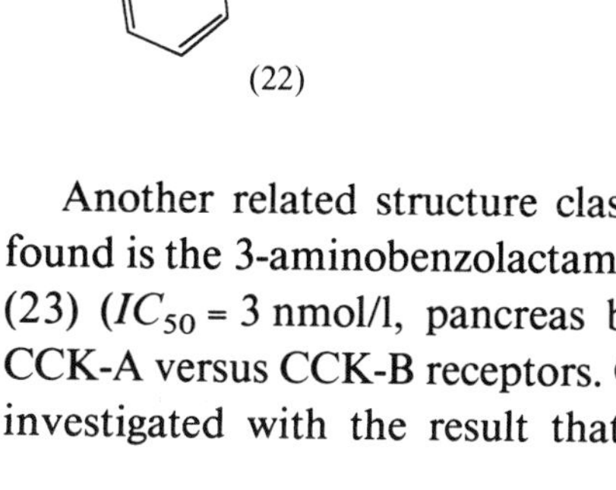

(22)

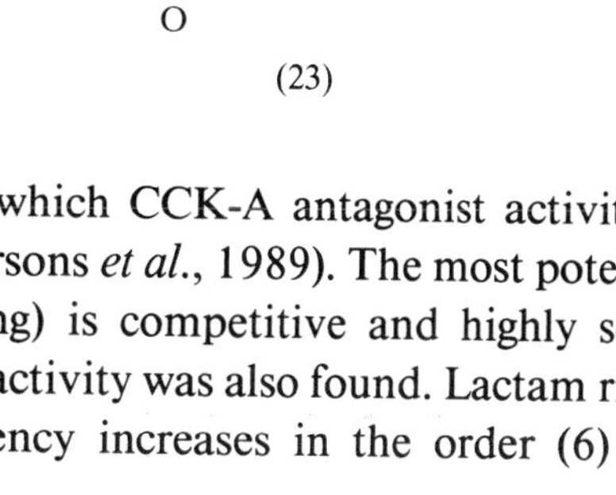

(23)

Another related structure class in which CCK-A antagonist activity has been found is the 3-aminobenzolactams (Parsons *et al.*, 1989). The most potent analogue (23) (IC_{50} = 3 nmol/l, pancreas binding) is competitive and highly selective for CCK-A versus CCK-B receptors. Oral activity was also found. Lactam ring size was investigated with the result that potency increases in the order (6) < (8) < (7).

Stereospecificity was demonstrated and a molecular modelling study showed good correspondence between structural features of the more potent *R*-enantiomer and MK-329. The authors concluded that the *tert*-butyloxycarboxymethyl group of (23) and the 5-phenyl of devazepide may share the same space on the receptor.

2.2. *Benzodiazepine CCK-B/gastrin-selective antagonists*

All of the antagonists discussed to this point have shown selectivity for the CCK-A receptor subtype. None of these compounds has been shown to bind selectively to

Table 16.3. Receptor binding stereospecificity of selected benzodiazepine derivatives

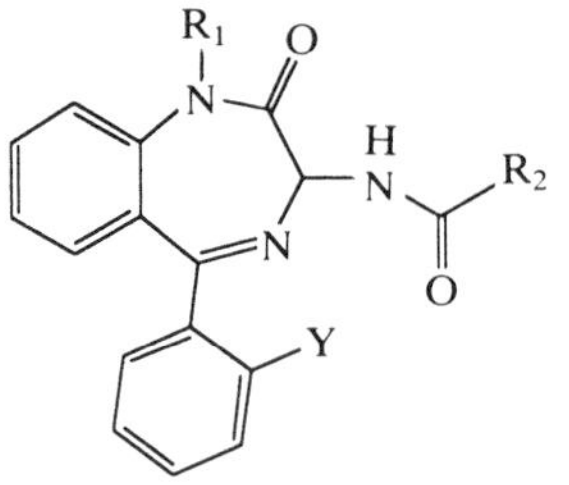

					IC_{50} (μmol/l)*	
Compound	R_1	Y	R_2	Stereo	Rat pancreas	Guinea-pig brain
17	CH_3	F	*p*-Cl-phenyl	*S*	0.002	29
24	CH_3	F	*p*-Cl-phenyl	*R*	0.049	11
25	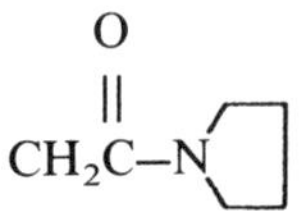	H	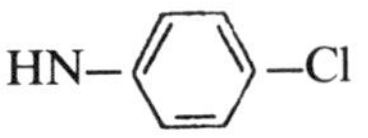	*R,S*	0.52	0.0003
26	CH_3	H	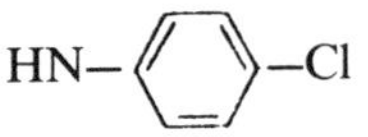	*R,S*	0.051	0.023
27	CH_3	H		*S*	0.026	0.41
28	CH_3	H	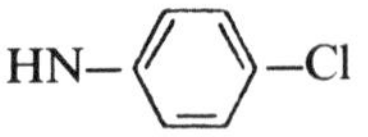	*R*	1.1	0.005
29	CH_3	H	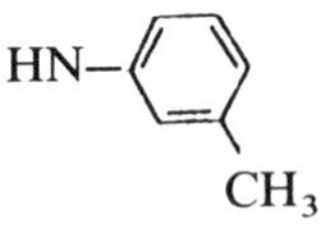	*R,S*	0.008	0.007
30	CH_3	H	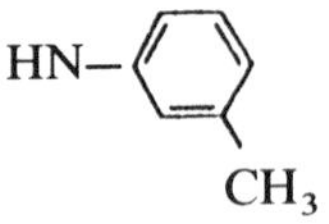	*S*	0.003	0.151
31	CH_3	H	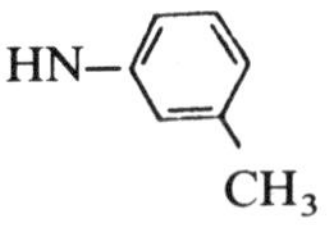	*R*	0.28	0.002

* Inhibition of ^{125}I-CCK binding.

the CCK-B or gastrin-type receptors. In 1989, 3-substituted benzodiazepine analogues were described which are potent and are the first non-peptide ligands to bind selectively to CCK-B and gastrin receptors versus CCK-A receptors (Bock *et al.*, 1989). These antagonists were designed from the earlier CCK-A-selective benzodiazepines such as (17) by combining individual structural modifications which led to loss of CCK-A selectivity. An example of this type of antagonist is (25) in which both the urea linkage and the large *N*-1 substituent are crucial for the indicated potency and selectivity (Table 16.3). Compound (25) is the most potent known antagonist with this type of selectivity.

An alternative method for achieving CCK-B/gastrin selectivity was found to be resolution of 1-methyl-3-arylurea benzodiazepines (Bock *et al.*, 1989). Compounds such as (26) and (29), which are racemic, are non-selective (Table 16.3). Separation into the *S*- and *R*-enantiomers revealed selective antagonists. The *S*-enantiomers such as (27) and (30) bind selectively to the CCK-A receptor subtype, while (28) and (31) are a new type of CCK-B/gastrin-selective ligand.

Investigation of a number of substituents on the phenyl showed *m*-methyl to be among the most potent, and, considering all properties, compound (31) (L-365,260) was the most preferred antagonist (Bock *et al.*, 1989; Lotti & Chang, 1989). Competitive binding inhibition was demonstrated for guinea-pig gastrin and brain CCK receptors, and no agonist activity was observed. Upon oral administration, L-365,260 potently antagonized gastrin-stimulated acid secretion in several animal species (mouse $ED_{50} = 30\ \mu g/kg$) with good duration of action. Good selectivity was demonstrated with respect to 13 other receptors (Lotti & Chang, 1989). Interestingly, none of these antagonists demonstrated significant separation of CCK-B and gastrin-receptor binding. L-365,260 is readily available through modification of the synthetic route to devazepide (Fig. 16.2). This CCK-B antagonist has been chosen for human clinical trials to evaluate its potential for therapy.

3 Conclusion

The development of devazepide, L-365,260, and related analogues represents pioneering basic drug development research on several fronts. The discovery of asperlicin was the first successful application of receptor-based screening for identifying a lead non-peptide antagonist for a peptide receptor. The subsequent medicinal chemistry effort utilized structural analogy of asperlicin to benzodiazepine receptor ligands in the first deliberate design of highly potent, selective, orally active non-peptide antagonists with appropriate properties for potential therapeutic agents. Key structural features were identified to produce selectivity for either receptor subtype. Devazepide and L-365,260 are now being widely used as tools in studies aimed at better understanding the biochemistry and pharmacology of CCK (Dourish *et al.*, 1991). These studies will undoubtedly lead to other potential applications for these drugs. Other successful examples of this overall approach are now beginning to appear, for example non-peptide angiotensin-II antagonist DuP 753 (32) (Timmermans *et al.*, 1991) and an alternative class of non-peptide CCK-B antagonist LY262691 (33) (Howbert *et al.*, 1991). Other novel structures such as cyclic hexapeptide oxytocin antagonists (34) have been discovered (Pettibone *et al.*, 1989). Very recently, the first potent, selective non-peptide substance P antagonist CP-96,345 (35) was reported (McLean *et al.*, 1991; Snider *et al.*, 1991). This methodology is likely to be very important into the 21st century in the development of novel structures with novel mechanisms of action as potential new therapeutic agents.

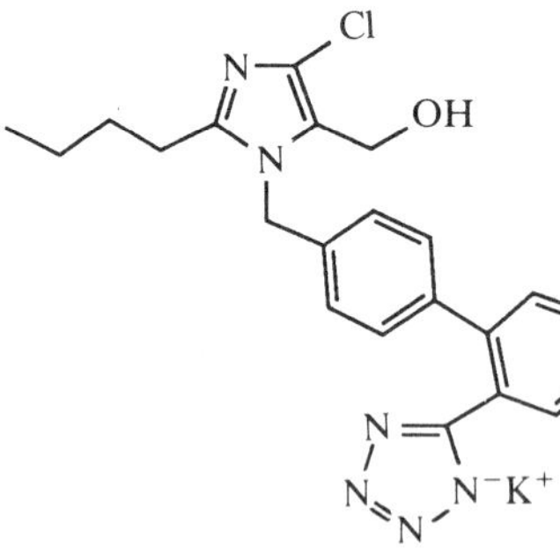

(32)

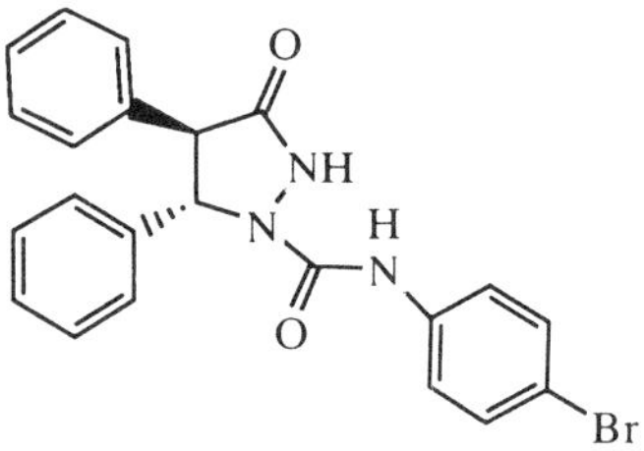

(33)

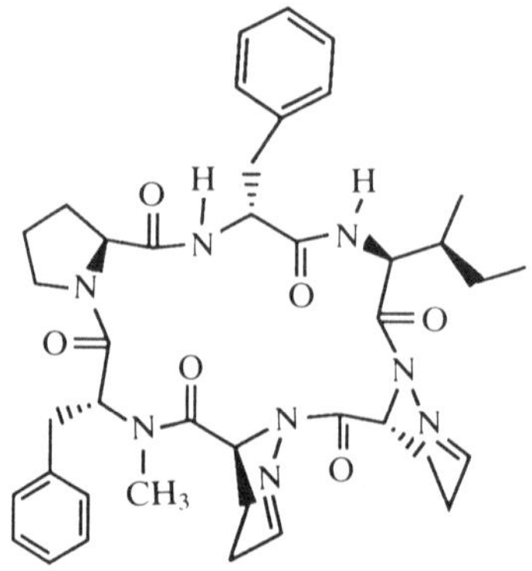

(34)

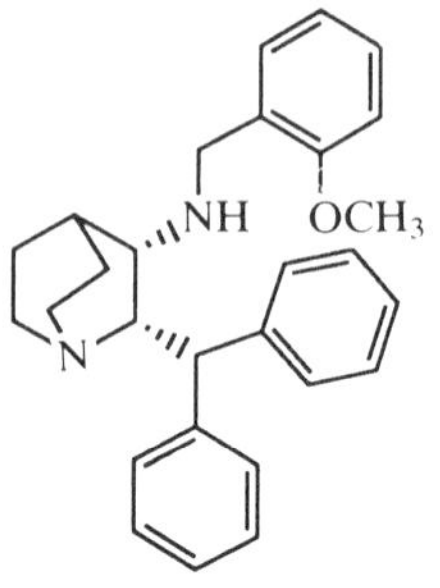

(35)

Interestingly, an alternative approach to lead generation and development has recently shown success in the CCK field. CCK-4 was simplified to the smallest fragment (a dipeptide) which retained an $IC_{50} \sim 10\ \mu mol/l$. This compound was then used as the lead for systematic structure/function studies which culminated in a novel class of non-peptide CCK-B antagonists exemplified by PD134308 (Hughes *et al.*, 1990; Horwell *et al.*, 1991).

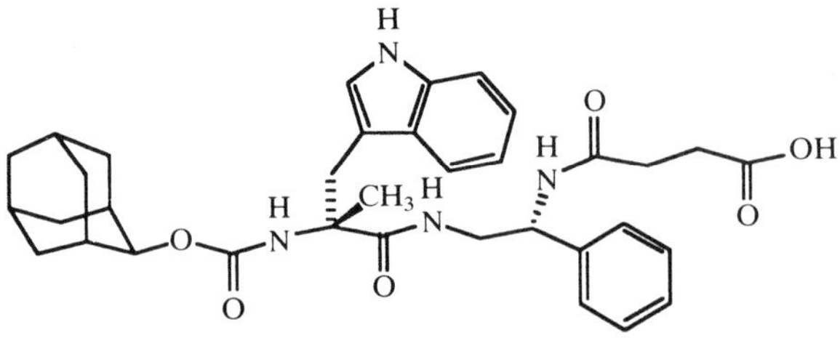

PD 134308

The rational design of non-peptide ligands for peptide receptors will come closer to reality over the next few years. As other successes emerge from receptor-based screening, the resultant structural database coupled with human and computerized abilities to draw analogies and model molecules should lead to success in other peptide systems. Improved capabilities for computerized modelling of peptides and small molecules should also be helpful.

4 Acknowledgements

It is a pleasure to acknowledge the contributions of my many colleagues at Merck who developed asperlicin, devazepide and L-365,260. They are listed as authors of the papers cited in those sections of this chapter. I also thank Ms Jean Kaysen for skilful help in preparing the manuscript.

5 References

Berlin, R.G. & Freidinger, R.M. (1991). Characterization of MK-329. In: G. Adler and C. Beglinger (Eds.), *Cholecystokinin Antagonists in Gastroenterology: Basic and Clinical Status*, pp. 70–9. Springer Verlag, Berlin.

Bock, M.G., DiPardo, R.M., Evans, B.E., Rittle, K.E., Whitter, W.L., Veber, D.F., Rittle, K.E., Evans, B.E., Freidinger, R.M., Veber, D.F., Chang, R.S.L., Chen T., Keegan, M.E. & Lotti, V.J. (1986). Cholecystokinin antagonists. Synthesis of asperlicin analogues with improved potency and water solubility. *J. Med. Chem.*, **29**, 1941–5.

Bock, M.G., DiPardo, R.M., Evans, B.E., Rittle, K.E., Veber, D.F., Freidinger, R., Hirshfield, J. & Springer, J.P. (1987). Synthesis and resolution of 3-amino-1,3-dihydro-5-phenyl-2H-1,4-benzodiazepin-2-ones. *J. Org. Chem.*, **52**, 3232–9.

Bock, M.G., DiPardo, R.M., Evans, B.E., Rittle, K.E., Veber, D.F., Freidinger, R.M., Chang, R.S.L. & Lotti V.J. (1988). Cholecystokinin antagonists. Synthesis and biological evaluation of 4-substituted 4H-[1,2,4]triazolo[4,3-a][1,4]benzodiazepines. *J. Med. Chem.*, **31**, 176–81.

Bock, M.G., DiPardo, R.M., Evans, B.E., Rittle, K.E., Whitter, W.L., Veber, D.F., Anderson, P.S. & Freidinger, R.M. (1989). Benzodiazepine gastrin and brain cholecystokinin receptor ligands: L-365,260. *J. Med. Chem.*, **32**, 13–16.

Bock, M.G., DiPardo, R.M., Evans, B.E., Rittle, K.E., Whitter, W.L., Veber, D.F., Freidinger, R.M., Chang, R.S.L., Chen, T.B. & Lotti, V.J. (1990). Cholecystokinin-A receptor ligands based on the K-opioid agonist tifluadom. *J. Med. Chem.*, **33**, 450–3.

Chang, R.S.L. & Lotti, V.J. (1986). Biochemical and pharmacological characterization of an extremely potent and selective nonpeptide cholecystokinin antagonist. *Proc. Natl. Acad. Sci. USA*, **83**, 4923–6.

Chang, R.S.L., Lotti, V.J., Martin, G.E. & Chen, T.B. (1983). Increase in ^{125}I-cholecystokinin receptor binding following chronic haloperidol treatment, intracisternal 6-hydroxydopamine or ventral tegmental lesions. *Life Sci.*, **32**, 871–8.

Chang, R.S.L., Lotti, V.J., Monaghan, R.L., Birnbaum, J., Stapley, E.O., Goetz, M.A., Albers-Schonberg, G., Patchett, A.A., Leisch, J.M., Hensens, O.D. & Springer, J.P. (1985). A potent nonpeptide cholecystokinin antagonist selective for peripheral tissue isolated from *Aspergillus alliaceus. Science*, **230**, 177–9.

Chang, R.S.L., Lotti, V.J., Chen, T.B. & Keegan, M.E. (1986). Tifluadom, a kappa opiate agonist, acts as a peripheral cholecystokinin receptor antagonist. *Neurosci. Lett.*, **72**, 211–14.

Dourish, C.T., Cooper, S.J., Iversen, L.I. & Iversen, S.D. (1991). *Multiple Cholecystokinin Receptors: Progress Towards CNS Therapeutic Targets*, Oxford University Press, Oxford. In press.

Evans, B.E., Bock, M.G., Rittle, K.E., DiPardo, R.M., Whitter, W.L., Veber, D.F., Anderson, P. S. & Freidinger, R.M. (1986). Design of potent, orally effective, nonpeptidal antagonists of the peptide hormone cholecystokinin. *Proc. Natl. Acad. Sci. USA*, **83**, 4918–22.

Evans, B.E., Rittle, K.E., Bock, M.G., DiPardo, R.M., Freidinger, R.M., Whitter, W.L., Gould,

N.P., Lundell, G.F., Homnick, C.F., Veber, D.F., Anderson, P.S., Chang, R.S.L., Lotti, V.J., Cerino, D.J., Chen, T.B., Kling, P.J., Kunkel, K.A., Springer, J.P. & Hirshfield, J. (1987). Design of nonpeptidal ligands for a peptide receptor: cholecystokinin antagonists. *J. Med. Chem.*, **30**, 1229–39.

Evans, B.E., Rittle, K.E., Bock, M.G., DiPardo, R.M., Friedinger, R.M., Whitter, W.L., Lundell, G.F., Veber, D.F., Anderson, P.S., Chang, R.S.L., Lotti, V.J., Cerino, D.J., Chen, T.B., Kling, P.J., Kunkel, K.A., Springer, J.P. & Hirshfield, J. (1988). Methods for drug discovery: development of potent, selective, orally effective cholecystokinin antagonists. *J. Med. Chem.*, **31**, 2235–46.

Freidinger, R.M. (1989a). Non-peptide ligands for peptide receptors. *Trends Pharmacol. Sci.*, **10**, 270–4.

Freidinger, R.M. (1989b). Cholecystokinin and gastrin antagonists. *Med. Res. Rev.*, **9**, 271–90.

Goetz, M.A., Lopez, M., Monaghan, R.L., Chang, R.S.L., Lotti, V.J. & Chen, T.B. (1985). Asperlicin, a novel non-peptidal cholecystokinin antagonist from *Aspergillus alliaceus*. Fermentation, isolation and biological properties. *J. Antibiot.*, **38**, 1633–7.

Horwell, D.C., Hughes, J., Hunter, J.C., Pritchard, M.C., Richardson, R.S., Roberts, E. & Woodruff, G.N. (1991). Rationally designed 'dipeptoid' analogues of CCK, α-methyltryptophan derivatives as highly selective and orally active gastrin and CCK-B antagonists with potent anxiolytic properties. *J. Med. Chem.*, **34**, 404–14.

Houck, D.R., Ondeyka, J., Zink, D.L., Inamine, E., Goetz, M.A. & Hensens, O.D. (1988). On the biosynthesis of asperlicin and the directed biosynthesis of analogs in *Aspergillus alliaceus*. *J. Antibiot.*, **41**, 882–91.

Howbert, J.J., Lobb, K.L., Brown, R.F., Reel, J.K., Neel, D.A., Mason, N.R., Mendelsohn, L.G., Hodgkiss, J.P. & Kelly, J.S. (1991). A novel series of non-peptide CCK and gastrin antagonists: medicinal chemistry and electrophysiological demonstration of antagonism. In: C.T. Dourish, S.J. Cooper, L.I. Iversen and S.D. Iversen (Eds.), *Multiple Cholecystokinin Receptors: Progress Towards CNS Therapeutic Targets*, Oxford University Press, Oxford. In press.

Hughes, J., Boden, P., Costall, B., Domeney, A., Kelly, E., Horwell, D.C., Hunter, J.C., Pinnock, R.D. & Woodruff, G.N. (1990). Development of a class of selective cholecystokinin type B receptor antagonists having potent anxiolytic activity. *Proc. Natl. Acad. Sci. USA*, **87**, 6728–32.

Innis, R.B. & Snyder, S.H. (1980). Distinct cholecystokinin receptors in brain and pancreas. *Proc. Natl. Acad. Sci. USA*, **77**, 6917–21.

Leisch, J.M., Hensens, O.D., Springer, J.P., Chang, R.S.L. & Lotti, V.J. (1985). Asperlicin, a novel non-peptidal cholecystokinin antagonist from *Aspergillus alliaceus*. Structure elucidation. *J. Antibiot.*, **38**, 1638–41.

Lotti, V.J. & Chang, R.S.L. (1989). A new potent and selective non-peptide gastrin antagonist and brain cholecystokinin receptor (CCK-B) ligand: L-365,260. *Eur. J. Pharmacol.*, **162**, 273–80.

Lotti, V.J., Pendleton, R.G., Gould, R.J., Hanson, H.M., Chang, R.S.L. & Clineschmidt, B.V. (1987). *In vivo* pharmacology of L-364,718, a new potent nonpeptide peripheral cholecystokinin antagonist. *J. Pharmacol. Exp. Ther.*, **241**, 103–9.

McLean, S., Ganong, A.H., Seeger, T.F., Bryce, D.K., Pratt, K.G., Reynolds, L.S., Siok, C.J., Lowe, J.A. & Heym, J. (1991). Activity and distribution of binding sites in brain of a nonpeptide substance P (NK_1) receptor antagonist. *Science*, **251**, 437–9.

Parsons, W.H., Patchett, A.A., Holloway, M.K., Smith, G.M., Davidson, J.L., Lotti, V.J. & Chang, R.S.L. (1989). Cholecystokinin antagonists. Synthesis and biological evaluation of 3-substituted benzolactams. *J. Med. Chem.*, **32**, 1681–5.

Pendleton, R.G., Bendesky, R.J., Schaffer, L., Nolan, T.E., Gould, R.J. & Clineschmidt, B.V. (1987). Roles of endogenous cholecystokinin in biliary, pancreatic and gastric function: studies with L-364,718, a specific cholecystokinin receptor antagonist. *J. Pharmacol. Exp. Ther.*, **241**, 110–16.

Petrillo, P., Amato, M. & Tavani, A. (1985). The interaction of the two isomers of the opioid benzodiazepine tifluadom with μ-, δ-, and κ-binding sites and their analgesic and intestinal effects in rats. *Neuropeptides*, **5**, 403–6.

Pettibone, D.J., Clineschmidt, B.V., Anderson, P.S., Freidinger, R.M., Lundell, G.F., Koupal, L.R., Schwartz, C.D., Williamson, J.M., Goetz, M.A., Hensens, O.D., Liesch, J.M. & Springer, J.P. (1989). A structurally unique, potent, and selective oxytocin antagonist derived from *Streptomyces silvensis. Endocrinology*, **125**, 217–22.

Plattner, J.J. & Norbeck, D.W. (1990). Obstacles to drug development from peptide leads. In: C.R. Clark and W.H. Moos (Eds.), *Drug Discovery Technologies*, pp. 92–126, Ellis Horwood, Chichester.

Pullen, R.G.L. & Hodgson, O.J. (1987). Penetration of diazepam and the non-peptide CCK antagonist, L-364,718, into rat brain. *J. Pharm. Pharmacol.*, **39**, 863–4.

Regoli, D. (1985). Peptide antagonists. *Trends Pharmacol. Sci.*, **6**, 481–4.

Reider, P.J., Davis, P., Hughes, D.L. & Grabowski, E.J.J. (1987). Crystallization-induced asymmetric transformation: stereospecific synthesis of a potent peripheral CCK antagonist. *J. Org. Chem.*, **52**, 955–7.

Romer, D., Buscher, H.H., Hill, R.C., Mauer, R., Petcher, T.J., Zeugner, H., Benson, W., Finner, E., Milkowski, W. & Thies, P.W. (1982). An opioid benzodiazepine. *Nature* (*London*), **298**, 759–60.

Saito, A., Goldfine, I.D. & Williams, J.A. (1981). Characterization of receptors for cholecystokinin and related peptides in mouse cerebral cortex. *J. Neurochem.*, **37**, 483–90.

Samanen, J. (1985). Biomedical polypeptides—a wellspring of pharmaceuticals. In: C.G. Gebelein and C.E. Carraher (Eds.), *Bioactive Polymeric Systems: An Overview,* pp. 279–382, Plenum Press, New York.

Setnikar, I., Bani, M., Cereda, R., Chiste, R., Makovec, F., Pacini, M. A., Revel, L., Rovati, L.C. & Rovati, L.A. (1987a). Pharmacological characterization of a new potent and specific nonpolypeptidic cholecystokinin antagonist. *Arzneim-Forsch.*, **37**(I), 703–7.

Setnikar, I., Bani, M., Cereda, R., Chiste, R., Makovec, F., Pacini, M. A. & Revel, L. (1987b). Anticholecystokinin activities of loxiglumide. *Arzneim-Forsch.*, **37**(II), 1168–71.

Snider, R.M., Constantine, J.W., Lowe, J.A., Longo, K.P., Lebel, W.S. Woody, H.A., Drozda, S.E., Desai, M.C., Vinick, F.J., Spencer, R.W. & Hess, H.-J. (1991). A potent nonpeptide antagonist of the substance P(NK_1) receptor. *Science*, **251**, 435–7.

Timmermans, P.M.W.M., Wong, P.C., Chiu, A.T. & Herblin, W.F. (1991). Nonpeptide angiotensin II receptor antagonists. *Trends Pharmacol. Sci.*, **12**, 55–62.

17 Peptidomimetic Research, Design and Synthesis

W.F. HUFFMAN
Peptidomimetic Research Department, SmithKline Beecham, 709 Swedeland Road, PO Box 1539, King of Prussia, PA 19406–0939, USA

1 Introduction

The design and synthesis of molecules which mimic key aspects of peptide pharmacophore, i.e. peptidomimetics, is emerging as an important contributor to medicinal chemistry in the 21st century. This is due in part to the fact that peptidomimetics have the potential to (i) provide an improved understanding of the biologically active conformation(s) of peptides of interest, (ii) provide improved bioavailability characteristics relative to the original peptide, and (iii) ultimately lead to the synthesis of small-molecule, non-peptide mimetics. This last objective, the rational design of a small-molecule, non-peptide 'peptidomimetic' from the peptide itself, represents the ultimate challenge in peptidomimetic research since it requires a rather complete understanding of the chemical and spatial elements of the pharmacophore of interest.

The scope of peptidomimetics extends from peptides where single amino acid residues (or peptide bonds) have been altered, all the way to the transformation of the peptide into a small, non-peptide molecule. Chemical modifications involved include the use of unusual amino acids, peptide bond replacements, mimics of secondary structural features of the peptide (e.g. helices, reverse turns), or non-peptide heterocycles as templates to mimic key aspects of the tertiary structure of the peptide. The result of these chemical changes is to impart altered characteristics to the peptide, for example protease resistance or conformational preference, as well as alterations in polarity, hydrogen bond potential, potency, or receptor selectivity.

Here the aim is to provide examples of peptidomimetic research to illustrate the scope and types of modifications in structure, conformation and biology that can be obtained, limiting discussion to transformations of an existing peptide lead. None the less, it should be apparent that some of the peptidomimetic approaches presented will also have applications to the equally intriguing problem of generating an initial peptide lead from a protein sequence of interest.

2 Unusual amino acids

Several examples from our vasopressin receptor antagonist studies serve to exemplify the synthetic challenges and the biological modifications which are typical for the use of unusual amino acids in peptidomimetic research.

SK&F 101926, a potent antagonist of the vasopressin antidiuretic hormone receptor (V_2) in a variety of animal species (both *in vitro* and *in vivo*) and in human renal tissue, behaved as an agonist, rather than an antagonist, when evaluated in humans. In order to prepare antagonists devoid of any partial agonist activity, we

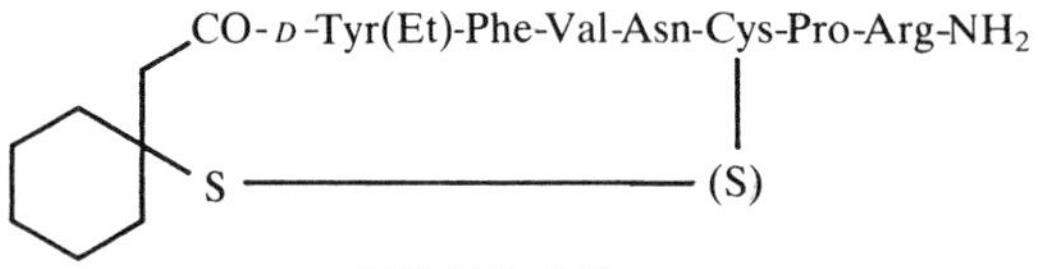

SK&F 101926

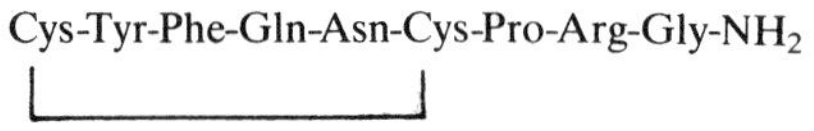

Arginine vasopressin

set out to determine the structure–activity relationships (SARs) for partial agonism of V_2-receptor antagonists. Attention was initially directed to those unusual amino acids which appeared to be essential for the conversion of agonists into antagonists; one such modification was the mercaptocyclohexylacetic acid moiety (1) (also referred to as pentamethylenemercaptopropionic acid or Pmp) which served as a mimetic for the cysteine residue at position-1 in arginine vasopressin (Manning *et al.*, 1981). In an attempt to understand the role of Pmp in the antagonist pharmacophore, a synthesis of Pmp was developed which allowed for the preparation of a wide variety of substitutions on the cyclohexane ring (Yim & Huffman, 1983). One such substitution, a *cis*-4′-MePmp (2), displayed remarkable features with respect to partial agonist activity.

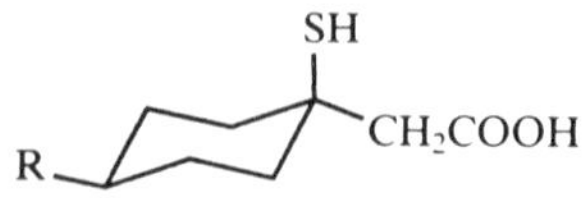

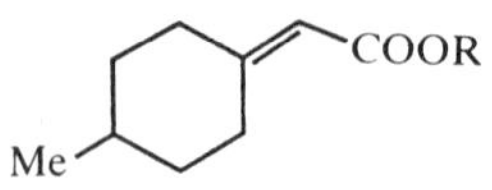

(1) R = H
(2) R = Me

(3) R = Et
(4) R = H

Michael-type addition of various benzyl mercaptans to ethyl (4-methylcyclohexylidene) acetate (3), followed by ester hydrolysis, afforded 4′-MePmp where the major product was the isomer with the methyl group and the sulphur atom on the same side of the cyclohexane ring (arbitrarily defined as the *cis*-isomer) (Newlander *et al.*, 1990). If the Michael addition was carried out on the corresponding acid (4) using a variety of amine bases as catalyst, the *trans*-isomer predominated; the stereochemical assignments of the *cis*- and *trans*-isomers were made based on X-ray crystallographic studies (Eggleston *et al.*, 1989). Peptide analogues of SK&F 101926 were prepared in which the Pmp residue at position-1 was replaced with either *cis*-4′-MePmp or *trans*-4′-MePmp. Unexpectedly, when tested in a canine model for vasopressin agonist activity, the peptide containing the *cis*-4′-MePmp was devoid of agonist activity while the peptide with the *trans*-isomer displayed substantial agonist activity similar to SK&F 101926 (Huffman *et al.*, 1989).

Cysteine mimetics like Pmp, that contain di-geminal substitution at the β-carbon of cysteine, have also been postulated to exert conformational effects on the adjacent disulphide bond when present in cyclic peptides such as the vasopressin antagonist SK&F 101926 (Meraldi *et al.*, 1977). To test the importance of the disulphide to partial agonist activity in antagonists, a disulphide mimetic was prepared in which the sulphur atoms were replaced with methylene groups; more precisely, the Pmp–Cys dipeptide unit (5) would be mimicked by the dicarba-dipeptide (6). Kolbe electrochemical coupling of a monoester of 1,1-cyclohexanediacetic acid (7) and a suitably protected glutamic acid afforded, after selective ester hydrolysis, a protected form of dipeptide (6) appropriate for peptide synthesis (Callahan *et al.*, 1988). Incorporation of dipeptide (6) into peptide vasopressin antagonists provided novel analogues which were also devoid of partial agonist activity when evaluated in a variety of animal models (Brooks *et al.*, 1988; Moore *et al.*, 1988).

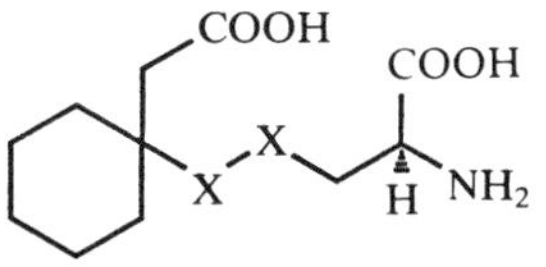

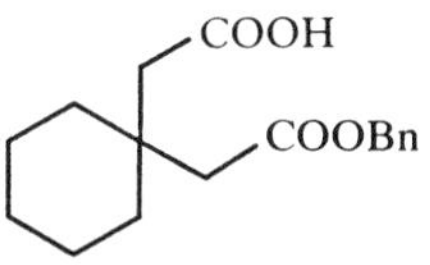

(5) X = S
(6) X = CH_2

(7)

Thus, two totally different types of unusual amino acids, representing what would appear to be minor chemical changes in the final peptides (Me for H; $-CH_2CH_2-$ for –SS–), resulted in potent vasopressin antagonists which were devoid of the partial agonist activity exhibited by the parent peptide. Accompanying this gratifying biological activity is a somewhat frustrating, albeit typical, lack of precise structural information.

Due to the conformational mobility inherent in molecules such as SK&F 101926, rigorous conformational analysis of the final peptides containing the unusual amino acids (2) and (6) was not possible. As a result, the constraints that these unusual amino acids place on total conformational mobility when incorporated into a peptide cannot be rigorously determined and one has to be content with achieving the biological objectives without really understanding the SAR. While this result is frustrating, it is not totally unexpected. The minor alterations in overall chemical structure achieved with incorporation of single amino acid modifications will rarely result in significant rigidification of highly flexible molecules. To reduce overall conformational flexibility, constrained mimics of larger portions of the peptide are needed.

3 Reverse-turn mimics

An important feature of peptide and protein secondary structure are those instances where the amino acid chain reverses direction. These ‘reverse turns’, as a consequence of their frequent appearance on the external surface of the molecule, are postulated to have a high probability of being part of receptor pharmacophores or antibody–epitope recognition (Rose *et al.*, 1985). Numerous organic mimetics of the most common types of reverse turns, β-turns and γ-turns, have been constructed in order to provide conformationally relevant templates on which to hang key pharmacophore elements. A recent example of the use of γ-turn mimetics in the design of enzyme inhibitors will serve to illustrate the value of these templates in peptidomimetic research.

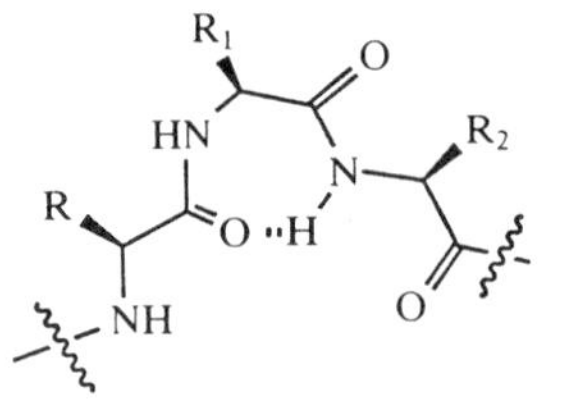

γ-turn

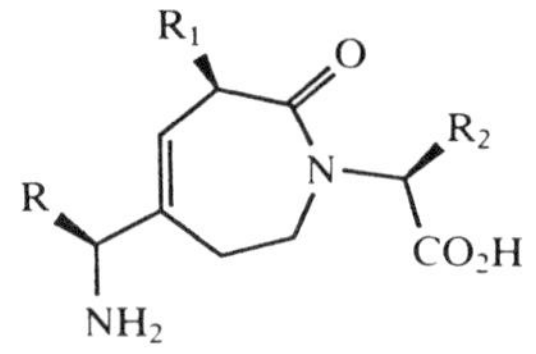

γ-turn mimic

The X-ray crystal structure of several aspartic proteinases complexed to a particular family of peptide inhibitors (endothiapepsin, Foundling *et al.*, 1987; rhizopuspepsin, Suguna *et al.*, 1987; HIV-1 protease, Miller *et al.*, 1989) provide the basis for an inhibitor pharmacophore hypothesis. The inhibitors studied (substrate-based peptide inhibitors containing a reduced amide bond at the scissile peptide bond) appeared to display a pharmacophore characterized by the reduced amide bond being a part of a γ-turn-like conformation.

H-Ala-Asn-Tyr-Pro-Val-Val-NH$_2$

(8)

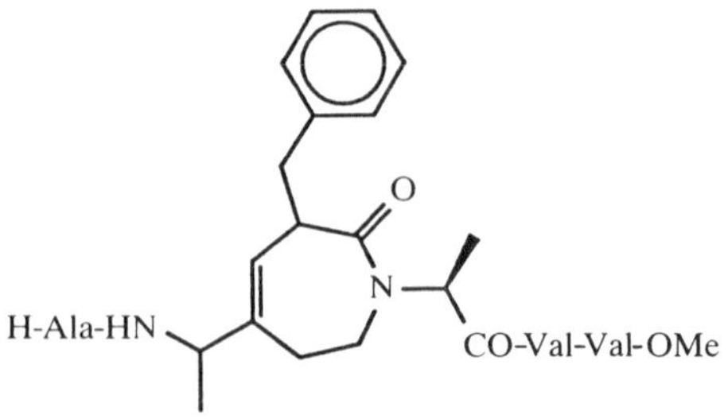

(9)

In experiments designed to test the hypothesis that a γ-turn is compatible with the active site of a representative aspartic acid protease, e.g. HIV protease, we employed our previously designed γ-turn mimic (Huffman *et al.*, 1988, 1989), shown above, which allowed for incorporation of side-chains in the key residues involved in the turn. A small peptide substrate of the HIV protease, compound (8) (Moore *et al.*, 1989), was used as the template in which a representative mimetic was placed about the putative cleavage site to afford compound (9). For synthetic ease, Ala was substituted for Asn at residue 2 and the carboxy terminus was the methyl ester instead of the primary carboxamide; both of these substitutions were known to be compatible with the HIV active site (Dreyer *et al.*, 1989; M.L. Moore *et al.*, 1991, unpublished results). A comparison of the K_m of substrate (8) (17.7 mmol/l) to the K_i of inhibitor (9) (100 μmol/l) indicates that incorporation of this particular turn mimic does indeed appear to be a modification which is acceptable to the active site of the enzyme. While it is tempting to speculate that the improvement in binding between compounds (8) and (9) validates the original design hypothesis, a key issue needs to be kept in mind. Substrates where the cleavage site is between Tyr and Pro have geometric and steric constraints which preclude the peptide backbone adopting the torsional angles required for a γ-turn;

indeed, the original modelling hypothesis arose from consideration of inhibitors of aspartic proteases. It is certainly possible that the γ-turn portion of (9) may not be binding in the S_1–S'_1 sites of the enzyme. This would not be unexpected since for many peptide-based ligand–receptor interactions, it is clear that agonists/substrates present quite different pharmacophores than antagonists/inhibitors. Therefore, in the case of the HIV protease under study, the appropriate comparison is between a reduced amide inhibitor and a reduced amide γ-turn mimic. In order to undertake this comparison, a method was found to remove the amide carbonyl group from the existing γ-turn mimic (K. Newlander *et al.*, 1991, unpublished results) which involved preparation of the thioamide, followed by conversion to a thioimmonium salt and subsequent reductive elimination. In this way, the reduced amide analogue of (9), compound (11), was obtained. Comparison of the K_i of a similarly substituted reduced amide heptapeptide (10), (K_i = 20 μmol/l) to the K_i of (11) (K_i = 650 nmol/l) again reveals biological activity that is consistent with the original hypothesis that a γ-turn mimic-containing molecule presents an acceptable inhibitor pharmacophore to the HIV protease.

H-Ser-Ala-Ala-Phe(R)Pro-Val-Val-NH_2

(10)

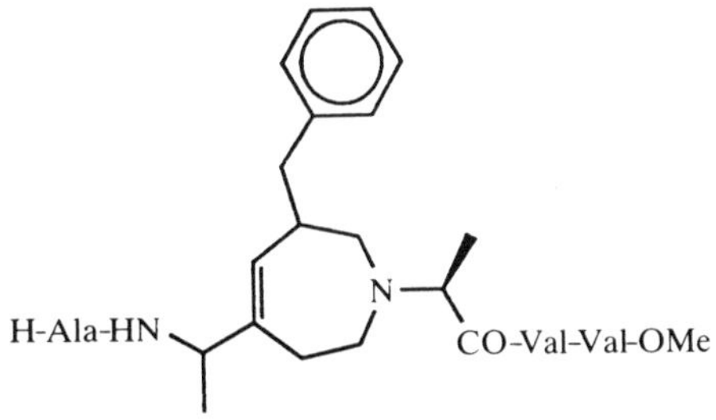

(11)

4 Rational design of small-molecule, non-peptide peptidomimetics

4.1 *Small-molecule lead optimization*

As stated above, the rational design of a small-molecule, non-peptide 'peptidomimetic' which presents the desired pharmacophore represents the ultimate challenge in peptidomimetic research since it requires a detailed understanding of the molecular features of the pharmacophore. Although small-molecule enzyme inhibitors have been rationally designed based on knowledge of the active site and substrate/inhibitor pharmacophore information (e.g. angiotensin-converting enzyme inhibitors), at present this has not been accomplished with peptide ligands for the more general class of membrane-associated receptors.

Progress in this area has usually been characterized by the initial discovery of a small-molecule lead from a receptor-based screening effort, followed by more traditional medicinal chemistry to improve affinity and selectivity in subsequent analogues, for example the cholecystokinin antagonists of Freidinger *et al.* (see Chapter 16). Recently several groups (Du Pont, SmithKline Beecham) have reported on the integration of peptide pharmacophore information into the design process to improve the potency and selectivity of mimetic leads discovered from screening. The successful application of this design paradigm, i.e. the use of peptide pharmacophore models to assist in small-molecule modification, represents a key step in the ultimate transformation of peptides into non-peptide mimics.

The Du Pont and SmithKline Beecham research groups used as a common starting point a compound (12) reported by Takeda scientists to be a weak, but selective angiotensin II antagonist (Furukawa *et al.*, 1982). Both Du Pont and SmithKline Beecham scientists have used models of the octapeptide angiotensin II pharmacophore to transform (12) into non-peptide antagonists, compounds (14) and (16), respectively.

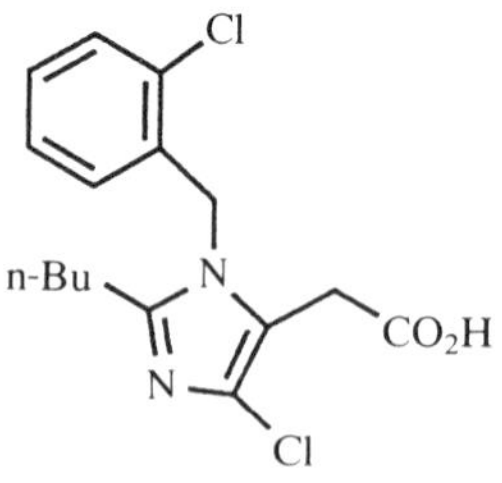

(12)

Asp[1]-Arg[2]-Val[3]-Tyr[4]-Ile[5]-His[6]-Pro[7]-Phe[8]

Angiotensin II

According to Johnson *et al.* (1990), the Du Pont scientists used a model of angiotensin II (A-II) developed by Smeby & Fermandjian (1978) to generate design hypotheses for the modification of the Takeda lead (12). The key aspects of their overlay hypothesis consisted of the alignment of (1) the carboxyl group in (12) with the C-terminal carboxyl group of A-II, and (2) the imidazole in (12) with the imidazole ring of the His[6] residue in A-II. This left the benzyl group in (12) pointed towards the N-terminus of the peptide and a site for further manipulation. After extensive SAR studies, the optimal substitution of the benzyl group was found to be the biphenyl carboxylic acid depicted in structure (13), which was believed, from modelling studies, to be in close proximity to the β-carboxyl group on the N-terminal Asp[1] residue.

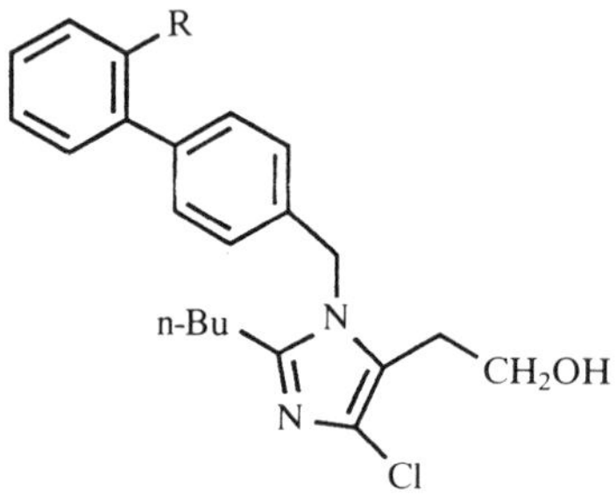

(13) R = COOH
(14) R = CN_4K

Further modification resulted in replacement of the aryl carboxylic acid group with its tetrazol-5-yl equivalent to afford DuP 753 (14) which is currently undergoing clinical evaluation as an antihypertensive agent. A final comparison between DuP 753 and the model of angiotensin II pharmacophore has not yet been presented by the Du Pont scientists, so the extent to which the original overlay

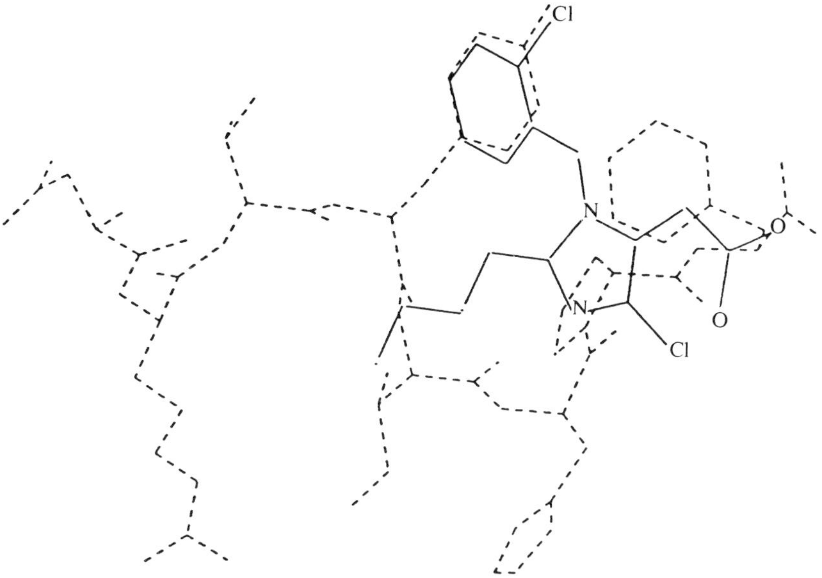

Figure 17.1. Model of A-II and (12) pharmacophores: ----, angiotensin II; ——, Takeda compound (12).

hypothesis still rationalizes the mimicry is unknown.

At SmithKline Beecham (Weinstock *et al.*, 1991), an internally generated model of angiotensin II pharmacophore (Samanen *et al.*, 1992) was used as a starting point for understanding the Takeda lead. A variety of overlays between (12) and A-II were modelled and finally an initial hypothesis was generated which placed (i) the carboxyl group of (12) near the C-terminal carboxyl group of A-II, (ii) the butyl side-chain of (12) near the lipophilic side-chain of Ile^5, and (iii) the aromatic ring of the benzy group in (12) near the aromatic ring of the Tyr^4 residue in A-II (Fig. 17.1). Two features of the model were apparent:

1 The fit between the carboxyl groups of the small-molecule lead and the peptide was not optimal; extension of the distance between the carboxyl group and the imidazole ring in (12) would be required to optimize the fit.

2 If the model were correct, functionality might be added to the small molecule which would allow access to the aromatic binding site occupied by the Phe^8 side-chain in the peptide pharmacophore.

A series of SAR studies directed at the carboxyl region of (12) culminated in the discovery of the thienyl-substituted acrylic acid moiety shown in compound (15).

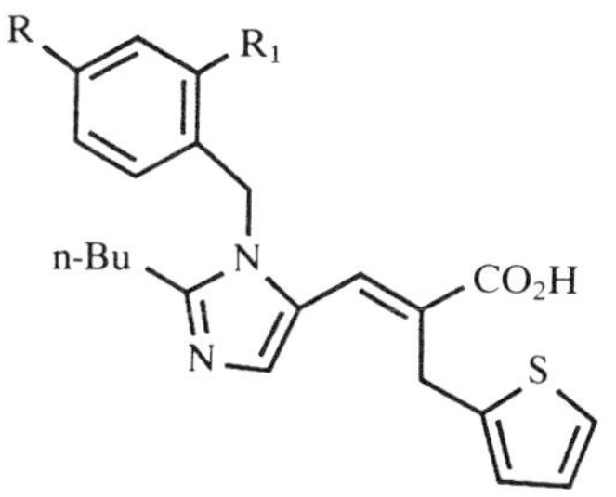

(15) R = H, R_1 = Cl
(16) R = COOH, R_1 = H

Table 17.1. *In vitro* SAR of angiotensin II antagonists

Compound	IC_{50} (nmol/l)*	K_B (nmol/l)†
Takeda lead (12)	43 000	2700
DuP 753 (14)	1.8	1.3
(15)	440	51
SK&F 108566 (16)	1.0	0.21

* Inhibition of ^{125}I-A-II specific binding to rat mesenteric membranes.
† Inhibition of A-II-induced vasoconstriction of rabbit aorta.

The angiotensin antagonist activity displayed by (15) was approximately 100-fold enhanced over that of the starting compound (12) (see Table 17.1). Further optimization of the benzyl ring substituent afforded the *p*-carboxylic acid analogue (16, SK&F 108566) which now exhibited a 10 000-fold enhancement in affinity and antagonist activity at the A-II receptor and seemed consistent with the modelling hypothesis (Fig. 17.2).

Regardless of the ultimate validity of the modelling hypotheses used by the Du Pont and SmithKline Beecham scientists, it is clear that their attempts to design improved small molecules using peptide pharmacophore models were very successful. Both groups have been able to transform a small-molecule lead into a potent receptor antagonist displaying a four log-fold increase in receptor affinity as shown in Table 17.1.

4.2 *Small-molecule lead generation*

The results with angiotensin II antagonists have validated the hypothesis that peptide pharmacophore models will assist the optimization of small-molecule leads. What remains is the ability to generate the lead directly from the same

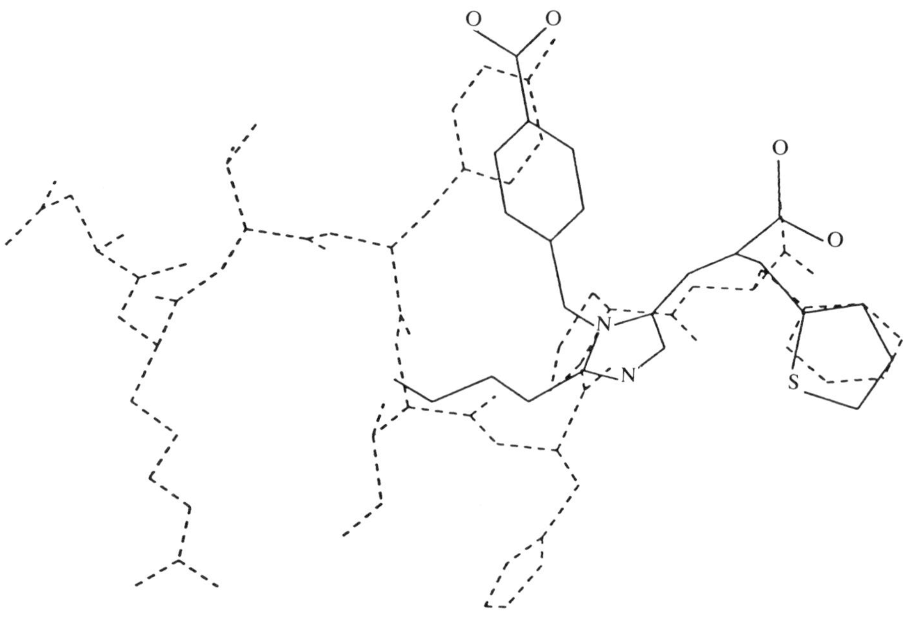

Figure 17.2. Model of A-II and (16): ----, angiotensin II; ——, SK&F 108566 (16).

models of peptide pharmacophore. This represents the penultimate (and most difficult?) step in the rational design process. An example of the type of system which might yield itself to the rational design of a small-molecule lead is the on-going work on the platelet GpIIb/IIIa receptor.

Platelet aggregation that is mediated via the glycoprotein (Gp) IIb/IIIa receptor can be modulated by small peptides which contain an 'RGD' adhesion site (Kloczewiak *et al.*, 1984; Gartner & Bennett, 1985; Plow *et al.*, 1987). RGD stands for that portion of amino acid sequence, Arg-Gly-Asp (designated in single amino acid code as RGD), discovered to be a common pharmacophore element in proteins involved in a variety of adhesion events (Pierschbacher & Ruoslahti, 1984). This key finding has prompted a large number of research groups to attempt to design platelet aggregation inhibitors by preparing GpIIb/IIIa receptor antagonists based on the RGD motif. Since the RGD sequence appears in a variety of unrelated proteins, our approach at SmithKline Beecham has been to hypothesize that a key element of RGD recognition at the GpIIb/IIIa receptor is based on the conformational array of the tripeptide. As a result, a variety of conformationally constrained peptides were prepared and evaluated for antiaggregatory activity (Ali *et al.*, 1990, Samanen *et al.*, 1991a,b). These SAR studies have culminated in the syntheses of two cyclic molecules (17 and 18) which are potent inhibitors of platelet aggregation both *in vitro* and *in vivo* (Nichols *et al.*, 1990; J. Samanen *et al.*, 1991, unpublished results).

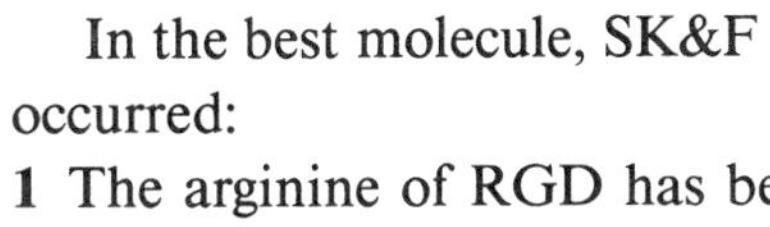

(17) (SK&F 106760)

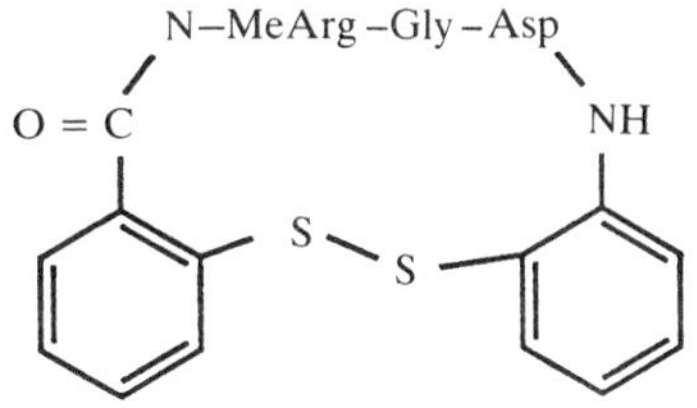

(18) (SK&F 107260)

In the best molecule, SK&F 107260, two peptidomimetic transformations have occurred:

1 The arginine of RGD has been replaced with N-methylarginine which imparts unexpectedly improved potency.

2 Conformational constraint has been introduced with a novel cystine, dipeptide mimetic, Mba–Man (*o*-mercaptobenzoic acid–*o*-mercaptoaniline).

Of particular relevance to this discussion is the fact that constraints present in (17) and (18) effectively limit the conformational array to the extent that solution nuclear magnetic resonance (NMR) spectroscopy of (17) and (18) provides a single conformational family (K. Kopple *et al.*, 1991, unpublished results). In addition, crystals of compound (18) have been obtained and an X-ray crystallographic structure has been determined (D.S. Eggleston *et al.*, 1991, unpublished results). Apart from some minor differences, the structure of (18) in the solid state closely resembles the solution structure determined by NMR; thus, a fairly reliable structural model of peptidomimetic (18) can be constructed (Fig. 17.3). Clearly the biologically active conformation of (18) may not be represented by either the X-ray

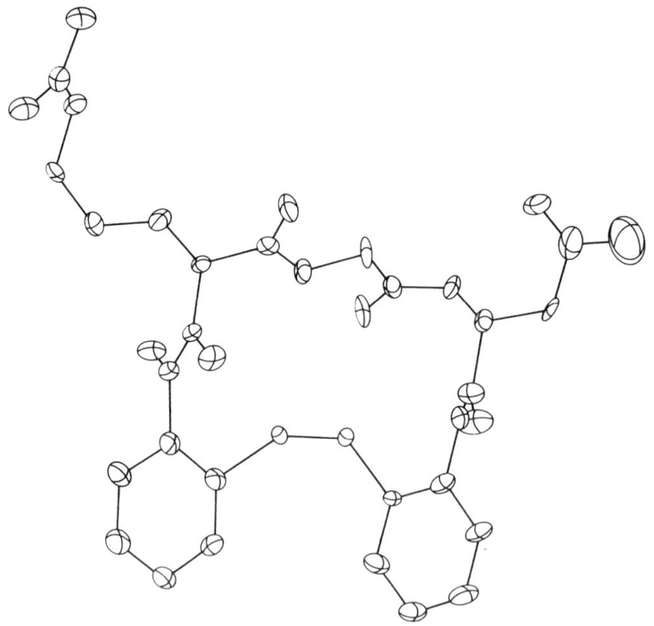

Figure 17.3. X-ray crystallographic structure of SK&F 107260 (18).

or solution-derived structure; however, the conformation depicted in Fig. 17.3 represents an excellent starting point for the design of a small-molecule, non-peptide lead structure which has affinity for the GpIIb/IIIa receptor. Once a suitable lead structure has been identified, the type of optimization strategies employed in the A-II antagonist efforts should afford a potent, selective antagonist. This would represent a successful conclusion to the overall objective rationally to design a non-peptide peptidomimetic from the peptide itself.

Peptide and protein molecules represent a rapidly expanding source of new therapeutic targets for drug discovery. The dilemma for the medicinal chemist is how to transform these large, proteinaceous materials into the types of molecules which have the best profile for widespread therapeutic benefit. Peptidomimetic research, as highlighted, provides a rational basis on which to attempt the transformation of large peptides into useful drug candidates.

5 References

Ali, F.E., Calvo, R., Romoff, T., Samanen, J., Nichols, A. & Storer, B. (1990). In: J. Rivier and G.R. Marshall (Eds.), *Peptides: Chemistry and Biology: Proceedings of the Eleventh American Peptide Symposium*, p. 94, ESCOM, Leiden.

Brooks, D.P., Koster, P.F., Albrightson-Winslow, C.R., Stassen, F.L., Huffman, W.F. & Kinter, L.B. (1988). *J. Pharm. Exp. Ther.*, **245**, 211.

Callahan, J.F., Newlander, K.A., Huffman, W.F., Bryan, H., Moore, M.L. & Yim, N.C.F. (1988). *J. Org. Chem.*, **53**, 1527.

Dreyer, G.B., Metcalf, B.W., Thomaszek, Jr., T.A., Carr, T.J., Chandler, III, A.C., Hyland, L., Fakhoury, S.A., Magaard, V.W., Moore, M.L., Strickler, J.L., Debouck, C. & Meek, T.D. (1989). *Proc. Natl. Acad. Sci. USA.*, **86**, 9752.

Eggleston, D.S., Yim, N.C.F., Silvestri, J.S., Bryan, H. & Huffman, W.F. (1989). *Acta Cyst.*, **C45**, 259–63.

Foundling, S.I., Cooper, J., Watson, F.E., Cleasby, A., Pearl, L.H., Sibanda, B.L., Hemmings, A.,

Wood, S.P., Blundell, R.L., Valler, M.J., Norey, C.G., Kay, J., Boger, J., Dunn, B.M., Leckie, B.J., Jones, D.M., Atrash, B., Hallett, A. & Szelke, M. (1987). *Nature* (*London*), **327**, 349.

Furukawa, Y., Kishimoto, S. & Nishikawa, K. (1982). U.S. Patents 4,340,598 and 4,355,040.

Gartner, T.K. & Bennett, J.S. (1985). *J. Biol. Chem.*, **260**, 11891.

Huffman, W.F., Callahan, J.F. Eggleston, D.S., Newlander, K.A., Takata, D.T., Codd, E.E., Walker, R.F., Schiller, P.W., Lemieux, C., Wire, W.S. & Burks, T.F. (1988). In: G.R. Marshall (Ed.), *Peptides: Chemistry and Biology: Proceedings of the Tenth American Peptide Symposium*, p. 105, ESCOM, Leiden.

Huffman, W.F., Callahan, J.F., Codd, E.E., Eggleston, D.S., Lemieux, C., Newlander, K.A., Schiller, P.W., Takata, D.T. & Walker, R.F. (1989). In: J. Tam and E.T. Kaiser (Eds.), *Synthetic Peptides: Approaches to Biological Problems*, p. 257, A.R. Liss, New York.

Johnson, A.L., Carini, D.J., Chiu, A.T., Duncia, J.V., Price, Jr., W.A., Wells, G.J., Wexler, R.R., Wong, P.C. & Timmermans, B.M.W.M. (1990). *Drug News Perspect.*, **3**, 337.

Kloczewiak, M., Timmons, S., Lukas, T.J. & Hawiger, J. (1984), *Biochemistry*, **23**, 17567.

Manning, M., Lammek, B., Kolodziejczyk, A.M., Seto, J. & Sawyer, W.H. (1981). *J. Med. Chem.*, **24**, 701.

Meraldi, J.-P., Hruby, V.J. & Brewster, A.I.R. (1977). *Proc. Natl. Acad. Sci. USA.*, **74**, 1373–7.

Miller, M., Schneider, J., Sathyanarayana, B.K., Toth, M.V., Marshall, G.R., Clawson, L., Selk, L., Kent, S.B.H. & Wlodawer, A. (1989). *Science*, **246**, 1149.

Moore, M.L., Albrightson, C., Brickson, B., Bryan, H.G., Callahan, J.F., Foster, J., Kinter, L.B., Newlander, K.A., Schmidt, D.B., Sorenson, E., Stassen, F.L., Yim, N.C.F. & Huffman, W.F. (1988). *J. Med. Chem.*, **31**, 1487.

Moore, M.L., Bryan, W.M., Fakhoury, S.A., Magaard, V.W., Huffman, W.F., Dayton, B.D., Meek, T.D., Hyland, L., Dreyer, G.B., Metcalf, B.W., Strickler, J.E., Gorniak, J.G. & Debouck, C. (1989). *Biochem. Biophys. Res. Commun.*, **159**, 420–5.

Newlander, K., Bryan, H.G., Callahan, J.F., Eggleston, D.S., Moore, M.L., Yim, N.C.F., Huffman, W.F. & Jackman, L.M. (1990). In: J. Rivier and G.R. Marshall (Eds.), *Peptides: Chemistry and Biology: Proceedings of the Eleventh American Peptide Symposium*, p. 283, ESCOM, Leiden.

Nichols, A., Vasko, J., Koster, P., Smith, J., Barone, F., Nelson, A., Stadel, J., Powers, D., Rhodes, G., Miller-Stein, C., Boppana, V., Bennet, D., Berry, D., Romoff, T., Calvo, R., Ali, F., Sorenson, E. & Samanen, J. (1990). *Eur. J. Pharmacol.*, **183**, 2019.

Pierschbacher, M.D. & Ruoslahti, E. (1984). *Nature* (*London*), **309**, 30.

Plow, E.F., Pierschbacher, M.D., Ruoslahti, E., Marguerie, G. & Ginsberg, M.H. (1987). *Blood*, **70**, 110.

Rose, G., Gierasch, L.M. & Smith, J.A. (1985). *Adv. Protein Chem.*, **37**, 1.

Samanen, J., Ali, F.E., Romoff, T., Calvo, R., Sorenson, E., Bennett, D., Berry, D., Koster, P., Vasko, J., Powers, D., Stadel, J. & Nichols, A. (1991a). In: E. Giralt and D. Andreu (Eds.), *Peptides 1990: Proceedings of the 21st European Peptide Symposium*, p. 781, ESCOM, Leiden.

Samanen, J., Ali, F.E., Romoff, T., Clavo, R., Sorenson, E., Vasko, J., Storer, B., Berry, D., Bennett, D., Strohsacker, M., Powers, D., Stadel, J. & Nichols, A. (1991b). *J. Med. Chem.*, **34**, 3114.

Samanen, J., Weinstock, J., Hempel, J., Keenan, R.M., Hill, D.T., Ohlstein, E.H., Weidley, E.F., Aiyar, N. & Edwards, R. (1992). In: J. Smith and J. Rivier (Eds.), *Peptides, Chemistry and Biology: Proceedings of the Twelfth American Peptide Symposium*, p. 386, ESCOM, Leiden.

Smeby, R.R. & Fermandjian, S. (1978). In: B. Weinstein (Ed.), *Chemistry and Biochemistry of Amino Acids, Peptides and Proteins*, p. 117, Marcel Dekker, New York.

Suguna, K., Padlan, E.A., Smith, C.W., Carlson, W.D. & Davies, D.R. (1987). *Proc. Natl. Acad. Sci. USA*, **84**, 7009.

Weinstock, J., Keenan, R.M., Samanen, J., Hempel, J., Finkelstein, J.A., Franz, R.G., Gaitanoupoulos, D.E., Girard, G.R., Gleason, J.G., Hill, D.T., Morgan, T.M., Peishoff, C.E., Aiyar, N., Brooks, D.P., Fredrickson, T.A., Ohlstein, E.H., Ruffolo, Jr, R.R., Stack, E.J., Sulpizio, A.C., Weidley, E.F. & Edwards, R.M. (1991). *J. Med. Chem.*, **34**, 1514.

Yim, N.C.F. & Huffman, W.F. (1983). *Int. J. Pept. Protein Res.*, **21**, 568.

18 Peptidomimetics as Tools for the Initiation and Analysis of Peptide and Protein Secondary Structure: The Prospects for Unnatural Proteins by Design

D.S. KEMP

Department of Chemistry Room 18–582, Massachusetts Institute of Technology, Cambridge, MA 01239, USA

1 Introduction

Molecular recognition leading to binding between a transportable agent and a receptor, usually a membrane-bound protein, underlies and indeed defines all aspects of medicinal chemistry. The endogenous hormones are often found to be small polypeptides, which are flexible molecules with short biological half-lives that can assume a wide range of conformations in solution. It is therefore not surprising that useful, highly active, clinical agonists and antagonists can be developed by a process of molecular re-design in which cyclic constraints are introduced into the natural hormones or in which portions of their amido backbones or the peptide side-chains themselves are replaced with non-peptide molecular functionalities that mimic the shapes and functions of regions of the natural hormone. Using these peptidomimetics, medicinal chemists have generated analogues of the natural hormone that extend their biological half-lives, enhance the efficiencies of binding or inhibition, and greatly increase their receptor specificity. These topics are not discussed in this review. Here an aspect of the peptidomimetic approach that is embryonic—but which has the potential to play a pivotal role as medicinal chemistry enters the 21st century—will be discussed: unnatural proteins by design.

Peptidomimetics operate largely by re-defining the local conformational bias of a polypeptide. Since the folding of globular proteins from the denatured to the native state depends in part on local structure in key regions of the primary amino acid sequence (Baldwin, 1990; Kim & Baldwin, 1990), the prospect exists for control of the folding process and of the properties of the ensuing protein through judicious introduction of peptidomimetic elements into the primary peptide sequence. This review focuses on the prospects for rational design and synthesis or semi-synthesis of small, medicinally relevant proteins that contain peptidomimetic elements. After a brief consideration of synthetic protein catagories and the likely medicinal applications of designer proteins, the prospects for increasingly efficient, reliable, and practical synthesis and semi-synthesis of proteins is discussed. Peptidomimetic control of protein folding can extend at least potentially beyond simple manipulation of secondary structure, and the alternatives to control of turns, sheets and helices are considered briefly. Finally, after a background presentation of the phenomenology of protein folding as it is understood at this time, a review is given of the available evidence for the nucleation of sheets and helices in solution.

2 Peptidomimetically derived proteins as medicinal agents: A prophecy

As well as revolutionizing modern biology, the biosynthesis of proteins by means of the recombinant DNA methodology has turned many hitherto rare proteins into practical pharmaceuticals, and rationally re-designed proteins can now be considered as therapeutic agents as a result of the power of site-specific mutagenesis, which permits ready modification of a protein amino acid sequence. No attempt has been made in this review to address the scope and potential of the biotechnology of protein biosynthesis. However, two of its consequences, one theoretical, one practical, directly impinge on the substance of this discussion and require brief comment.

Site-specific mutagenesis has permitted correlations between the stabilities of proteins and their primary amino acid sequence (Matthews, 1987; Alber, 1989). These correlations prove the structural code governing protein folding to be robust in the sense that most small structural changes are found to perturb stability but not to disrupt the global folding pattern of a protein or to destroy the basic capability of folding. Thus it is likely that circumspect incorporation of peptidomimetics into otherwise normal peptide sequences also can yield structures that fold non-capriciously. Recent efforts from several groups have demonstrated that it is indeed possible starting with the 20 natural amino acids to design and synthesize polypeptide sequences not found in nature that fold in water to small helical bundles (Hodges *et al.*, 1988; Oas & Kim, 1988; DeGrado *et al.*, 1989), to packed sheets (Richardson & Richardson, 1986), and to helix-sheet arrays (Gutte *et al.*, 1979). These preparations exhibit at least some of the properties of small proteins. Following a different approach, Mutter has advanced the concept of TASPs (template assembled synthetic proteins) in which polypeptides of uniform secondary structural bias are linked to a polyvalent flexible spacing element that serves as an entropic constraint. Recently his group has demonstrated the TASP concept with several cogent examples (Mutter & Vuilleumier, 1989).

Given the availability of site-mutated proteins from recombinant sources and the likely improvements in our capacity to design predictably foldable sequences consisting solely of the natural amino acids, the obvious practical question is whether the properties of peptidomimetic-derived unnatural proteins can prove to be sufficiently unusual and useful to justify the cost and synthetic effort required to generate them. What properties might be expected from a peptidomimetic-derived protein that would be difficult to achieve with a properly engineered natural protein or from its simple covalent modifications? In 1991 there are no peptidomimetic-derived proteins with pharmacologically interesting properties! Presumptuous as it may be to ask this question, let us be frankly speculative.

Why proteins as pharmacological agents? Viewed as a non-covalently bonded molecular recognizer of a targeted cell surface and compared with a smaller molecule, a protein offers the potential advantage of tailored fit to a much greater area of mutual interaction, resulting in a larger range of binding constants and greater specificity of binding. Yet conventional proteins are high-molecular-weight species for which the recognition site is usually only a small fraction of the total surface area. Peptidomimetic-derived protein analogues might indeed be attractive clinical candidates if they could be re-designed to achieve long biological half-lives and binding specificity in a molecular weight range significantly below that found for small proteins. Feasibility seems highly probable, given the properties of certain tightly structured medium-sized peptides. Thus the conotoxins achieve conformational homogeneity together with extremely potent and selective neurotoxicity as a result of the conformational constraint of multiple crosslinking by internal cysteine disulphides (Gray *et al.*, 1988).

A reduction in the accessible molecular weight range for chimeric species with the structured homogeneity and versatility of proteins is likely to generate other attractive features. The side-chains of the 20 natural amino acids provide a limited palette for targeting receptor molecules. An unnatural protein analogue could provide a potentially limitless range of functionalities at the binding region, increasing the sophistication of covalent attachment chemistry, allowing highly directed photochemical or nuclear energy transfer into a cell, and permitting tailoring of cooperative conformational responses to cell-membrane components. Low-molecular-weight protein analogues that are also efficient and specific molecular catalysts could find significant medicinal applications if they were also synthetically accessible and metabolically clean.

Whatever the lower limit on the practical molecular weights of agents capable of binding versatility, the feasibility of such re-design efforts hinges on a functional understanding of the rules of the design and folding processes that generate native protein structure, and all such speculations are of course premature until protein analogues can be more reliably synthesized.

3 Prospects for the ready synthesis and semi-synthesis of peptidomimetic-derived protein analogues

Synthesis of small proteins poses a major unsolved problem. The current methodologies for the synthesis of medium to large-sized polypeptides must be improved both in ease, reliability and cost, and the prospects for preparing peptidomimetic-derived protein analogues as medicinals are slender without a breakthrough in both strategy and tactics of amide bond formation.

Conventional polypeptide synthesis uses partially protected amino acids as binding blocks and generates all amide bonds of the product by peptide coupling reactions. These may be carried out sequentially, starting at one end of the molecule, leading to linear synthetic strategies of which the most convenient and widely used are the solid-phase methods, pioneered by Merrifield (Barany & Merrifield, 1979). Alternatively, side-chain protected peptide fragments can be prepared by solid-phase protocols and consolidated by modern peptide-bond-forming chemistry (Felix *et al.*, 1985; Seyer *et al.*, 1989). The strengths and limitations of these alternative methodologies differ sufficiently to require separate discussions.

After two decades of remarkable methodological refinements, the limitations of the solid-phase methodology when applied to very large peptides now appear to be intrinsic, arising from inexorable attrition and accumulation of impurities during the lengthy synthetic sequence, unpredictable catastrophic reduction in coupling rate owing to peptide association (Kent, 1985) and, above all, introduction of impurities during the final removal of multiple side-chain protective groups. Reported solid-phase syntheses of polypeptides in the size range of small proteins (Clark-Lewis *et al.*, 1986) are indeed impressive, but offer little hope for resolution of the intrinsic problems.

Fragment condensation strategies, despite their potential versatility and efficiency, share with solid-phase procedures the intractable problem of impurity introduction at the final de-blocking step. They compound it with capricious coupling behaviour during the late stages of fragment consolidation, which is seen for all conventional amide-bond-forming procedures. The development of novel methods for generating blocked, linkable peptide fragments by solid-phase procedures (DeGrado & Kaiser, 1982) has not resolved or circumvented these problems. The synthesis by Yajima of RNase A using fragment consolidation (Fujii & Yajima, 1981) remains the definitive achievement with this strategy, but it offers no prospect for routine or facile synthesis of small proteins.

These methodological deficiencies are compounded by the demands of semi-synthesis, in which completely unblocked peptide fragments derived from natural proteins are linked to synthetic fragments. Despite the obvious merits of this strategy (Offord, 1987), it has been infrequently applied to protein synthesis, primarily because of the inadequacies of conventional reagents for generating the polypeptide amide linkage, all of which require complete protection of side-chain functionalities. A methodology is urgently needed that permits consistent, reliable couplings of unblocked polypeptide fragments at high dilution in polar, preferably aqueous, solvents. The availability of a proven, reliable methodology of this type would allow synthetic or semi-synthetic access to a wide class of peptidomimetic-derived protein analogues.

Modification and tailoring of enzymatic peptide-bond formation provides one fertile avenue of exploration. Another lies in capture methodologies (Kemp, 1981), in which a linkage is established between a pair of peptide fragments prior to amide-bond formation, which then occurs intramolecularly, as seen schematically in Fig. 18.1. Provided the intramolecular amide-bond formation at the second step of the scheme occurs with high efficiency, the method is rendered compatible with reactive side-chain functions of unblocked peptide fragments. A practical capture

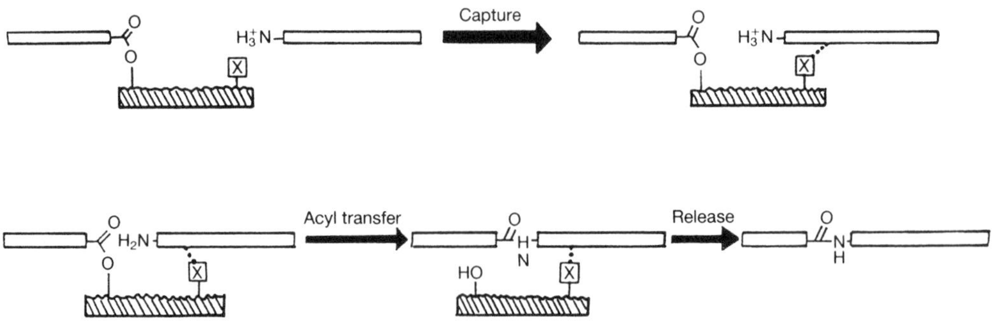

Figure 18.1. Schematic summary of a capture strategy for amide-bond formation. An unspecified affinity between a pair of peptides, symbolized by the group X, links the fragments and permits amide formation to occur intramolecularly (Kemp, 1981).

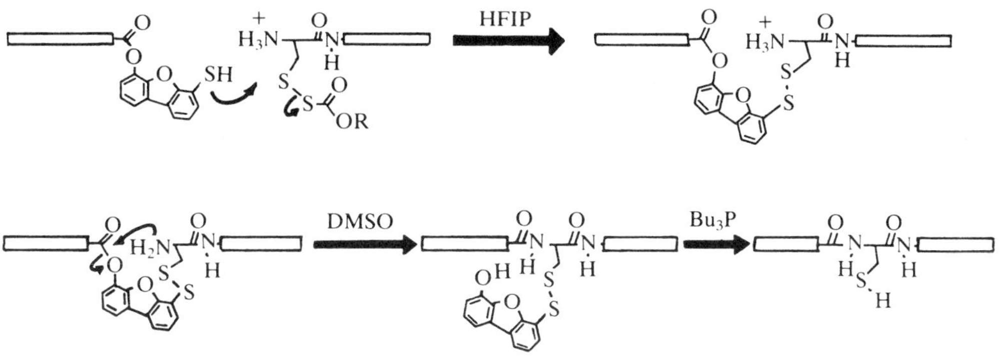

Figure 18.2. The thiol capture strategy. Formation of an S–S bond between an N-terminal Cys-SH and a thiol of an C-terminal dibenzofuran spacer results in efficient intramolecular O,N-acyl transfer (Fotouhi *et al.*, 1989).

methodology must exhibit three key features. First, the capture step must occur cleanly and rapidly at high dilution in aqueous solvents. Second, the amide bond-forming step must occur rapidly and efficiently without the enthalpic driving force of a highly activated acylating agent. Third, the elements of this relatively complex scheme must operate *synergetically* to maximize generality, reliability and convenience, with internal analytical controls that facilitate monitoring of each step of the sequence as well as separation and purification of intermediates. At present, thiol capture provides the working example of such methodologies. The capture step (Fig. 18.2) involves nucleophilic substitution at a sulphur function of an N-terminal cysteine residue to form a pair of peptides linked by a disulphide bond. The Scm (–S–CO–OCH_3) sulphur leaving group of the scheme permits rapid, clean formation of the required disulphide bond at neutral pH in aqueous medium at 25°C using reactants at submillimolar concentrations (Fotouhi *et al.*, 1989).

Efficient intramolecular acyl transfer to form an amide bond in the second step occurs with an effective molarity of 5–10 M using a dibenzofuran functionality (Fig. 18.3) which was selected to provide optimal molecular fit to the geometry of the transition state of the reaction (Kemp *et al.*, 1989). Achieving synergetic complementarity of the tactical and strategic elements of the capture scheme has involved design of a new protocol for solid-phase peptide synthesis for generation of peptide fragments with dibenzofuran functionalization (Kemp & Galakatos, 1986), development of practical procedures for synthesis of requisite blocked amino acid derivatives (Kemp *et al.*, 1988), demonstrations of control of racemization (McBride & Kemp, 1987) and disulphide interchange (Kemp & Fotouhi, 1987), and design of new protective groups optimized for the scheme (Kemp & Carey, 1989; Fotouhi & Kemp, 1992). Molecular mechanics calculations indicate that the transition state of Fig. 18.3 has significant van der Waals strain, raising the prospect that molecular templates substantially more efficient than dibenzofuran can be achieved by a rational design process. Progress toward this goal has included development of a molecular template that shows 150-fold enantioselectivity for acyl transfer of L-alanine to L- and D-cysteines (Kemp & Buckler, 1991).

The proof of the thiol capture scheme must lie in its capacity to ligate peptides bearing minimal side-chain blocking groups. We have recently demonstrated clean,

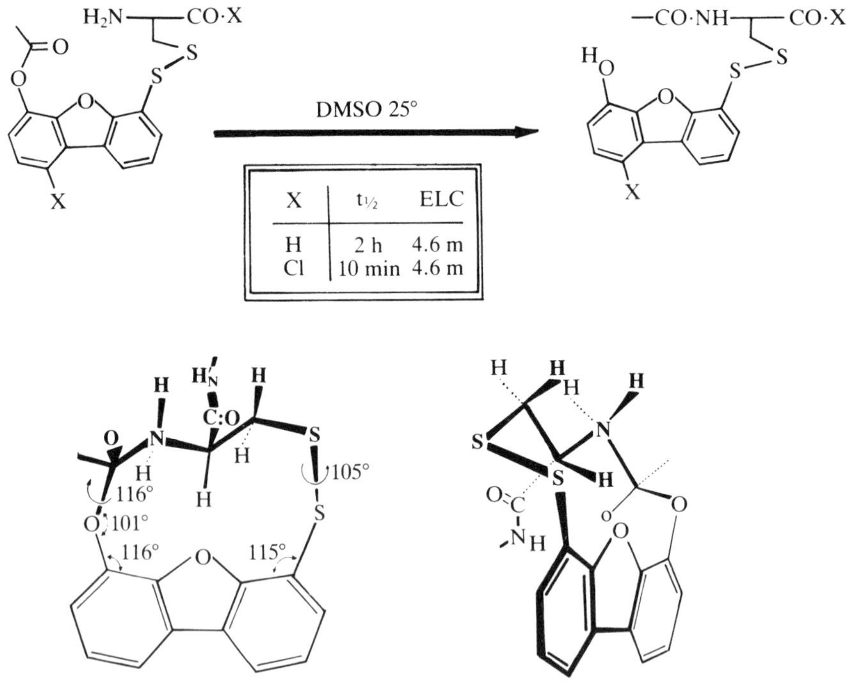

X	$t_{1/2}$	ELC
H	2 h	4.6 m
Cl	10 min	4.6 m

Figure 18.3. The dibenzofuran spacing element exhibits optimal distance complementarity to one conformation of the transition state for O,N-acyl transfer (Kemp *et al.*, 1989).

high-yield syntheses of peptides in the size range of 20–40 amino acid residues by ligation of pairs of peptides bearing protective groups only on Cys–SH. With proper attention to reaction conditions, the side-chain functionalities of lysine and histidine do not interfere (Fotouhi *et al.*, 1989; Kemp & Carey, 1991). Although many more examples are needed before the reliability of the method can be confidently asserted, these results offer the hope that a workable and general semi-synthetic protocol for the generation of small proteins bearing chimeric or unnatural peptidomimetic elements may be realizable.

What further chemical achievements are needed to attain this goal? Ultimately the applicability of schemes such as those of Fig. 18.2 must rest on cost and generality, although it must be remembered that the molar quantities of reagents required for synthesis of significant quantities of semi-synthetic proteins are likely to be small, and the apparent loss of generality by the ligation requirement at cysteine may be compensated by the potential for adaptation to unnatural amino acid residues derived by alkylation at the Cys–SH after the ligation and SS cleavage steps. It is premature at this point to predict the long-range utility of the thiol capture protocol as we have developed it, or of some second-generation variant whose development may be built upon its example, or of strategic alternatives. The important point is that ligation of fragments derived from cleavages of natural proteins is for the first time feasible.

A major missing element required for practical semi-synthesis is a process for selective cleavage of a protein backbone at Xxx–Cys amide bonds, followed by activiation of the liberated Xxx–acyl function in a manner that permits its inclusion in the thiol capture scheme. Such processes can be readily envisaged that could extend the scope of known but inefficient chemistry (Spande *et al.*, 1970), and only time and laboratory effort should be required to bring them to realization.

4 The protein-folding process

In a seminal review written more than 30 years ago, Kauzmann described the denaturation–renaturation of proteins as, 'one of the most interesting and complex classes of reactions that can be found either in nature or in the laboratory' (Kauzmann, 1959). Data accumulated in the intervening years have lessened neither the interest nor the complexity. Even though the major issues in the field of protein folding remain unresolved, the rich structural, thermodynamic and kinetic data now available provide a secure footing from which to plan re-design of protein structure based on peptidomimetics.

Although the broad class of proteins properly includes all biomolecules consisting of a ribosomally synthesized primary sequence of amide-linked α-amino acid residues, the folding properties and structures of the water-soluble globular proteins have received the most attention. A protein characterized by a single folding domain has structure at three levels. The simplest is the primary amino acid sequence along the peptide backbone. Secondary structure refers to local conformational interactions between neighbouring amino acid residues along that chain and is usually categorized as comprising bends or loops (Rose *et al.*, 1985), extended structures including parallel or antiparallel β-sheets, and helices. Tertiary structure refers to the relatively long-range interactions that occur as atoms belonging to amino acid side-chains or backbone pack to form the protein interior. The folded or native protein conformation and the folding process itself have been found to exhibit the following striking regularities:

1 First, the folding to a native state from an unfolded, conformationally diverse, denatured state which can approximate the polymer chemist's random coil appears in many, if not most, cases to be a reversible process in which the native state lies at a free-energy minimum, at least among states accessible on a reasonable time scale (Dill, 1990). The structure of the folded state and very likely the path by which it is formed are completely specified by the sequence of amino acid residues along the peptide backbone.

2 Second, for many proteins, under at least some conditions that permit folding from the denatured to the native state, the folding process is highly cooperative in the sense that intermediates are present only at low concentrations, and the readily detectable species are the limiting denatured and native structural manifolds (Dobson & Evans, 1984). Some proteins exhibit a substantial deviation from the thermochemical behaviour expected for a two-state folding model, and additional partially folded non-native states exhibiting a high degree of secondary structure can be demonstrated at equilibrium under some conditions. Features of these 'molten globule' states have been characterized by ^{1}H-nuclear magnetic resonance (NMR) spectroscopy in favourable cases (Baum *et al.*, 1989). Moreover after controlled proton exchange under conditions of very rapid exergonic re-folding from a denatured state, NMR spectroscopy has been used to demonstrate the structures of short-lived early intermediates along protein-folding pathways (Roder *et al.*, 1988; Udagaonkar & Baldwin, 1988). The present status of detectable intermediates along protein-folding pathways and their significance for elucidating the folding mechanism has been recently reviewed (Kim & Baldwin, 1990).

3 Third, the overall free-energy change for conversion of a denatured to a native protein state is modest, in most cases lying in the range of only − 5 to − 20 kcal/mol (Privalov & Gill, 1988). Strikingly, despite a stabilizing free energy per residue of only hundredths of a kcal/mol, native proteins do not for the most part behave as conformationally frayed structures, although a substantial amount of cooperative motion occurs (Karplus & McCammon, 1981).

4 Fourth, correlations among the extensive database of X-ray structures of crystalline proteins establish that the globular proteins fall into well-defined secondary structural families (Richardson, 1981), that backbone and side-chain torsional angles of protein interiors lie at local energy minima (McGregor *et al.*, 1987), that the preferred packing orientations of sheets and helices within proteins are well defined and achieve maximum density (Janin & Chothia, 1980; Chothia *et al.*, 1981), and that the packing density within the solvent-shielded core of the protein equals, or exceeds, that of the typical crystal packing observed for small molecules (Richards, 1977). A recent computational modelling exercise implies that the observed core packing density is an exceptionally stringent structural criterion (Ponder & Richards, 1987). Although the principal driving force governing protein folding appears to be the transfer of hydrophobic amino acid side-chains from a solvent-exposed state to the shielded interior packing of the native state (Kauzmann, 1959; Privalov & Gill, 1988; Dill, 1990), one cannot model proteins accurately as pure oil drops surrounded by water-exposed polar groups. In a survey of 37 representative proteins (Miller *et al.*, 1987), the charged Asp, Glu, Lys and Arg residues are indeed found to comprise 27% of the surface and only 4% of the interior, but non-charged polar residues constitute 39% of the interior and only 24% of the surface. Non-polar residues constitute equal fractions (57–58%) of surface and interior regions of folded proteins. However, the large non-polar side-chains of Val, Leu, Ile and Phe are found to constitute 44% of the interior, but only 14% of surface. Strikingly, even in large proteins, only 15% of amino acids are totally buried, and for small proteins 30–50% of these are the methylene disuphide functions of cystine residues which provide interchain crosslinking. These empirical observations constitute a practical paradigm that can guide the construction of unnatural protein structures.

Although the weighting of the factors that determine protein folding remains controversial (Dill, 1990; Kim & Baldwin, 1990), rational approaches to the design of foldable polypeptide sequences are best modelled from the stages of the 'building up' principle of Scheraga (1973) in which formation of elements of secondary structure is presumed to initiate and underlie the folding process. Regions of stablized secondary structure are therefore a primary design factor.

Of equal importance is the need to ensure a structurally well-defined state for unnatural protein structures. Isolated sheets or helices in solution are almost invariably frayed manifolds of large numbers of partially structured conformations of nearly equal stability. The strikingly different, very high conformational homogeneity exhibited in solution by many natural proteins must result from their special topological features (Richardson, 1981) together with the conformation-locking effects of their high-core density. Achieving dense core packing appears to be essential for construction of peptide assemblages that exhibit the non-frayed properties of native proteins (Saunders & Terwilliger, 1989).

Evidence from site-specific mutagenesis experiments as well as from the properties of semi-synthetic RNAse S derivatives suggests that a change in stability for a small, locally folded element of a protein is likely to be reflected in enhanced stability of the global structure in which this element is imbedded (Mitchinson & Baldwin, 1986). In general, this correlation between local and global stabilization provides a design principle of great power, since it allows independent energetic tuning of separable elements of peptidomimetic-derived proteins.

5 Types of peptidomimetic control of protein folding

According to the framework models of protein folding (Kim & Baldwin, 1990) and the older model of Anfinsen and Scheraga (1976), the first significant intermediates on the folding pathway are likely to be secondary structural elements consisting of turns, sheets or helices. A natural approach for the introduction of structure and stability-influencing peptidomimetics into a protein amino acid sequence that is based on this model would therefore involve unnatural backbone replacements that greatly increase the tendency of linked peptides to generate these structures in a local region. Here we will deal exclusively with the prospects and available information concerning this approach.

At least two alternatives to this strategy exist. As symbolized schematically in Fig. 18.4, an initially structureless or random-coil state which characterizes the denatured protein is envisaged to generate under conditions optimal for the folding process local elements of secondary structure which coalesce by a process of mutual molecular recognition to form compact regions of solvent-shielded tertiary structure. This latter process has attributes similar to the docking of a natural hormone with its receptor, as first noted by Hofmann (Hofmann & Bohm, 1966) for the interaction of the RNAse S-peptide and S-protein. Construction of peptidomimetics that are complementary in fit and van der Waals envelope to a developing secondary-structural region elsewhere in the peptide backbone offers a novel and potentially fertile prospect for the construction of tailored protein-like molecules. It should thus be possible to design molecular templates for the formation of tertiary protein structure.

The third prospect for peptidomimetic matching involves tuning of the energetics of the random-coil state. Drawing upon evidence from mutagenesis experiments, Shortle has argued that folding can be furthered by selective destabilization of the denatured state (Shortle *et al.*, 1990). Moreover, Sauer has recently observed that the results of site-specific mutagenesis experiments on the thermal stability of λ Cro (Pakula & Sauer, 1990) are best explained if a local region of the peptide backbone is more exposed to solvent in the native, folded conformation than in the denatured state. Use of peptidomimetics selectively to destabilize the unfolded state may therefore be a productive strategy.

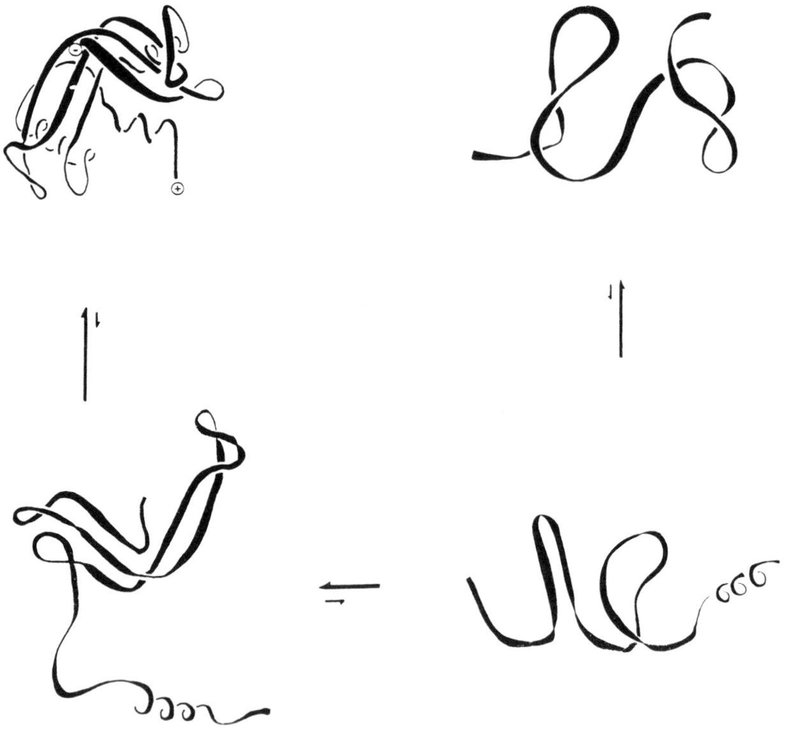

Figure 18.4. A schematic diagram of the framework model for protein folding, showing denatured (top right) and native (top left) states, with hypothetical intermediates containing local elements of secondary structure.

Very likely the medicinal chemist of the future who aims at tailoring protein properties must consider all three avenues of structural modification as appropriate tools for stabilizing a desired protein conformation.

6 Peptidomimetics and the nucleation of helices and sheets

Many small peptides in solution assume a wide range of conformational states of similar energy, and it was once generally assumed that all simple peptides approximate random-coil structures in water. The availability of high-field ^{1}H-NMR spectroscopy as a conformational probe has changed that picture.

Of the natural amino acids, proline is the most conformationally constrained, and for amino acids in globular protein structures it is the least randomly distributed among the three major types of secondary structures, being found with high frequency in turns (Chou & Fasman, 1974; Levitt, 1978). Not surprisingly, small proline-containing peptides in aqueous solution are frequently found by NMR analysis to have significantly populated β-turn conformations (Dyson *et al.*, 1988) or other non-random conformational features (Grathwohl & Wüthrich, 1976). Building upon and extending the conformational ring-locking of proline, many workers have actively pursued the synthesis and pharmacological properties of conformationally restricted β-turn mimetics during the past decade (Morgan & Gainor, 1989).

Unlike loops or turns, which are strictly local, helices and sheets are global structures, and therefore mimetics must direct the conformational bias of the backbones of linked peptide sequences over moderate distances. We define a *conformational template* as a functionality that acts under favourable conditions to enforce a single conformation on an otherwise conformationally inhomogeneous chain of atoms to which it is linked. Since helices and sheets correspond respectively to locally compact and to extended conformations of the peptide backbone, the structural features of conformational templates for these two classes of global secondary structures are likely to be very different. A helical template must act in a specific manner to increase the compactness of its contiguous tri- or tetra-peptide sequence; a sheet-inducing template must augment and regularize the natural tendency of peptide backbones to form extended conformations, most probably by providing a parallel, linear sequence of hydrophobic or hydrogen-bondng interactions.

In principle, a helix can be nucleated from the N-terminus, from the C-terminus, or by stabilizing constraints introduced into the barrel region. The last can be covalent linkages, as with the potent Roche growth hormone-releasing factor analogues prepared with lactam linkages between side-chain functions of aspartic acid and lysine groups (Felix *et al.*, 1988) or metal chelates, as studied by Ghadiri (Ghadiri & Choi, 1990; Ghadiri & Fernholz, 1990). Alternatively, hydrophobic interactions on one side of an amphipathic helical barrel can result in dramatic stabilization (DeGrado, 1988; Hodges *et al.*, 1988; Oas & Kim, 1988; DeGrado *et al.*, 1989; Kaiser, 1989). The stabilities of helices can be modified by salt-bridge effects and by interactions of charges with the significant helix dipole (Schoemaker *et al.*, 1987).

Conformationally restricted nucleation from the N-terminus has been achieved by substitution of a hydrazone–ethylene bridge for an (i,i + 4) backbone hydrogen bond (Arrhenius *et al.*, 1987) as well as by an acetyl-L-prolyl-L-proline peptidomimetic in which a conformational constraint in the form of a thiamethylene bridge has been introduced, as seen in Fig. 18.5 (Kemp & Curran, 1988; Kemp *et al.*,

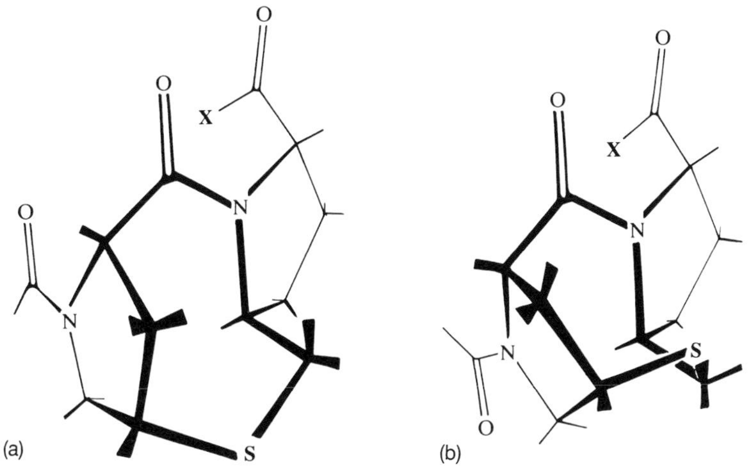

Figure 18.5. Nucleating (a) and non-nucleating (b) conformations of Ac-Hel_1-X, a conformationally constrained mimetic for Ac-L-Pro-L-Pro-X (Kemp & Curran, 1988).

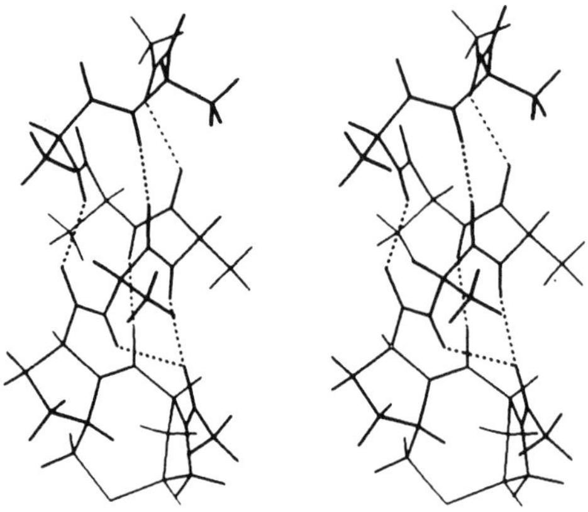

Figure 18.6. Computer-generated stereogram of a helical local-energy minimum for Ac-Hel$_1$-(L-Ala)$_5$-NHCH$_3$.

1991a,b). From ^{1}H-NMR evidence a conjugate of this functionality at the N-terminus of a hexa-L-alanine peptide is found in water and other solvents to exist as a frayed helix (Fig 18.6). Owing to an equilibration between nucleating and non-nucleating conformations that is slow on the NMR time scale, this conformational template has reporter properties that can be used to measure helical stability, as noted in Figs 18.7 and 18.8 (Kemp *et al.*, 1991a). These results establish that helical conformations of polypeptides can be nucleated and stabilized by properly designed conformational templates. Studies of peptide–template conjugates should allow the definition of a complete list of quantitative rules that govern the formation of isolated short helices as well as helical bundles, and with the aid of

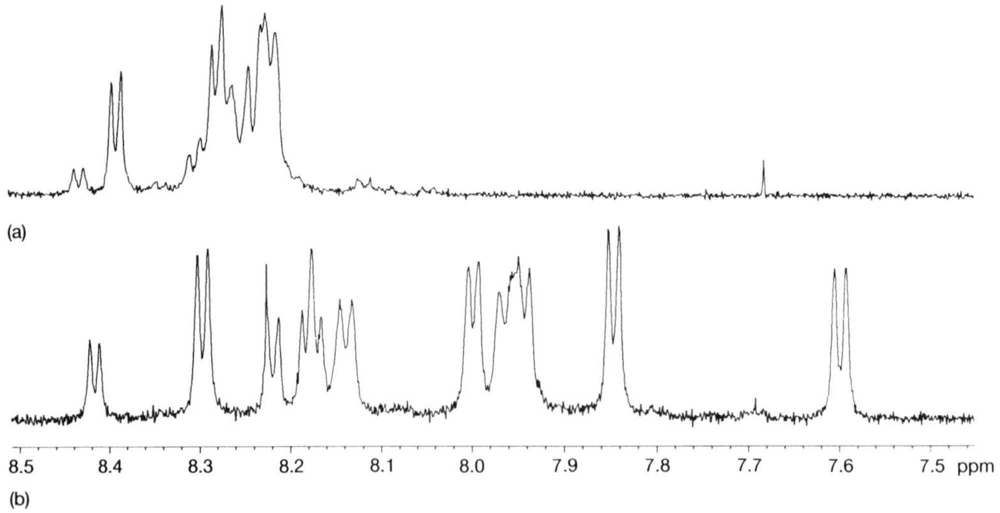

Figure 18.7. ^{1}H-NMR spectra of NH resonances in H_2O, 23°C, pH 1. (a) Spectrum of Ac-L-Pro-L-Pro-(L-Ala)$_6$-OH shows six unresolved resonances in the random coil region, indicative of no detectable helical structure. (b) Spectrum of Ac-Hel$_1$-(L-Ala)$_6$-OH shows the presence of two conformational states corresponding to the non-nucleating template conformation (Figure 18.5b) with six NH resonances in the random-coil region between 8.1 and 8.5 δ and to the nucleating template conformation (Figure 18.5a) with six NH resonances in the helical region between 7.5 and 8.0 δ.

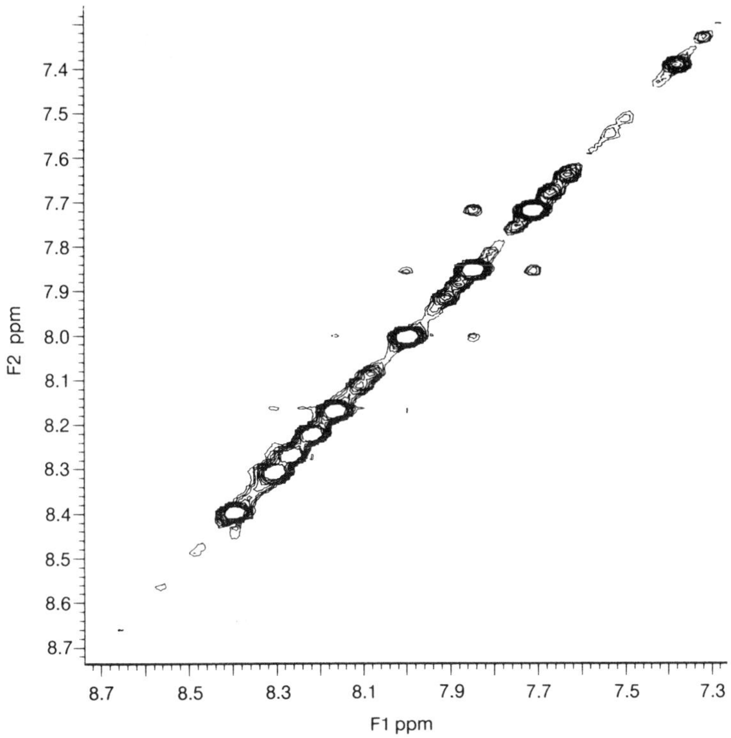

Figure 18.8. ^{1}H-NMR ROESY spectrum in the NH region of deuterium-labelled conjugate Ac-Hel$_1$-(L-Ala-d_4)4OH in water, 23°C, pH 1. The four random-coil NH resonances between 8.1 and 8.4 δ show no crosspeaks; the four helical NH resonances between 7.5 and 8.0 δ show crosspeaks indicative of nuclear Overhauser interactions between adjacent amide NH protons, consistent with a frayed helix.

these rules, peptidomimetically stabilized helical structures should be constructable essentially on demand.

Are templates necessary for the nucleation of helices? Classical Zimm–Bragg models imply an affirmative answer (Scheraga, 1978), but recent studies have shown that certain small alanine-rich peptides can assume substantial strongly temperature-dependent helicity in water (Marqusee *et al.*, 1989; Padmanabhan *et al.*, 1990). The generalizability of these remarkable results to other peptides is at present unclear.

Until recently, acylic, unassociated models for the formation of β-sheets have been unavailable. A sheet is less compact than a helix, and peptides with extended backbone structures that can associate edgewise to form sheet-like hydrogen bonds display a strong tendency toward aggregation which has hindered characterization. However, conjugates of specific peptides with properly selected mimetics carrying hydrogen-bonding sites that correspond to those of a sheet signature form unaggregated antiparallel and parallel sheet structures in dilute solutions (Kemp & Bowen, 1988, 1990; Kemp *et al.*, 1990a,b).

Owing to aqueous insolubility, studies of the antiparallel sheet models of Fig. 18.9 have largely been carried out in organic solvents such as dimethyl sulphoxide (DMSO), and conformations have been established by chemical shift

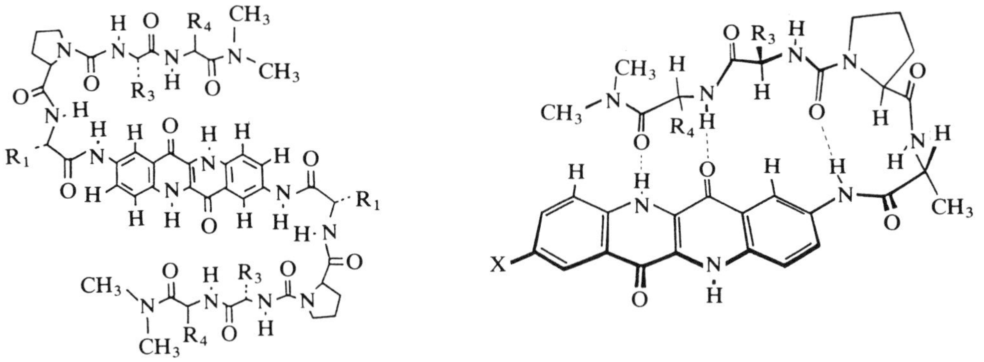

Figure 18.9. General structure of epindolidione–peptide conjugates shown to form antiparallel β-sheet structures in DMSO-d_6 solution. (The X group duplicates the peptide function.) (Kemp & Bowen, 1990.)

correlations, temperature and solvent dependencies of chemical shifts, and nuclear Overhauser effects. Data for nearly 50 conjugates have revealed the following structural patterns. Presence of a strong β-turn-forming element is required for detectable sheet structure, and the NMR data can be quantitatively fitted to a model in which the significant solution conformations have either one, two or three contiguous hydrogen bonds, starting with the turn region. This result is consistent with a model of sheet nucleation along the peptide backbone. Variation of the amino acid side-chains at sites 3 and 4 of the structure of Fig. 18.9 results in very large, linearly correlated chemical shift changes in two proton resonances (Figs 18.10 and 18.11), which are most readily explained as corresponding to an increased dominance and stabilization of the sheet conformations. An experimental sheet stabilization order for antiparallel sheets in DMSO of Ile = Val > Phe > Leu, Met > Ala > Gly is observed.

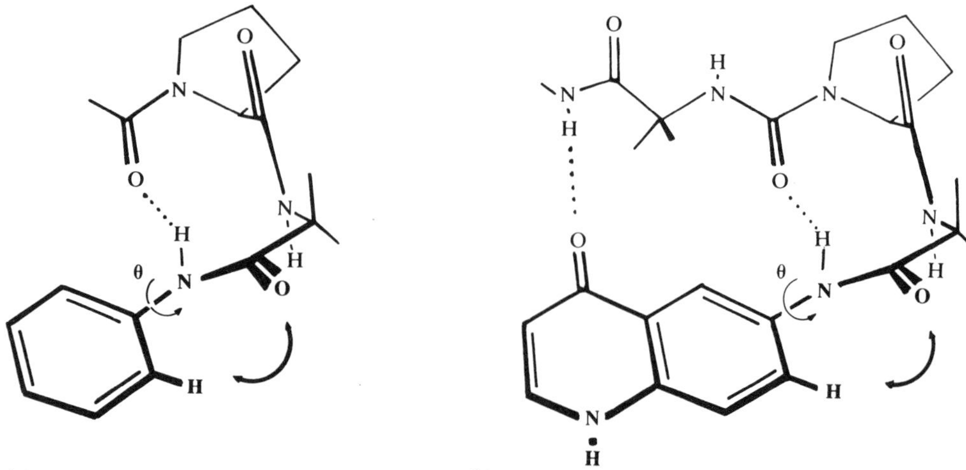

(a) (b)

Figure 18.10. The origin of the large amino acid sequence-dependent chemical shift changes shown in Figure 18.11. (a) Formation of a β-turn hydrogen bond does not constrain the angle θ. (b) Formation of a second hydrogen bond fixes θ, positions the amide carbonyl oxygen near H-3, and results in large correlated chemical shift changes at the NH of amino acid −4 and the aryl hydrogen H-3.

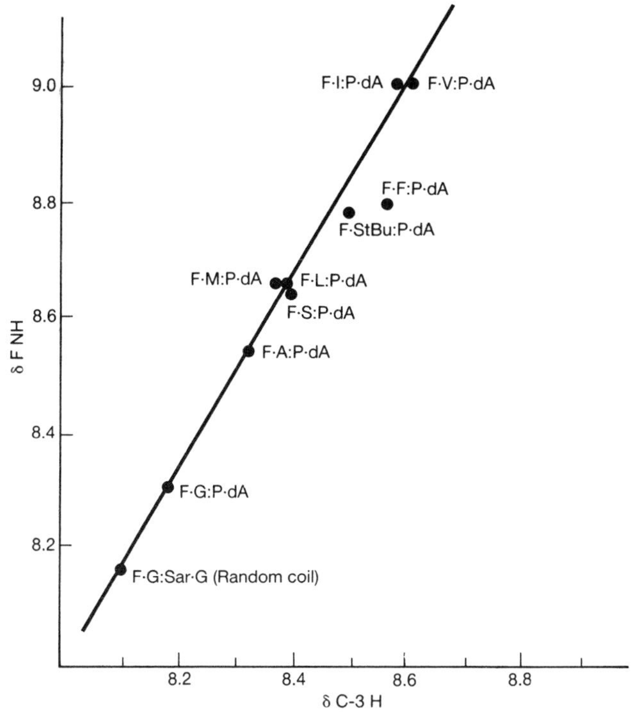

Figure 18.11. Correlation of ^{1}H-NMR chemical shifts for the Phe NH and CH-3 of the antiparallel sheet structure of Figure 18.9. The fourth amino acid residue is Phe; the third amino acid residue is varied as indicated by the amino acid one-letter codes (S-tBu ≡ L-serine *tert*-butyl ether). The linear correlation is best explained by an increasing stabilization of sheet structures with change in amino acid at site 3.

By ^{1}H-NMR spectroscopy, typified by the data of Fig. 18.12, the parallel sheet structure of Fig. 18.13 has been demonstrated to be the dominant conformation in water, and the solvent stabilization order for sheets has been found to be $CDCl_3$ ≫ DMSO > H_2O. This work is in its relative infancy, yet striking generalizations

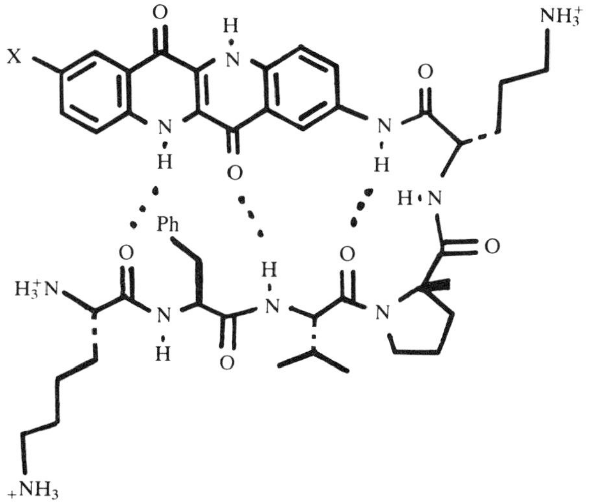

Figure 18.12. The structure of the epindolidione–peptide conjugate shown by NMR evidence to form a parallel β-sheet structure in water, pH 2, 23°C.

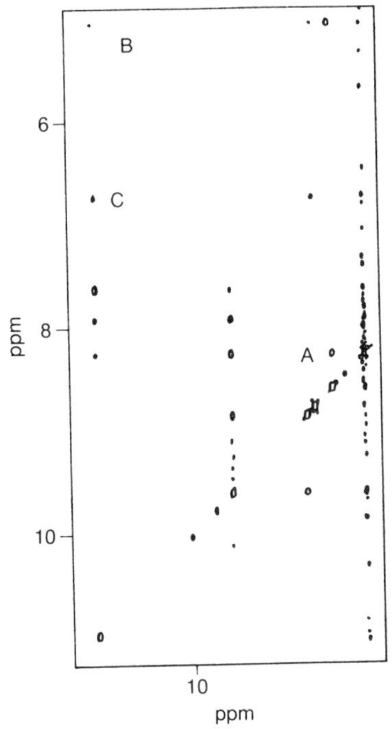

Figure 18.13. A portion of the ^{1}H-NMR NOESY spectrum of the conjugate of Figure 18.12, 6 mmol/l, 9 : 1 H_2O–D_2O, pH 2, 25°C. Some of the sheet-defining NOE interactions are apparent: (A) Val NH to CH-1; (B) heterocyclic NH to Phe α-CH; (C) heterocyclic NH to Phe phenyl.

can be offered with some confidence. Both the parallel- and antiparallel-sheet studies share the following features: first, the degree of stabilization of these templated secondary structural elements is relatively modest and, in water, all structures we have studied are to some degree conformationally frayed. Second, modification of the amino acid sequence can result in substantial variation in conformational stability. Third, temperature changes in DMSO and water result in only modest changes in the stabilities of these templated secondary structures.

All elements of protein secondary structure have now been shown to be amenable to nucleation and control in aqueous solution by rigid conformational templates that interact with adjacent peptide functionalities. Although aspects of the problems of conformational fraying and achieving optimum core packing density remain to be resolved, the availability of pathfinding examples of *ab initio* protein design (Richardson & Richardson, 1986; DeGrado *et al.*, 1989) makes this a propitious moment for development of rationally designed peptidomimetic-based protein analogues.

The second and unresolved half of the genetic code governs the protein-folding process. A complete understanding of protein folding will resolve the riddles of how nature, using mere linear sequences of amino acids, achieves molecular recognition, exquisite binding selectivity and discrimination, transduction of chemical to mechanical or electrical energy, and enzymatic catalysis. Without question, a complete and satisfying understanding of protein function will require duplication of each of these properties with analogues of proteins in which regions of structure are replaced by rationally designed mimetics and surrogates. New medicinal agents with exceptional potency and selectivity are not the least of the dividends that will result from these ventures.

7 Acknowledgements

Financial support from the National Institutes of Health Grants GM 13453 and GM 40547, from the National Science Foundation Grant 8813429-CHE, and from Pfizer Inc., is gratefully acknowledged.

8 References

Albert, T., (1989). In: G. Fasman (Ed.), *Predictions of Protein Structure and the Principles of Protein Conformation*, pp. 161–92, Plenum Press, New York.

Anfinsen, C.B. & Scherage, H.A. (1976). *Adv. Protein Chem.*, **29**, 205–99.

Arrhenius, T., Lerner, R.A. & Satterthwait, A.C. (1987). *UCLA Symp. Mol. Cell. Biol. New Ser.*, **69**, 453–6.

Baldwin, R.L. (1990). *Nature* (*London*), **346**, 409.

Barany, G. & Merrifield, R.B., (1979). In: J. Meienhofer and E. Gorss (Eds.), *The Peptides*, Vol. 2, pp. 1–284, Academic Press, New York.

Baum, J., Dobson, C.M., Evans, P.A. & Hanley, C. (1989). *Biochemistry*, **28**, 7–13.

Chothia, C., Levitt, M. & Richardson, D. (1981). *J. Mol. Biol.*, **145**, 215–50.

Chou, P.Y. & Fasman, G.D. (1974). *Biochemistry*, **13**, 211–29.

Clark-Lewis, I., Abersold, R., Ziltemer, H., Schrader, J., Hood, L. & Kent, S.B.H. (1986). *Science*, **231**, 134–9.

DeGrado, W.F. (1988). *Adv. Protein Chem.*, **39**, 51–124.

DeGrado, W.F. & Kaiser, E.T. (1982). *J. Org. Chem.*, **47**, 3258–61.

DeGrado, W.F., Wasserman, A.R. & Lear, J.D. (1989). *Science*, **243**, 622–8.

Dill, K.A. (1990). *Biochemistry*, **29**, 7133–55.

Dobson, C.M. & Evans, P.A. (1984). *Biochemistry*, **23**, 4267–70.

Dyson, H.J., Rance, M., Houghten, R.A., Lerner, R.A. & Wright, P.E. (1988). *J. Mol. Biol.*, **201**, 161–200.

Felix, A.M., Heimer, E.P., Wang, C-T, Lambros, T.J., Swistok, J., Roszkowski, M., Ahmad, M., Confalone, D., Scott, J.W., Tarker, D., Meienhofer, J., Trzeciak, A. & Gillessen, D. (1985). *Int. J. Pept. Protein Res.*, **26**, 130–48.

Felix, A.M., Heimer, E.P., Wang, C-T., Lambros, T.J., Fournier, A., Mowles, T.F., Mainer, S., Campbell, R.M., Wegrazynski, B.B., Toome, V., Fry, D. & Madison, V.S. (1988). *Int. J. Pept. Protein Res.*, **32**, 441–54.

Fotouhi, N. & Kemp, D.S. (1992). In: J.A. Smith and J.E. Rivier (Eds.), *Peptides: Chemistry and Biology*, pp. 635–6. ESCOM, Leiden.

Fotouhi, N., Galakatos, N.G. & Kemp, D.S. (1989). *J. Org. Chem.*, **54**, 2803–17.

Fujii, N. & Yajima, H. (1981). *J. Chem. Soc. Perkin I*, 831–41.

Ghadiri, M.R. & Choi, C. (1990). *J. Am. Chem. Soc.*, **112**, 1630–2.

Ghadiri, M.R. & Fernholz, A.K. (1990). *J. Am. Chem. Soc.*, **112**, 9633–5.

Grathwohl, C. & Wüthrich K. (1976). *Biopolymers*, **15**, 2025–41.

Gray, W.R., Olivera, B.M. & Cruz, I.J. (1988). *Ann. Rev. Biochem.*, **57**, 665–700.

Gutte, B., Daeumingen, M. & Wittschieber, E. (1979). *Nature* (*London*), **281**, 650–2.

Hodges, R.S., Semchuk, R.D., Taneja, A.K., Kay, C.M., Parker, J.M.R. & Mant, C.T. (1988). *Peptide Res.*, 19–33.

Hofmann, K. & Bohm, H. (1966). *J. Am. Chem. Soc.*, **88**, 5914–17.

Janin, J. & Chothia, C. (1980). *J. Mol. Biol.*, **143**, 95–128.

Kaiser, E.T. (1989). In: G. Fasman (Ed.), *Predictions of Protein Structure and the Principles of Protein Conformation*, pp. 761–75, Plenum Press, New York.

Karplus, M. & McCammon, J.A. (1981). *CRC Crit. Rev. Biochem.*, **9**, 293–349.
Kauzmann, W. (1959). *Adv. Protein Chem.*, **14**, 1-63.
Kemp, D.S. (1981). *Biopolymers*, **20**, 1793–804.
Kemp, D.S. & Bowen, B.R. (1988). *Tetrahedron Lett.*, **40**, 5077–80 & 5081–2.
Kemp, D.S. & Bowen, B.R. (1990). In: L.M. Gierasch and J. King (Eds.), *Protein Folding—Deciphering the Second Half of the Genetic Code*, pp. 293–303, AAAS, Washington, D.C.
Kemp, D.S. & Buckler, D.R. (1991). *Tetrahedron Lett.*, **32**, 3009–16.
Kemp, D.S. & Carey, R.I. (1989). *J. Org. Chem.*, **54**, 3640–6.
Kemp, D.S. & Carey, R.I. (1991). *Tetrahedron Lett.*, **32**, 2845–8.
Kemp, D.S. & Curran, T.C. (1988). *Tetrahedron Lett.*, **39**, 4931–4 & 4935–8.
Kemp, D.S. & Fotouhi, N. (1987). *Tetrahedron Lett.*, **28**, 4637–41.
Kemp, D.S. & Galakatos, N.G. (1986). *J. Org. Chem.*, **51**, 1821–929.
Kemp, D.S., Bowen, B.R. & Muendel, C.C. (1990a). *J. Org. Chem.*, **55**, 4650–7.
Kemp, D.S., Boyd, J.G. & Muendel, C.C. (1991a). *Nature* (*London*), **352**, 451–4.
Kemp, D.S., Fotouhi, N., Boyd, J.G., Carey, R.I., Ashton, C. & Hoare, J. (1988). *Int. J. Pept. Protein Res.*, **31**, 359–72.
Kemp, D.S., Carey, R.I., Dewan, J.C., Galakatos, N.G., Kerkman, D. & Leung, S.-L. (1989). *J. Org. Chem.*, **54**, 1589–603.
Kemp, D.S., Muendel, C.C., Blanchard, D.E. & Bowen, B.R., (1990b). In: J.E. Rivier and G.R. Marshall (Eds.), *Peptides, Chemistry, Structure, and Biology*, pp. 675–6, ESCOM, Leiden.
Kemp, D.S., Curran, T.C., Boyd, J.G. & Allen, T.J. (1991b). *J. Org. Chem.*, **56**, 6683–97.
Kemp, D.S., Curran, T.C., Davis, W.M., Boyd, J.G. & Muendel, C. (1991c), *J. Org. Chem.*, **56**, 6672–82.
Kent, S.B.H. (1985). In: C. Deber, V. Hruby and K. Kopple (Eds.), *Peptides—Structure and Function, 9th American Peptide Symposium*, pp. 407–14, Pierce, Rockford, IL.
Kim, P.S. & Baldwin, R.L. (1990). *Ann. Rev. Biochem.*, **59**, 631–60.
Levitt, M. (1978). *Biochemistry*, **17**, 4277–85.
McBride, B. & Kemp, D.S. (1987). *Tetrahedron Lett.*, **28**, 3435–9.
McGregor, M.J., Islam, S.A. & Sternberg, M.J.E. (1987). *J. Mol. Biol.*, **198**, 295–310.
Marqusee, S., Robbins, V.H. & Baldwin, R.L. (1989). *Proc. Natl. Acad. Sci. USA*, **86**, 5286–90.
Matthews, B.W. (1987). *Biochemistry*, **26**, 6885–7.
Miller, S., Janin, J., Lesk, A. & Chothia, C. (1987). *J. Mol. Biol.*, **196**, 641–56.
Mitchinson, C. & Baldwin, R.L. (1986). *Proteins*, **1**, 23–33.
Morgan, B.A. & Gainor, J.A. (1989). *Ann. Rep. Med. Chem.*, **24**, 243–52.
Mutter, M. & Vuilleumier, S. (1989). *Agnew. Chem.* (*Int Edn. Engl.*), **28**, 535–54.
Oas, T.G. & Kim, P.S. (1988). *Nature* (*London*), **336**, 42–4.
Offord, R.E. (1987). *Protein Eng.*, **1**, 151–7.
Padmanabhan, S., Marqusee, S., Ridgeway, T., Laue, T.M. & Baldwin, R.L. (1990). *Nature* (*London*), **344**, 268–70.
Pakula, A.A. & Sauer, R.T. (1990). *Nature* (*London*), **344**, 363–4.
Ponder, J. & Richards, F.M. (1987). *J. Mol. Biol.*, **34**, 775–91.
Privalov, P.L. & Gill, S.J. (1988). *Adv. Protein Chem.*, **39**, 191–234.
Richards, F.M. (1977). *Ann. Rev. Biophys. Bioeng.*, **6**, 151–76.
Richardson, J.S. (1981). *Adv. Protein Chem.*, **34**, 167–339.
Richardson, J.S. & Richardson, D.C. (1986). In: D.L. Oxender and C.F. Fox (Eds.), *Protein Engineering*, pp. 149–57, A.R. Liss, New York.
Roder, H., Elove, G.A. & Englander, S.W. (1988). *Nature* (*London*), **335**, 700–4.
Rose, G.D., Gierasch, L.M. & Smith, J.A. (1985). *Adv. Protein Chem.*, **37**, 1–109.
Saunders, W.S. & Terwilliger, T.C. (1989). *Science*, **23**, 54–7.
Scheraga, H. (1973). *Pure Appl. Chem.*, **36**, 1–8.
Scheraga, H. (1978). *Pure Appl. Chem.*, **50**, 315–24.
Schoemaker, K.R., Kim, P.S., York, E.J., Stewart, J.M. & Baldwin, R.L. (1987). *Nature* (*London*), **326**, 563–7.

Seyer, R., Aumelas, A., Marie, J., Bonnafous, J.C., Jard, S. & Castro, B. (1989). *Helv. Chim. Acta*, **72**, 678–89.

Shortle, D., Stites, W.E. & Meeker, A.K. (1990). *Biochemistry*, **29**, 8033–41.

Spande, T.F., Witkop, B.W., Dagani, Y. & Patchornik, A. (1970). *Adv. Protein Chem.*, **24**, 117–25.

Udagaonkar, J.B. & Baldwin, R.L. (1988). *Nature*, **335**, 694–9.

Optimization

19 On the Future of QSAR

C. HANSCH
Department of Chemistry, Pomona College, Claremont, CA 91711, USA

1 Introduction

The QSAR (quantitative structure–activity relationship) paradigm is an extension of the developments in physical organic chemistry, in particular the use of models developed by Hammett, Taft and others. The necessary condition for its rise was general access to computers which began rapidly to gain momentum in about 1965 with the advent of the IBM 360 series. The QSAR label has generally been taken to connote biological applications (rather than those in the area of organic chemistry) which had been almost non-existent before 1962. Today, almost 30 years after the first general approach was outlined, most such research is associated with drug development, so much so that it is often considered to be simply a means for optimizing drug potency. It is surprising that biochemists have paid relatively little attention to QSAR (purified enzymes are the best subjects for QSAR) despite the fact that in the period 1950–60 many attempts were made to employ the simple Hammett equation to interpret enzymic mechanisms. It is our view that to consider QSAR simply as a means for drug optimization is short-sighted: In fact, it is a means for enhancing our understanding of how organic chemicals interact with living systems (microbes, plants, insects, animals) or their parts (macromolecules, enzymes, organelles, membranes, cells). Just how much enlightenment QSAR will produce will not be clear for a few more decades.

The intense excitement that many of us experienced in the 1960s of finding it possible to correlate mathematically the biological activity of a set of congeners with their physicochemical properties has now worn off. With the many new parameters as well as new approaches an equation can be had for almost any good set of data. In fact, the word 'good' is probably an unnecessary qualifier! It has become clear that there is no statistical test which can assure us that a given QSAR has any useful meaning. Worse, since there is no fundamental law or set of laws from which to start rationalizing the relationship between structure and activity, even for a system as simple as a purified enzyme reacting with a set of simple inhibitors, we are left to steer through a seemingly endless fog, with many research groups contending for the best approach. The situation is confusing even to those who have spent most of their career trying to keep up with the new ideas continually being advanced. (Kamlet *et al.*, 1988; Cohen *et al.*, 1990; Ramsden, 1990). Often the biological data are so noisy and the mathematical models so rudimentary that one hardly knows where to begin to look for errors. To imagine how QSAR will continue to develop it is instructive to consider the development of the Hammett σ constant which was introduced in 1935. By 1968, Swan & Lupton had counted 43 variations! But today, over half a century later, there is rather broad agreement on the need for four: σ, σ^{o}, σ^{-} and σ^{+}. However, the agreement on how to factor σ into inductive and resonance components is still not settled. The problems connected with defining suitable hydrophobic and steric parameters is vastly more complex and will certainly require a longer time span before some kind of consensus begins to form. Moreover, in the development of σ, *de novo* computational methods were not advanced enough for consideration. Today computational methods contend with experimentally based methods for defining a plethora of numerical values—seemingly more than can be evaluated. Since there is no hope in sight for a fundamental approach to rationalize how a set of drugs perturbs an enzyme, much less a mouse, how are we to search for order in this immensely confusing game in the next 10–20 years? One even wonders if the search is worthwhile. All we do know is that there is enormous pressure and vast amounts of money being spent in a huge effort to understand the selective toxicity of organic compounds in the quest for better health via better drugs and a cleaner environment. The present author believes that understanding will slowly begin to emerge through what might be called *lateral validation of QSAR*, i.e. tying a new QSAR into a matrix of self-consistent structure–activity relationship. Not only self-consistent from the point of view of physicochemical parameters, but also from the point of view of basic biochemistry. It is no longer enough to derive a new QSAR with $r = 0.9$ and all terms statistically justified. Such a QSAR only gains meaning as it can be shown to fit into a self-consistent pattern when compared to many other QSARs from as widely divergent areas as possible.

What is meant by lateral validation can be illustrated from some examples from our laboratory. The application of quantum chemical parameters in QSAR can also be illustrated, something which is becoming increasingly important (Lewis, 1990). Just as computers opened up the possibility for QSAR in the 1960s, supercomputers are opening up the application of improved molecular orbital (MO) programs to others beside the professional quantum chemists. Indeed, undergraduate students now attempt such studies. The Hammett equation has been limited in its development to the study of rather conservative substituent changes on a parent structure.

The potential for the use of MO calculations can be illustrated with the following examples.

2 Mutagenicity of X–C_6H_4–N=NN(CH_3)R with TA92 in the Ames test (Shusterman *et al.*, 1989)

3 1

$$\log 1/C = 1.04 \log P - 1.63\,\sigma^+ + 3.06 \qquad (19.1)$$

$$n = 17,\ r = 0.974,\ s = 0.315$$

$$\log 1/C = 0.97 \log P - 7.76\,q\,\text{HOMO} + 5.96 \qquad (19.2)$$

$$n = 21,\ r = 0.931,\ s = 0.585$$

In eqn. (19.1) C is the molar concentration of mutagen required to produce 30 mutations above background in 10^8 bacteria. The negative ρ with σ^+ shows that electron donation by X favours mutagenic potency and since σ^+ yields somewhat better results than σ through resonance is involved. For mutagenesis by the triazenes, activation by the S9 microsomal fraction is essential. Microsomal oxidation yields carbocations (CH_3^+ or R^+) which attack DNA resulting in mutations (Scheme 1) and it is with this step that it is assumed that σ^+ is associated. The log *P* term may model movement of the mutagen through the cellular components, binding to P450 and possibly binding with the DNA.

SCHEME 1

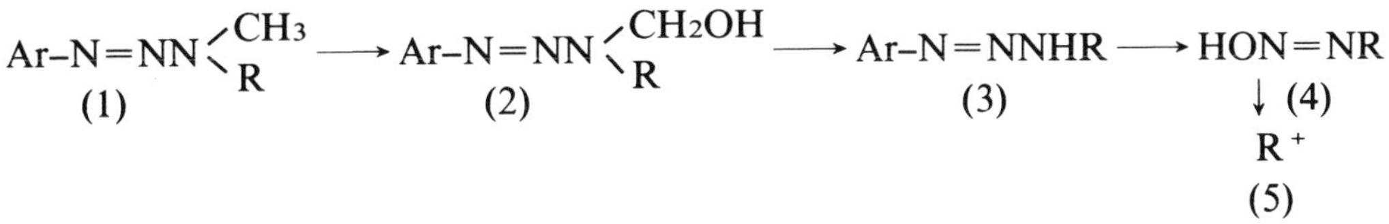

About 10 years after the derivation of eqn. (19.1), eqn. (19.2) was derived where *q* HOMO is the electron density on N_1 in the HOMO. In addition to the 17 data points on which eqn. (19.1) is based, eqn. (19.2) covers four examples where the simple $X–C_6H_4–$ has been replaced by one of the following heteroaromatic moieties:

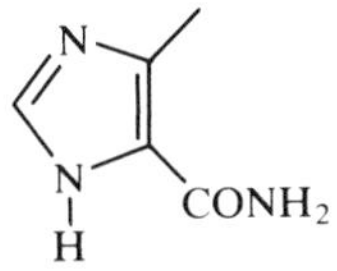

(6)

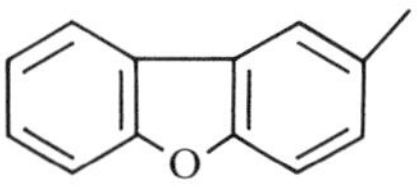

(7)

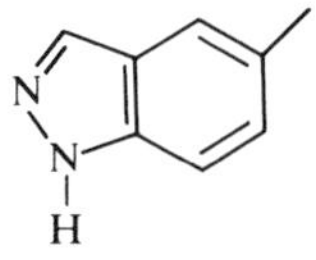

(8)

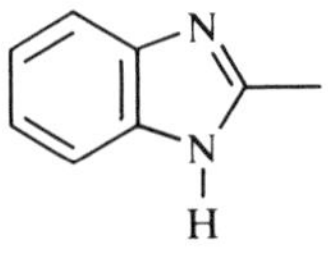

(9)

In addition to the above, eqn. (19.2) predicts that

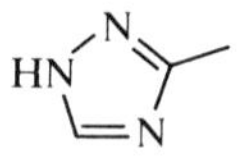

should be very weakly active and, in fact, it is of doubtful activity and, hence, was not included in the derivation of eqn. (19.2). The activity of one heterocyclic structure

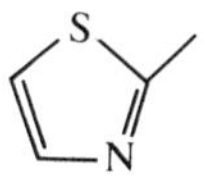

was poorly predicted for reasons which are not clear but may be, in part, due to the MNDO methodology used to calculate *q* HOMO.

An important aspect of the above two QSARs is that *h* (coefficient with log *P*) is essentially identical for both examples which shows the independent nature of log *P* and that *q* HOMO is a good substitute for σ^+. While σ^+ does give a considerably better correlation (compare values of the standard deviation *s*) *q* HOMO does open up a wide range of structural possibilities denied to the Hammett equation. Of course, the Hammett approach could be extended to the heterocycles by synthesizing the structure from which σ^+ could be obtained and using these values in eqn. (19.1). However, going to the additional trouble to obtain a sharper correlation very likely would not provide greater insight into the basic mechanism behind mutation. It would be useful to see whether heterocyclic triazenes were then fit by eqn. (19.1) as well as the simpler phenyl analogues. Work of this type needs to be done in order to understand better the limitations of the programs for doing semi-empirical MO calculations. One of the known limitations is that when through resonance is involved, MNDO and AM1 methodologies do not perform as well. In the above example the use of σ^+ in place of σ produces only a modest improvement in correlation indicating the inductive effect is more important than the resonance effect, and no doubt this is conducive to the reasonable result.

The potential illustrated by eqn. (19.2) is most exciting and we believe portends great opportunities for the use of MO parameters in QSAR in the future. Lewis's extensive bibliography (Lewis, 1990) contains rather few convincing examples of biological QSARs and the question arises as to why this is so. One of the major causes, besides the lack of proper biological data, has been the reluctance of theoretical chemists to combine hydrophobicity terms in their equations.

Equation (19.2) may be close to the best correlation attainable with the present methodology using MO parameters. Most of the biological testing was done in a single laboratory and that done for the heterocycles was identical to that used for eqn. (19.1) Log *P* for all of the compounds was obtained experimentally in our laboratory. However, it is unlikely that if all congeners contained different heterocyclic rings the fit would have been as good. Turning to the following, more problematical, example is reassuring.

Equation (19.3) correlates mutation rates for 188 nitroaromatic and heteroaromatic compounds in the Ames test using the TA98 organism (Debnath *et al.*, 1991).

$$\log \mathrm{TA98} = 0.65 \log P - 2.90 \log (\beta \cdot 10^{\log P} + 1) - 1.38\varepsilon_{\mathrm{LUMO}} + 1.88\,\mathrm{I}_1 - 2.89\mathrm{Ia} - 4.15 \qquad (19.3)$$

$$n = 188,\ r = 0.900,\ s = 0.886,\ \log P_o = 4.93$$

In this bilinear (on log *P*) model TA98 represents revertants per nannomole of nitro compound and $\varepsilon_{\mathrm{LUMO}}$ is the energy of the lowest unoccupied MO calculated using the AM1 program. The correlation is not nearly as good as that of eqn. (19.2) nor would one expect it to be. Almost 40 different aromatic systems are covered ranging in complexity from benzene to chrysene, benzopyrene and coronene. A dozen types of nitroheterocycles were also included ranging in complexity from indole to phenazines. A wide variety of substituents was present including as many as four nitro groups per aromatic ring and only about one-third of the log *P* values were experimentally determined; the rest were calculated. Probably the most serious difficulty is that the testing was done in many different laboratories with compounds which were not always of the highest purity. Nevertheless, eqn. (19.3) is a robust equation based on an activity range of 10^8 revertants/nmol.

As in the case of eqns (19.1) and (19.2) the most important variable in eqn. (19.3) is log *P* which accounts for 47% of the variance in log TA98. The $\varepsilon_{\mathrm{LUMO}}$ term accounts for 19%, I_1 for 10% and Ia for 5%.

I_1 is an unusual indicator variable which takes the value of 1 for all congeners having three or more fused rings (phenanthrene, acridine, etc.) and zero for those having only one or two fused rings (benzene, quinoline, etc.). Other factors being equal, the larger molecules are about 75 times more active; however, this holds for TA98 but not for TA100 organisms. It is possible that the larger molecules intercalate more effectively with the DNA, but is it not clear from graphics analysis just how this might occur. The Ia indicator variable is for a set of five acenthrylenes which are about 1000-fold less active than expected; deleting these has essentially no effect on the QSAR.

Dropping Ia and the five data points associated with it, yields eqn.(19.4) which is essentially the same as eqn.(19.3).

$$\log \mathrm{TA98} = 0.65 \log P - 2.90 \log (\beta \cdot 10^{\log P} + 1) - 1.38\, \varepsilon_{\mathrm{LUMO}} + 1.88\, \mathrm{I}_1 - 4.16 \qquad (19.4)$$

$$n = 183,\ r = 0.901,\ s = 0.893,\ \log P_o = 4.93$$

That is, these very poorly active compounds do not distort the QSAR.

The above two examples highlight some of the successes and some of the limitations of QSAR as it is now practised and illustrate in broader terms the potential for quantum chemistry in QSAR.

In the case of eqn. (19.1), the standard deviation is clearly higher than one would expect from experimental error alone so that even with a good value of *r* we cannot remain complacent. Focusing on eqn. (19.1) one wonders about the problems the octanol–water log *P* is attempting to cover. Activation by S9 occurs according to Scheme 1. While there is abundant evidence that microsomal oxidation is log *P* dependent (Hansch *et al.*, 1990) little is known about electronic or steric effects. Once product (2) is obtained the other steps take place spontaneously, but product (2) must make its way from the site of its formation to the site of reaction on DNA before decomposition to (5). Intermediate (5) would be so highly reactive that it would react with any nucleophile including water. Presumably log *P* also accounts to some extent for hydrophobic effects in the random walk process as well as those which might be involved in interactions with the DNA. Which of these processes is dominant in setting the value of *h*? There are more complexities. It is well known that hydrophobic compounds are toxic to all kinds of cells (Lien, *et al.*, 1968; Hansch *et al.*, 1989). Hence, how much of the unexplained variance of eqn (19.1) might be due to toxic side-reactions especially of the more hydrophobic members of the set? Are steric factors involved? The most important problem, what is the relationship between mutagenicity and cancer, which to be meaningful must be studied in animals, must also be addressed. Each of these problems merits a carefully planned programme of study in a single laboratory which may require years of patient effort. An academic setting would be preferable for such a study. At present, a large fraction of biological structure–activity research is done in industrial organizations where the pressure for immediate profit is great. This has resulted in much money being spent on computer and graphic hardware and software with relatively little spent on gathering the kind of tractable data which there is some hope of dealing with by current methodology. The problem is that academic laboratories where research can be done without pressure for immediate profit lack financial support for meaningful long-term multidisciplinary projects. The result is that many of us in academe have violated the old adage of biochemists not to 'waste clean thinking on dirty data'. We have had to resort to taking whatever data (often of poorly devised sets of chemicals with rampant collinearity problems) could be found in the literature and making the best of it.

Equations (19.2) and (19.3) do not stand alone; lateral support comes from several directions in addition to that of the hydrophobic terms common to both equations. With respect to eqns (19.1) and (19.2) it is expected that activation by microsomal oxidation would be promoted by electron-releasing substituents, and the negative value for ρ^+ supports this as does the relative electron density on N_1.

From various kinds of experimental evidence, the mutagenicity of the nitroaromatics (eqn. 19.3) is assumed to occur by the following mechanism:

SCHEME 2

$$\text{Ar–NO}_2 \xrightarrow[(10)]{} \text{Ar–NHOH} \xrightarrow[(11)]{} \begin{bmatrix} \text{Ar–NHOSO}_3^- \\ \text{Ar–NHOCOCH}_3 \\ \text{Ar–NH}^+ \end{bmatrix} \xrightarrow[(12)]{} \text{DNA–NHAr}$$

Reduction of the nitro group would be favoured by low-level unoccupied MOs and, indeed, eqn. (19.3) supports this. There is evidence that the hydroxylamine is converted to esters which react with DNA or the nitrenium may form and react with DNA. The nitro compounds do not require S9 but are reduced within the TA98 cell by cytosolic reductases. Information about this step can be had from QSAR 5 and 6 which are formulated from the reduction of a set of substituted nitrobenzenes by xanthine oxidase.

$$\log k = -1.53\ \varepsilon_{\text{LUMO}} - 0.06 \qquad (19.5)$$

$$n = 21,\ r = 0.897,\ s = 0.242$$

$$\log k = 1.09\ \sigma^- + 1.73 \qquad (19.6)$$

$$n = 21,\ r = 0.936,\ s = 0.192$$

It is suspected that xanthine oxidase may be the cellular enzyme which accomplishes step (10) although other reductases may be involved. Lateral support for eqn. (19.3) comes from the similarity between the LUMO terms of eqns (19.3) and (19.5). Again we see that the Hammett equation (19.6) gives a better correlation but does not provide more insight than eqn. (19.5). It is of interest that adding a hydrophobic term to eqn. (19.5) or (19.6) does not improve the correlation. That is, hydrophobicity does not play a role in the reduction step. However, hydrophobicity does play a role in the phenol sulphotransferase enzyme suggesting that hydrophobicity may be important for step (11) for the formation of sulphate esters (Campbell *et al.*, 1987).

The matrix of related correlations can be extended in another direction by QSAR 7 for the mutagenesis by aromatic and heteroaromatic amines with TA98 cells (Debnath *et al.*, 1992).

$$\log \text{TA98} = 1.09 \log P + 2.23\ \text{I}_1 - 3.57 \qquad (19.7)$$

$$n = 88,\ r = 0.875,\ s = 0.938$$

In the case of the amines, in contrast to the nitro compounds, S9 is required for activation. With the more hydrophilic amines, as with the triazenes, only the linear relationship is found. As for the other examples, the log *P* is highly important and I_1 term is lateral validation for eqn. (19.3). After oxidation of the amine to the hydroxylamine the mutagenic mechanism of the amine proceeds according to Scheme 2 for the nitro compounds. Hence, I_1 cannot be associated with activation but must be related to the attack on DNA. A surprising feature of QSAR 7 is that it does not contain an electronic term. If two terms are added ($\varepsilon_{LUMO} + \varepsilon_{LUMO}$), a significant but not important improvement results. This has been interpreted to mean that a positive ε_{HOMO} term important for microsomal oxidation is counterbalanced by a negative ε_{LUMO} term important for the stability of the nitrenium ion and possibly associated with the esterification step. In the case of the nitro compounds the electronic effects operate in the same direction, but with amines they are in opposition (Debnath *et al.*, 1992). The use of a common language allows us to interrelate the different studies of mutagenesis.

In the coming decades, whatever approach is taken to QSAR development, a serious effort must be made to show the interrelatedness, not only among the QSARs, but also with what we know about organic and biochemical reaction mechanisms. A new QSAR standing alone means little. Appreciation of this point of view has been shown by Kamlet *et al.* (1988) and by Magee (1990) in the context of other approaches to QSAR.

The difficulty of making sense from an isolated QSAR can be illustrated from a study of nitrosobenzylamines producing mutations in TA1535 in the presence of S9.

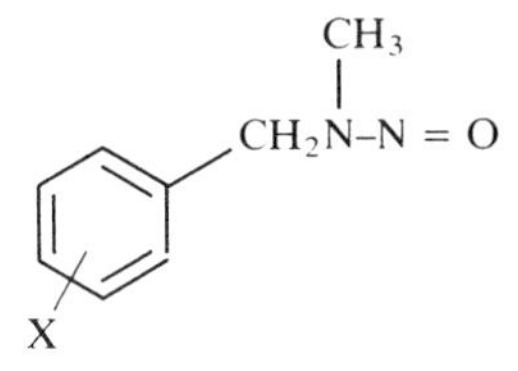

(13)

$$\log 1/C = 3.55\sigma - 3.88\sigma^2 + 1.62\,{}^3\chi_p^v - 5.11 \quad (19.8)$$

$n = 13$, $r = 0.873$

$$\log 1/C = 0.92\pi + 2.08\sigma - 3.26 \quad (19.9)$$

$n = 12$, $r = 0.891$, $s = 0.314$

C is the concentration of mutagen in nannomoles per plate needed to produce 50 revertants per plate. Equation (19.8) was developed by Singer *et al.* (1986) who determined π constants for the nitrosamines by measuring their log *P* values in octanol–water. They concluded from their values that hydrophobic forces were not important and that eqn. (19.8) was ‘intuitively satisfying’. From an inspection of their π values it was clear that they did not all agree with π from the benzene system (Hansch & Leo, 1979) which would seem appropriate for the benzylamine amines. Accordingly, we formulated eqn. (19.9) using π from the benzene system. There is no objective way to choose between these two QSARs. One data point (3–OCH_3)

has been omitted in deriving eqn. (19.9) which makes it less attractive than eqn. (19.8); however, eqn. (19.8) uses one more parameter. Although the quality of the correlation with eqn. (19.9) is slightly better the difference is not enough to tip the balance toward eqn. (19.9). The only reason for favouring eqn. (19.9) is that many other examples (see, for example, eqns 19.1–19.3) of the QSAR of mutagenesis contain log *P* terms. Moreover, the value of *h* in eqn. (19.9) is close to that of eqns (19.1) and (19.7) where microsomal oxidation is involved. The $^3\chi_\rho^v$ parameter is the three-path connectivity index devised from graph theory which obviously must be in some way collinear with π. Of course, it can be said that either, or both, of the above correlations are simple coincidence. The only answer to this is the building up of a massive, self-consistent bank of QSARs which has general as well as specific predictive value.

To some extent, log *P* and π have achieved this in that it is rare that such terms are missing from a biological QSAR; however, this does not mean that they are the best of all possible parameters for what we loosely call hydrophobic interactions.

The successful use of MO parameters in the above QSAR presages the coming importance of quantum chemistry in QSAR. This can also be seen from a recent study of Sotomatsu *et al.* (1989) in which they derived eqn. (19.10).

$$\sigma = 2.9166[q(=\text{O}) + q(-\text{O}-) + q(\text{H})] + 12.681 \quad (19.10)$$

$$n = 27,\ r = 0.971,\ s = 0.083$$

In this expression, *q* represents the charge (calculated by the AM1 program) on the oxygen and hydrogen atoms of the carboxyl group for a set of 27 meta- and para-substituted benzoic acids. While the standard deviation is larger than the experimental error in measuring σ (which may be about 0.02), it does approach that found in good correlations in pure physical organic chemistry.

A still further extension of our QSAR matrix can be illustrated with studies with microsomal P450 (Hansch *et al.*, 1990). Equations (19.11–19.13) correlate the induction of cytochrome P450 (50% increase) by three quite different types of compounds.

2.1 *Induction of P450 by ROH in chick embryo hepatocytes*

$$\log 1/C = 0.78 \log P + 1.46 \quad (19.11)$$

$$n = 7,\ r = 0.988,\ s = 0.095$$

2.2 *Induction of P450 by 4-X-pyrazoles in chick embryo hepatocytes*

$$\log 1/C = 0.85 \log P + 1.93 \quad (19.12)$$

$$n = 8,\ r = 0.970,\ s = 0.305$$

2.3 *Induction of P450 by 5,5-X_2-barbiturates in chick embryo hepatocytes*

$$\log 1/C = 1.02 \log P + 2.75 \quad (19.13)$$

$$n = 9,\ r = 0.984,\ s = 0.186$$

Binding to P450 is very similar and raises the question: Does binding = induction (Shusterman *et al.*, 1989)?

2.4 *Binding of 4-X-pyrazoles to rat liver microsomes*

$$\log 1/K_s = 1.06 \log P + 3.05 \tag{19.14}$$

$n = 10$, $r = 0.977$, $s = 0.347$

2.5 *Binding of 4-X-pyrazoles to chick embryo hepatocytes*

$$\log 1/K_s = 1.01 \log P + 2.93 \tag{19.15}$$

$n = 8$, $r = 0.989$, $s = 0.215$

Safe, Fujita and their colleagues (1986) have also shown that for more complex compounds binding to, and induction of, P450 is closely associated with π.

It has long been known that microsomal oxidation was related to hydrophobicity, but this was first demonstrated in QSAR terms in the early 1970s (Martin & Hansch, 1971; Hansch 1972) as illustrated by QSAR 16 (Martin & Hansch, 1971).

$$\log 1/K_{m(corr)} = 0.69 \log P + 2.90 \tag{19.16}$$

$n = 14$, $r = 0.920$, $s = 0.330$

QSAR 16 correlates the demethylation and oxidation of a variety of compounds where $K_{m(corr)}$ applies only to the un-ionized form of the chemical. For the examples covered by eqn. (19.16) k_{cat} was essentially constant so that binding = oxidation.

Carcinogenicity has also been shown to be closely tied to log P (Hansch & Fujita, 1964) as one would, of course, anticipate from eqns (19.1–19.3, 19.7 and 19.9). Thus, we have seen in dim outline the interrelatedness of mutagenicity, carcinogenicity and P450 induction via log P. The generalization can be extended. Many hundreds of QSARs correlating toxicity and hydrophobicity take the form:

$$\log 1/C = a \log P + b \tag{19.17}$$

Where C is the molar concentration producing some kind of toxic effect with enzymes, organelles, cells, membranes or whole organisms (Dunn & Hansch, 1974; Hansch *et al.*, 1989). Many different classes of compounds are covered by eqn. (19.17). Hence, it is not surprising that P450 appears to have evolved, at least in part, as a means to protect organisms from hydrophobic molecules. It is a counterpart, for small molecules, to the immune system which protects against macromolecules and cells. Oxidation by the P450 system results in less toxic, more hydrophilic substances which can be more easily eliminated, at least from animals.

Oxidation alone is not enough to establish a complete defence system against small, hydrophobic molecules which are toxic. It requires a number of back-up systems, two of which will now be considered.

The sulphation of OH and NH_2 moieties by phenolsulphotransferase (PST) is a means for greatly increasing the hydrophilicity of compounds and thus reducing their toxicity. Glucuronidation by the enzyme uridine-5′-diphospho-α-D-glucuronic acid (UDPGA) is another.

(14) (15)

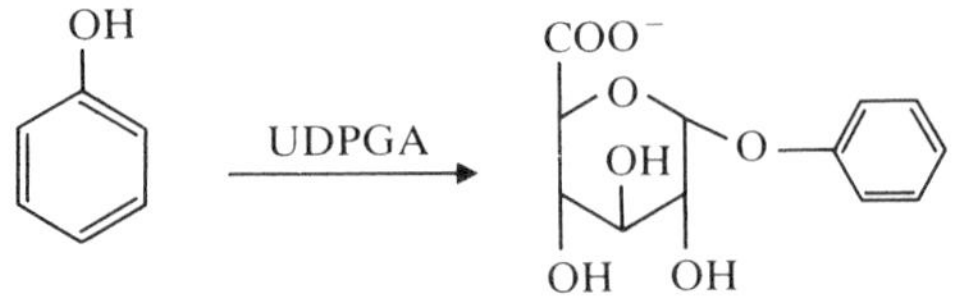

(16) (17)

That both of these enzymes evolved to operate on hydrophobic compounds is apparent from the following QSAR.

2.6 *Sulphation of phenols by human PST* (Campbell *et al.*, 1987)

$$\log 1/K_m = 0.92 \log P - 1.48\ MR_4' - 0.64\ MR_3 + 1.04\ MR_2 + 0.67\sigma^- + 4.03 \quad (19.18)$$

$n = 35$, $r = 0.950$, $s = 0.477$

2.7 *Glucuronidation of ROH in rabbits* (Kamil *et al.*, 1953)

$$\log A = 0.90 \log P - 1.17 \log(\beta \cdot 10^{\log P} + 1) + 0.72\ \mathrm{I} - 0.01 \quad (19.19)$$

$n = 24$, $r = 0.936$, $s = 0.233$, $\log P_o = 1.98$

While steric and electronic effects also play a role in eqn. (19.18), log P is the most important term. Its coefficient is similar to those we have seen in mutagenesis and P450 processes. Equation (19.19) is similar. The indicator variable I takes the value of 1 for secondary alcohols. That these are more effectively glucuronidated probably stems from the fact that they are more stable to oxidation. These two enzymes can operate directly on xenobiotics or in concert with microsomal oxidation.

A rather different example of lateral validation involving electronic factors comes from a study of the enzymic hydrolysis of the following two classes of esters (Selassie *et al.*, 1988).

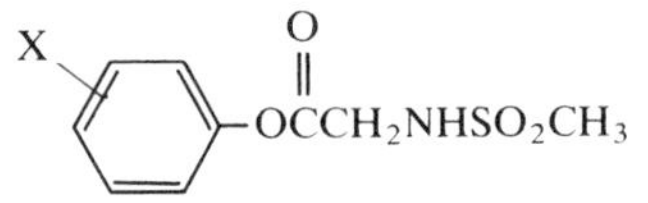

I

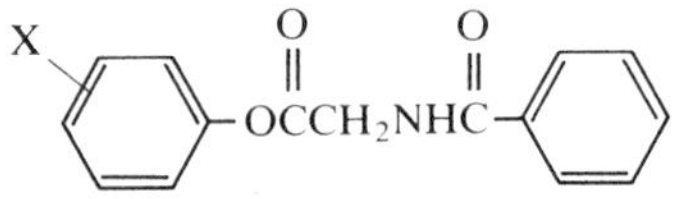

II

In Table 19.1 are listed ρ values for the hydrolysis of esters I and II for the K_m step. For most of the above examples, k_{cat} is essentially constant so the ρ represents the overall importance of the electronic effect of X on hydrolysis. There are three

Table 19.1. Values of ρ (coefficient with σ) in QSAR for hydrolysis of esters I and II. The average value is 0.58

ρ	Enzyme	Ester
0.42	Chymotrypsin	II
0.46	Trypsin	II
0.49	Subtilisin (Carlsberg)	II
0.50	Subtilisin (BPN)	II
0.55	Papain	I
0.57	Papain	II
0.62	Ficin	I
0.63	Bromelain D	II
0.70	Bromelain B	II
0.74	Ficin	II
0.74	Actinidin	II

points of interest. The different side-chains of I and II have little or no effect even though this might cause slightly different binding modes for the substrates. The constancy of ρ is surprising, although there seems to be a small difference between the serine proteases (subtilisin, chymotrypsin and trypsin) and the other cysteine hydrolases.

The small values of ρ show that enzymes play the major role in the hydrolytic process, that is, the substituents play only a small part. The values of ρ can be compared to those obtained from hydrolysis of esters in general. For the basic hydrolysis of aromatic esters ρ is around 2; while for acid-catalysed hydrolysis it is close to 0. The low value for acid hydrolysis is the result of very strong polarization of the carbonyl group which promotes attack by water.

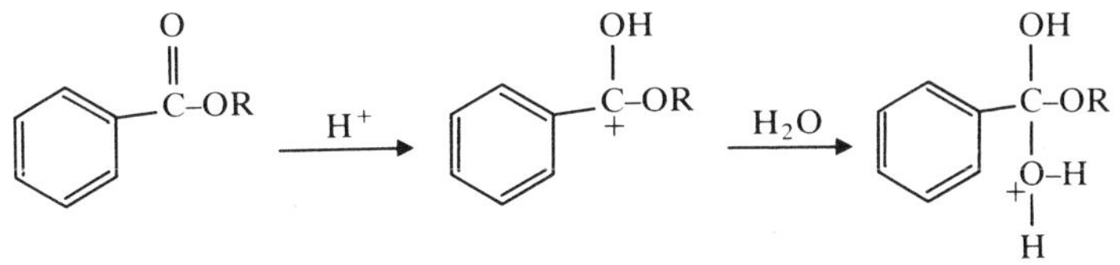

(20) (21) (22)

Thus the enzymic process resembles acidic hydrolysis with the enzyme setting up the substrate for attack by serine or cysteine.

The above examples illustrates how QSARs can, in a number of small ways, provide generalizations about chemico-biological interactions. Indeed there are many other ways which are outside of the usual goal of optimization of potency in a particular set of congeners. It has long been recognized by those in the field that the prediction of potency from a QSAR for new congeners lying beyond explored data space is a very risky undertaking (Hansch *et al.*, 1977). This is not to say it is without value since it has been demonstrated to be of direct value in the design of bioactive compounds now in the market (Fujita, 1984) and there are hundreds of examples where accurate predictions from a QSAR have been realized (see Kim *et al.*, 1979, for a few examples). However, it is our thesis that this is not the most interesting or important role for QSARs, and increasingly in the next few decades more attention will be directed to defining a clearer understanding of the major processes in chemico-biological interactions.

Some examples of this line of thought from other points of view can be suggested.

• Early in QSAR studies, it was recognized that optimum hydrophobicity (log P_o) for the penetration of the CNS by drugs was about 2. Actually, this may be the ideal for penetration to many parts of the body (note eqn. 19.19). A similar value has been found for log P_o in the penetration of plant tissue by herbicides. (Kakkis *et al.*, 1984). Thus, in the early phases of drug development, striving for maximum potency, for say a heart drug or an antihistamine, might lead one to rather lipophilic compounds. But, a few moments of reflection now informs us that such drugs might have CNS side-effects. In fact, this seems to be true (Hansch *et al.*, 1987).

• The discovery of mutagenicity in the early phases of drug development is always a disturbing event. However, QSAR has a role to play in the design of less toxic substances. In the development of eqn. (19.2), it was shown that mutagenic potency of the triazenes could be raised almost to the level of that of aflatoxin or wiped out by the manipulation of the electronic and hydrophobic properties of the substituents. Hence, if efficacy and mutagenicity are not highly collinear (as they are in many anticancer drugs), the latter can be eliminated with the help of QSAR. There are few examples where QSARs have been developed for various toxicities along with those for potency.

• Selective toxicity is at the heart of drug design. Ideally, one should be concerned with this from the very start of a programme at the receptor level. In fact, this has been done by Selassie *et al.* (1991) by deriving QSARs for the inhibition of dihydrofolate reductase from bacterial and vertebrate sources.

• It has been pointed out that when there are no conformational restrictions a phenyl ring can by 180° rotation place a meta-substituent in two potentially different environments. Thus a substituent might be placed in either hydrophobic or polar space (Hansch & Klein, 1986; Selassie *et al.*, 1991) as, for example, in binding into a cleft. A polar substituent would orient outside of a cleft while a hydrophobic one would orient into the cleft. Of course, this would not be limited to phenyl rings, but could occur with other structures when the elements of symmetry are suitable. Advantage can be taken of this in drug design by utilizing one meta-position to enhance binding at the receptor level, but controlling overall hydrophobicity by placing a polar substituent in the other meta-position or at some other point in the molecule (Hansch & Klein, 1986; Selassie *et al.*, 1991).

• In conclusion, QSARs have reached a level of maturity where there is no longer any doubt that mathematical relationships can be established correlating structure with activity. In the next few decades, we must struggle to find the best parameters for use in QSARs. It is clear that statistics alone, although very necessary, will be of little help in this process, nor will the fundamental laws of physical chemistry. To describe accurately, mathematically, how a set of drugs reacts with a receptor in a mouse would require so many variables that the effort would not be worth the return. What we need to strive for is a relatively small set of descriptors which will provide a general picture of what the major features of the process are. The value of the descriptors will only become clear as they are seen to bring order into a large amount of SAR data. Already it seems clear that quantum chemistry will be an important means for extending the range of QSARs which can be compared and contrasted in our search for order in the understanding of how xenobiotics affect life.

3 References

Campbell, N.R.C., van Loon, J.A., Sundaram, R.S., Ames, M.M., Hansch, C. & Weinshilboum, R. (1987). *Mol. Pharmacol.*, **32**, 813.

Cohen, N.C., Blaney, J.M., Humblet, C., Gund, P. & Barry, D.C. (1990). *J. Med. Chem.*, **33**, 883.

Debnath, A.K., de Compadre, R.L.L., Debnath, G., Shusterman, A.J. & Hansch, C. (1991). *J. Med. Chem.*, **34**, 786.

Debnath, A.K., Debnath, G., Shusterman, A.J. & Hansch, C. (1992). *Environ. Mol. Mutagenesis*, **19**, 35.

Dunn III W. J. & Hansch, C., (1974). *Chem. Biol. Interact.*, **9**, 75.

Fujita, T. (1984) In: G. Jolles and K.R.H Wooldridge (Eds.), *Drug Design: Fact or Fantasy?* p. 19, Academic Press, London.

Hansch, C. (1972). *Drug Metab. Rev.*, **1**, 1.

Hansch, C. (1977). In: J.A.K. Buisman (Ed.), *Biological Activity and Chemical Structure*, p. 47, Elsevier, Amsterdam.

Hansch, C. & Fujita, T. (1964). *J. Am. Chem. Soc.*, **86**, 1616.

Hansch, C. & Klein, T. (1986). *Acc. Chem. Res*, **19**, 392.

Hansch, C. & Leo, A. (1979). *Substituent Constants for Correlation Analysis in Chemistry and Biology*, Wiley Interscience, New York.

Hansch, C., Björkroth, J.P. & Leo, J. (1987). *J. Pharm. Sci.*, **76**, 663.

Hansch, C., Sinclair, J.F. & Sinclair, P.R. (1990). *Quant. Struct.-Act. Relat.*, **9**, 189.

Hansch, C., Kim, D., Leo, A.J., Novellino, E., Silipo, C. & Vittoria, A. (1989). *CRC Crit. Rev. Toxicol.*, **19**, 185.

Kakkis, E., Palmire Jr., V.C., Strong, C.D., Bertsch, W., Hansch, C. & Schirmer, U. (1984). *J. Agr. Food Chem.*, **32**, 133.

Kamil, I.A., Smith, J.N. & Williams, R.T. (1953). *Biochem. J.*, **53**, 129.

Kamlet, M.J., Doherty, R.M., Abraham, M.H. & Taft, R.W. (1988). *Quant. Struct.-Act. Relat.*, **7**, 71.

Kim, K.H., Hansch, C., Fukunaga, J.Y., Steller, E.E., Jow, P.Y.C., Craig, P.N. & Page, J. (1979). *J. Med. Chem.*, **22**, 366.

Lewis, D.F.V. (1990). *Prog. Drug Met.*, **12**, 206.

Lien, E.J., Hansch, C. & Anderson, S.M. (1968). *J. Med. Chem.*, **11**, 430.

Magee, P.S. (1990). *Quant. Struct.-Act. Relat.*, **9**, 202.

Martin, Y.C. & Hansch, C. (1971). *J. Med. Chem.*, **14**, 777.

Ramsden, C.A. (Ed.) (1990). *Comprehensive Medicinal Chemistry*, Vol. 4. C, Pergamon Press, Oxford.

Safe, S.H. (1986). *Ann. Rev. Pharmacol. Toxicol.*, **26**, 371.

Selassie, C.D., Chow, M. & Hansch, C. (1988). *Chem.-Biol. Interactions*, **68**, 13.

Selassie, C.D., Li, R.-L., Poe, M. & Hansch, C. (1991). *J. Med. Chem.*, **34**, 46.

Schusterman, A.J., Debnath, A.K., Hansch, C., Horn, G.W., Fronczek, F.R., Greene, A.C. & Watkins, S.F. (1989). *Mol. Pharmocol.*, **36**, 939.

Singer, G.M., Andrews, A.W. & Guo, S.-M. (1986). *J. Med. Chem.*, **29**, 40.

Sotomatsu, T., Murata, Y. & Fujita, T. (1989). *J. Comp. Chem.*, **10**, 94.

20 Computer-assisted Drug Design in the 21st Century

Y.C. MARTIN, K.-H. KIM and M.G. BURES

Computer Assisted Molecular Design Project, Pharmaceutical Products Division, Abbott Laboratories, Abbott Park, IL 60064, USA

1 Introduction

Two dramatic developments have recently extended the methodology of computer-aided drug design. First, the computer can now design novel compounds and predict new biological properties of known compounds by using three-dimensional (3-D) substructure searching (Van Drie, *et al.*, 1989; Martin, 1990; Martin *et al.*, 1990; Bures, *et al.*, 1991; Martin, 1992). Second, the versatile and robust comparative molecular field analysis (CoMFA) method quantitatively forecasts potency from 3-D structure (Cramer III *et al.*, 1988; Kim, 1991; Kim & Martin, 1991a,b; Y.C. Martin & G. Greco, unpublished results). Ever-increasing computer power, ever faster and more reliable networking between computers at different locations, ever larger databases, and ever more intelligent and comprehensive software will remove much of the empiricism in new drug design.

Before medicinal chemists or computers can design novel bioactive molecules, they must know the molecular characteristics associated with the desired biological activity, undesirable side-effects, toxicity and metabolic inactivation. Structure–activity relationships usually provide this information. Figure 20.1 shows the relationships between the structure–activity-based computer methods discussed

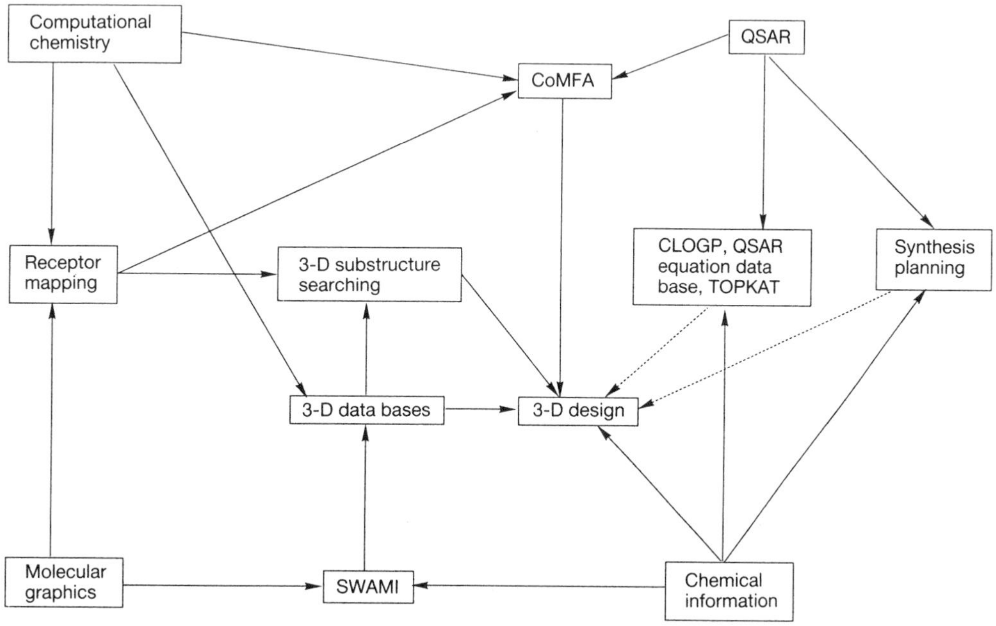

Figure 20.1. The relationships between different computer methods of examining structure–activity relationships.

here. Four disparate uses of computers in chemistry (QSAR, computational chemistry, molecular graphics and chemical information/synthesis planning) rarely intersected before 1980. However, in the last decade, affordable graphics and more powerful and friendly computers have increased both the numbers of workers in these fields and the synergism between these fields.

Many needed concepts in computer-aided molecular design have been developed and the power of computer hardware will continue to outpace software. Two factors limit the optimum use of computers in medicinal chemistry: the various techniques require extensive human expertise for their correct application, and they are accessed in different programs. We anticipate that the diverse computer techniques used in drug design will be consolidated to produce more powerful computer programs. As the experts become more adept, this will be translated into computer software that will supply expertise now required of users. Such sophisticated software will remove much of the human labour and subjectivity from the various approaches. We also foresee that databases will continue to grow in size and ease of acccess and that new database techniques will produce answers to medicinal chemistry questions more directly.

Advances in the computer prediction of the biological effects of compounds and their by-products in the body will not only lead us more quickly to safer, potent, new drugs but will lead to the need for less pharmacological and toxicological testing in animals.

2 Computer information tools for the perception of substructures, chemical database searching, and synthesis planning based on 2-D structures

Searching structure–activity databases is a computer application familiar to medicinal chemists. Biological and physical data may be stored with standard database techniques. However, it requires special chemical information techniques to store chemical structures so it is easy for the user either to find a specific compound (one would not want to supply the IUPAC name for this, for example) or to search for molecules that contain a particular substructure (Willett, 1987; Martin *et al.*, 1990)

Databases of millions of structures are now practical. Also, one can search for molecules that are substructures of the search structure or are structurally similar to it (Willett, 1987). Recently, new methods for the organization, clustering and analysis of the data found in the searches and integration with systems that better handle biological data have further increased the power of chemical information systems.

We use the 'Daymenus' system (Daylight, 1991). Its GENIE program is a particularly powerful utility: it provides a language to process chemical structures based on the substructures present. The user programs the actions to be taken when a substructure match is found. For example, in medicinal chemistry GENIE has been used for the following: to predict carcinogenicity (Enslein *et al.*,1987) from just the 2-D chemical structure and a known QSAR equation; to describe programming rules for the generation of a 3-D structure from templates (Martin *et al.*, 1986); to assign atom types for a molecular mechanics program (O'Donnell *et al.*, in preparation); and to identify atoms in our molecular graphics program (J. DeLazzer *et al.*, unpublished results). Because of the sophisticated capabilities that chemical information tools offer computer chemistry, the following sections will discuss applications of these methods to specific computer-aided drug-design problems.

Another area in which computers will continue to affect medicinal chemistry and compound design is in synthesis planning. Programs that propose synthetic sequences for organic compounds (retrosynthetic analysis) or evaluate pathways or products of organic reactions (synthetic analysis) are available. Retrosynthetic analysis programs, such as LHASA (Long *et al.*, 1983) and REACCS (Moock *et al.*, 1988), suggest intermediates and reaction transformations en route to proposed synthesis targets. Programs such as CAMEO (Laird & Jorgensen, 1990) and EROS (Gasteiger *et al.*, 1987) provide a detailed analysis of the products and mechanistic pathways of organic reactions. Retrosynthetic analysis in conjunction with synthetic analysis can give a thorough evaluation of intermediates, products and side products expected during a planned synthetic scheme.

3 QSAR—the statistical analysis of the relationship between bioactivity and properties of molecules calculated from 2-D structure

QSAR is especially useful for optimizing the groups that modulate the potency of a molecule (Martin, 1978, 1981). Its beginning a generation ago (Hansch *et al.*, 1962; Fujita *et al.*, 1964) opened up the field of quantitative potency forecast by a computer. This experience can now be applied to more diverse applications of computers to drug design. QSAR also alerted medicinal chemists to the power of statistical analysis of complex data and the value of planning the series of molecules before synthesis. Since there are thousands of known QSAR relationships, any new method that can show a relationship to QSAR will itself be strengthened. QSAR has been used successfully in many cases of new compound design (Martin, 1981; Hopfinger, 1985; Fujita, 1990).

A QSAR begins with a data table in which each row contains the properties of a particular compound and each column a different molecular property, physical or biological. The early QSARs generally considered the hydrophobicity, the electronic nature, and the size of variable substituents on a common parent molecule. In traditional QSAR, multiple regression analysis of the correlation between biological potency and one or more physical properties produces the best-fitting equation. An investigator makes three choices in running a QSAR: the properties, descriptors of the molecules; the mathematical form of relationship between the potency and the descriptors; and the statistical method to analyse it.

A successful QSAR equation can provide information on the mechanism of action of the compounds or the atomic environment of the biological binding site. For example, one might find that substituents at one position bind in hydrophobic space whereas those at another position do not contact the binding site. Figure 20.2 shows an example. QSARs are usually derived without knowledge of the 3-D structure of the macromolecular target. For cases where the 3-D structure is known, the conclusions from QSAR and molecular graphics displays of the target with a bound ligand show that QSAR agrees with molecular graphics (Smith *et al.*, 1982).

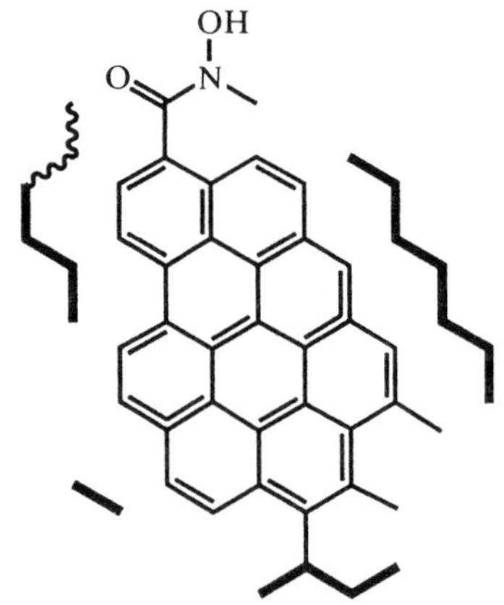

Figure 20.2. A 2-D binding-site map derived from QSAR analysis of hydroxamate inhibitors of 5-lipoxygenase (Summers *et al.*, 1990). Solid lines indicate lipophilic boundaries; curved line indicates steric hindrance.

A QSAR provides an estimate of the highest potency expected of a molecule from the series. Additionally, QSAR can sometimes establish whether all parts of the molecule are in close contact with the binding site (Kim *et al.*, 1989).

The relationship between potency and hydrophobicity is often not linear but can be fit with a parabola. However, Martin & Hackbarth (1976) and Kubinyi (1976, 1977) suggested that instead one should fit a different equation. When the compounds also differ in degree of ionization, the situation can become difficult to understand. With compartmental modelling one can distinguish the physical properties that affect binding to the target from those that affect accessibility to the target (Martin & Hackbarth, 1976, 1977).

Colinearity arises when two or more properties are correlated with each other. To prevent colinearity of physical properties in a data set, Hansch *et al.* (1973) suggested that it should be designed from a cluster analysis on all possible substituents. A well-designed series includes one member from each of the clusters of similar substituents. We showed that series designed from cluster analysis show more independent variation in physical properties than do traditional medicinal chemistry series (Martin & Panas, 1979). There is roughly a doubling in information gained per compound tested. Thus, QSAR concepts can make medicinal chemistry more efficient.

However, certain properties of molecules are intrinsically correlated. For example, chemists expect that partition coefficients of compounds between water and various organic solvents are correlated and contain more information than any one property. Because of such intrinsic correlations, Wold and colleagues suggested that one analyse QSARs with principal components (Wold *et al.*, 1984) or partial least squares (PLS) (Berntsson, 1986). Principal components and PLS methods are similar since they extract from the data table a set of independent vectors that are linear combinations of the individual properties. Each physical property contributes to each eigenvector. Generally, many fewer than the number of input columns explain most of the variance in the data. PLS does not overfit data, but it might miss a true relationship. In contrast, regression analysis will not miss a true relationship but it might also regard a false one as true.

PLS uses cross-validation to test the robustness of the results. In a cross-validation, one repeats the analysis several times, each time leaving out a portion of the observations from the fit and predicting them from it. This is repeated until each observation has been predicted once. The cross-validated standard deviation is higher than the fitted standard deviation, but it is a more reliable estimate of the precision of a true prediction.

One may use other statistical methods to analyse the relationships between potency and molecular properties. For example, if the activity is a simple 'yes/no' response or is graded, the technique of discriminant analysis is useful (Martin *et al.*, 1973).

A limitation of QSAR is that it is often difficult to calculate the property of interest for some molecules. Another is that traditional properties do not fit all data sets. The 3-D QSAR method of CoMFA (see Section 6) overcomes these problems in many cases.

What will be the role of traditional QSAR in the 21st century? Is it being phased out by the 3-D methods? Our answer is no. First, 3-D methods require one to propose a bioactive conformation whereas QSAR does not. Thus, for a series of flexible molecules QSAR may be the only unambiguous method. Second, the thousands of known QSAR relationships attest to its versatility and are useful in forecasting bioactivity of new compounds. Third, QSARs based on sets of diverse structures have been derived for mutagenicity, carcinogenicity and toxicity (Craig, 1990). The use and further optimization of such QSARs will help the medicinal chemist or the computer to design safe compounds. Last, QSARs are often easily calculated for large series. The special advantages of QSAR mean that it will continue to be a useful tool. Friendlier and more intelligent programs discussed in the next section will increase the use of QSAR.

4 The application of chemical information tools to QSAR

The Pomona College Medicinal Chemistry Project has integrated QSAR and chemical information tools. They have compiled a chemical information database of measured solvent–water partition coefficients and pK_a values. Their CLOGP program uses chemical information tools to calculate the octanol–water log P from an input 2-D structure of the molecule. When done manually, this calculation requires considerable expertise. CLOGP recognizes the substructures present in the molecule and looks up the log P increments and corrections due to these substructures (Leo, 1990; Daylight, 1991). It also reports the measured value of the log P if this is in the database.

This group also has a system for creating and searching a database of QSAR equations and the raw data behind them (Hansch, 1989). One can search the database of equations by:

- the type or value of a biological or physical property;
- the type of biological system such as *in vitro*, isolated enzyme, or whole animal;
- the parent compound of the series;
- the source of the data;
- the properties that correlate with bioactivity;
- the regression coefficients for these properties and their ideal values.

The database of raw data is a traditional chemical database. Currently there are 1854 equations based on 25 000 data points from more than 10 000 distinct compounds. Chemists around the world are adding all published QSARs.

We imagine a future version of this system that would report for any input structure its forecast potencies from the equations for which it matches the parent structure. For example, the program would warn a scientist of potential toxicity or side-effects if a new compound fits the criteria for a previously derived relationship. If all the world's medicinal chemistry literature were summarized in QSARs, then this would be an efficient way to forecast the biological properties of molecules proposed for synthesis.

We think that the programs just described will affect the way that QSAR will be practised in the 21st century. One can imagine that the computer could be fed a list of chemical structures and it would retrieve the biological data, calculate or look up the physical properties, and do the statistical analysis. While this is possible in some programs today, the advanced skills of human experts have not been included. If a program were really expert, it would decide which compounds and physical properties to consider and might use the results of a preliminary analysis to change the parameters or data set considered. Some aspects of these tasks might involve searching the equation database for relationships with the same type of compound or same type of bioactivity. A more advanced expert system would perceive when there is enough information for a QSAR in a structure–activity database, run the QSAR, and report the results to the scientist. It might even recognize when the synthesis and testing of a few specific compounds would produce a good QSAR data set.

5 Modelling the 3-D structure of binding sites from structure–activity relationships of ligands

The binding sites for small molecules on biomacromolecules are of defined shape and properties. The aim of molecular modelling of bioactive small molecules is to explore the characteristics of this binding site— a process called 'pharmacophore mapping'. The importance to medicinal chemistry of the move from 2-D to 3-D structure–activity relationships cannot be overemphasized.

Pharmacophore mapping has been available nearly as long as has QSAR. However, it affected medicinal chemistry later because it requires more extensive computer hardware and because much of the needed software has matured only in the last decade. Today the medicinal chemist has the choice of many molecular modelling and computational chemistry programs. These programs run on hardware ranging in power from personal computers to supercomputers.

Computational chemistry programs (Counts, 1987) are based on quantum (Clark, 1985) or molecular mechanical principles (Burkert & Allinger, 1982; Allinger *et al.*, 1989). They explore the relative energy of the various conformations of a molecule or configurations of a molecular ensemble and calculate chemical or physical properties of individual molecules or ensembles (Martin, 1992). The Quantum Chemistry Program Exchange, which supplies any of several hundred standard computational chemistry programs for a modest fee (QCPE, 1990), has catalysed the growth of the field.

The commercially available computer programs CONCORD and COBRA generate 3-D structures from 2-D input. CONCORD generates, in a few seconds, one 3-D molecular structure (Pearlman *et al.*, 1990). COBRA generates many low-energy conformations in approximately the same time per conformation (Leach *et al.*, 1990).

One uses a molecular modelling program to visualize and manipulate the information obtained from computational chemistry calculations. The displays are typically line drawings with dot surfaces. Physical properties of molecules, such as molecular or solvent-accessible surface, electrostatic potentials, molecular orbital coefficients, or partial atomic charge, can be displayed with colour coding, contours, or colour-coded 3-D contours. Some programs organize modelling data in tables or add it to databases. Most can generate 3-D structures from 2-D input such as sketching or downloading from a chemical information database, calculate and minimize the steric energy of structures, examine the conformations available to a compound, and export data to computational chemistry programs.

Molecular modelling systems are advancing rapidly. The number of atoms and the complexity of the display constantly increases. The programs are more natural to use with spaceball or data glove to control the rotation and translation of the molecule. Stereoscopic viewing devices have also improved in recent years. Lastly, high-quality copies of the screen images are easily produced.

Specific computational chemistry and molecular modelling techniques aid pharmacophore mapping (Martin, 1992). For example, one expects to superimpose molecules by overlapping essential atoms, molecular features, or proposed receptor binding points. Molecular modelling software for pharmacophore mapping thus has facilities to calculate the location of preferred binding points from structures and to use these points in further manipulation. Versatile superposition functions are also essential. The space occupied by the ensemble of active compounds represents a model of the shape of the macromolecular binding site. If there are inactive molecules that can adopt the required conformation, the extra volume that they occupy represents space occupied by the macromolecule, i.e. space that is forbidden to ligands. Therefore, pharmacophore mapping software can display the union or intersection volume of several structures and the unique volume of one set of structures compared to another set.

For pharmacophore mapping one must decide both the superposition rule and the bioactive conformation for the compounds. 2-D structure–activity relationships provide the data for making these decisions. Specifically, the superposition rule may be clear if analogue studies have established the groups essential for a bioactivity. Also, just one conformationally constrained analogue may establish the bioactive conformations of many molecules in a set. Systematic search or distance geometry methods generate the consistent conformations of a set of molecules. For example, we synthesized a set of conformationally constrained analogues of diverse structure to establish the bioactive conformation of the selective D_1 agonist (1) (Martin *et al.*, 1991). It is that shown in Fig. 20.3. This led to the design, forecast and synthesis of (2) as a potent and selective D_1 agonist.

From similar studies on α_1, α_2, D_1 and D_2 agonists we concluded that dopamine has the same bioactive conformation at these four receptors (Martin *et al.*, 1987).

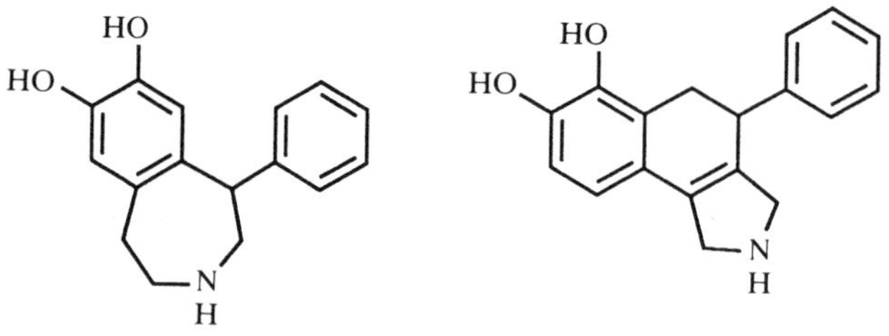

(1) (2)

Figure 20.4 shows the shapes of the binding pockets deduced from these studies. Later, molecular biologists have shown that these receptors appear to be homologous proteins that differ only in the details of their 3-D structure (Sunahara, 1990).

In another study we extended a known pharmacophore model for plant growth regulators (Katekar *et al.*, 1987) to include more diverse structures. Figure 20.5 shows the structures of representative inhibitors and their proposed pharmacophore points and bioactive orientation.

The search for the bioactive conformation from structure–activity information is not always successful. We could not establish the bioactive conformation of erythromycin analogues even with the aid of nuclear magnetic resonance (NMR) and X-ray crystal structures of several erythromycin analogues (Kim & Martin, 1989a). The difficulty is not with the conformational search methods nor energy minimization, since molecular modelling correctly forecasted the crystal structures of several erythromycin analogues. For example, the minimum energy conformation of A-23788 is essentially identical to the later determined X-ray crystal structure as shown in Fig. 20.6. We could not establish the bioactive conformation because every active erythromycin analogue can assume all three low-energy conformations shown in Fig. 20.7. Because of the synthetic challenge, no analogue that eliminates even one of these conformations has been tested.

The many molecules, conformations and superpositions to be integrated with bioactivity information led us to organize our work with chemical information tools (Martin *et al.*, 1988). Specifically, we store the 3-D coordinates of the molecules in the same database as we store the observed, fit and forecast biological properties of the molecules. The sections on SWAMI and ALADDIN, below, will describe this. We have also prepared a chemical information database of parameters for

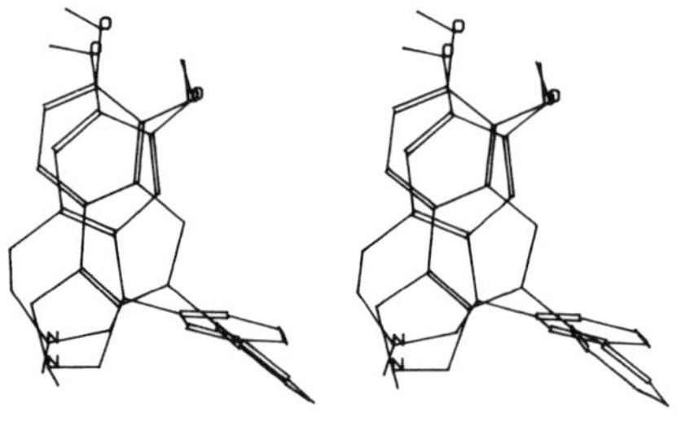

(1) (2)

Figure 20.3. A superposition of the bioactive conformation of (1) as established by synthesis of a variety of analogues and of (2), the designed compound.

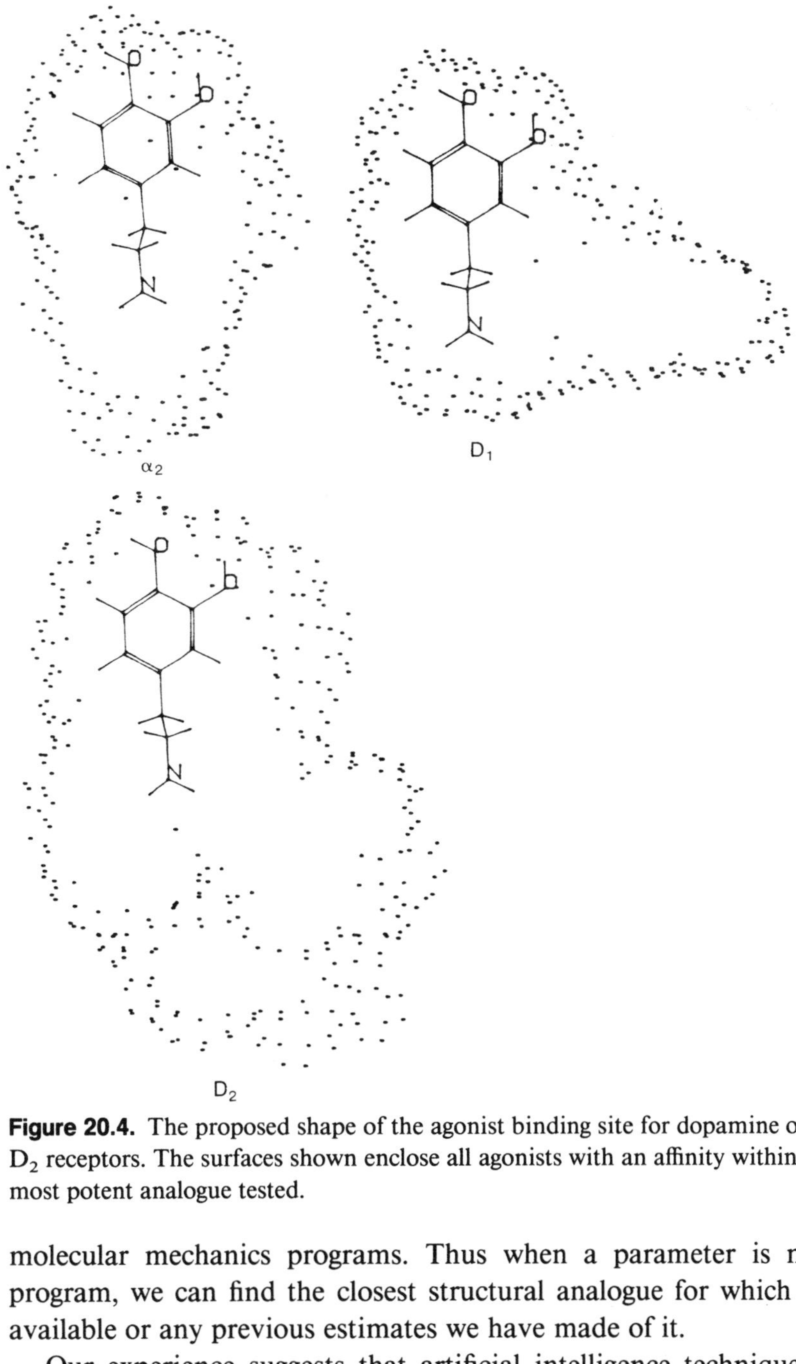

Figure 20.4. The proposed shape of the agonist binding site for dopamine on the α_2, D_1 and D_2 receptors. The surfaces shown enclose all agonists with an affinity within 100-fold of the most potent analogue tested.

molecular mechanics programs. Thus when a parameter is missing from the program, we can find the closest structural analogue for which the parameter is available or any previous estimates we have made of it.

Our experience suggests that artificial intelligence techniques would be very helpful to anyone starting a pharmacophore mapping study. One major problem is to pick the compounds to be modelled. There might be 1000 in-house structures available or dozens of literature structure–activity studies. Which contain the information likely to yield a solution? Does it appear there is enough information or must more be generated? What structures would provide the needed information? Beyond this, it takes much skill and experience to choose the proper methods to use with a set of data and then to use them properly. Should ensemble distance geometry, template forcing, or systematic search be used to find the bioactive conformations? How will the structures be optimized? Is this problem related to

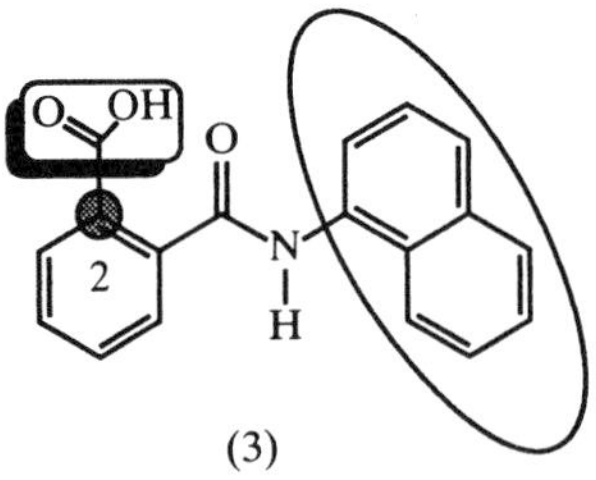

(3)

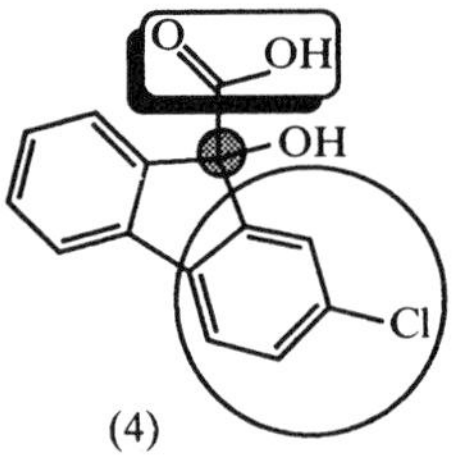

(4)

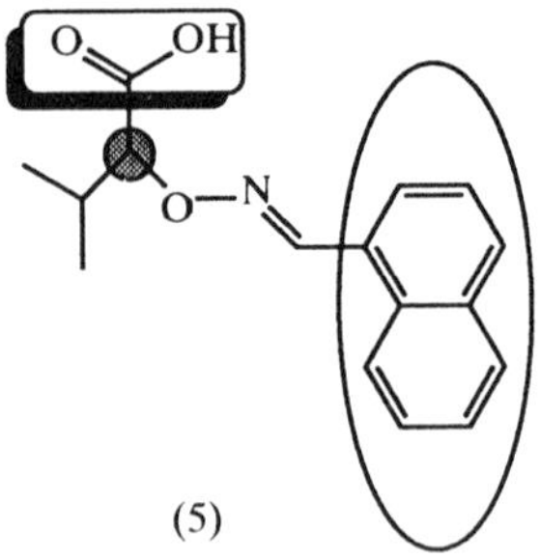

(5)

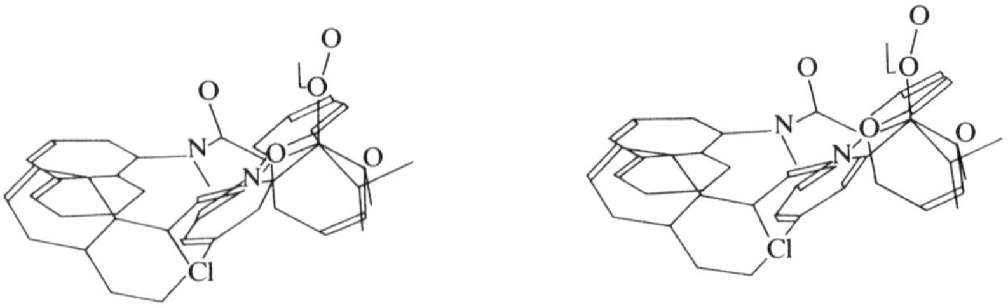

Figure 20.5. A pharmacophore model for auxin transport inhibitors. The proposed pharmacophore points of a standard inhibitor, compound 1, and those of two recently discovered inhibitors, 2 and 3. At the bottom of the figure is a stereoview of the proposed superposition of these compounds.

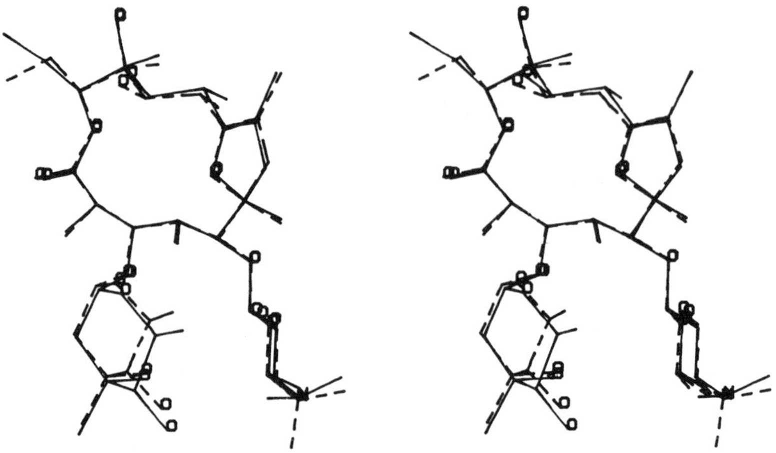

Figure 20.6. A superposition of the forecast and observed crystal conformation of an erythromycin analogue.

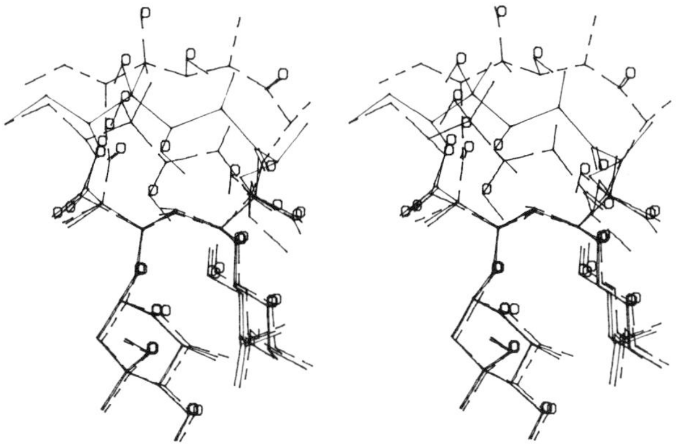

Figure 20.7. Stereo pair showing the three low-energy conformations of erythromycin.

one already solved? Is QSAR a better approach for this information? The variety and complexity of the software needed for pharmacophore mapping make the idea of an expert guide attractive.

6 CoMFA—the statiscal analysis of the relationship between bioactivity and 3-D properties of molecules to produce quantitative pharmacophore maps

Clearly, QSAR practitioners would like to be able to include 3-D properties in their equations and molecular modellers would like to make quantitative forecasts of potency. Of the various approaches to 3-D QSAR, we think that CoMFA (Cramer *et al.*, 1988) is superior in ease and unambiguity of application and in its statistical and theoretical basis.

CoMFA identifies the quantitative influence on potency of specific chemical features at particular regions in space. Input to CoMFA is the relative potency of molecules and their superimposed bioactive conformations. For each molecule one uses probe atoms to calculate the steric, electrostatic and (in our implementation) hydrogen-bonding interaction energies at hundreds of points on a lattice surrounding the molecules. PLS identifies the relationship between these energies and biological potency. CoMFA equations directly forecast the affinity of proposed molecules. Additionally, the coefficients of the energy values in a CoMFA can be displayed as contours to aid new compound design. CoMFA has the advantage over QSAR because only a limited number of properties is available for exploration.

The basic assumption of CoMFA is that the molecules interact with the target biomolecule in a manner reflected in the 3-D structures as superimposed. For example, one implication of a CoMFA fit is that if one were to superimpose the ligand–biomolecule complexes, the location of protein atoms would be constant and the ligand locations would be as modelled in the pharmacophore mapping. However, it is also consistent with a CoMFA if the protein moves in a way that is correlated with the 3-D structure of the ligand. To date these assumptions have not been experimentally verified in any series.

Our implementation of CoMFA differs from the one reported by Cramer *et al.* (1988) in that we use a program that includes explicit functions for hydrogen bonding (Goodford, 1985; Boobbyer *et al.*, 1989).

Traditional QSARs can be re-cast as CoMFA relationships. For example, CoMFA electrostatic properties reproduce the traditional QSAR electronic parameter σ, i.e the effect of substituents on the pK_a values of benzoic acids (Kim & Martin, 1991a,c). The CoMFA equation predicts the pK_a values of 21 of 23 additional compounds to within 0.27 kcal/mole. The pK_a values of clonidine-like imidazolines and 2-substituted and 1-methyl-2-substituted imidazoles are also well described by CoMFA electrostatic properties. Also, the formation constants of the hydrogen bond between acyclic ethers and *p*-chlorophenol (Bellon *et al.*, 1980) were well described by the GRID hydrogen-bond donor probe.

In QSAR, the substituent effect on the relative rate of acid-catalysed hydrolysis of ethyl esters, E_s, has been used as a measure of the steric effect of a substituent. CoMFA analysis showed that both steric and electrostatic properties significantly contribute to E_s, (Fig. 20.8) (Kim & Martin, 1991b). CoMFA steric energies also reproduce the traditional QSAR bulk parameters, molar volume and surface area. Additionally, the binding affinities of glycosides to concanavalin-A, the molar sweetening potencies of nitroanilines and cyanoanilines, and the inhibitory potencies of alkylphosphonic acid esters on cholinesterase (biological properties that have been reported to be correlated with molar refractivity, a Sterimol parameter, or E_s, respectively) are all well fit and cross-validated with CoMFA steric energies (Kim, 1992a).

CoMFA also fits, with good cross-validation, data in which QSAR shows a linear correlation of biological potency with log P (Kim, 1992a). For example, we found a good CoMFA fit using the water probe for the following: the potencies of aliphatic alcohols to inhibit the effect of acetylcholine on guinea-pig ileum, the concentration of phenols at which cytochrome P450 is half reduced to P420 at 10°C, and the K_m at 25°C, for the papain-catalysed hydrolysis of phenyl hippurates. Each series is

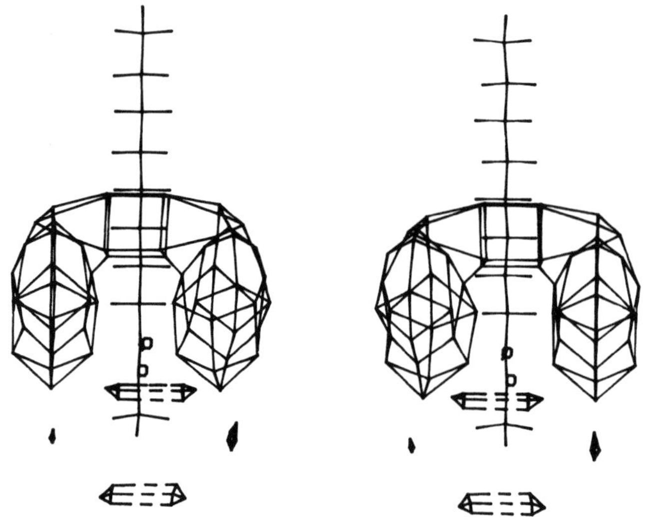

Figure 20.8. The regions in space which, if occupied, retard the rate of hydrolysis of esters (——) and those at which electron-donating substituents increase the rate of hydrolysis (----). More than 99% of the variance is accounted for by the CoMFA properties.

better fit with a water probe that has both hydrogen bonding and steric properties than by a probe with no hydrogen-bonding properties. The CoMFA fit is never worse than traditional QSAR and sometimes better.

CoMFA also fits data for which the traditional QSAR is parabolic or bilinear (Kim, 1992b). Examples are the bactericidal potencies of alkylnikethamide chlorides or bromoalkylcarboxylic acids (Fig. 20.9) and the acute intravenous toxic concentrations of aliphatic alkanes in 50 or 100% of male mice after bolus injection. The standard deviations and correlation coefficients from both the fitted models and the cross-validations are excellent and usually superior to the corresponding parabolic or bilinear models. The optimum log *P* value can be estimated from the coefficient contour map. We expect that non-linear steric or electrostatic effects should also be handled by CoMFA.

Three data sets in which traditional QSAR showed the contribution of both hydrophobicity and electrostatic effects to potency were also studied (K.H. Kim, unpublished results): the local anaesthetic action of diethylaminoethyl benzoates, the enzymatic hydrolysis of 4-substituted phenyl β-D-glucosides, and the inhibition of carbonic anhydrase by substituted benzene sulphonamides. We observed that the extraction of PLS components from a matrix of all descriptors simultaneously may not be the ideal method for such mixed descriptor data. Nevertheless, CoMFA analysis showed a clear influence of both hydrophobic and electronic properties on potency. In all three examples the correlations were superior to the literature QSARs.

We used a second method of validating CoMFA. That is, we asked whether steric contours found by CoMFA from the structure–activity relationships of ligands reflect the true shape of the binding site on the protein. For this purpose, we studied a series of quinones and quinone analogues that had been tested as inhibitors of electron transport in the bacterial photoreaction centre (L. Dutton, private commun.), a protein of known 3-D structure (M. Schiffer, 1988, unpublished

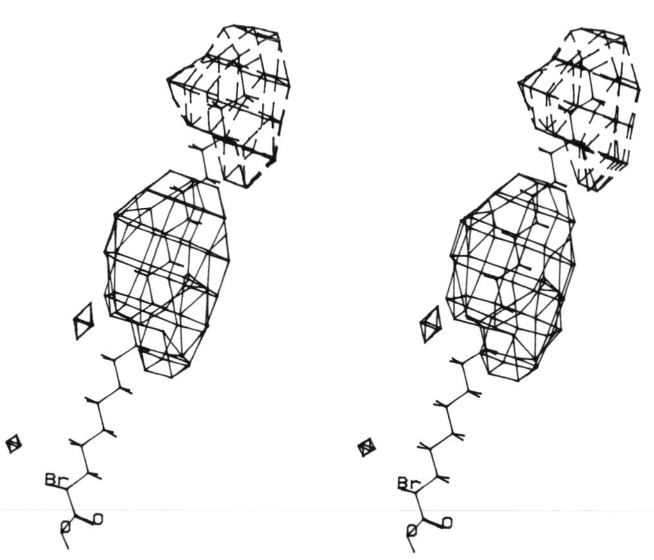

Figure 20.9. The CoMFA contours that result from using a water probe to investigate the bactericidal potencies of bromoalkylcarboxylic acids.

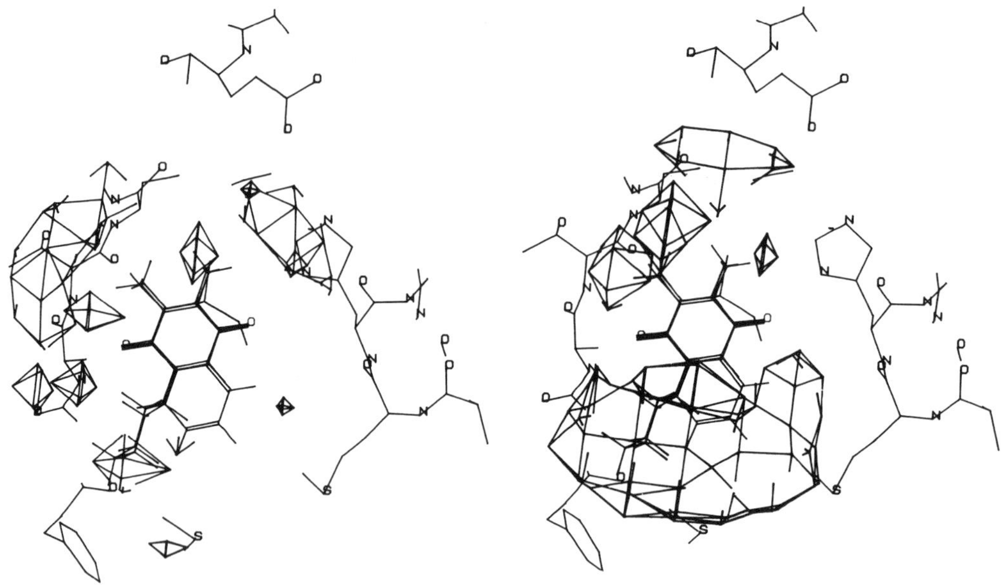

Figure 20.10. A comparison of the CoMFA steric contours and the crystallographic structure of the reaction-centre binding site for ubiquinone. (Left) the negative steric contours, regions in space which if occupied by the quinone decrease affinity. (Right) the positive steric contours. Note that the positive steric contours do not contact the protein atoms whereas the negative ones do.

observations). The methyl probe produced a significant CoMFA (G. Greco & Y.C. Martin, unpublished results). More importantly, Fig. 20.10 shows that the CoMFA contours do reflect the shape of the binding site as established by protein crystallography.

The ultimate validation of such a method is its ability to forecast the potency of analogues not included in its derivation. Figure 20.11 shows the forecast and observed affinities of eight non-quinone analogues for the reaction-centre protein. Clearly, the CoMFA has excellent forecasting ability. Other examples include the correct predictions of the p*K*a values of the benzoic acids and clonidine analogues mentioned above and several sets of biological potencies from our laboratory.

Since running a CoMFA analysis is quite straightforward once one has the molecules superimposed, we expect that CoMFA will become a standard part of the pharmacophore mapping toolkit. As we continue to generate CoMFA relationships, we would like to store them in the same database as traditional QSARs. Then they could be used as discussed above to forecast biological properties of proposed or newly synthesized molecules.

We expect that increases in computer power will allow the theoretical medicinal chemist to tackle ever more complex problems and to treat problems more realistically by including solvent, for example. Additionally, the spin-off of the greater computer power will be an increased understanding of the complexity needed in any computation. It is possible that new strategies for simplification of the calculations might emerge as computational chemists gain more experience.

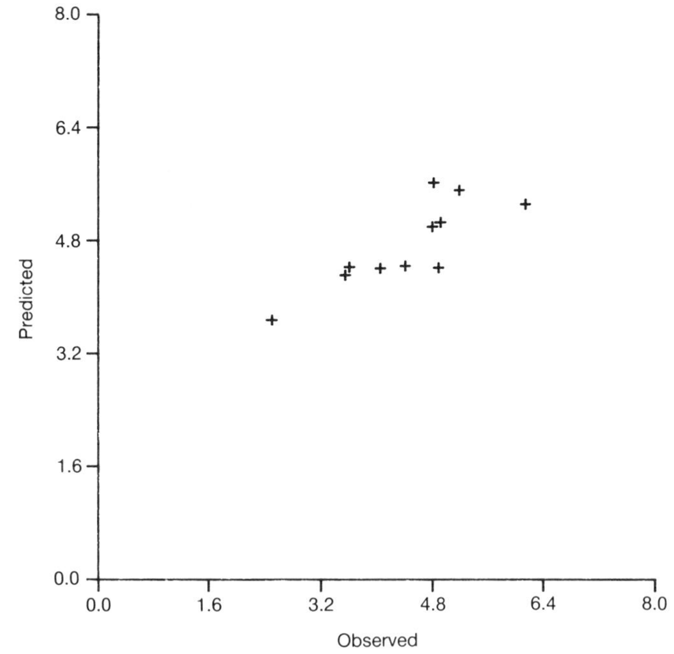

Figure 20.11. The forecast and observed affinities of compounds for the reaction-centre protein.

Advances in pharmacophore mapping will include advances in the concepts and strategies and in specific software. We expect that the great increase in the number of experimental 3-D structures of protein–ligand complexes will produce valuable information for mapping binding sites of macromolecules of unknown 3-D structure. For example, in pharmacophore mapping we implicitly assume the binding site is identical in all protein–ligand complexes with the various ligands of the series. We seldom superimpose to optimize shape overlap as well as binding-point overlap because we do not know what parts of the shape are relevant and which are not. Is this reasonable? How much will proteins move to accommodate a slightly bad-fitting ligand? How will such movements affect affinity? How high in energy can the structure of the bound ligand be? The experience with experimental data will spur refinement of pharmacophore mapping techniques and corresponding changes in the software.

One can model the 3-D structure of a protein from its sequence if one knows the 3-D structure of a homologous protein (Greer, 1990). These strategies will be more valuable as more 3-D structures of proteins are determined. Improvements in the methods to calculate protein energetics will lead to further improvements in the precision of prediction of 3-D protein structure. Additionally, the very active investigation of perturbation free-energy methods to predict the energetics of ligand–protein interaction will both refine the methods for calculating energetics and provide tools to forecast both the structure and the affinity of a new ligand–protein complex.

We expect that in the 21st century it will be more common for a medicinal chemist to design a molecule using some knowledge of the 3-D structure of the macromolecular target. This might remove the empiricism from the choice of a superposition rule and a bioactive conformation. A synergism between pharmacophore mapping and 3-D modelling of macromolecular structure from sequence is possible. For example, proposed 3-D details of the binding site should fit the 3-D structure–activity relationships of ligands as well as the amino acid sequence of the target and the 3-D structure of a homologous protein. Structure–activity information comes from experiment—any method that adds it to other information will benefit in accuracy. Additionally, CoMFA requires modest computer resources to accomplish forecasts within the precision of the biological data. These characteristics ensure that CoMFA will be used even when 3-D information on the target macromolecule is available.

7 SWAMI—the integration of molecular modelling and chemical information techniques to provide flexibility and power for the scientist

We have incorporated chemical information tools into molecular modelling with our unpublished program SWAMI (Structural Wisdom And Modelling Integration; J. DeLazzer, M.B. Bures & Y.C. Martin) with the result that molecular modelling is both easier and more powerful than in traditional systems. Although as far as we know these capabilities are unique to Abbott, we expect that commercial vendors will see the advantages of such a powerful system.

As noted above, we store the results of our molecular modelling work, such as 3-D structures, energies, partial atomic charges and notes on the calculations, in a chemical information database (Martin *et al.*, 1988). This database also contains the observed and forecast biological properties of the molecules. We include both literature and proprietary compounds. This organizes data that are otherwise scattered in many files and simplifies sharing information.

Storing 3-D coordinates in a database is not new (the Cambridge Structural Database (CSD, 1990) contains 3-D structures of ca. 85 000 small molecules). The integration of a 3-D database with pharmacophore mapping, CoMFA and molecular modelling is unique and extremely powerful. SWAMI can find compounds by their 2-D structure or by a variety of names including Abbott number. Once SWAMI finds a compound, it lists all conformations stored and identifies the proposed bioactive conformation. This makes it easy to retrieve for display only the conformation of interest. Also, SWAMI can easily retrieve all conformations of a molecule or all XYZ sets that formed a particular pharmacophore map or CoMFA analysis. Since we store all the low-energy conformations that result from a conformational energy search, we can quickly build the corresponding conformations of analogues. Additionally, since we identify the proposed bioactive conformations of molecules, we can quickly model the bioactive conformation of close analogues. Lastly, SWAMI can use database information to guide the display, for example to colour the molecules displayed by the value of their potencies.

The database can be searched in traditional ways. For example, we can search for potent molecules for which we have not proposed a bioactive conformation and additionally require that the molecules identified have a certain pharmacophore and less than some number of rotatable bonds.

Internally, SWAMI uses chemical information facilities (Weininger *et al*, 1989; Weininger & Weininger, 1990; Daylight, 1991) to perceive chemical structures from input coordinate sets. Figure 20.12 shows an example of how chemical information tools are used in SWAMI. As shown in Fig. 20.12, SWAMI can recognize atoms by their substructural environment. The result of this is that one can apply the same command to a variety of structures in one operation—an important time-saving automation. Furthermore, SWAMI can use substructural information for its own operations such as to identify the ring atoms for a polar map display of ring-torsion angles.

We would like to have every modelling group store their results in a similar database so that everyone could directly search and retrieve literature models. This database would be a unified repository for the results of molecular modelling and computational chemistry studies.

SWAMI is the first step in the development of an expert artificial intelligence system that would guide molecular modelling and pharmacophore mapping investigations. The ultimate system could help researchers pick the proper structure–activity data and computational chemistry method, start up computations to generate needed information or submit samples for experimental study, correct problems encountered during execution of calculations, interpret results, and design new molecular modelling studies. As with the ideal expert system for QSAR, an expert system for pharmacophore mapping would monitor a structure–activity database to detect when there is enough information to start a modelling study,

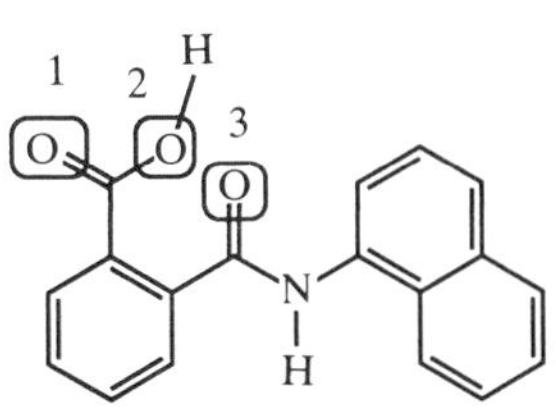

A

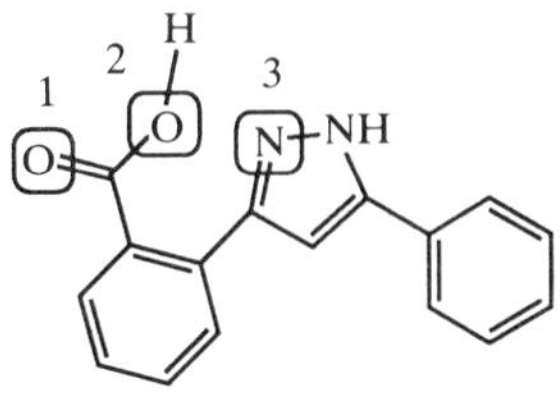

B

RMS A,B 1 O=C(O),O=C(O) 2 [O;H1]C=O, [O;H1]C=O 3 O=CN, [n;H0]c

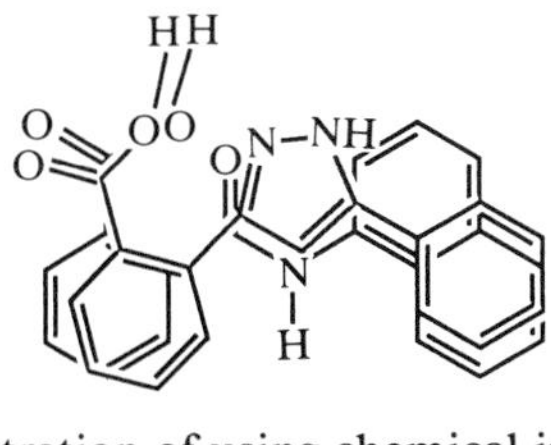

Figure 20.12. An illustration of using chemical information tools in SWAMI to superimpose two compounds from Figure 20.5. The arguments of the root mean squared (RMS) command are substructural descriptions of the atoms to be overlaid. (This contrasts with more conventional descriptions of atoms by an atom symbol and a number.) The superposition that results from this command is also shown.

suggest biological testing of certain compounds, and even design a few key analogues for synthesis and testing.

8 ALADDIN—3-D searching databases of small molecules to find new bioactivities in old molecules or to design new molecules

It is not easy for a human to design novel molecules that match a pharmacophore map or that bind to a macromolecule of known 3-D structure. We wrote ALADDIN to do this. ALADDIN searches 3-D databases to find molecules that match geometric criteria between atoms in a specified substructural environment (Van Drie *et al.*, 1989). Other substructural, steric and database requirements may also be specified. Finally, it can test whether the database molecules fit into the binding site. ALADDIN uses SWAMI databases and writes macros that allow the easy viewing of ALADDIN results with SWAMI. Since it functions as part of the Daylight Chemical Information System, full 2-D database facilities are also available to identify the compounds. Several other 3-D searching programs are available (Martin *et al.*, 1990); however, ALADDIN is unique since it also can automatically design new molecules that meet 3-D and substructural criteria (Martin, 1990; Martin and Van Drie, 1992).

To supplement our molecular modelling databases, we and others (Rusinko *et al.*, 1989) have used CONCORD to generate databases of 3-D structures of large and diverse sets of compounds. We routinely search 3-D databases of the Abbott collection of compounds and the *Fine Chemicals Directory* (*FCD*).

The first specification of an ALADDIN search is the substructural environment of the atoms of interest. One next tells ALADDIN the points, lines and planes to calculate from these atoms. The location of a point may be at an atom or lone pair, at the centre of mass of a substructure, or calculated from other points, lines and planes. Lastly, one specifies the needed distances, angles, or torsion angles between these points, lines or planes. For example, Fig. 20.13 is a scheme of the D_2

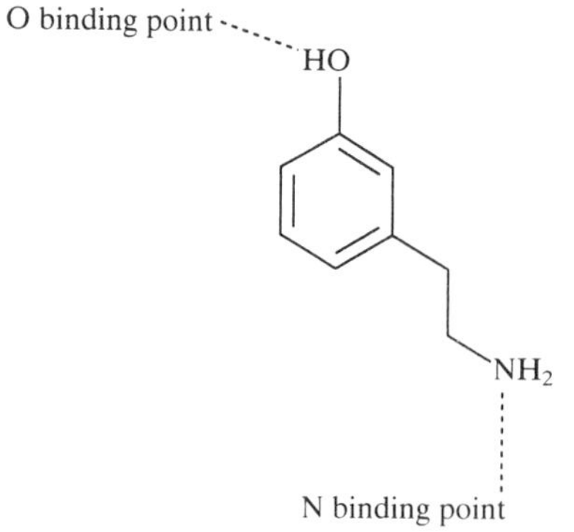

Figure 20.13. A geometric description of our model of the ligand-binding site on the D_2 dopaminergic receptor. The ligand binding point for the O is placed 2.4 Å from the O in the plane of the aromatic ring, in the direction shown along the O–H line. That for the N is placed along the N–lone-pair direction 2.4 Å from the N. The following distances apply to the model: O(H) to N, 5.8–7.5 Å; O binding point to N binding point, 8.4–10.0 Å; O to N binding point, 6.5–8.0 Å; N to O binding point, 7.5–9.5 Å. Additionally the torsion angle (O binding point)–O–N–(N binding point) is 85–130°.

pharmacophore used in our ALADDIN design of dopaminergics.

ALADDIN has identified compounds that have previously unsuspected biological properties. For example, from pharmacophore mapping of auxin transport inhibitors as described above and in Fig. 20.5, ALADDIN identified (6)–(9) as new inhibitors (Bures *et al.*, 1991). These compounds were shown to be actually active inhibitors.

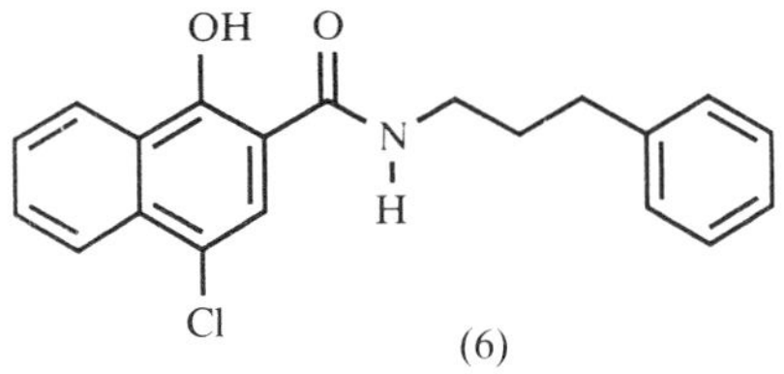

(6)

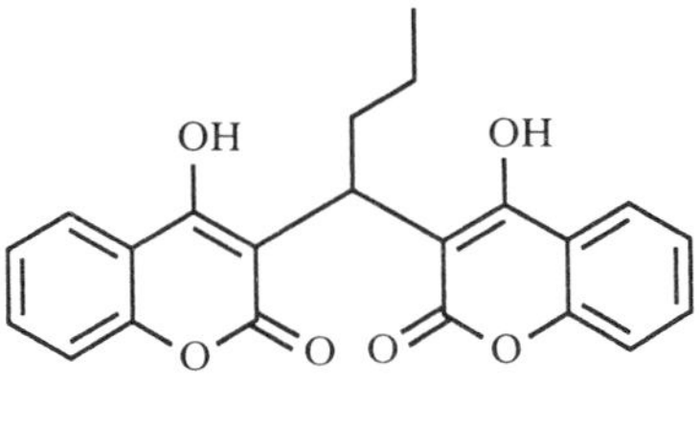

(7)

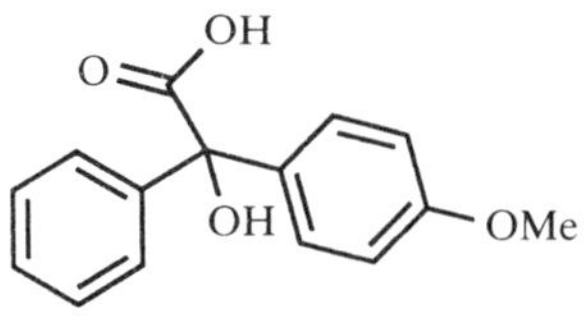

(8)

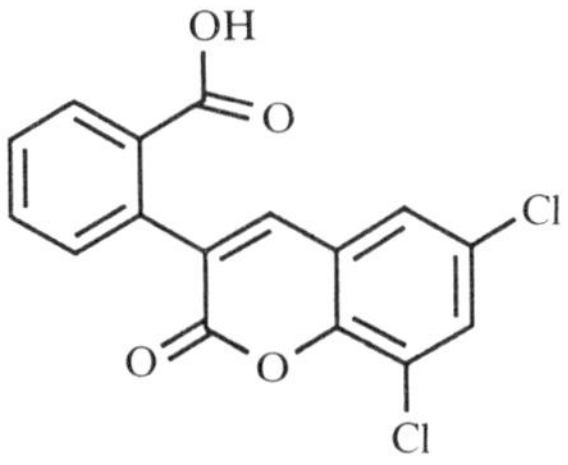

(9)

To design novel compounds with ALADDIN (Martin, 1990) one first searches for molecules that meet the geometric constraints without regard to the type of atom involved in the match. The next step is to change the database molecule into that to be synthesized by replacing the identified atoms with those needed for the pharmacophore. The MODSMI language does this (Martin & Van Drie, 1992). Also, atoms irrelevant to the 3-D fit are removed so that each compound has a unique backbone. Next, the 3-D structures of these modified molecules are generated with CONCORD. Lastly, a second ALADDIN test ensures that the designed molecules do truly meet the pharmacophore requirements. We then forecast the biological activities of both enantiomers of the designed compounds with CoMFA.

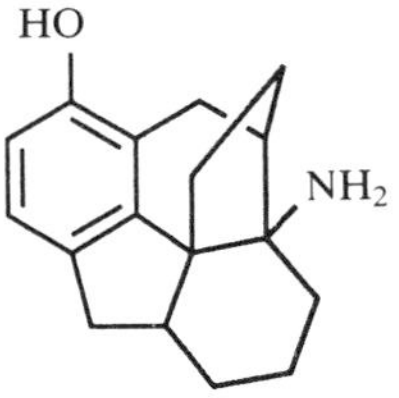

(10)

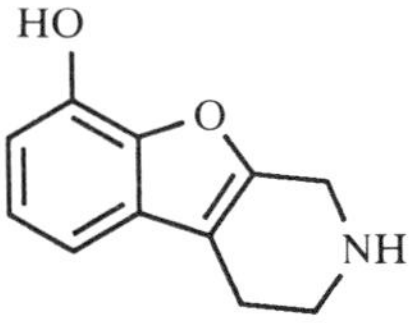

(11)

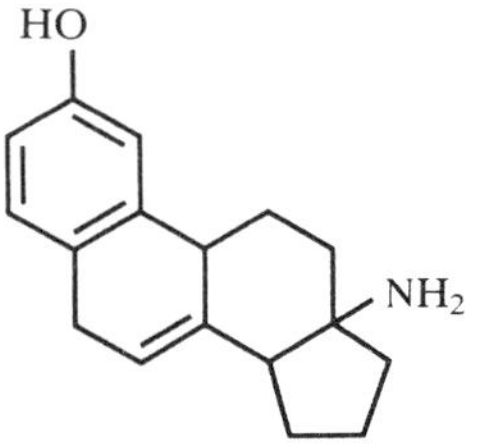

(12)

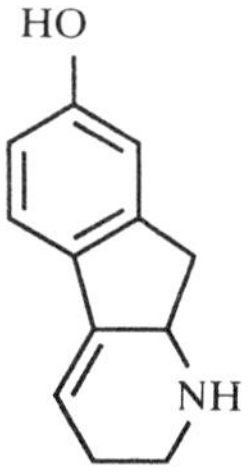

(13)

This strategy has generated a wide variety of compounds that meet our geometric criteria for dopaminergic activity (Martin, 1990). We searched three different databases to generate approximately 500 structurally unique molecules. Compounds (10)–(13) are examples of new compounds designed. As well as identifying eight or nine classes of known fused-ring phenolic dopaminergic compounds, the search identified 62 other classes of fused-ring compounds that match the pharmacophore. Since so many research groups have synthesized dopaminergic compounds and yet ALADDIN suggested so many unique compounds, we concluded that 3-D searching is useful in mature medicinal chemistry areas as well as unexplored ones. The low frequency of finding the same ring class more than once suggests that more searches will design other molecules. Such design does not need the building of special libraries of templates.

To use ALADDIN one must describe a pharmacophore in terms a computer can understand. We expect that such descriptions of pharmacophores will be incorporated into databases such as the QSAR database (Hansch, 1989). Since one can now quickly generate 3-D structures, a program that tests new molecules for matches to stored pharmacophore requirements would be possible.

Automated drug-design procedures are still in their infancy. One important improvement will be to incorporate CoMFA descriptions into the design criteria so that only potent molecules would be proposed. Integration with literature and patent searching and with computer design and evaluation of synthetic pathways would help ensure that only compounds that can be synthezised and patented would be suggested. Lastly, selective compounds would be identified by forecasting affinities of the compounds for any QSAR or other ALADDIN or CoMFA models for which they meet the substructural requirements.

9 What's in the future?

We are already witnessing an explosion in the number of known amino acid sequences of receptors and in the number of subtypes of some receptors. It is hard to predict how many subtypes of receptors there might be and what biological role they play. Medicinal chemistry design and synthesis of ligands to probe the biological function of these multiple subtypes will be an exciting contribution to biology.

Although computer tools design new compounds with forecast-desired biological activities, it is important to remember that these compounds must be synthesized and tested to assess the success of the design goals. Therefore, an essential aspect of any computer-aided molecular design strategy is the ability to evaluate the synthetic feasibility of the compounds proposed. Interaction between the synthetic and theoretical chemists is a start, but we feel that coupling the tools of computer-aided compound design and those of synthesis planning is an important goal.

While exciting advances have been made, medicinal chemistry is at the early stages of computer automation of its thought processes. The integration of computer methodologies is already observable in methodologies such as CLOGP, CoMFA, SWAMI and ALADDIN. Continued integration of computer tools will require that scientists and scientific programmers be broadly trained so that opportunities are recognized. Integration will happen more quickly when authors of individual programs recognize the importance of integration to others. For example, many chemistry programs now recognize the SMILES description of a 2-D chemical structure. Any of these can be added to a system such as SWAMI that already uses SMILES notation. However, if a new program uses an independently invented notation of 2-D chemical structure, then a conversion between the two notations would also need to be written. This might or might not be easy.

The integration of different computer tools means that scientists can more easily apply the new techniques to their problems. This new group of scientists using an 'old' method will not only benefit themselves but also they in turn will improve the old method by their particular experience and insights. Thus, both the donor and recipient gain by integration of tools.

On the other hand, as methodologies are integrated, it becomes more likely that the user is not an expert in using all of them. This increases the need for the computer program to make intelligent decisions for the user. While this can be done with default or usual values, a better system would make choices depending on the context of the modelling situation. Such artificial intelligence systems would ideally incorporate all the thinking that an expert (or better, a collection of experts) would apply to the situation. Artificial intelligence has been only modestly applied to medicinal chemistry interests: however, anyone who has used the CLOGP program appreciates how useful it is for one to have an expert system that will do complex operations. A special feature of the CLOGP program is the facility for users easily to change the rules temporarily, permanently, or for their own private use. This separation of the knowledge from the detailed programming makes the program powerful for users and yet maintains its integrity. There is great opportunity for expert systems in QSAR, molecular modelling, pharmacophore mapping and CoMFA. Expert systems will continue to grow in power as they become easier to program and as scientists become themselves more expert.

We expect that the future will see more extensive and accessible databases of medicinal chemistry information. We have noted the opportunity to store pharmacophore mapping, CoMFA and ALADDIN information in the same type of database just as it is now being used for QSAR equations. However, we need databases of structure–activity information that has not been successfully studied with computer tools. If we want to make analogues, we would like to know the chemical synthesis used and the scope of any patents or publications on the compounds. It is unclear whether such databases need to be maintained on one computer with special librarian functions. One vision of the future sees databases of the same type scattered over many computers with the contents of each database maintained by the person responsible for generating the information. Searching would be by some variant of current electronic mail techniques.

This review shows that exciting computer tools for medicinal chemists are now available and that the optimum use of these tools is only beginning.

References

Allinger, N.L., Yuh, Y.H. & Lii, J.-H. (1989). *J Am. Chem. Soc.*, **111**, 8551, 8566 & 8576.
Bellon, T., Taft, R.W. & Abboud, J.-L. M. (1980). *J. Org. Chem.*, **45**, 1166.
Boobbyer, D.N.A., Goodford, P.J., McWhinnie, P.M. & Wade, R.C. (1989). *J. Med. Chem.*, **32**, 1083.

Bures, M.G., Black-Schaefer, C. & Gardner, G. (1991), *J. Comput.-Aided Mol. Design*, **5**, 323–34.
Burkert, U. & Allinger, N.L. (1982). *Molecular Mechanics, ACS Monograph 177*, American Chemical Society, Washington D.C.
Berntsson, P. & Wold, S. (1986). *Quantitative Structure–Activity Relationships*, **5**, 45.
Clark, T. (1985). *A Handbook of Computational Chemistry*, John Wiley, New York.
Counts, R.W. (1987). *J. Comput.-Aided Mol. Design*, **1**, 95.
Craig, P.N. (1990). In: C. Hansch, P. G. Sames and J.B. Taylor (Series Eds.), C.A. Ramsden (Vol. Ed.), *Comprehensive Medicinal Chemistry*, p. 645, Pergamon, Oxford.
Cramer III, R.D., Patterson, D.E. & Bunce, J.D. (1988). *J. Am. Chem. Soc.*, **110**, 5959.
CSD (1990). Cambridge Structural Database, Cambridge Crystallographic Data Centre, University Chemical Laboratory, Lensfield Road, Cambridge CB2 1EW, UK.
Daylight Chemical Information Systems, Inc. (1991). 3951 Claremont St., Irvine, CA 92714, USA.
Enslein, K., Borgstedt, H.H., Tomb, M.E., Blank, B.W. & Hart, J.B. (1987). *Toxicol. Ind. Health*, **3**, 267.
FCD: The Fine Chemicals Directory is distributed as a MACCS database by Molecular Design, San Leandro.
Fujita, T. (1990). In: C. Hansch, P.G. Sammes and J. B. Taylor (Series Eds.,), C.A. Ramsden (Vol. Ed.), *Comprehensive Medicinal Chemistry*, p. 497, Pergamon, Oxford.
Fujita, T., Iwasa, J. & Hansch, C (1964). *J. 19m. Chem. Soc.*, **86**, 5175.
Gasteiger, J., Hutchings, M.G., Christoph, B., Gann, L., Hiller, C., Low, P., Marsili, M., Saller H. & Yuki, K. (1987). *Topics Curr. Chem.*, **137**, 19.
Goodford, P.J. (1985). *J. Med. Chem.*, **28**, 849.
Greer, J. (1990). *Proteins*, **7**, 317.
Hansch, C. (1989). In: J.L. Fauchere (Ed.), *QSAR: Quantitative Structure–Activity Relationships in Drug Design*, p. 23, Alan R. Liss, New York.
Hansch, C., Unger, S.H. & Forsythe, A.B. (1973). *J. Med. Chem.*, **16**, 1217.
Hansch, C., Maloney, P. P., Fujita, T. & Muir, R.M. (1962). *Nature* (*London*), **194**, 178.
Hopfinger, A.J. (1985). *J. Med. Chem.*, **28**, 1133.
Katekar, G.F., Winkler D.A. & Geissler, A.E. (1987). *Phytochemistry*, **26**, 2881.
Kim, K.-H. (1991). *Med. Chem. Res.*, **1**, 259.
Kim, K.-H. (1992a). *Quant. Struct.-Act. Relat.* In press.
Kim, K.-H (1992b). *Quant. Struct.-Act. Relat.* In press.
Kim, K.-H. & Martin, Y.C. (1989a). In: J.L. Fauchere (Ed.), *QSAR: Quantitative Structure–Activity Relationships in Drug Design*, p. 325, Alan R. Liss, New York.
Kim, K.-H. & Martin, Y.C. (1991a). *J. Org. Chem.*, **5b**, 2723.
Kim, K-H. & Martin, Y.C. (1991b). In: C. Silipo and A. Vittoria (Eds.), *QSAR: Rational Approaches on the Design of Bioactive Compounds: Proceedings of the 8th European Symposium, Sorrento (Napoli), Italy, 9–13 September 1990*, p. 151, Elsevier, Amsterdam.
Kim, K.-H. & Martin, Y.C. (1991c). *J. Med. Chem.*, **34**, 2056.
Kim, K.-H, Martin, Y.C., Otis, E. & Mao, J. (1989). *J. Med. Chem.*, **32**, 84.
Kubinyi, H. (1976). *Drug Res.*, **26**, 1991.
Kubinyi, H. (1977). *J. Med. Chem.*, **20**, 625.
Laird, E. R. & Jorgensen, W.L. (1990). *J. Chem. Inform. Comput. Sci.*, **30**, 458.
Leach, A.R., Dolata, D.P. & Prout, K. (1990). *J. Chem. Inform. Comput. Sci.*, **30**, 316.
Leo, A. (1990). In: C. Hansch, P.G. Sames and J.B. Taylor (Series Eds.), C.A. Ramsden (Vol. Ed.), *Comprehensive Medicinal Chemistry*, p. 295, Pergamon, Oxford.
Long, A.K., Rubeinstein, S.D. & Joncas, L. J. (1983). *Chem. Engin. News*, **61**(19), 22.
Martin, Y.C. (1978). *Quantitative Drug Design*, Marcel Dekker, New York.
Martin, Y.C. (1981). *J. Med. Chem.*, **24**, 229.
Martin, Y.C. (1990). *Tetrahedron Comput. Methodol.*, **3**, 15.
Martin, Y.C. (1991). *Methods Enzymol*, 587.
Martin, Y.C. (1992). *J. Med. Chem.*, **33**.

Martin, Y.C. & Hackbarth, J.J. (1976). *J. Med. Chem.*, **19**, 1033.
Martin, Y.C. & Hackbarth, J.J. (1977). In: B. Kowalski (Ed.), *Chemometrics: Theory and Application*, p. 153, American Chemical Society, Washington D.C.
Martin, Y.C. & Panas, H.N. (1979). *J. Med. Chem.*, **22**, 784.
Martin, Y.C. & Van Drie, J.H. (1992). In: W. Warr (Ed.), *Proceedings of the 2nd International Conference on Chemical Information, Noordwijkerhout 1990.* In press.
Martin, Y.C., Bures, M.G. & Willett, P. (1990). In: K.B. Lipkowitz and D.B. Boyd (Eds.), *Reviews in Computational Chemistry*, VCH Publishers, New York.
Martin, Y.C., Jarboe, C.H., Krause, R.A., Lynn, K.R., Dunnigan, D. & Holland, J.B. (1973). *J. Med. Chem.*, **16**, 147.
Martin, Y.C. Danaher, E.B., Weininger, A.M. & Weininger, D. (1986). *Abstracts for Molecular Graphics Society Meeting.*
Martin, Y.C., Danaher, E.B., Kauslauskas, L. & Kim, K.-H. (1987). In: M.J. Rand and C. Raper (Eds.), *Pharmacology*, p. 611, Elsevier, Amsterdam.
Martin, Y.C., Danaher E.B., May, C.S. & Weininger D. (1988). *Comput. Aided Mol. Design*, **2**, 15.
Martin, Y.C., Kebabian, J.W., MacKenzie, R. & Schoenleber, R. (1991). In: C. Silipo and A. Vittoria (Eds.), *QSAR: Rational Approaches on the Design of Bioactive Compounds: Proceedings of the 8th European Symposium, Sorrento (Napoli), Italy, 9–13 September 1990*, p. 469, Elsevier, Amsterdam.
Moock, T.E., Grier, D.L., Hounshell, W.D., Grethe, G., Cronin, K., Nourse, J.G. & Theodosiou, J. (1988). *Tetrahedron Comput. Methodol.*, **1**, 117.
O'Donnell, T.J., Rao, S.N., Martin, Y.C. & Eccles, B. In preparation.
PDB (1990). Brookhaven Protein Databank, Brookhaven National Laboratory, Upton, New York 11973, USA.
Pearlman, R.S., Ruskinko III, A., Skell, J.M., Balducci, R. & McGarity, C.M. (1990). CONCORD, distributed by Tripos Associates, Inc., 1699 S. Hanley Road, Suite 303, St Louis, Missouri 63144, USA.
Quantum Chemical Program Exchange (1990). Department of Chemistry, Indiana University, Bloomington, Indiana 47405, USA.
Ruskinl III, A., Sheridan, R.P., Nilakantan, R., Haraki, K.S., Bauman, N. & Venkataraghavan, R. (1989). *J. Chem. Inform. Comput. Sci*, **29**, 251.
Smith, R.N., Hansch C., Kim, K.-H., Omiya, B., Fukumura, G., Selassie, C.D., Jow, P.Y.C., Blaney, J.M. & Langridge, R. (1982). *Arch. Biochem. Biophys.*, **215**, 319.
Summers, J. B. & Kim, K.-H., Mazdiyasni, H., Holms, J.H., Ratajzcyk, J.D., Stewart, A.O., Dyer, R.D. & Carter, G.W. (1990). *J. Med. Chem.*, **33**, 992.
Sunahara, R.K., Niznik, H.B., Weiner, D.M., Stormann, T.M., Brann, M.R., Kennedy, J.L., Gelernter, J.E., Roxmahel, R., Yang, Y., Israel, Y., Seeman, P. & O'Dowd, B.F. (1990). *Nature (London)*, **347**, 80.
Van Drie, J. H., Weininger, D. & Martin Y.C., (1989). *J. Comput.-Aided Mol. Design*, **3**, 225.
Weininger, D. & Weininger, J.L. (1990). In: C. Hansch, P.G. Sammes and J.B. Taylor (Series Eds.), C.A. Ramsden (Vol. Ed.), *Comprehensive Medicinal Chemistry*, p. 59, Pergamon, Oxford.
Weininger, D., Weininger, A. & Weininger, J.L. (1989). *J. Chem. Inform. Comput. Sci.*, **29**, 97.
Willett, P. (1987). *Similarity and Clustering in Chemistry Information Systems*, Research Studies Press, Letchworth, Hertfordshire, England.
Wold, S., Albano, C., Dunn III, W.J., Edlund, U., Esbensen, K., Geladi, K., Hellberg, S., Johansson, E., Lindberg, W. & Sjostrom, M. (1984). In: B.B. Kowalski (Ed.), *Chemometrics: Mathematics and Statistics in Chemistry, NATO ASI Series C, 138*, p. 4, D. Reidel, Dordrecht, Holland.

Part 4
Bioavailability Manipulations
Prodrugs

21 Trends in Design of Prodrugs for Improved Drug Delivery

H. BUNDGAARD

The Royal Danish School of Pharmacy, Department of Pharmaceutical Chemistry, 2 Universitetsparken, DK-2100 Copenhagen, Denmark

1 Rationale and applications of the prodrug concept

Prodrug design comprises an area of drug research that is concerned with the optimization of drug delivery. A prodrug is a pharmacologically inactive derivative of a parent drug molecule that requires spontaneous or enzymatic transformation within the body in order to release the active drug, and that has improved delivery properties over the parent drug molecule (Albert, 1958).

A molecule with optimal structural configuration and physico-chemical properties for eliciting the desired therapeutic response at its target site does not necessarily possess the best molecular form and properties for delivery to its point of ultimate action. Usually, only a minor fraction of doses administered reach the target area and since most agents interact with non-target sites as well, an inefficient delivery may result in undesirable side-effects. This fact of differences in transport and *in situ* effect characteristics for many drug molecules is the basic reason why bioreversible chemical derivatization of drugs, i.e. prodrug formation, is a means by which a substantial improvement in the overall efficacy of drugs can often be achieved.

Prodrugs are designed to overcome pharmaceutically and/or pharmacokinetically based problems associated with the parent drug molecule that would otherwise limit the clinical usefulness of the drug. The prodrug approach can be illustrated as shown in Fig. 21.1. The usefulness of a drug molecule is limited by its suboptimal physico-chemical properties, e.g. it may show poor biomembrane permeability. By attachment of a pro-moiety to the molecule or otherwise modifying the compound, a prodrug is formed that overcomes the barrier for the drug's usefulness. Once past the barrier, the prodrug is reverted to the parent compound

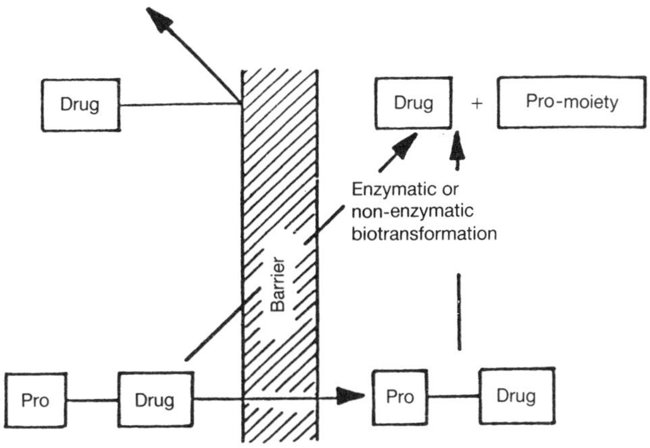

Figure 21.1. Schematic illustration of the prodrug approach.

by a postbarrier enzymatic or non-enzymatic process. Prodrug formation can thus be considered as conferring a transient chemical cover to alter or eliminate undesirable properties of the parent drug molecule.

A number of barriers may limit the clinical usefulness of a drug. In the pharmaceutical phase, i.e. the incorporation of a potential drug entity into a practically useful drug-delivery system or dosage form, the barriers may be presented by formulation problems due to the physico-chemical properties of the drug, such as poor water solubility, and by aesthetic properties of the drug, such as taste and tissue irritation.

In the pharmacokinetic phase, i.e. the absorption, distribution, metabolism and excretion of the drug, major barriers which may limit the usefulness of a drug are:

1 Incomplete absorption of the drug across biological membranes such as the gastrointestinal mucosal cells and the blood–brain barrier.

2 Incomplete systemic bioavailability of a drug due to presystemic metabolism (first-pass metabolism).

3 Too rapid absorption or excretion of the drug when a longer duration of action is desired.

4 Toxicity problems related to local irritation or distribution into tissues other than the desired target organ.

5 Poor site specificity of the drug.

The design of prodrugs in a rational manner requires that the underlying causes which necessitate or stimulate the use of the prodrug approach be defined and clearly understood. It may then be possible to identify the means by which the difficulties can be overcome. The rational design of prodrugs can thus be divided into three basic steps:

1 Identification of the drug-delivery problem.

2 Identification of the physico-chemical properties required for optimal delivery.

3 Selection of a prodrug derivative that has the proper physico-chemical properties and that will be cleaved in the desired biological compartment.

In the past, the prodrug approach has been successfully applied to a wide variety of drugs. Most of the applications have involved:

- enhancement of bioavailability and passage through various biological barriers;
- increased duration of pharmacological effects;
- increased site-specific delivery;
- decreased toxicity and adverse reactions;
- improvement of organoleptic properties;
- improvement of stability and solubility properties.

The achievements attained have been described in a number of comprehensive reviews or monographs (Sinkula & Yalkowsky, 1975; Stella, 1975; Roche, 1977, 1987; Yalkowsky & Morozowich, 1980; Notari, 1981; Wermuth, 1981, 1984; Bundgaard, 1985a, 1989b; Stella *et al.*, 1985; Lee & Li, 1989; Sloan, 1989; Balant *et al.*, 1990; Lee & Bundgaard, 1992). The prodrug approach to optimization of drug delivery has undergone substantial expansion over the past two decades and more than 50 chemical entities are now used clinically in the form of prodrugs.

Here some views on future developments in this field are presented and a number of challenges and possibilities for prodrug research are indicated. Based on recent significant achievements, the following areas should progress substantially in the next decade: improvement of drug bioavailability, site-specific delivery or targeting via prodrugs, delivery of peptide drugs using the prodrug approach and the development of novel chemical approaches in prodrug design.

2 Bioavailability

2.1 *Absorption*

The most important factors in drug absorption occurring by passive diffusion are aqueous solubility and lipophilicity. Drugs that are too polar or hydrophilic often exhibit poor transport properties, whereas those that are too non-polar or lipophilic frequently have low bioavailabilities because of their poor aqueous solubility and dissolution characteristics. Since the prodrug principle involves transient modification of these properties, it is readily evident that this principle is perfectly suitable to improve drug absorption through epithelial tissue. Numerous examples of the use of prodrug derivatives to improve the oral, rectal, ocular or dermal delivery of various drugs can be found in the reviews mentioned above. Without doubt, this area of application of prodrugs will continue to be of substantial importance. The major limitation in the application of the prodrug approach to improve bioavailability is often of a chemical character, i.e. the lack of a bioreversible derivative type applicable to the given drug molecule. Thus, attempts to improve the corneal absorption of carbonic anhydrase inhibitors like acetazolamide by the prodrug approach have hitherto not met with success due to the difficulty in obtaining optimal bioreversible derivatives for the primary sulphonamide moiety occurring in these compounds (Larsen & Bundgaard, 1987, 1989). Similarly, the problem of the very poor peroral absorption of several hydrophilic phosphorus compounds such as inositol triphosphates and bisphosphonates can only be solved by the prodrug approach if suitable bioreversible derivatives for the phosphate or phosphonate group are developed.

It is also interesting to note that although the antiherpetic agent acyclovir is widely used dermally, its physico-chemical properties, i.e. poor aqueous solubility and lipophilicity, greatly limit efficient access of the drug to its site of action in the skin. The prodrug approach should theoretically offer an ideal solution to this problem but, again, the limitation may be the lack of a suitable bioreversible derivatization principle for this molecule. The same applies to both the oral and dermal delivery of the analogous ganciclovir.

Another aspect of prodrugs for improved bioavailability that should receive increased attention in the future is the problem of presystemic metabolism.

2.2 *Prevention of first-pass metabolism*

Several drugs are efficiently absorbed from the gastrointestinal tract, but show limited systemic bioavailability due to presystemic (first-pass) metabolism or inactivation before reaching the systemic circulation. This metabolism can occur in the intestinal lumen, at the brush border of the intestinal cells, in the mucosal cells lining the gastrointestinal tract or in the liver. In addition to decreasing the percentage of dose reaching its intended site of action, extensive first-pass metabolism often results in significant variability in bioavailability, necessitating careful monitoring of drug blood levels in the patient. Thus, if first-pass metabolism of a certain drug in individuals varies between 85 and 95%, there will be a threefold difference in the amount of available drug.

A major class of drugs undergoing extensive first-pass metabolism is those containing phenolic hydroxyl groups. The rapid inactivation of these drugs (e.g. salicylamide, morphine, isoprenaline, fenoldopam, dopamine, L-dopa, naloxone, nalbuphine, naltrexone and β-oestradiol) is mainly due to sulphation, glucuronidation and methylation of the phenolic moieties, the conjugation reactions being catalysed by enzymes present in the gut and liver.

The prodrug approach can sometimes be very useful to reduce first-pass metabolism. Most attempts performed in the past have been concerned with phenolic drugs. The traditional approach has been to mask the metabolizable moiety, i.e. to derivatize the phenolic group to yield an ester prodrug. This approach will only be useful if the prodrug-to-drug conversion occurs mainly in an organ other than the intestine or liver. If the demasking of the protective group occurs in the intestinal wall or liver, the active parent drug will subsequently be metabolized within the same organ. In fact, an increased extent of first-pass metabolism of the diisobutyryl ester (ibuterol, (2)) of terbutaline (1) as compared with the parent drug has been observed (Hörnblad *et al.*, 1976). Apparently the more lipophilic prodrug ester enters the lipoidal microsomal media in the liver to a greater extent than the parent drug, and is subsequently hydrolysed and conjugated.

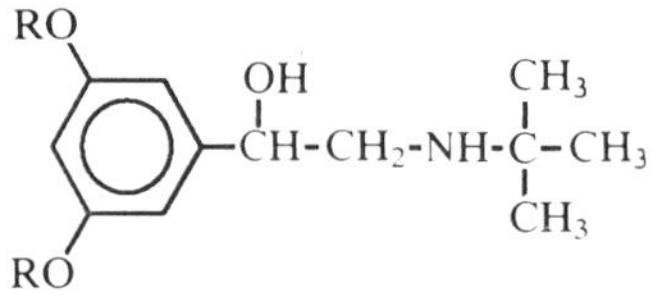

(1) R = H

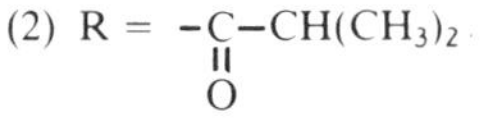

Nevertheless, a number of recent examples show the potential usefulness of protecting the metabolizable phenolic moiety in a bioreversible derivative. Thus, anthranilate and acetylsalicylate (or salicylate) esters of nalbuphine (Aungst *et al.*, 1987), naltrexone and β-oestradiol have been found to result in greatly increased systemic bioavailability of the parent drugs following oral administration in dogs due to depressed first-pass metabolism. For example, the anthranilate ester (4) and the salicylate ester (5) of naltrexone (3) resulted in a bioavailability of the parent drug of 49 and 31%, respectively. Naltrexone itself had a bioavailability of only 1% whereas the unsubstituted benzoate ester (6) did not improve the bioavailability.

Hydrolysis data indicated that the esters (4) and (5) are more stable toward enzymatic hydrolysis than the benzoate ester, indicating that these esters survive presystemic hydrolysis to a greater extent (Hussain *et al.*, 1987; Hussain & Shefter, 1988). In the case of β-oestradiol, its oral bioavailability after administration to dogs of the 3-anthranilate and 3-acetylsalicylate ester was 5- and 17-fold higher, respectively, than after oral β-oestradiol (Hussain *et al.*, 1988). Whether such esters may be generally useful to depress first-pass metabolism of various phenolic drugs is not known as yet. One should also be aware of potential species differences. Thus, recent studies in this laboratory have shown that the anthranilate and acetylsalicylate esters of β-oestradiol do not depress first-pass metabolism in rats as opposed to the findings reported with dogs (Lokind *et al.*, 1991).

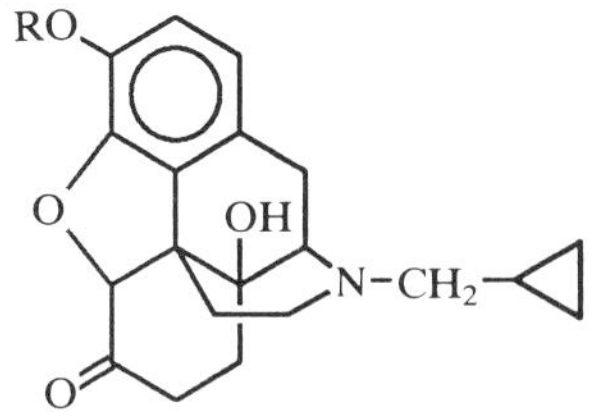

(3) R = H

(4) R = 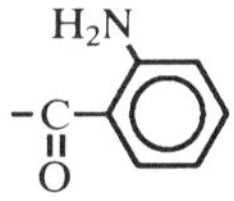

(5) R = 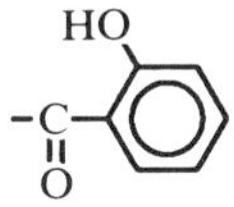

(6) R = 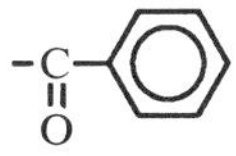

Another way to protect a phenolic moiety against presystemic metabolism is to prepare an ester with an built-in esterase inhibiting function, so that the prodrug can slow down its own rate of hydrolysis, and thereby pass intact through the gut wall and liver. A nice example of this strategy is the bronchodilator prodrug bambuterol (7) which is the bis-*N*,*N*-dimethylcarbamate of terbutaline (1). *N*,*N*-disubstituted carbamate esters are generally very stable against both chemical and enzymatic hydrolysis and have, in addition, esterase-inhibiting properties. Bambuterol has been found to possess these properties, being a potent inhibitor of pseudocholinesterase. Upon oral administration, the compound is readily absorbed and passes unmetabolized through the gut wall so that most of the dose reaches the liver and the systemic circulation in intact form. The regeneration of active terbutaline from the prodrug takes place by a multistep reaction involving an initial enzyme-mediated oxidation at the methyl groups in the carbamate moiety to give *N*-hydroxymethyl carbamates (8) which are subsequently decomposed spontaneously to farmaldehyde and monomethyl carbamates (9). The latter are then enzymatically hydrolysed by virtue of pseudo-cholinesterase (Scheme 1). This enzyme is selectively inhibited by bambuterol, i.e. the prodrug inhibits its own hydrolysis, and the result is a slow formation of the parent drug and a sustained action (Svensson & Tunek, 1988).

In addition to these desirable gains in bioavailability and duration of action, bambuterol has been shown to afford enhanced delivery of the parent drug to its site of action, the lungs, with concomitant reduction of side-effects such as muscle tremor due to the lower plasma levels of terbutaline (Svensson & Tunek, 1988). Apparently bambuterol and its hydroxylated metabolites have greater affinity for lung tissue than terbutaline and are retained in this tissue. Following lung uptake, terbutaline is regenerated from the prodrug and its primary metabolites.

The favourable results seen with bambuterol have since been used as a basis for the preparation of *N,N*-dimethylcarbamate esters of other phenolic drugs susceptible to undergo first-pass metabolism, and in a number of cases significant depression of this bioavailability-limiting factor has been obtained (Hansen *et al.*, 1991).

A third approach to depress presystemic metabolism is to derivatize the susceptible drug molecule at some other position in the molecule so that the prodrug obtained is no longer a substrate for the presystemic metabolizing enzyme, i.e. even though the functional group originally attacked is not directly masked. The diminished affinity of such a prodrug for the metabolizing enzyme may be due to electronic, steric, hydrophilic or other effects. An example illustrating this approach is propranolol which undergoes extensive first-pass metabolism in the liver, in particular through cytochrome P450-mediated oxidation. Esterification of the

SCHEME 1

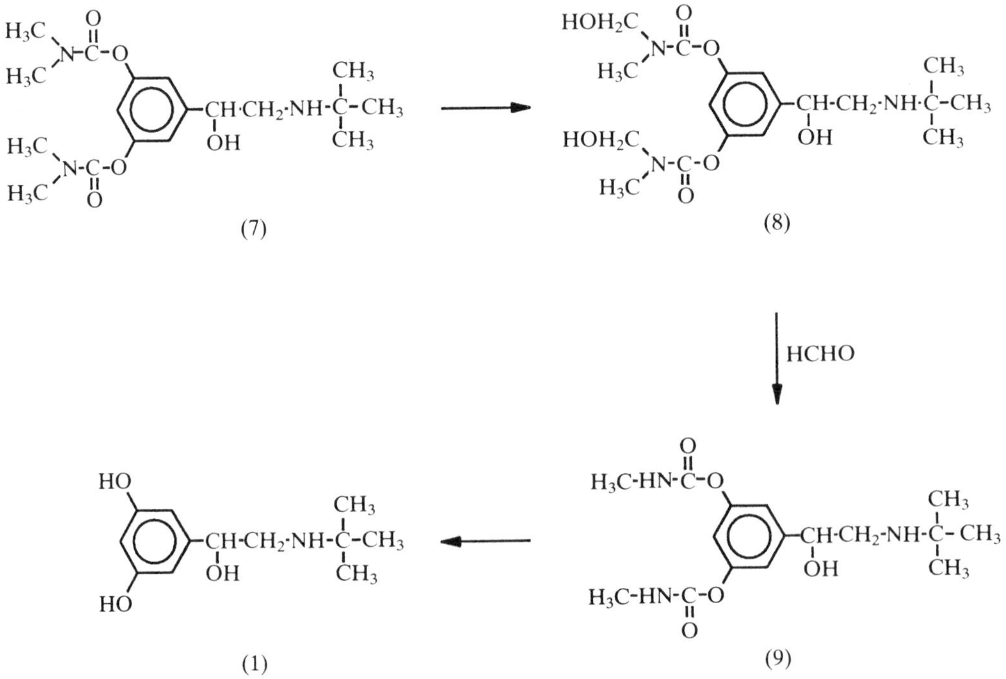

hydroxyl group in the molecule, which is attacked to only a minor extent during first-pass metabolism, gives the acetate and hemisuccinate esters and has thus been shown to afford a significant protection against first-pass metabolism in rats and dogs (Garceau *et al.*, 1978; Anderson *et al.*, 1988). Similarly, the N-Mannich base (10) of salicylamide (11) has been found to reduce the extent of presystemic metabolism somewhat in rabbits even though the phenolic group in salicylamide is conjugated by more than 95% following peroral administration (D'Souza *et al.*, 1986).

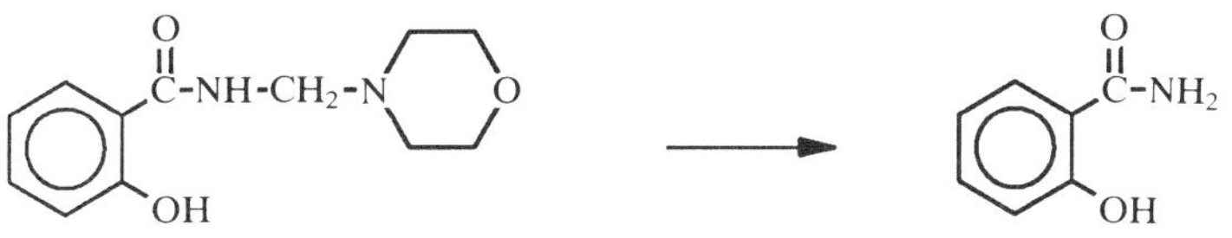

(10) (11)

A third example is *N*-acetylcysteine. The systemic bioavailability of this drug following oral administration is only about 5–10% due to extensive first-pass metabolism by deacetylation, primarily in the gut (Sjödin *et al.*, 1989). By bioreversible esterification of the thiol or the carboxylic acid group of *N*-acetylcysteine, the compound is no longer a substrate for an intestinal α-*N*-acyl-L-amino acid hydrolase responsible for the deacetylation and, accordingly, such prodrug modification may be a useful approach to circumvent the problem of intestinal first-pass metabolism (Kahns & Bundgaard, 1990).

Rational application of this approach will require a greater knowledge of the structure–reactivity requirements of the metabolizing enzymes such as phenol sulphotransferase.

As stated above, derivatization of a phenolic function in the form of a prodrug requiring enzymatic cleavage may often be a futile approach since the demasking of the protective group occurs before the drug reaches the sites of metabolism. A most promising, and perhaps generally applicable, approach to depress the first-pass metabolism of a phenolic drug may be the development of prodrug derivatives in which conversion to the active parent drug occurs by non-enzymatic means, e.g. by chemical hydrolysis at the physiological pH of 7.4 or by an intramolecular reaction occurring with an appropriate rate at pH 7.4 and 37°C. A number of derivatives of phenols undergoing an intramolecularly facilitated and enzyme-independent cleavage with release of the parent phenol can readily be envisaged. For example, various basic carbamates of 4-hydroxyanisole have recently been shown to undergo a facile base-catalysed cyclization in aqueous solution with formation of an imidazolidinone and the parent phenol (Saari *et al.*, 1990) (Scheme 2). At pH 7.4 and 37°C, half-lives for the conversion were in the range of 0.5–12 h and no catalysis was effected by plasma.

No studies have yet been performed to exploit such a prodrug approach involving a non-enzymatic prodrug cleavage for depressing first-pass metabolism of phenols or other susceptible entities, but such studies are certainly warranted.

SCHEME 2

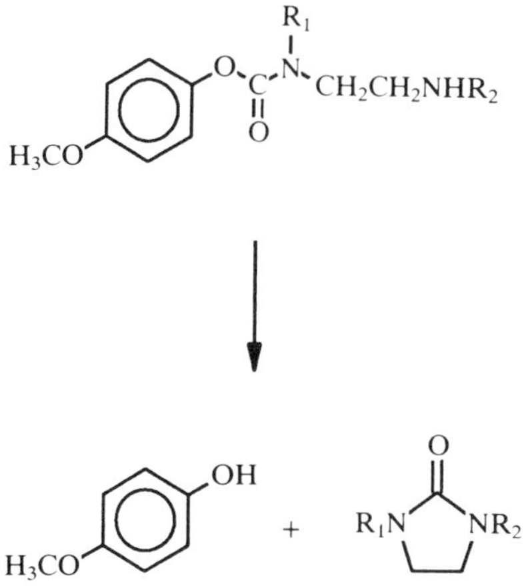

3 Site-specific delivery

Two approaches to the design of site-specific delivery of drugs via the prodrug concept can be visualized:

1 Aim to design a bioreversible derivative that affords an increased or selective transport of the parent drug to the site of action (*site-directed drug delivery*).

2 Design of a derivative that is ubiquitous, but undergoes bioactivation only at the target (*site-specific bioactivation*).

The site-directed drug delivery can further be divided into *localized drug delivery* and *systemic site-specific delivery.* Below, recent examples are given to illustrate the utilization of these principles to achieve site-specific delivery or targeting of drug molecules. In several cases, site-specific delivery has been obtained by a combination of site-directed delivery and bioactivation.

3.1 *Site-directed drug delivery*

3.1.1 LOCALIZED SITE-SPECIFIC DELIVERY

Up to now, most success in achieving site-directed drug delivery via prodrugs has been through localized drug delivery, i.e. where drug input is direct to the target organ such as the skin or the eye. The therapeutic success achieved stems primarily from increased absorption or transport of the prodrug across the biological membrane to which it is applied. Thus, improved site-specific delivery of the anti-glaucoma agents epinephrine, timolol and phenylephrine has been achieved with prodrugs possessing improved corneal permeability characteristics. This, in turn, leads to decreased concentrations of the parent drug at sites where it is unwanted such as the systemic circulation. The net result obtained is a reduction of the dose necessary for eliciting the pharmacological effect and hence a reduction in side-effects (Lee & Bundgaard, 1992).

In the case of epinephrine, the clinically used prodrug dipivefrin has led to a markedly improved ocular delivery. This dipivalate ester prodrug is much more lipophilic than epinephrine and the esterification of the metabolically susceptible phenolic hydroxyl groups affords a delay in metabolic destruction. These properties, coupled with a sufficiently high susceptibility to undergo enzymatic hydrolysis in the eye during and after absorption, are responsible for the approximately 20-fold greater antiglaucoma activity of the prodrug in comparison with the parent drug upon local administration in humans. In addition, because lower doses of the prodrug can be used, untoward cardiac side-effects due to epinephrine absorption from the tear duct overflow are greatly diminished. Dipivefrin has also a longer duration of action than epinephrine, because the metabolism of the latter, which involves a methylation of the phenolic OH-groups, is delayed or prolonged as it cannot occur until the prodrug has undergone conversion to epinephrine (Mandell *et al.*, 1978).

As shown with timolol ester prodrugs, the unwanted systemic absorption of a topically applied drug can be significantly reduced without affecting the ocular absorption (Chang *et al.*, 1988a,b). The basis for this achievement is most probably the differential lipophilic characteristics of the membranes responsible for ocular absorption (corneal) and systemic absorption (conjunctival and nasal mucosae). Thus, the therapeutic index, defined as the ratio of aqueous humour to plasma timolol concentrations, can be improved as much as 15-fold with the butyryl ester (13) prodrug of timolol (12) (Chang *et al.*, 1988b; Bundgaard, 1989c).

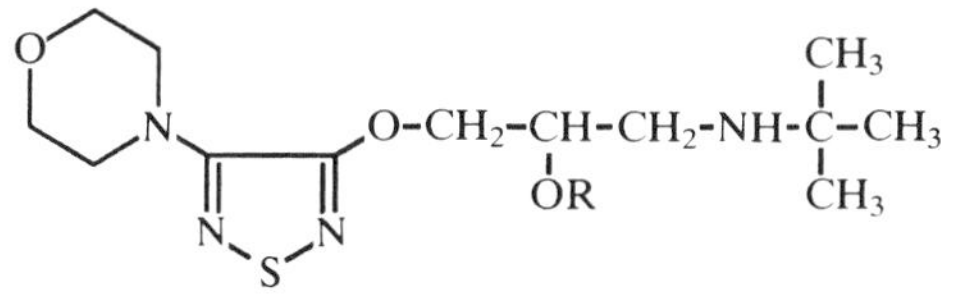

(12) R = H

(13) R = $-C(=O)-CH_2CH_2CH_3$

The ability of prodrugs to achieve relative oculoselectivity, i.e. selective drug action in the eye without the risk of systemic actions, is a concept which may find wider applications in the design of new ocular agents. The key to this concept is the ability of prodrugs to enhance the potency of a drug candidate originally designed for systemic use but which was deliberately chosen because of its lack of significant systemic potency. By this means the incidence of systemic side-effects can be minimized whereas the loss in potency can be offset by enhancing the ocular

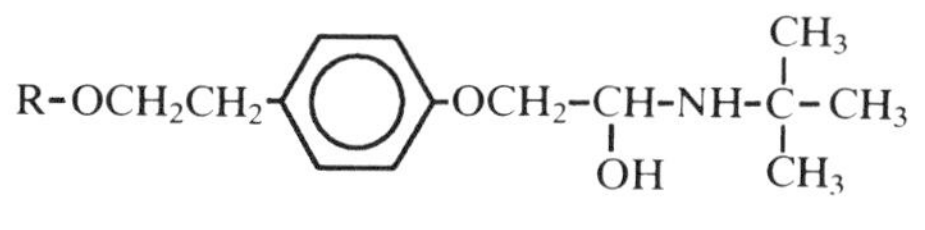

(14) R = H

(15) R = $-C(=O)-CH_3$

absorption via the use of prodrugs. Such an approach has been described by Sugrue *et al.* (1988). These investigators reported that L-653,328 (15), an acetate ester of the β-blocker L-652,698 (14), was as potent as timolol in lowering the intraocular presure in a rabbit model but was at least 100-fold less potent than timolol against the heart. This lack of systemic potency was consistent with the modest affinity of L-652,698, the active moiety of L-653,328, for extraocular β-receptors.

Further examples on the use of a selectively acting drug combined with the prodrug approach to improve its delivery to its site of action are certainly going to appear during the next decade.

3.1.2 SYSTEMIC SITE-SPECIFIC DELIVERY

Delivery to a specific internal site or organ through selective drug transport is more difficult to achieve than localized site-specific delivery, since the drug must be transported in the blood to the desired organ or tissue, passing various complex barriers on the way. Despite the difficult goal some successful examples have appeared.

Of particular interest is the concept developed by Bodor and co-workers for site-specific delivery of drugs to the brain (Bodor & Brewster, 1983; Bodor, 1987). In this method (Scheme 3), the drug (D) which is aimed to be delivered to the brain is coupled to a quaternary carrier (e.g. *N*-methylnicotinic acid) $(QC)^+$ and the obtained $(D{-}QC)^+$ is reduced chemically to the neutral, lipophilic dihydro form (dihydrotrigonelline) (D–DHC). After administration of this compound, it is distributed quickly throughout the body, including the brain. The lipophilic form is then enzymatically oxidized back to the original quaternary salt $(D{-}QC)^+$ which, because of its ionic, hydrophilic character, is eliminated quickly from the body except from the brain. Its hydrophilic character prevents it from passing through the blood–brain barrier and thus it is locked in the brain. Slow enzymatic cleavage of $(D{-}QC)^+$ in the brain will then result in a steady release of the parent drug there. Because of the facile elimination of $(D{-}QC)^+$ from the general circulation, only small amounts of free drug are released in the blood and the overall result is a concentration of the drug at its site of action within the brain. This dihydropyridine–pyridinium salt redox delivery system has been applied for brain-specific or brain-enhanced delivery of several agents including phenethylamine, dopamine, phenytoin, testosterone, β-oestradiol and penicillins. The chemical linkage connecting the drug with dihydrotrigonelline can be an amide or ester

SCHEME 3

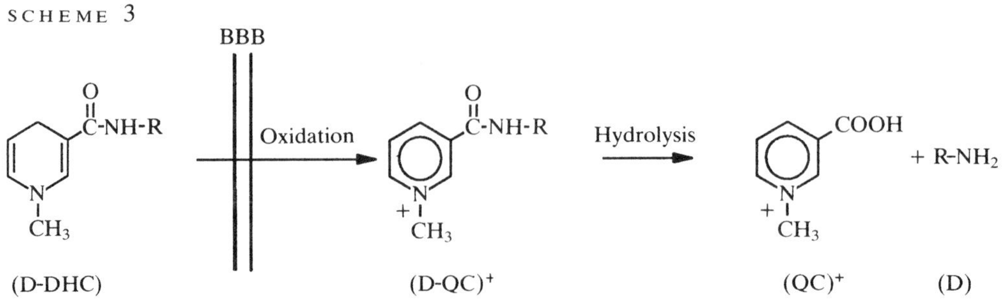

bond if the drug contains an amino or hydroxyl group, respectively. For a carboxylic acid drug, acyloxymethyl derivatives of the type (16) have been used (Pop *et al.*, 1989).

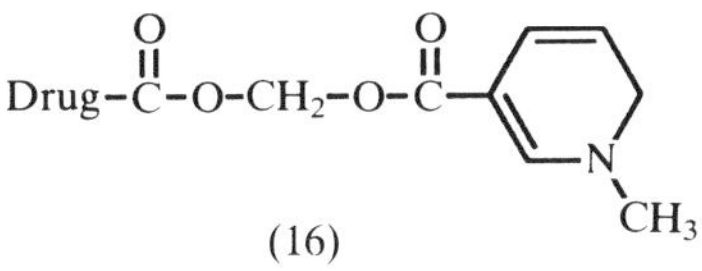

(16)

Another example of site-directed drug delivery obtained by the prodrug approach is provided by lovastatin (17). This drug has recently been approved as the first of its therapeutic class for the biochemical control of hypercholesterolaemia. It is not active itself but is converted *in vivo*, particularly in the liver, upon peroral administration to the active hydroxyl acid form (18) (Duggan *et al.*, 1989). The latter is a potent inhibitor of hydroxymethylglutaryl-CoA-reductase (HGR) and thus of cholesterol synthesis. Since the liver is the major site of cholesterol synthesis and regulation, it is the principal target organ for HGR inhibitors. *In vivo*, the prodrug lovastatin is in reversible metabolic equilibrium with the active ring-opened form (Scheme 4). Although lovastatin is less well (about 30%) absorbed than compound (18) (about 90%) following peroral administration, the absorbed fraction reaches the portal circulation largely unchanged, and is efficiently extracted

SCHEME 4

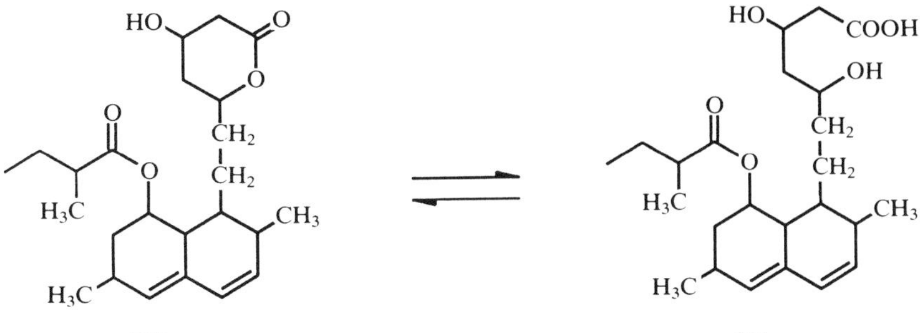

(17) (18)

by the liver after which it is reversibly hydrolysed to the active hydroxy acid. The prodrug shows a much higher first-pass extraction by the liver than the hydroxy acid and provides adequate levels of this agent within the liver, with subsequent minimization of systemic drug levels. Thus, the better therapeutic index for the prodrug (17) as compared to the active species is due to its specific uptake by the target organ combined with adequate conversion to and retainment of the active form at the target. Similar liver-specific uptake and bioactivation is seen with simvastatin, an analogue of lovastatin (Todd & Goa, 1990; Vickers *et al.*, 1990).

3.2 *Site-specific drug release*

Site-specific drug delivery through site-specific prodrug activation may be accomplished by the utilization of some specific property at the target site, such as altered

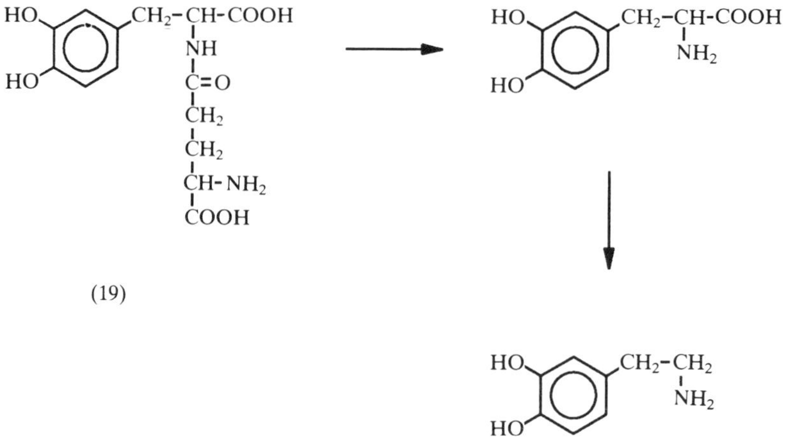

(19)

Figure 21.2. Selective generation of dopamine in the kidney by sequential action of γ-glutamyl transpeptidase and aromatic L-amine acid decarboxylase on γ-glutamyl dopa.

pH or high activity of certain enzymes relative to non-target tissues, for the prodrug–drug conversion.

It has been shown that the kidney is highly active in the uptake and metabolism of γ-glutamyl derivatives of amino acids and peptides. This property is due to the high concentration in the kidney of γ-glutamyl transpeptidase, an enzyme capable of cleaving γ-glutamyl derivatives of amino acids and other compounds containing an amino function. Based on this finding the possibility of using γ-glutamyl derivatives of pharmacologically active substances as kidney-specific prodrugs has been explored by various groups. Thus, γ-glutamyl derivatives of dopamine and L-dopa have been developed as kidney-specific prodrugs (Wilk *et al.*, 1978). Administration of the derivatives to animals leads to selective generation of the active dopamine in the kidney as a consequence of the high activity of γ-glutamyl transpeptidase in this organ. The γ-glutamyl-L-dopa derivative (19) is initially converted to L-dopa which then decarboxylates to dopamine by the action of L-amino acid decarboxylase, an enzyme that is also highly concentrated in the kidney (Fig. 21.2). Tissues other than the kidney showed only low free drug concentrations, resulting in a separation of the renal dopaminergic from the systemic adrenergic effects with the prodrugs. The results suggest that these prodrugs of L-dopa and dopamine may be useful as specific renal vasodilators.

Although the amino acid (20) is a potent inhibitor of CMP-KDO synthetase, a key enzyme in the biosynthesis of the lipopolysaccharide of Gram-negative bacteria, it is unable to reach its cytoplasmic target and is therefore inactive as an antibacterial agent. Simple lipophilic esters are not useful to enhance the delivery of the amino acid (20) since they are not cleaved by the bacteria. The double prodrug (21) has, on the other hand, recently been found to solve the problem (Norbeck *et al.*, 1989). Upon entry into bacterial cells the disulphide bond in compound (21) is reduced by sulphydryl compounds in the bacteria, resulting in the formation of the thiol (22). This is highly unstable and the active amino acid (20) is formed by a rapid, intramolecular displacement (Scheme 5).

There are several examples of prodrugs developed to exploit an atypically high level of an enzyme occurring in neoplastic cells. 5′-Deoxy-5-fluorouridine (doxi-

SCHEME 5

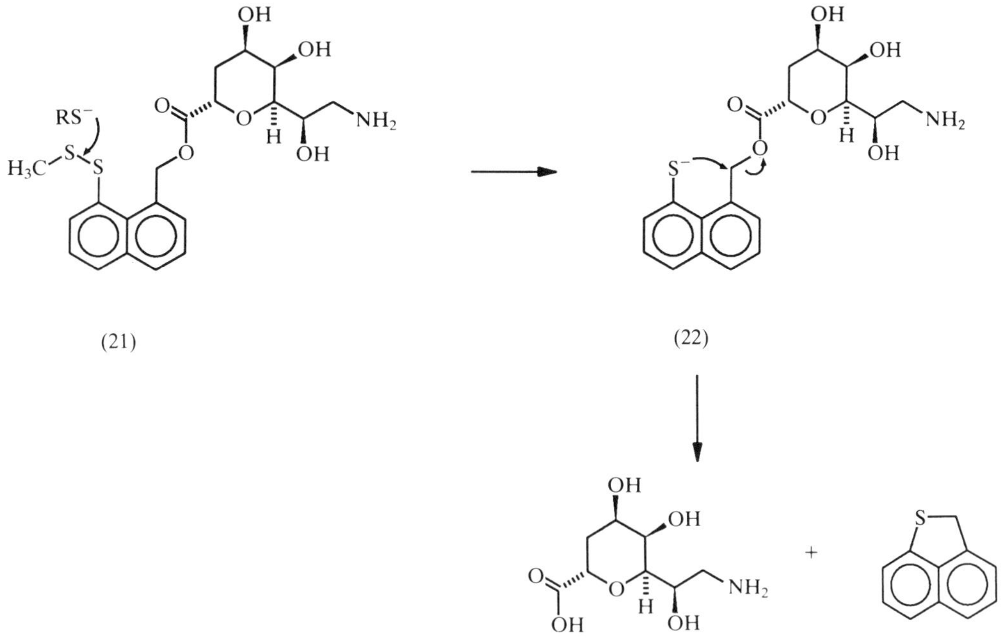

(21) (22) (20)

fluridine) is a prodrug of 5-fluorouracil that is used in the treatment of human malignancies. Compared with 5-fluorouracil, doxifluridine has a similar spectrum of activity but a higher therapeutic index. The antiproliferative activity of doxifluridine is due to its intracellular conversion in tumour tissues to 5-fluorouracil by uridine phosphorylase, and its higher therapeutic index is due to lack of this enzyme in bone marrow cells (Suzuki *et al.*, 1980).

An alternative approach is to exploit the differences in ratio of two or more enzymes that participate in activating the prodrug, which by necessity must be a sequentially labile prodrug. Bodor & Visor (1984a) exploited the marked differences in the ratio of ketone reductase to esterase levels among the anterior segment tissues to achieve site-specific delivery of diisovaleryl adrenalone (23), a bioreversible derivative of epinephrine (25), in the ciliary body, where epinephrine is regenerated and where this drug acts. Epinephrine is regenerated when the diester is reduced to (27) before hydrolysis occurs; no epinephrine is regenerated from the adrenalone (26) formed from diester hydrolysis (Scheme 6). The proportion of the diester undergoing hydrolysis only versus the reduction–hydrolysis sequence, and hence its duration of action, is a function of susceptibility of the ester linkage to hydrolysis (Bodor & Visor, 1984b). Thus, di(ethylsuccinyl)adrenalone (24), with a plasma half-life of 1 min, is anticipated to have a shorter duration of action than its diisovaleryl counterpart, which has a plasma half-life of 19 min. It is reasonable to expect that a better understanding of the enzymatic systems in the various anterior segments would enhance the usefulness of this approach to achieve drug targeting in the eye.

The recently introduced antiulcer agent omeprazole (28) is an excellent example of a prodrug showing a high degree of site-specific bioactivation resulting in site-specific drug delivery. The drug is an effective inhibitor of gastric acid secretion

SCHEME 6

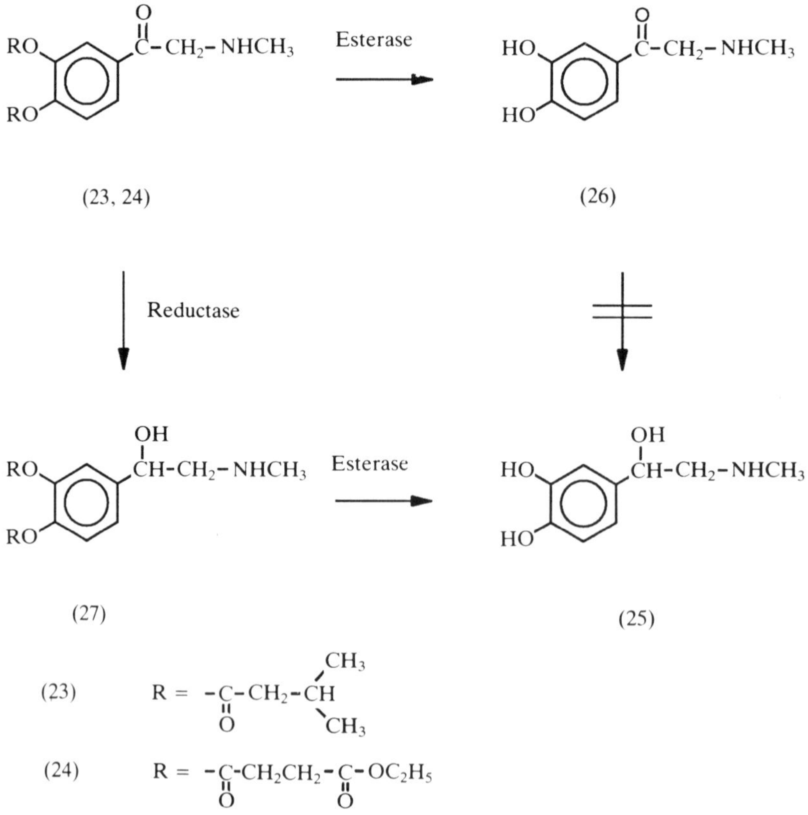

by inhibiting the gastric H^+,K^+-ATPase. This enzyme is responsible for gastric acid production, and is located in the secretory membranes of parietal cells. Omeprazole itself is not an active inhibitor of this enzyme, but is transformed within the acid compartments of the parietal cells into the active inhibitor, a cyclic sulphonamide (29). This reacts with the thiol groups in the enzyme with the formation of a disulphide complex (30), thus inactivating the H^+,K^+-ATPase (Scheme 7) (Lindberg *et al.*, 1986). The high specificity in the action of omeprazole is due to a combination of factors (Lindberg *et al.*, 1990):

SCHEME 7

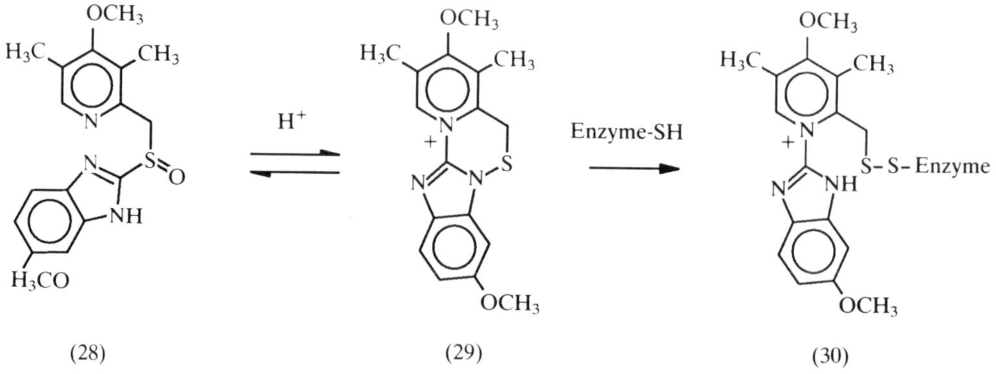

1 Omeprazole is a weak base (the pyridine nitrogen has a pK_a = 4.0) and therefore concentrates in acidic compartments, i.e. in parietal cells which have the lowest pH of the cells in the body.
2 The low pH value of the parietal cells causes the conversion of omeprazole into the active inhibitor close to the target enzyme.
3 The active inhibitor (29) is a permanent cation with limited possibilities to penetrate the membranes of the parietal and other cells, and thus will be retained at its site of action.
4 In the neutral part of the body omeprazole has good stability and only slight conversion to the active species occurs.

The examples cited above show that the prodrug concept can be useful to achieve drug targeting or drug localization. Any attempts to use prodrugs to direct drugs selectively to their site of action should take into consideration that the following basic criteria are to be met by the prodrugs and drugs if the approach is to be successful:

- the prodrug should be able to reach the site for drug action;
- the prodrug should be converted efficiently to the drug at that site;
- the parent active drug should to some extent be retained or trapped at the target site for a sufficient period of time to exert its effect.

The reason why attempts to promote site-specific delivery via prodrugs have failed in many past cases is certainly that not all these criteria have been met (Stella & Himmelstein, 1980, 1982). Thus, although a prodrug will release the parent drug at its target site in a highly selective manner due to a target-specific cleavage mechanism (e.g. due to an atypical enzyme activity), it will not be successful if the prodrug is not able to reach the target tissue. Both conditions should be fulfilled at the same time. Similarly, the selective delivery of prodrug forms of known active drugs may fail simply because of the fact that for the drug to be active it must already have at least some access to the site of action. Therefore, it will also readily equilibrate with other tissues and selective delivery of that agent to the site of action by a prodrug will result in only a transient increase of parent drug concentration at the site, before equilibrium is attained with the rest of the body. Thus, drugs most likely to benefit from site-specific delivery to target sites are those agents that are still in the phase of design and development. An exception to this is the situation where the prodrug becomes trapped or retained at the target, followed by a slow conversion to the parent drug as illustrated above.

4 Prodrug derivatives of peptides

A major obstacle to the application of peptides as clinically useful drugs is their poor delivery characteristics. Most peptides are rapidly metabolized by proteolysis at most routes of administration, they are in general non-lipophilic compounds showing poor biomembrane penetration characteristics, and they possess short biological half-lives due to rapid metabolism and clearance (Humphrey & Ringrose, 1986; Lee & Yamamoto, 1990).

Several peptides also suffer from systemic transport problems in that they do not readily penetrate cell membranes to reach the receptor biophase or cross the blood–brain barrier (Meisenberg & Simmons, 1983).

A possible approach to solve these delivery problems is derivatization of the bioactive peptides to produce prodrugs or transport forms which possess enhanced physico-chemical properties in comparison to the parent compounds with regard to delivery and metabolic stability. Thus, such derivatization may on one hand protect small peptides against degradation by enzymes present at the mucosal barrier, and on the other hand render hydrophilic peptides more lipophilic and hence facilitate their absorption. To be a useful approach, however, the derivatives should be capable of releasing the parent peptide spontaneously or enzymatically in the blood following their absorption.

This prodrug approach has only recently been pursued (Bundgaard, 1986, 1992) but, as illustrated below, it may in fact be a highly useful means of improving the delivery of peptide drugs.

4.1 *Bioreversible derivatization of the peptide bond*

It is generally recognized that *N*-alkylation of peptide bonds usually makes them resistant to enzymatic attack (Veber & Freidinger, 1985; Fauchère, 1986). However, since *N*-methyl and similar alkyl derivatives are not bioreversible, the approach of simple *N*-alkylation implies the design of a new peptide (the analogue approach). A possible strategy in a prodrug approach (Fig. 21.3) may be to create an *N*-α-hydroxyalkyl derivative of the peptide bond since such derivatives of primary and cyclic amides are known to be spontaneously converted to the parent amide and the corresponding aldehyde in aqueous solution, the rate of conversion being dependent on the nature of the alkyl group and the acidity of the amide (Bundgaard and Johansen, 1984; Bundgaard, 1985b). A major obstacle to this approach has been the difficulty in performing *N*-α-hydroxyalkylation of secondary, acyclic amides such as peptide bonds, but as reported by Buur & Bundgaard (1988) this difficulty can be overcome by making 5-oxazolidinones. Such compounds are readily formed by condensing an α-amino acid or an *N*-acylated amino acid with an aldehyde. The lactone ring in *N*-acyl-5-oxazolidinones is highly reactive and is

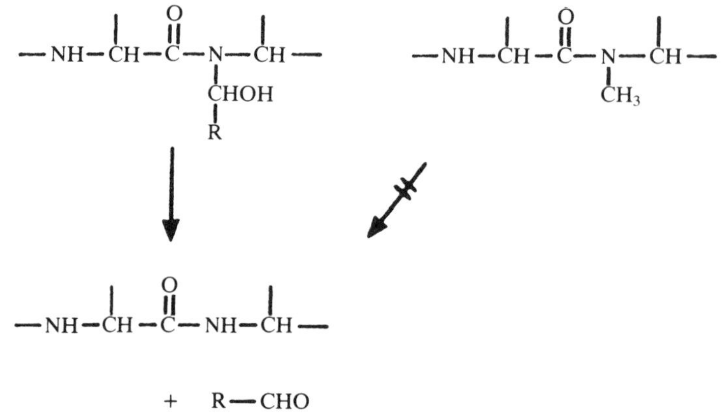

Figure 21.3. Illustration of a possible prodrug approach to protect a peptide bond against enzymatic cleavage. Whereas an *N*-methyl derivative usually remains stable *in vivo*, the *N*-α-hydroxyalkyl derivative is spontaneously decomposed at physiological pH with release of the parent peptide and an aldehyde.

easily opened by hydrolysis or aminolysis, resulting in the intermediate formation of an *N*-α-hydroxyalkyl derivative. Such derivatives of various peptides and *N*-acylated amino acids are quantitatively decomposed in aqueous solution to their parent peptide and aldehyde at rates solely dependent on pH (Bundgaard & Rasmussen, 1991a). The stability of a given peptide derivative can be controlled by selection of appropriate aldehydes for the initial 5-oxazolidinone formation. Thus, by using acetaldehyde or chloral instead of formaldehyde, a higher rate of prodrug conversion at physiological pH can be achieved (Bundgaard & Rasmussen, 1991a).

That stabilization of a peptide bond by such *N*-α-hydroxyalkylation can in fact be achieved has been documented with the pancreatic proteolytic enzymes carboxypeptidase A and α-chymotrypsin. Thus, whereas the *N*-benzyloxycarbonyl (Z) derivatives of the dipeptides Gly-L-Leu and Gly-L-Ala are readily hydrolysed by carboxypeptidase A, the *N*-hydroxymethylated compounds, i.e. Z-Gly(CH_2OH)-Leu and Z-Gly(CH_2OH)-Ala are totally resistant to cleavage by the enzyme as revealed by their similar rates of decomposition in the presence or absence of the enzyme at pH 7.4 and 37°C (Bundgaard & Rasmussen, 1991a). Similarly, whereas *N*-acetyl-L-phenylalaninamide (31) is readily degraded by α-chymotrypsin, the *N*-hydroxymethyl derivative (32) is completely inert toward the enzyme (Bundgaard & Kahns, unpublished results). It is interesting to note that in these compounds the susceptible C-terminal peptide bond is stabilized by *N*-hydroxymethylation of the second peptide bond. This implies that such bioreversible derivatization may not only protect the peptide bond directly derivatized but also adjacent underivatized peptide bonds.

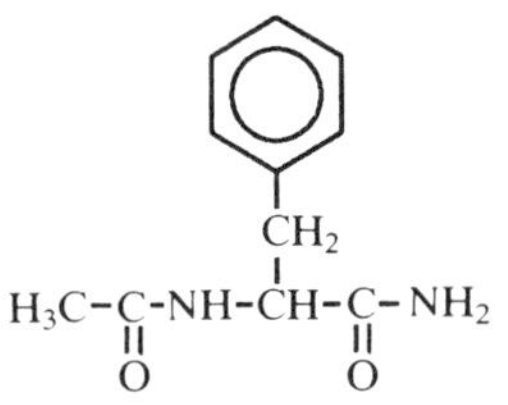

(31)

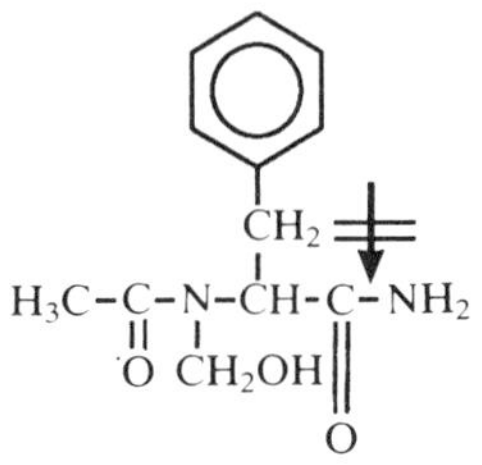

(32)

Another interesting aspect of this prodrug approach to protect a peptide bond against rapid proteolytic cleavage is further derivatization of the hydroxyl group in the *N*-α-hydroxyalkyl derivatives, e.g. by esterification to afford *N*-α-acyloxyalkyl derivatives which are cleavable by non-specific esterases. In this manner not only the stability but also the lipophilicity can be further modified (Bundgaard & Rasmussen, 1991b).

4.2 *4-Imidazolidinone derivatives*

Potentially useful and broadly applicable prodrug types for the α-amino amide moiety, which occurs in a large number of peptides, are 4-imidazolidinones (33). Such derivatives are readily formed by condensing compounds containing an α-amino amide moiety with aldehydes or ketones (Scheme 8) (Hardy & Samworth, 1977). In aqueous solution the derivatives are readily decomposed with release of

SCHEME 8

(33)

the parent peptide, the stability being dependent on the structure of the carbonyl component and the steric properties within the amino acid residue next to the N-terminal residue (Klixbüll & Bundgaard, 1984; Rasmussen & Bundgaard, 1991a).

An important feature of the 4-imidazolidinones is their resistance to undergo enzymatic hydrolysis. The rates of degradation of the compounds are not increased in the presence of human plasma but are either the same or even slower than the rates of degradation in buffer solutions. The compounds are also inert toward various aminopeptidases capable of degrading the parent di- and tri-peptides (Rasmussen & Bundgaard, 1991a). In contrast to the 4-imidazolidinones, the parent peptides are readily hydrolysed by plasma enzymes. Thus, by forming a 4-imidazolidinone derivative it is feasible to protect the peptide bond involved in the derivatization against enzymatic cleavage. By spontaneous hydrolysis at physiological pH the parent peptide is then released at a rate dependent on the structure of the imidazolidinone.

The nature of the carbonyl component can have a large influence on the rate of hydrolysis. Thus, the 4-imidazolidinone derivatives of Leu-enkephalin formed with acetaldehyde, acetone or cyclohexanone decompose at pH 7.4 and 37°C with half-lives of 30, 11 and 5.5 h, respectively (Rasmussen & Bundgaard, 1991b).

4.3 *Prodrugs of TRH*

Thyrotropin-releasing hormone (TRH, pyro-Glu-L-His-L-Pro-NH_2) is the hypothalamic peptide that regulates the synthesis and secretion of thyrotropin from the anterior pituitary gland. Since its discovery in 1969, TRH has been shown to have not only a variety of endocrine and CNS-related biological activity, but also potential as a drug in the management of various neurological and neuropsychiatric disorders including depression, brain injury, acute spinal trauma, Alzheimer's disease and schizophrenia (for reviews see Griffiths (1987) and Metcalf & Jackson (1989)).

The clinical utilization of the neuropharmacological properties of TRH is, however, greatly hampered by its rapid metabolism and clearance as well as by its poor access to the CNS (Griffiths, 1987). Following parenteral administration in humans, TRH shows a plasma half-life of only 6–8 min (Iversen, 1988) which is mainly due to rapid enzymatic degradation of the peptide in the blood, in particular by the so-called TRH-specific pyroglutamyl aminopeptidase serum enzyme (PAPase II) (Møss & Bundgaard, 1990a, and references cited therein). This enzyme catalyses the hydrolysis of TRH at the pGlu-His bond yielding pyroglutamic acid and His-Pro-NH_2 (Fig. 21.4). The lipophilicity of TRH is very low

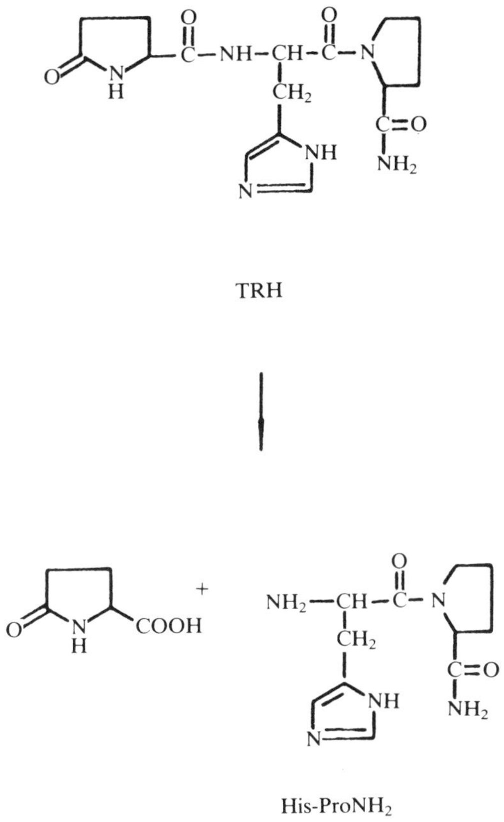

Figure 21.4. Enzymatic hydrolysis of TRH by virtue of a pyroglutamyl aminopeptidase to pyroglutamic acid and L-histidyl-L-prolineamide.

(Bundgaard & Møss, 1989) and this may be a primary reason for the limited ability of the peptide to penetrate the blood–brain barrier (Nagai *et al.*, 1980).

A possible approach to solve these delivery problems may be derivatization of the peptide to produce prodrugs or transport forms which are markedly more lipophilic than the parent peptide and which are resistant toward the TRH-degrading serum enzyme but cleavable by other means such as by chemical or non-specific enzyme-catalysed hydrolysis to release the parent TRH *in vivo*.

We have recently found that it is feasible to achieve these aims by modifying the imidazole group of the histidine residue in TRH (Bundgaard & Møss, 1990). By reacting TRH with various chloroformates, a number of *N*-alkoxycarbonyl derivatives of TRH (34) were obtained. These derivatives proved to be totally resistant to cleavage by the TRH-degrading serum enzyme. On the other hand, the derivatives are readily bioreversible as the parent TRH is formed quantitatively by spontaneous or plasma esterase-catalysed hydrolysis (Fig. 21.5) (Table 21.1).

Besides being potentially useful to prolong the duration of action of TRH *in vivo*, the *N*-alkoxycarbonyl prodrug derivatives possess greatly increased lipophilicity relative to TRH as assessed by octanol–buffer partition coefficients (Table 21.1). This property may render the prodrug forms more capable of penetrating the blood–brain barrier or various other biomembranes than the parent peptide. In fact, recent experiments have shown that it may be possible to achieve transdermal

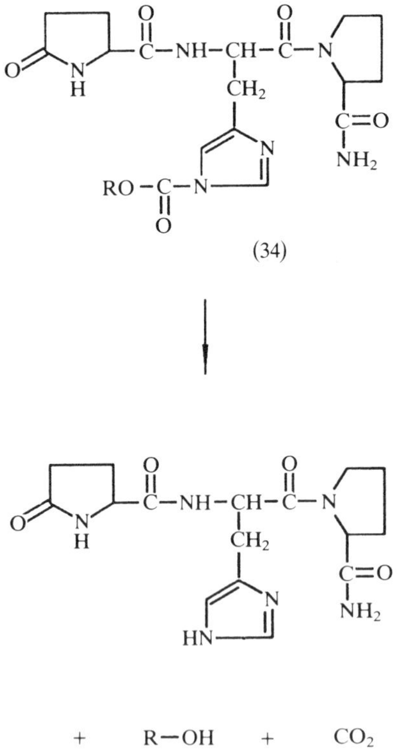

Figure 21.5. Spontaneous and plasma esterase-catalysed hydrolysis of *N*-alkoxycarbonyl derivatives of TRH (34) to the parent peptide. By *N*-alkoxycarbonylation of the imidazole moiety the pyroglutamyl peptide bond has become totally resistant to degradation by the TRH-inactivating pyroglytamyl aminopeptidase enzyme.

Table 21.1. Rate data for the hydrolysis and lipophilicity of TRH* and its various *N*-alkoxycarbonyl prodrug derivatives (Bundgaard & Møss, 1990)

Compound	Half-lives at 37°C (h)		
R in (34)	pH 7.4 buffer	Human plasma	log *P*†
CH_3	9.3	2.8	- 1.88
C_2H_5	14.2	3.8	- 1.30
i-C_3H_7	35.8	6.6	- 0.80
C_4H_9	19.0	4.3	- 0.47
C_6H_{13}	17.9	1.2	0.71
C_8H_{17}	17.5	0.4	1.88
c-C_6H_{11}	36.8	6.4	0.60

* Half-life of hydrolysis of TRH at a concentration of $< 5 \times 10^{-6}$ mol/l is 9.4 min; log $P = -2.46$.
† *P* is the partition coefficient between octanol and 0.02 mol/l phosphate buffer of pH 7.4.

delivery of TRH by using a lipophilic TRH prodrug such as *N*-octyloxycarbonyl-TRH (34, $R = C_8H_{17}$). In contrast to TRH itself, this prodrug was found to penetrate human skin *in vitro* quite efficiently and to be converted to the parent peptide during the transport through the skin (Møss & Bundgaard, 1990b).

The example with TRH demonstrates how a simple prodrug derivatization at a site remote from the site of enzymatic cleavage of the peptide nevertheless can be utilized to protect the peptide against its specific enzymatic inactivation. The *N*-alkoxycarbonyl prodrug derivatives may well be applicable to other peptides containing a histidine residue as well as to other drugs containing an imidazole moiety.

4.4 *Future developments*

Although it has been somewhat overlooked in the past, the prodrug approach may be a highly useful means of improving the absorption of peptides. By this bioreversible derivatization technique it is readily feasible to obtain derivatives with increased lipophilicity and, in some cases, metabolic stability. An illustrating example has been given with the tripeptide TRH. The prodrug technique may find its greatest use for peptide drugs containing not more than 7–10 amino acids, but a large part of future peptide drugs will certainly also fall in this range. For example, the various renin inhibitors presently under development (Greenlee, 1990) belong to this category.

While our knowledge of the design of prodrug derivatives for various functional groups occurring in peptides and proteins has increased greatly in recent years, more work is needed to enable us rationally to utilize this knowledge in the design of prodrugs capable of protecting a peptide bond against enzymatic cleavage and hence peptide inactivation. Thus, several peptides are rapidly cleaved by the gastrointestinal enzymes trypsin, α-chymotrypsin and carboxypeptidase A. As described above, a most efficient way to protect the vulnerable peptide bond(s) against proteolytic cleavage by the prodrug approach may be *N*-α-hydroxyalkylation or *N*-α-acyloxyalkylation of the NH group in this bond. However, the possibility of obtaining the same results by bioreversible modification of the peptide elsewhere in the molecule should also be considered. It can thus be imagined that a pentapeptide with a C-terminal amide group (which is often seen for endogenous peptide hormones), cleavable by the pancreatic enzymes at, for example, the bond between the second and third amino acid, can be stabilized against these enzymes just by modifying the C-terminal amide function in a bioreversible

SCHEME 9

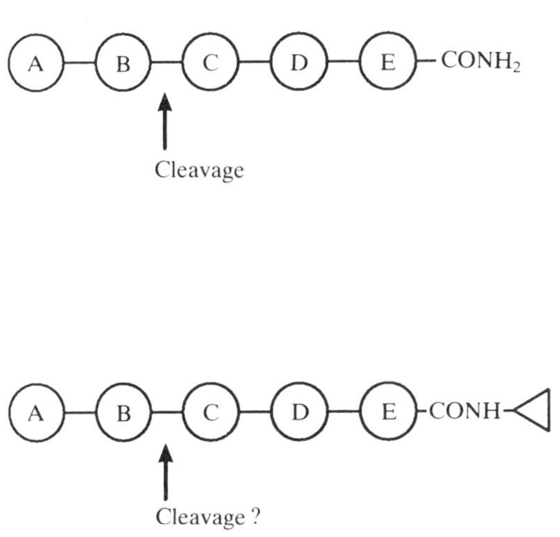

manner (Scheme 9). To this end, one should take advantage of the great knowledge existing about the substrate specificities of the various proteolytic enzymes.

5 Chemical approaches in prodrug design

A basic requisite for the prodrug approach to be useful in solving drug-delivery problems is the ready availability of chemical derivative types satisfying the prodrug requirements, the most prominent of these being reconversion of the prodrug to the parent drug *in vivo*. This prodrug–drug conversion may take place before absorption (e.g. in the gastrointestinal tract), during absorption, after absorption or at the specific site of drug action in the body, all dependent upon the specific goal for which the prodrug is designed. Ideally, the prodrug should be converted to the drug as soon as the goal is reached.

The necessary conversion or activation of prodrugs to the parent drug molecules in the body can take place by a variety of reactions. The most common prodrugs are those requiring a hydrolytic cleavage mediated by enzymatic catalysis. Active drug species containing hydroxyl or carboxyl groups can often be converted to prodrug esters from which the active forms are regenerated by esterases within the body (e.g. in the blood or liver). In other cases, active drug substances are regenerated from their prodrugs by biochemical reductive or oxidative processes.

Besides use of the various enzyme systems of the body to carry out the necessary activation of prodrugs, the buffered and relatively constant value of the physiological pH (7.4) may be useful in triggering the release of a drug from a prodrug. In these cases, the prodrugs are characterized by a high degree of chemical lability at pH 7.4 while preferably exhibiting a higher stability at for example pH 3–4. A serious drawback of prodrugs requiring chemical (non-enzymatic) release of the active drug is the inherent lability of the compounds, raising some stability-formulation problems at least in cases of solution preparations. Such problems have in several cases been overcome by using a more sophisticated approach involving pro-prodrugs or double prodrugs where use is made of an enzymatic release mechanism prior to the spontaneous reaction (Fig. 21.6). The utility of this *double prodrug concept* to optimize the properties of prodrugs has been demonstrated with pilocarpine prodrugs (Bundgaard *et al.*, 1985, 1986) and in several other cases as described in a recent review (Bundgaard, 1989a).

During the past decade several new types of bioreversible derivatives including double or triple prodrugs have been exploited for utilization in designing prodrugs

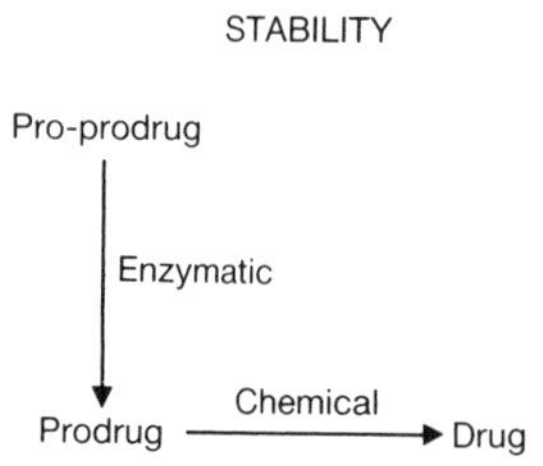

Figure 21.6. Stabilization of a spontaneously decomposing prodrug by further derivatization to yield a chemically stable but enzymatically labile double prodrug.

of a variety of drug molecules (Bundgaard, 1985b, 1989a; Bundgaard *et al.*, 1989). However, more research in this area is still needed in order to obtain an arsenal of derivatives to make use of. Thus, although esterification of molecules containing a carboxyl group appears as an obvious prodrug approach, it may sometimes fail because of an insufficiently high rate and extent of ester prodrug hydrolysis *in vivo*. In such cases, *N,N*-disubstituted glycolamide esters (Bundgaard & Nielsen, 1987; Nielsen & Bundgaard, 1988) or O-α-acyloxyalkyl esters may be useful because of their high enzymatic lability. The latter type has, for example, been used successfully to obtain various prodrugs of penicillins, notably ampicillin (Ferres, 1983). As mentioned above, no generally applicable and optimal prodrug principle has been exploited for the sulphonamide group (present, e.g., in various carbonic anhydrase inhibitors). The same holds true for the phosphate and phosphonate groups although α-acyloxyalkyl esters may be useful in some cases (Farquhar *et al.*, 1983; Bundgaard, 1989a). Experiments demonstrating a facile conversion of bis(acyloxymethyl) esters of phosphates, originally proposed by Farquhar *et al.* (1983) (Srivastva & Farquhar, 1984), under *in vivo* conditions to release the parent phosphate are, however, still lacking. Likewise, there is a lack of a broadly applicable prodrug type for the carbonyl group such as that present in various steroids and for the amino group. Some double prodrug types for primary and secondary amines involving an initial enzymatic hydrolysis step followed by a rapid, spontaneous release of the parent amine have, however, recently been discovered (Gogate & Repta, 1987; Alexander *et al.*, 1988; Amsberry & Borchardt, 1990) and may be attractive candidates.

6 Conclusions

Over the past two decades, the prodrug concept has been widely practised to optimize the delivery characteristics of a great number of existing drugs, largely as a result of an increased awareness and understanding of the physico-chemical factors that affect the efficacy of drug delivery and action. Several drugs are now used clinically in the form of prodrugs, and as the prodrug approach is becoming an integral part of the new drug design process one may expect that the new drugs in many cases will appear as prodrugs. As examples can be mentioned the ethyl esters of several angiotensin-converting enzyme inhibitors such as enalapril, the antiulcer agent omeprazole which possesses a high degree of site-specific bioactivation, and the cholesterol-lowering agents lovastatin and simvastatin which are prodrugs showing a liver-specific uptake and bioactivation following oral administration. Besides being an obligatory part of basic drug design, the prodrug approach will certainly still be used to solve delivery problems associated with older drugs (e.g. in situations where an alternative route of administration is desirable). An example of this is the feasibility of obtaining transdermal delivery of morphine (Drustrup *et al.*, 1991) and various contraceptives (Friend, 1990) using various ester prodrugs.

7 References

Albert, A. (1958). Chemical aspects of selective toxicity. *Nature*, **182**, 421–3.
Alexander, J., Cargill, R., Michelson, S.R. & Schwamm, H. (1988). (Acyloxy)alkyl carbamates as

novel bioreversible prodrugs for amines: Increased permeation through biological membranes. *J. Med. Chem.*, **31**, 318–22.

Amsberry, K.L. & Borchardt, R.T. (1990). The lactonization of 2′-hydroxyhydrocinnamic acid amides: A potential prodrug for amines. *J. Org. Chem.*, **55**, 5867–77.

Anderson, B.D., Chu, W.W. & Galinsky, R.E. (1988). Reduction of first-pass metabolism of propranolol after oral administration of ester prodrugs. *Int. J. Pharm.*, **43**, 261–5.

Aungst, B.J., Myers, M.J., Shefter, E. & Shami, E.G. (1987). Prodrugs for improved oral nalbuphine bioavailability: Interspecies differences in the disposition of nalbuphine and its acetylsalicylate and anthranilate esters. *Int. J. Pharm.*, **38**, 199–209.

Balant, L.P., Doelker, E. & Buri, P. (1990). Prodrugs for the improvement of drug absorption via different routes of administration. *Eur. J. Drug Metab. Pharmacokin.*, **15**, 143–53.

Bodor, N. (1987). In: E.B. Roche (Ed.), *Bioreversible Carriers in Drug Design. Theory and Application*, pp. 95–120, Pergamon Press, New York.

Bodor, N. & Brewster, M.E. (1983). Problems of delivery of drugs to the brain. *Pharmacol. Therap.*, **19**, 337–86.

Bodor, N. & Visor, G. (1984a). Formation of adrenaline in the iris-ciliary body from adrenalone diesters. *Exp. Eye Res.*, **38**, 621–6.

Bodor, N. & Visor, G. (1984b). Improved delivery through biological membranes. XVII. A site-specific chemical delivery system as a short-acting mydriatic agent. *Pharm. Res.*, **1**, 168–73.

Bundgaard, H. (Ed.) (1985a). *Design of Prodrugs*, Elsevier, Amsterdam.

Bundgaard, H. (1985b). In: H. Bundgaard (Ed.), *Design of Prodrugs*, pp. 1–92, Elsevier, Amsterdam.

Bundgaard, H., (1986). In: S.S. Davis, L. Illum and E. Tomlinson (Eds)., *Delivery Systems for Peptide Drugs*, pp. 49–68, Plenum Press, New York.

Bundgaard, H. (1989a). The double prodrug concept and its applications. *Adv. Drug Delivery Rev.* **3**, 39–65.

Bundgaard, H. (1989b). In: L.F. Prescott and W.S. Nimmo (Eds.), *Novel Drug Delivery and its Therapeutic Application*, pp. 193–208, John Wiley, London.

Bundgaard, H. (1989c). Improved ocular delivery of pilocarpine and timolol through prodrugs. *Proc. 5th Int. Conf. Pharm. Technol., Paris*, pp. 52–60.

Bundgaard, H. (1992). Prodrugs as a means to improve the delivery of peptide drugs. *Adv. Drug Delivery Rev.*, **8**, 1–38.

Bundgaard, H. & Johansen, M. (1984). Hydrolysis of *N*-(α-hydroxybenzyl)benzamide and other *N*-(α-hydroxyalkyl)amide derivatives: implications for the design of *N*-acyloxyalkyl-type prodrugs. *Int. J. Pharm.*, **22**, 45–56.

Bundgaard, H & Møss, J. (1989). Prodrug derivatives of thyrotropin-releasing hormone and other peptides. *Trans. Biochem. Soc.*, **17**, 947–9.

Bundgaard, H. & Møss, J. (1990). Prodrugs of peptides. 6. Bioreversible derivatives of thyrotropin-releasing hormone (TRH) with increased lipophilicity and resistance to cleavage by the TRH-specific serum enzyme. *Pharm. Res.*, **7**, 885–92.

Bundgaard, H. & Nielsen, N.M. (1987). Esters of N,N-disubstituted 2-hydroxyacetamides as a novel highly biolabile prodrug type for carboxylic acid agents. *J. Med. Chem.*, **30**, 451–3.

Bundgaard, H. & Rasmussen, G.J. (1991a). Prodrugs of peptides. 9. Bioreversible *N*-α-hydroxyalkylation of the peptide bond to effect protection against carboxypeptidase or other proteolytic enzymes. *Pharm. Res.*, **8**, 313–22.

Bundgaard, H. & Rasmussen, G.J. (1991b). Prodrugs of peptides. II. Chemical and enzymatic hydrolysis kinetics of *N*-acyloxymethyl derivatives of the peptide-like bond. *Pharm. Res.*, **8**, 1238–42.

Bundgaard, H., Falch, E. & Jensen, E. (1989). A novel solution stable, water-soluble prodrug type for drugs containing a hydroxyl or an NH-acidic group *J. Med. Chem.*, **32**, 2503–7.

Bundgaard, H., Falch, E., Larsen, C., Mosher, G.L. & Mikkelson, T.J. (1985). Pilocarpic acid esters as novel sequentially labile pilocarpine prodrugs for improved ocular delivery. *J. Med. Chem.*, **28**, 979–81.

Bundgaard, H., Falch, E., Larsen, C., Mosher, G.L. & Mikkelson, T.J. (1986). Pilocarpine prodrugs II. Synthesis, stability, bioconversion and physicochemical properties of sequentially labile pilocarpine acid diesters. *J. Pharm. Sci.*, **75**, 775–83.

Buur, A. & Bundgaard, H. (1988). Prodrugs of peptides. III. 5-Oxazolidinones as bioreversible derivatives for the α-amido carboxy moiety in peptides. *Int. J. Pharm.*, **46**, 159–67.

Chang, S.C., Chien, D.S., Bundgaard, H. & Lee, V.H.L. (1988a). Relative effectiveness of prodrug and viscous solution approaches in maximizing the ratio of ocular to systemic absorption of topically applied timolol. *Exp. Eye Res.*, **46**, 59–69.

Chang, S.C., Bundgaard, H. Buur, A. & Lee, V.H.L. (1988b). Low dose *O*-butyryl timolol improves the therapeutic index of timolol in the pigmented rabbit. *Invest. Ophthalmol. Vis. Sci.*, **29**, 626–9.

D'Souza, M., Venkataramanan, R., D'Mello, A. & Niphadkar, P. (1986). An alternative prodrug approach for reducing presystemic metabolism of drugs. *Int. J. Pharm.*, **31**, 165–7.

Drustrup, J., Fullerton, A., Christrup, L. & Bundgaard, H. (1991). Utilization of prodrugs to enhance the transdermal absorption of morphine. *Int. J. Pharm.*, **71**, 105–16.

Duggan, D.E., Chen, I.-W., Bayne, W.F., Halpin, R.A., Duncan, C.A., Schwartz, M.S., Stubbs, R.J. & Vickers, S. (1989). The physiological disposition of lovastatin. *Drug Metab. Disp.*, **17**, 166–73.

Farquhar, D., Srivastva, D.N., Kuttesch, N.J. & Saunders, P.P. (1983). Biologically reversible phosphate-protective groups. *J. Pharm. Sci.*, **72**, 324–5.

Fauchère, J.-L. (1986). Elements for the rational design of peptide drugs. *Adv. Drug Res.*, **15**, 29–69.

Ferres, H. (1983). Pro-drugs of β-lactam antibiotics. *Drugs of Today*, **19**, 499–538.

Friend, D.R. (1990). Transdermal delivery of contraceptives. *Crit. Rev. Therap. Drug Carrier Syst.*, **7**, 149–86.

Garceau, Y., Davis, I. & Hasegawa, J. (1978). Plasma propranolol levels in beagle dogs after administration of propranolol hemisuccinate ester. *J. Pharm. Sci.*, **67**, 1360–3.

Gogate, U.S. & Repta, A.J. (1987). *N*-(Acyloxyalkoxycarbonyl) derivatives as potential prodrugs of amines. II. Esterase-catalyzed release of parent amines from model prodrugs. *Int. J. Pharm.*, **40**, 249–55.

Greenlee, W.J. (1990). Renin inhibitors. *Med. Res. Rev.*, **10**, 173–236.

Griffiths, E.C. (1987). Clinical applications of thyrotropin-releasing hormone. *Clin. Sci.* **73**, 449–57.

Hansen, K.T., Faarup, P. & Bundgaard, H. (1991). Carbamate ester prodrugs of dopaminergic compounds: Synthesis, stability and bioconversion. *J. Pharm. Sci.*, **80**, 793–8.

Hardy, P.M. & Samworth, D.J. (1977). Use of *N-N'*-isopropylidene dipeptides in peptide synthesis. *J. Chem. Soc. Perkin Trans. I.*, 1954–60.

Hörnblad, Y., Ripe, E., Magnusson, P.O. & Tegner, K. (1976). The metabolism and clinical activity of terbutaline and its prodrug ibuterol. *Eur. J. Clin. Pharmacol.*, **10**, 9–18.

Humphrey, M.J. & Ringrose, P.S. (1986). Peptides and related drugs: A review of their absorption, metabolism and excretion. *Drug Metab. Rev.*, **17**, 283–310.

Hussain, M.A. & Shefter, E. (1988). Naltrexone-3-salicylate (a prodrug of naltrexone): Synthesis and pharmacokinetics in dogs. *Pharm. Res.*, **5**, 113–15.

Hussain, M.A., Aungst, B.J. & Shefter, E. (1988). Prodrugs for improved oral β-estradiol bioavailability. *Pharm. Res.*, **5**, 44–7.

Hussain, M.A., Koval, C.A., Myers, M.J., Shami, E.G. & Shefter, E. (1987). Improvement of the oral bioavailability of naltrexone in dogs: A prodrug approach. *J. Pharm. Sci.*, **76**, 356–8.

Iversen, E. (1988). Intra- and extravascular turn over of thyrotropin-releasing hormone in normal man. *J. Endocrinol.*, **118**, 511–16.

Kahns, A.H. & Bundgaard, H. (1990). Prodrugs as drug delivery systems. 107. Synthesis and chemical and enzymatic hydrolysis kinetics of various mono- and diester prodrugs of *N*-acetylcysteine. *Int. J. Pharm.*, **60**, 193–205.

Klixbüll, U. & Bundgaard, H. (1984). Prodrugs as drug delivery systems. 30. 4-Imidazolidinones as potential bioreversible derivatives for the α-amino amide moiety in peptides. *Int. J. Pharm.*, **20**, 273–84.

Larsen, J.D. & Bundgaard, H. (1987). Prodrug forms for the sulfonamide group. I. Evaluation of *N*-acyl derivatives, *N*-sulfonylamidines, *N*-sulfonylsulfilimines and sulfonylureas as possible prodrug derivatives. *Int. J. Pharm.*, **37**, 87–95.

Larsen, J.D. & Bundgaard, H. (1989). Prodrug forms for the sulfonamide group. III. Chemical and enzymatic hydrolysis of various *N*-sulfonyl imidates—a novel prodrug form for a sulfonamide group or an ester function. *Int. J. Pharm.*, **51**, 27–38.

Lee, V.H.L., & Bundgaard, H. (1992). In: K.B. Sloan (Ed.), *Prodrugs Topical and Ocular Drug Delivery*, pp. 221–97, Marcel Dekker, New York.

Lee, V.H.L. & Li, H.K. (1989). Prodrugs for improved ocular drug delivery. *Adv. Drug Delivery Rev.*, **3**, 1–38.

Lee, V.H.L. & Yamamoto, A. (1990). Penetration and enzymatic barriers to peptide and protein absorption. *Adv. Drug Delivery Rev.*, **4**, 171–207.

Lindberg, P., Nordberg, P., Alminger, T., Brändström, A. & Wallmark, B. (1986). The mechanism of action of the gastric acid secretion inhibitor omeprazole. *J. Med. Chem.*, **29**, 1327–9.

Lindberg, P., Brändström, A., Wallmark, B., Mattsson, H., Rikner, L. & Hoffmann, K.-J. (1990). Omeprazole: The first proton pump inhibitor. *Med. Res. Rev.*, **10**, 1–54.

Lokind, K.B., Lorenzen, F.H. & Bundgaard, H. (1991). Oral bioavailability of 17β-estradiol and various ester prodrugs in rats. *Int. J. Pharm.*, **76**, 177–82.

Mandell, A.I., Stentz, F. & Kitabchi, A.B. (1978). Dipivalyl epinephrine: A new prodrug in the treatment of glaucoma. *Ophthalmology*, **85**, 268–75.

Meisenberg, G. & Simmons W.H. (1983). Peptides and the blood–brain barrier. *Life Sci.*, **32**, 2611–23.

Metcalf, G. & Jackson, I.M.D. (Eds.) (1989). Thyrotropin-releasing hormone. Biomedical significance. *Ann. N.Y. Acad. Sci.*, **553**, 1–631.

Møss, J. & Bundgaard, H. (1990a). Kinetics and pattern of degradation of thyrotropin-releasing hormone (TRH) in human plasma. *Pharm. Res.*, **7**, 751–5.

Møss, J. & Bundgaard, H. (1990b). Prodrugs of peptides. 7. Transdermal delivery of thyrotropin-releasing hormone (TRH) via prodrugs. *Int. J. Pharm.*, **66**, 39–45.

Nagai, Y., Yokohama, S., Nagawa, Y. Hirooka, Y. & Nihei, N. (1980). Blood level and brain distribution of thyrotropin-releasing hormone (TRH) determined by radioimmunoassay after intravenous administration in rats. *J. Pharm. Dyn.*, **3**, 500–6.

Nielsen, N.M. & Bundgaard, H. (1988). Glycolamide esters as biolabile prodrugs of carboxylic acid agents: Synthesis, stability, bioconversion, and physicochemical properties. *J. Pharm. Sci.*, **77**, 285–98.

Norbeck, D.W., Rosenbrook, W., Kramer, J.B., Grampovnik, D.J. & Lartey, P.A. (1989). A novel prodrug of an impermeant inhibitor of 3-deoxy-D-manno-2-octulosonate cytidylyl-transferase has antibacterial activity. *J. Med. Chem.*, **32**, 625–9.

Notari, R.E. (1981). Prodrug design. *Pharmacol Therap.*, **14**, 25–53.

Pop, E., Wu, W.-M., Shek, E. & Bodor, N. (1989). Brain-specific chemical delivery systems for β-lactam antibiotics. Synthesis and properties of some dihydropyridine and dehydroisoquinoline derivatives of benzylpenicillin. *J. Med. Chem.*, **32**, 1774–81.

Rasmussen, G.J. & Bundgaard, H. (1991a). Prodrugs of peptides. 10. Protection of di- and tripeptides against aminopeptidase by formation of bioreversible 4-imidazolidinone derivatives. *Int. J. Pharm.*, **71**, 45–53.

Rasmussen, G.J. & Bundgaard, H. (1991b). Prodrugs of peptides. 15. 4-Imidazolidinone prodrug derivatives of enkephalines to prevent aminopeptidase-catalysed metabolism in plasma and absorptive mucosae. *Int. J. Pharm.*, **76**, 113–22.

Roche, E.B. (Ed.) (1977). *Design of Biopharmaceutical Properties Through Prodrugs and Analogs*, American Pharmaceutical Association, Washington D.C.

Roche, E.B. (Ed.) (1987). *Bioreversible Carriers in Drug Design. Theory and Application.* Pergamon Press, New York.

Saari, W.S., Schwering, J.E., Lyle, P.A., Smith, S.J. & Engelhardt, E.L. (1990). Cyclization – activated prodrugs. Basic carbamates of 4-hydroxyanisole. *J. Med. Chem.*, **33**, 97–101.

Sinkula, A.A. & Yalkowsky, S.H. (1975). Rationale for design of biologically reversible drug derivatives: Prodrugs. *J. Pharm. Sci.*, **64**, 181–210.

Sjödin, K., Nilsson, E., Hallberg, A. & Tunek, A. (1989). Metabolism of *N*-acetyl-L-cysteine. Some structural requirements for the deacetylation and consequences for the oral bioavailability. *Biochem. Pharmacol.*, **38**, 3981–5.

Sloan, K.B. (1989). Prodrugs for dermal delivery. *Adv. Drug Delivery Rev.*, **3**, 67–101.

Srivastva, D.N. & Farquhar, D. (1984). Bioreversible phosphate protective groups: Synthesis and stability of model acyloxymethyl phosphates. *Bioorg. Chem.*, **12**, 118–29.

Stella, V. (1975). In: T. Higuchi and V. Stella (Eds.), *Pro-drugs as Novel Drug Delivery Systems*, pp. 1–115, American Chemical Society, Washington, D.C.

Stella, V.J & Himmelstein, K.J. (1980). Prodrugs and site-specific drug delivery. J. Med. Chem., **24**, 1275–82.

Stella, V.J. & Himmelstein, K.J. (1982). In: H. Bundgaard, A.B. Hansen and H. Kofod (Eds.), *Optimization of Drug Delivery*, pp. 134–53, Munksgaard, Copenhagen.

Stella, V.J. Charman, W.N.A. & Naringrekar, V.H. (1985). Prodrugs: Do they have advantages in clinical practice? *Drugs*, **29**, 445–73.

Sugrue, M.F., Gautheron, P., Grove, J., Mallorga, P., Viader, M.-P., Baldwin, J.P., Ponticello, G.S. & Varga, S.L. (1988). L-653,328: an ocular hypotensive agent with modest beta-receptor blocking activity. *Invest. Ophthalmol. Vis. Sci.*, **29**, 776–84.

Suzuki, S., Hongu, Y., Fukazawa, H., Ichihara, S. & Shimizu, H. (1980). Tissue distribution of 5′-deoxy-5-fluorouridine and derived 5-fluorouracil in tumor-bearing mice and rats. *Gann*, **71**, 238–45.

Svensson, L.-Å. & Tunek, A. (1988). The design and bioactivation of presystemically stable prodrugs. *Drug Metab. Rev.*, **19**, 165–94.

Todd, P.A. & Goa, K.L. (1990). Simvastatin. A review of its pharmacological properties and therapeutic potential in hypercholesterolaemia. *Drugs*, **40**, 583–607.

Veber, D.F. & Freidinger, R.M. (1985). The design of metabolically stable peptide analogs. *Trends Neurosci.*, **8**, 392–6.

Vickers, S., Duncan, C.A., Chen, I.-W., Rosegay, A. & Duggan, D.E. (1990). Metabolic disposition studies on simvastatin, a cholesterol-lowering prodrug. *Drug Metab. Disp.*, **18**, 138–45.

Wermuth, C.G. (1981). Modulation of natural substances in order to improve their pharmacokinetic properties. In: J.L. Beal and E. Reinhard (Eds.), *Natural Products as Medicinal Agents*, pp. 185–216, Hippokrates Verlag, Stuggart.

Wermuth, C.G. (1984). Designing prodrugs and bioprecursors. In: G. Jollès and K.R.H. Wooldridge (Eds.), *Drug Design: Fact or Fantasy?*, pp. 47–72, Academic Press, London.

Wilk, S., Mizoguchi, H. & Orlowski, M. (1978). γ-Glutamyl dopa: A kidney specific dopamine precursor. *J. Pharmacol. Exp. Therap.*, **206**, 227–32.

Yalkowsky, S.H. & Morozowich, W. (1980). In: E.J. Ariëns (Ed.), *Drug Design*, pp. 121–85, Academic Press, New York.

Drug Targeting

22 Antibody-based Drug Targeting Approaches: Perspectives and Challenges

D.J.A. CROMMELIN, J. BERGERS and J. ZUIDEMA
Department of Pharmaceutics, University of Utrecht, PO Box 80.082, 3508 TB Utrecht, The Netherlands

1 Drug targeting

With our present arsenal of drugs, humankind can treat or cure a large number of diseases that were life threathening in the past. However, the success with chemotherapy in the treatment of cancer and viral diseases, like AIDS, has been limited. The therapeutic index of the current drugs developed for these diseases is too small. Toxic and therapeutic doses are too close together. The causes of failure of the present generation of drugs in combating diseases like cancer and certain microbial, parasitic and viral diseases, can be summarized as follows:

1 The active compound is rapidly eliminated in intact form from the body through the kidneys or the liver, or inactivated through metabolic action and never reaches the target site.

2 Accumulation of the drug at the target site is the exception and not the rule; most of the drug is distributed over other organs exerting its (toxic) effects there.

3 The tendency of a drug to be taken up into the target cells is often rather low, which poses a problem if intracellular delivery is required for its action.

Drug targeting is the approach to increase the therapeutic index of drugs by delivering the active compound specifically at its site of action and to keep it there until it has been inactivated, so maximizing the therapeutic effects and avoiding toxic effects elsewhere. Although the basics of the concept of drug targeting were defined in the early days of this century by Paul Ehrlich (Baumler, 1984), only in the last decade has substantial progress been made. Knowledge about the pathophysiology of diseases at the cellular and molecular level has been growing rapidly. Insights about the anatomical and physiological barriers to be overcome to reach target sites have increased considerably. Cell-specific receptors have been identified and homing devices developed. And—last but not least—the number of technological options (e.g. carriers) for drug delivery has grown considerably.

Site-specific and 'timed' delivery is particularly important for the new generation of protein-based drugs. Proteins produced via recombinant DNA or hybridoma technology represent a group of potentially powerful therapeutic substances. Many of them act as endocrine or paracrine-like mediators (e.g. interferon-2α). For optimum delivery of these compounds information about the timing is required as they often participate in a polymediator cascade of events. Besides, the dose–response relationships of these compounds can deviate from the traditional S-shape and show, for instance, a bell-shape. Therefore, these drugs offer a great challenge for the designer of the delivery system: access, retention and timing are the key issues for a successful therapeutic effect (Tomlinson, 1987).

Several approaches can be discerned to achieve site-specific delivery of drugs. The prodrug concept has gained a lot of interest and has proven to be successful.

Prodrugs are defined as drug molecules modified in such a way that the inactivated derivative is reactivated *in vivo*, producing the active parent compound. Either site-directed drug delivery or site-specific drug release or activation can be the purpose for the design of a prodrug (Bundgaard, 1983). An interesting and elegant example of prodrug development focusing on site-directed delivery is the so-called 'lock-in principle' for delivery of drugs to the brain. Briefly, lipophilic prodrugs, which can pass through the blood–brain barrier are converted in the brain into a hydrophilic form. This hydrophilic form is 'locked in' inside the brain and slowly releases the active compound. The 'lock-in' effect is mainly restricted to the brain, because of the properties—only permeable for lipophilic compounds—of the blood–brain barrier surrounding it (e.g. Bodor & Loftsson, 1987).

Alternatively, a combination of a carrier system (protection of the drug, manipulation of the release kinetics) with a homing device can be used to deliver the drug at the site of its action and to keep it there over the desired period of time. In these delivery systems regularly three units can be discerned: (i) an active moiety (drug), (ii) a carrier and (iii) a homing device. Sometimes two or even all three functions can be assigned to one basic molecular structure (e.g. non-modified antibody).

This review will deal only with the potential and limitations of these carrier-based site-specific drug-delivery systems; it will focus on the approaches where monoclonal antibodies are being used—either as such or in modified form: conjugated with an active moiety or in the form of drug-laden liposomes.

2 Transport considerations for site-specific delivery: Anatomical, physiological and pathological hurdles

Upon injection carrier transport in the body depends on the physico-chemical properties of the carrier: its size/molecular weight, surface hydrophobicity, charge and the presence of ligands for interaction with surface receptors (Crommelin & Storm, 1990). It is important fully to appreciate the nature of the barriers that have to be passed by the homing device–carrier–drug combination. A schematic picture of the capillary wall structures and their pores is given in Fig. 22.1 which shows the intact endothelium. The fenestrae in the liver are about 100–200 nm; the pores between the endothelial cells and in the basement membrane outside liver, spleen and bone marrow are much smaller. Under pathological conditions (such as tumours and inflammation sites) a higher permeability of the endothelium can exist allowing larger particles to enter tumour tissue (e.g. Poste, 1985; Gabizon & Papahadjopoulos, 1988). Long circulating, small and stable liposomes (see below) are required for optimum localization in tumour tissue. On the other hand, tumour tissue can be poorly perfused and difficult to enter, because of the presence of necrotic material. Tumours often lack lymphatic drainage causing a pressure build-up and induction of an outward flow (Jain, 1987a,b). This pressure may interfere with diffusional uptake of targeted material into tumour tissue.

Indices have been defined in the literature to express the degree of drug targeting. The Drug Targeting Index (DTI) is the ratio of drug delivered to the response and toxicity sites in the case of targeted delivery divided by this ratio in

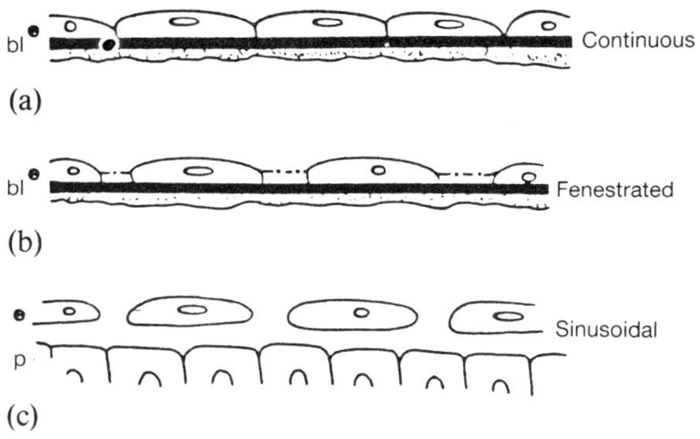

Figure 22.1. Schematic illustration of the structure of different classes of blood capillaries. (a) Continuous capillary. The endothelium is continuous with tight junctions between adjacent endothelial cells. The subendothelial basement membrane is also continuous. (b) Fenestrated capillary. The endothelium exhibits a series of fenestrae which are sealed by a membranous diaphragm. The subendothelial basement membrane is continuous. (c) Discontinuous (sinusoidal) capillary. The overlying endothelium contains numerous gaps of varying size enabling materials in the circulation to gain access to the underlying parenchymal cells. The subendothelial basement is either absent (liver) or present as a fragmented interrupted structure (spleen, bone marrow). (From Poste, 1985.)

the case of free drug (Hunt *et al.*, 1986; Boddy *et al.*, 1989). The therapeutic availability (TA) is defined as the rate of input of free drug over the input rate of targeted drug to achieve the same maximal therapeutic effect. Another targeting index can be defined by substituting 'maximal therapeutic effects' for toxic effects for the compartment where the toxicity is localized.

On the basis of the kinetic models developed so far, a number of conclusions can be drawn for situations where targeted delivery is, in principle, advantageous. Drugs with high total clearance are good candidates for targeted delivery. Response sites with a relatively small blood flow require carrier-mediated transport. Increases in the rate of elimination of free drug from either central or response compartments tend to increase the need for targeted drug delivery; this also implies a higher input rate of the drug–carrier conjugate to maintain the therapeutic effect. To maximize the targeting effect, the release of drug from the carrier should be restricted to the response compartment. Boddy *et al.* (1989) also included pharmacodynamics (dose–response curve) in their considerations. Gupta & Hung (1989) emphasized the importance of taking tissue weight into account in the calculation of drug targeting parameters for a proper evaluation of the targeting potential. In practice, the targeting potential is not regularly evaluated through the above-mentioned indices. Data on levels of free drug in target tissue after administration of a targeted delivery system are rare since data on the intact conjugate and released free drug in the response and toxicity compartment are not available.

Theoretical studies on the requirements for optimum delivery of antibodies to tumour tissue were performed by Weinstein and collaborators (Weinstein *et al.*, 1986; Fujimori *et al.*, 1989). After passage through the endothelial wall and the basement membrane, a targeted macromolecule (e.g. a monoclonal antibody (MAb) directed against tumour-specific antigens) has to penetrate into the tumour tissue. Convection is considered the most important process over macroscopic distances; the basis for transport across microscopic distances into densely packed tumour material is presumably diffusion. An interesting conclusion derived from their work is that for deep penetration of the antibody into the tumour tissue under certain conditions, a lower-affinity antibody may be preferable to a high-affinity antibody as the latter is kept at the outside of the tumour mass and does not distribute evenly over it. Antigen–antibody interaction imposes a 'binding-site barrier' retarding MAb percolation into the tumour; consequently the lower molecular weight and affinity of Fab′ (fragments of the MAb) would favour—theoretically—a more uniform distribution over the tumour mass than the full immunoglobulin G (IgG) molecule.

In conclusion, anatomical and physiological constraints may limit the distribution of conjugates over the body and therefore access to the target site. There are still questions concerning the effect of pathological conditions on distribution characteristics. Presently, pharmacokinetic studies provide theoretical guidelines for the rational selection of the drug, homing device and carrier. Experimental attempts to validate these guidelines have not been published yet.

3 (Monoclonal) antibodies (MAb) as targeted therapeutic agents

Antibodies can act simultaneously as a homing device and active agent. After attachment to their target cell, MAb may affect the functioning of the cell (e.g. kill it). Via its Fc receptor, complement can be bound which may cause lysis of the target cell. Alternatively, contact with macrophages can be established, or certain Fc-receptor bearing killer (K) cells can induce 'antibody-dependent cell-mediated cytotoxicity' (ADCC) (Roitt *et al.*, 1987). Besides, a blockade of certain essential cell surface receptors by MAb can cause metabolic deficiencies (Lowder, 1986; Crommelin & Storm, 1990).

OKT3, a murine MAb of the IgG_{2a} class, is marketed for the treatment of allograft rejection. It is claimed that OKT3 can induce reversal of transplant rejection through different mechanisms: it blocks the killer function of sessile T-cells and, besides, it interacts with the CD3–T-cell receptor complex on late thymocytes and mature T-cells. After opsonization the MAb–T-cell complex is taken up by macrophages. Finally, it modulates the CD3–T-cell receptor complex of circulating T-cells, resulting in a cell with all phenotypic characteristics with the exception of the CD3–T-cell receptor complex (Goldstein, 1987). Antimouse antibody production after administration of OKT3 is a serious problem preventing re-treatment with many murine MAb. It can be partly avoided by concomitant administration of immunosuppressive agents (e.g. Shield *et al.*, 1987). The use of $F(ab')_2$ or F(ab′) fragments is a way of reducing the likelihood raising an immune response against the Fc part. Another, more challenging, approach is the development of human or humanized MAb (Burton *et al.*, 1990). Two different approaches for the humanization of MAb are presently under development: chimeric molecules consisting of a human Fc part and a murine Fab part, or complementary determining regions (CDRs), grafted molecules in which the genetic elements encoding for the six CDRs of the murine antibody are ligated into a human immunoglobulin gene. In principle the CDR approach minimizes the exposure to murine material. Completely human MAb can be produced via heterohybridomas where human antibody genes are transfected in mouse cells, which subsequently produce the human MAb. Alternatively, transgenic mice can be used. Although these measures reduce the immunogenicity of the existing generation of murine MAb, the problem of anti-idiotypic immune responses still exists in all these human or humanized MAb.

Bispecific MAb provide an interesting therapeutic potential. Bispecific MAb expose two different antigen-binding sites combined in one antibody structure. Bispecific MAb offer the opportunity to bring target cells or tissue (one antigen-binding site) in contact with other structures (second antigen-binding site). An example is the re-targeting after intraperitoneal injection of rIl2-stimulated, autologous T-lymphocytes with the bispecific MAb exposed on their surface into patients with intraperitoneally located ovarian carcinoma cells. This MAb combines specificity for T-cells with a carcinoma antigen and the bispecific MAb are *in vitro* hooked up to the stimulated T-lymphocytes prior to intraperitoneal injection (De Leij *et al.*, 1990). Promising results in a limited group of patients with a malignant glioma were recently published. A chemically combined anti-CD3 antibody and an antiglioma antibody was incubated with lymphokine-activated killer (LAK) T-cells *in vitro* and subsequently injected (Nitta *et al.*, 1990). Another example is the enhanced binding of tissue-plasminogen activator (t-PA) to fibrin using bispecific monoclonal antibodies exposing antigen-binding sites to both fibrin and t-PA (Bos *et al.*, 1990).

4 Immunoconjugates as targeted therapeutic agents

To improve the therapeutic potency of MAb, combinations of a MAb with a drug have been developed. The efforts described in the literature mainly focus on the treatment of cancer (Crommelin & Storm, 1990; Mitra & Ghosh, 1990; Ramakrishnan, 1990). Many different drugs have been used for conjugation with MAb (Hellström *et al.*, 1987; Pietersz *et al.*, 1988). Covalent binding strategies have gained in sophistication over the years (Crommelin & Storm, 1990; Mitra & Gosh, 1990) and several effector molecules can be bound to one MAb. The best candidates are considered to be cytostatics with a high intrinsic cytotoxicity, as only a limited number of antibodies can be bound to a tumour cell. Crommelin & Storm (1990) dealt with the potential problems encountered with the constructed conjugates. Covalent binding can change the cytostatic potential of the drug and decrease the affinity of the MAb for the antigen. The stability of the conjugate *in vivo* can be insufficient; fragmentation will lead to loss of targeting potential. And, finally, the immunogenicity of the MAb and toxicity of the drug involved can change dramatically.

By far the most popular group of cytotoxic agents conjugated with MAb are the toxins. Ricin, a toxin with a protein structure and a molecular weight of 66 kD, has been the prime target for MAb conjugation. The unmodified compound is extremely cytotoxic as it enzymatically blocks intracellular protein synthesis at the ribosomal level. Unmodified ricin conjugated with MAb still has a limited target-cell specificity; sequestration in the liver is a major problem. Reportedly, only about 1% of the dose accumulates in the tumour tissue (Buchsbaum & Lawrence, 1990; Ramakrishnan, 1990). Besides, phase I clinical trials show that the present conjugates are quite immunogenic. Therefore, at the present time attempts are being made to reduce non-target site delivery and immunogenicity by adapting the ricin molecule (e.g. by genetic engeneering). Human or humanized MAb instead of murine MAb and modified toxins (e.g. blocking or decoupling the galactose receptor sites, production of non-glycosylated toxins or polyethyleneglycol (PEG) attachment) might reduce the problem of immunogenicity and non-target site delivery. More details about the 'state of the art' of immunotoxins can be found in Crommelin & Storm (1990) and Ramakrishnan (1990).

MAb may thus play a critical role as they act as homing devices to tumour tissue. The problem of reaching the target tissue and penetration into it has already been dealt with. The questions now arise: how well can MAb discriminate target from non-target cells? And, do all tumour cells expose the tumour-associated antigen?

The literature dealing with these questions is abundant (e.g. Hellström *et al.*, 1987). Many MAb raised against tumours are directed against differentiation antigens. Differentiation antigens are not fully unique for tumour cells; the specificity of differentiation antigens for tumour tissue compared with normal tissue is quantitative rather than qualitative. This means that for these surface antigens, no absolute tumour-specific targeting can be expected. Apart from these differentiation antigens 'clone-specific' antigens occur. They are unique to the clone forming the tumour. The use of clone-specific antigens probably implies that for each patient MAb have to be tailor-made.

Tumour heterogeneity, antigen shedding and antigen modulation are phenomena that can interfere with the 'homing' process. Tumours tend to be heterogeneous as they consist of many subpopulations of cells differing in expressing antigens. This means that not all cells in a tumour will interact with the targeted conjugate. Besides, tumour cells can avoid recognition through antigen shedding or antigen modulation. Shedding of antigens to the extracellular compartment interferes with the homing function of the circulating MAb through MAb–antigen complex formation. Binding of MAb to the cell-surface antigen can cause antigen modulation, implying that the cell no longer exposes this antigen.

The problem of tumour-cell heterogeneity may be solved by using 'cocktails' of MAb, or by the use of drugs that are released from the conjugate after reaching a target cell (cf. Fig. 22.6), or by attaching radio isotopes with a range of action beyond the dimensions of only one cell. Shedding and modulation of antigens does not occur with all antigens. The antigen/MAb combinations that do not demonstrate these effects should be selected. Another approach to circumvent early inactivation of MAb by shedding antigens presently under investigation is the injection of pure MAb prior to the conjugate to bind shedded material.

In conclusion, MAb and MAb-conjugates can, in principle, contribute to our therapeutic arsenal to fight diseases like cancer. However, the optimum conditions and structure of the drug molecule and the (modified) MAb have not been properly defined yet.

5 Liposomes and immunoliposomes

Many colloidal particulate systems varying in size, charge, release mechanism and biodegradability have been tested as carriers for site-specific drug delivery. A full overview of the literature has been published by Tomlinson (1987). As particulate systems tend to be over 50 nm in diameter, they are, in general, unable to pass through epithelial and endothelial membranes (see above). Their fate *in vivo* depends on their stability in the blood, their size, charge and surface-exposed molecules. Unless special measures are taken stable colloidal particulates are taken up by the cells of the mononuclear phagocyte system (MPS) (Davis & Illum, 1986; Tomlinson, 1987; Crommelin & Storm, 1990).

Among the colloidal particulate systems proposed for site-specific drug delivery, liposomes have gained considerable attention (e.g. Gregoriadis, 1984, 1988). Liposomes are water-filled, vesicular structures based on lipid bilayers surrounding an aqueous core (Fig. 22.2). Dependent on the preparation technique, they vary in size, charge and bilayer rigidity. They can carry a drug either in the lipid core of the bilayer through partitioning or by physical entrapment in the aqueous phase; homing devices can be attached to the outside by well-established covalent coupling techniques (Toonen & Crommelin, 1983; Green & Widder, 1987). Liposomes stand out among particulate carriers because of their relatively low toxicity (Zonneveld & Crommelin, 1988; Storm *et al.*, 1991) and the versatility of their release characteristics and disposition *in vivo* by changing preparation techniques and bilayer constituents (Senior, 1987). Dependent on their size, charge, bilayer rigidity —and possibly some still unknown factors—liposomes circulate only for a short time (minutes) in the circulation before degradation or uptake by macrophages of the MPS occurs; alternatively, their residence time in blood can be hours or even days if they are stable and not recognized by macrophages as 'foreign body-like' structures. The exact surface characteristics required for long-circulating liposomes are still under discussion. Monosialoganglioside (GM_1), phosphatidylinositol (PI)

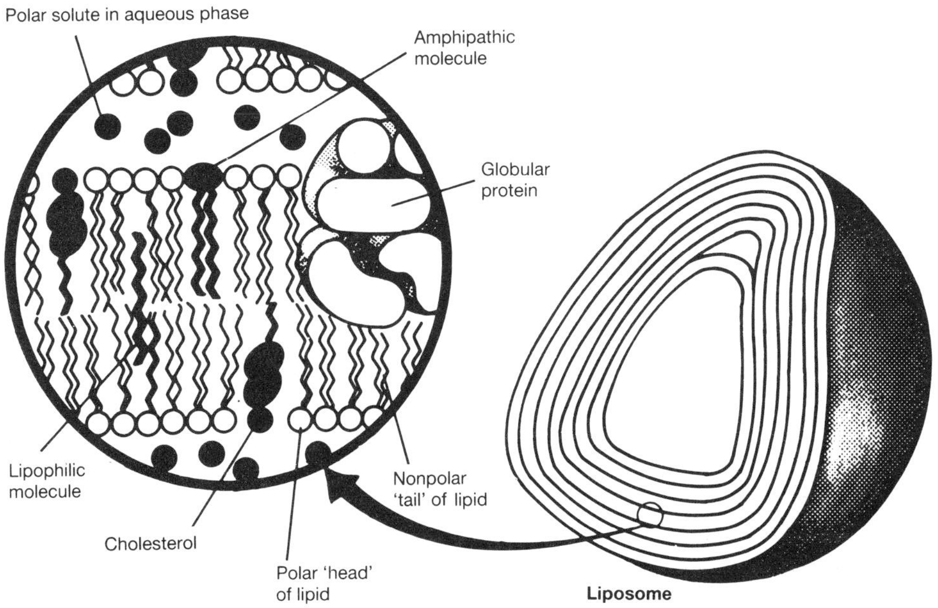

Figure 22.2. A simplified representation of the sites of interactions of hydrophilic, lipophilic and amphipathic molecules, cholesterol and proteins with multilamellar liposomes. (From Fendler, 1980.)

or polyethyleneglycol–phosphatidylethanolamine incorporation into bilayers prolongs blood residence times (Gabizon & Papahadjopoulos, 1988; Allen, 1989; Klibanov *et al.*, 1990; Woodle *et al.*, 1990).

The accumulation of drug-laden liposomes in macrophages offers interesting opportunities to fight macrophage-located microbial, viral or bacterial diseases effectively (Emmen & Storm, 1987). Alternatively, strong macrophage activitation can be induced by liposome-encapsulated lymphokines and 'microbial' products, e.g. interferon-γ or muramyltripeptide–phosphatidylethanolamine (MTP–PE) (Fidler, 1986; Storm *et al.*, 1990).

Liposome targeting to other sites of the body requires homing devices. Several attempts have been made to accumulate antibody (fragment)–liposome combinations (immunoliposomes) at predetermined sites in the body. Usually, the antibody or antibody fragment is covalently bound to the surface of liposomes via an anchor molecule in the bilayer (Toonen & Crommelin, 1983; Peeters *et al.*, 1987; Wright & Huang, 1989). The experience with immunoliposomes obtained so far indicates that after intravenous injection, target sites outside the blood circulation are difficult to reach, presumably because of their relatively short circulation time and the poorly permeable endothelial lining at target sites (e.g. Matzku *et al.*, 1990). A different situation exists for blood cells like red blood cells (RBC) or the cells of the endothelial lining (Hughes *et al.*, 1989; Peeters *et al.*, 1989). An example of targeting of Fab′-immunoliposomes to mouse RBC (Cr-labelled mRBC) circulating in rats is depicted in Fig. 22.3. Briefly, Cr-mRBC were injected into rats. These Cr-mRBC circulated in the blood compartment and over about 1 h only a small fraction disappeared from the blood. Subsequently, immunoliposomes (exposing anti-mRBC Fab′ fragments) with different Fab′ density were injected. It is clear that these immunoliposomes caused a sudden and—dependent on the Fab′ density—dramatic drop in circulating Cr-label (mRBC). In Fig. 22.4, it is demonstrated that about 2 h after injection, a major part of the label accumulated in the spleen and liver. Presumably, the macrophages of the rat recognized the immunoliposome-coated mRBC as 'foreign body-like' structures and phagocytosed them. This concept of 'dragging' circulating target cells to the macrophages was used to cure rats infected with mRBC containing the malaria parasite *Plasmodium berghei*. A dramatic increase in survival rate (number of rats radically cured) was observed with chloroquine (CQ)-loaded immunoliposomes (anti-mRBC Fab′ fragments) compared to a number of control experiments (Table 22.1). In this experimental model (parasitized mRBC in rats) evidence was collected that intravenously injected immunoliposomes with a drug: (i) could reach their target cells (mRBC), (ii) manipulate the target-cell fate (macrophage uptake) and (iii) attack the intracellularly located parasite. This approach offers promising opportunites to treat other diseases where the target site is located in the blood compartment (e.g. infected or dysfunctioning lymphocytes, thrombi).

In certain conditions, the diseased tissue is confined to a particular cavity in the body. This is true, for instance, for ovarian carcinoma, which usually stays throughout its clinical life mainly within the peritoneal cavity. After diagnosis the prognosis for ovarian carcinoma patients is poor. The primary tumours and larger tumour nodules can be removed surgically, but micronodules and floating tumour

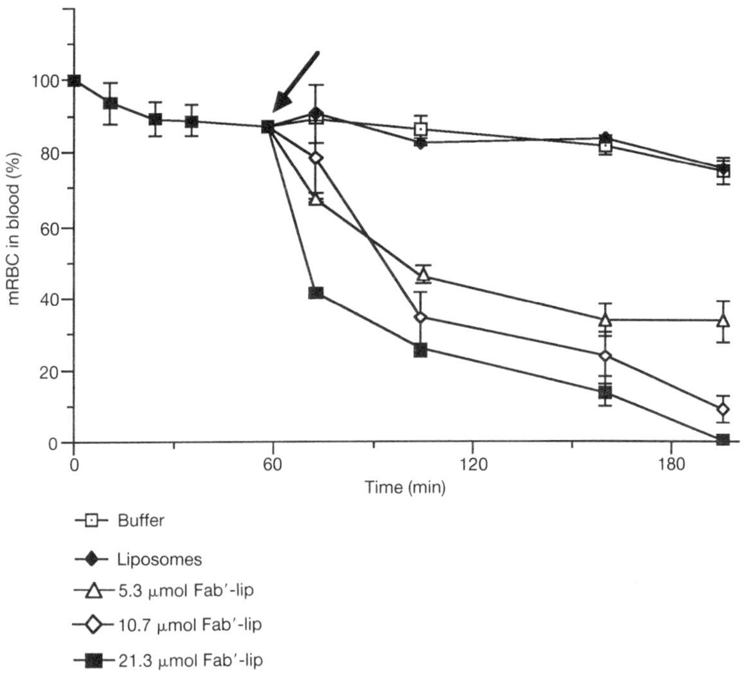

Figure 22.3. Blood levels of Cr-mRBC (see text) in rats after intravenous administration of liposomes or anti-mRBC Fab′ liposomes with different liposomal Fab′ densities: 1.5×10^9 Cr-labelled mouse RBC (Cr-mRBC) were injected per rat. In all cases a total amount of 10 μmol phospholipid was injected in a total volume of 0.5 ml, 1 h after injection of Cr-mRBC (arrow). Each data point represents the mean and standard deviation (error bars) of four animals. Small standard deviations are not shown. The liposome composition was cholesterol: phosphatidylcholine: phosphatidylserine: anchor-PE: 10:9.5:1:0.5 (molar ratio); mean particle size was 0.32 μm. (From Peeters *et al.*, 1989.)

cells cannot be treated adequately by surgery. Intraperitoneal injection of cytostatic solutions is being used with considerable success, but the residence time in the cavity is short providing insufficient opportunity for deep intratumoral penetration and a major fraction still reaches the systemic circulation (Los *et al.*, 1989). Sustained release of cisplatin, a cytostatic regularly used in the treatment of ovarian

Table 22.1. Effect of treatment with Fab′ liposomes containing CQ (anti-mRBC Fab′-lipCQ) on the number of rats radically cured after intravenous infection with parasitized (*Pl. berghei*) mRBC

Treatment	Number of rats radically cured*
Anti-mRBC Fab′-lipCQ†	6 (12)
Lip-CQ	0 (8)
CQ	0 (8)
Buffer	0 (8)

* No parasites were detected during a 28-day period in rats infected with 10^5 parasitized mRBC. Number in parentheses indicates total number of rats per group.

† CQ (chloroquine) dose was 0.8 mg per rat. The data are summarized from two independently performed experiments. Other relevant control experiments, e.g. administration of anti-mRBC Fab′-lip without CQ showed no radically cured rats. (From Peeters *et al.*, 1989, figure 5.)

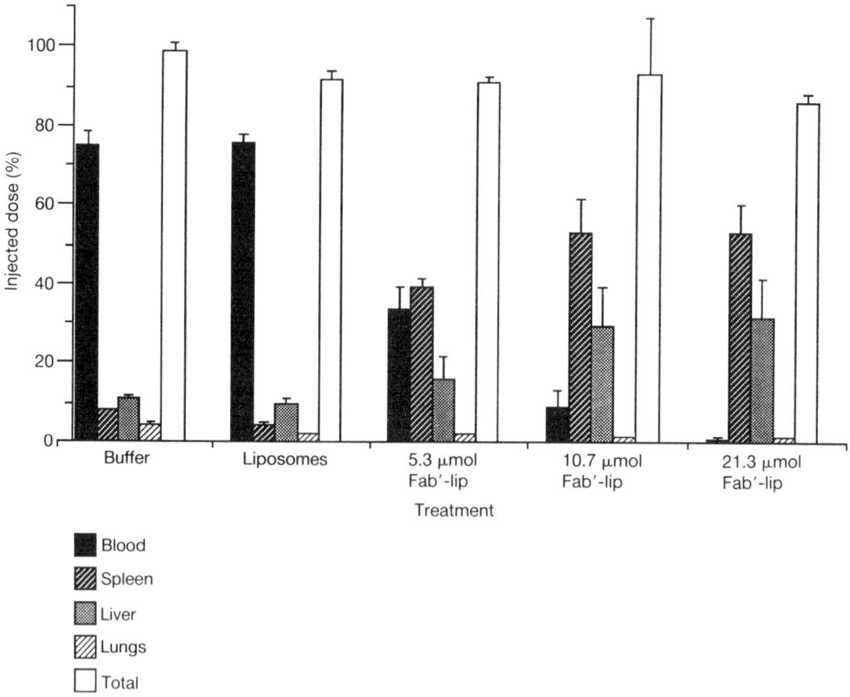

Figure 22.4. Recovery of Cr-mRBC in spleen, liver, lungs, blood and their sum. Animals were killed 2 h after intravenous administration of liposomes or antimouse RBC immunoliposomes (Fab′ liposomes) with different liposomal Fab′ densities (indicated as μg/μmol phospholipid). Presented data are the mean and standard deviation (error bars) from four animals. For further details see Figure 22.3. (From Peeters *et al.*, 1989.)

carcinomas, from hydrogel rods implanted in the peritoneal cavity reduced the dose-limiting renal toxicity of the drug, but also negatively affected the antitumour effect (Los *et al.*, 1991). As sustained release in the peritoneal cavity alone did not improve the therapeutic index of the drug, targeted delivery of cytostatics to the ovarian carcinoma cells with immunoliposomes was a logical next step in the optimization of chemotherapy. Here the aim is to deliver drug-laden liposomes directly at the surface of—or preferably in—the ovarian carcinoma cells (Straubinger *et al.*, 1988). *In vitro* immunoliposomes bearing methotrexate-γ-aspartate showed an eightfold increase in potency against human ovarian carcinoma cells. After intraperitoneal injection, a specific interaction between these human ovarian carcinoma cells and the immunoliposomes was observed qualitatively. Nässander & colleagues (1990) performed similar experiments and quantified *in vitro* the binding of immunoliposomes with a murine Fab′ fragment (OV-TL-3) to the human ovarian carcinoma-3 (OVCAR-3) cells. In Fig. 22.5 the effect of increasing amounts of phospholipid (PL) added to the wells on the PL bound to the cells is shown. Binding efficacy and specificity could be clearly demonstrated. In athymic nude mice, a highly efficient and long-lasting adherence to tumour cells was observed (U.K. Nässander, pers. commun.).

After establishment of an immunoliposome–cell interaction the drug has to exert its action on the cell. There are several pathways proposed to reach this goal (Fig. 22.6) (Peeters *et al.*, 1987). Macrophages can phagocytose the cell plus

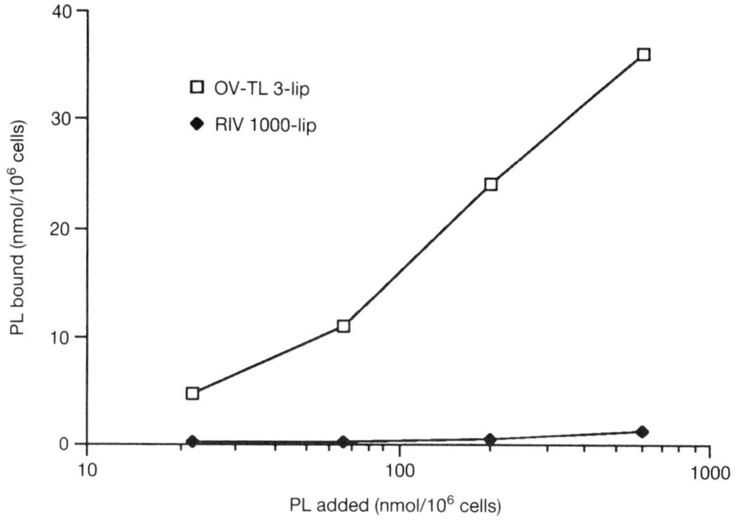

Figure 22.5. Binding of immunoliposomes to OVCAR-3 cells. Various amounts of OV-TL-3-liposomes (11 μg Fab′μmol PL) and RIV1000 liposomes (7 mg Fab′/μmol/PL) were incubated with OVCAR-3 cells (2×10^6 cells/ml). RIV1000 is a non-relevant antibody. The cells were washed three times to separate bound Fab′ liposomes from the unbound vesicles. The amount of cell binding was determined after lysing the liposomes, thereby releasing the entrapped carboxyfluorescein, a marker for the liposome integrity (see Nässander *et al.*, 1990).

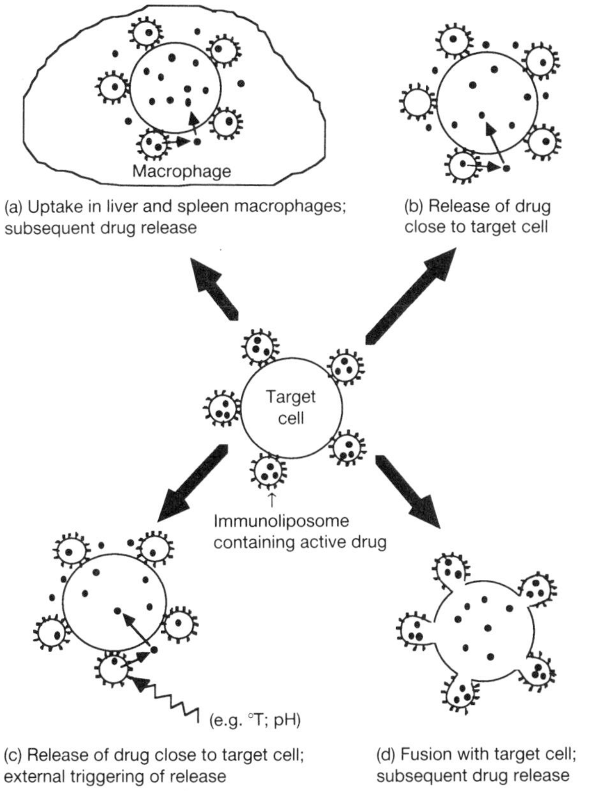

Figure 22.6. Several pathways of drug internalization after immunospecific binding of the immunoliposomes to the appropriate target cell (Peeters *et al.*, 1987).

adhering liposome and the drug can subsequently be released from the degrading liposome inside this macrophage in the close proximity of the co-phagocytosed target cell (discussed above with RBC). In the situation depicted in Fig. 22.6(b) the drug is released from the adhering immunoliposomes in the close proximity of the target cell. This can be achieved by constructing liposomal bilayers with delayed or sustained drug-release characteristics. External triggering of drug release from liposomal bilayers by external stimuli is a third option. Temperature-sensitive liposomes have been proposed to be used for this purpose. Finally (Fig. 22.6d) immunoliposomes carrying fusogenic proteins could deliver the immunoliposomes directly to the target-cell cytosol. In particular, the last option requires more insight into the basics of the interaction between liposomes and cells, particularly in combination with the homing devices used for targeting of the liposomes (Felgner, 1990).

6 Conclusions and perspectives

The concept of drug targeting has been developing at a rapid pace in the last decade. In particular, the availability of a new generation of homing devices in the form of monoclonal antibodies and a better understanding of the anatomy and physiology of the human body, also under pathological conditions, has been critical for the impressive progress made in the last decade. These new insights also revealed the potential—and even more important—the limitations of the different concepts currently under investigation.

The fate of macromolecules after intravenous injection strongly depends on molecular weight and charge (Takakura *et al.*, 1990). Too-small molecules are rapidly cleared; high-molecular-weight and negatively charged compounds tend to be excreted slowly, unless they are recognized by MPS cells. More work should be done to monitor the pharmacokinetics of carriers and associated drugs more closely to gain a better insight into the critical parameters controlling the fate of drug and carrier *in vivo*. This would allow the design of drug–carrier–homing device combinations on the basis of rationale instead of 'trial and error'.

At present only a minor fraction of the intravenously injected dose accumulates at the target site. An exception is the targeting to MPS cells in the liver and spleen; they can be reached almost quantitatively without loss to other tissues or via excretion. Particles in the colloidal size range avoid MPS phagocytosis only through the selection of special ingredients which leads to long circulation times (cf. Stealth® concept as mentioned above) or through rapid degradation in the blood compartment.

Apart from the obviously crucial design of advanced drug delivery systems like immunotoxins and immunoliposomes on laboratory scale, scaling up and quality assurance issues like reproducibility, purity and shelf-life require considerable attention, as experience so far shows that these issues pose many problems and hamper the final introduction of these advanced and sophisticated systems in therapy.

The first prototypes of these achievements have now reached the clinical stage. Hopefully, new, more specific and potent drug-delivery systems will follow. The approaches to drug targeting under development nowadays regularly include macromolecules, like proteins, and colloidal particles. Immune reactions often develop after the first injection, forcing the cessation of therapy. Then, the exploration of pathways to reduce the immunogenicity (e.g. by humanization of MAb or the use of human MAb) should be a priority. Another point to be made is that the drug involved in drug targeting should not necessarily be a drug that is already on the market and whose safety has been proven in an existing clinical protocol. Targeting of a drug implicity means that the disposition of the drug is different from the free drug and so is its spectrum of side-effects. This means that by developing the site-specific delivery concept a whole group of extremely potent drug molecules, like toxins, that were—because of their side-effects—never before considered for therapy have a chance to be used. Heath *et al.* (1985a,b) explored another approach: the development of carrier-dependent drugs. This concept concerns potentially potent drugs, which would affect the cell function through interference with intracellular processes; however, these compounds have a poor ability to enter cells. If carrier systems deliver these drugs directly (and specifically) into the cytosol of the target cell, they can exert their therapeutic action with a minimum of side-effects. A high degree of specificity and a high level of endocytosis of the carrier–drug combination is a prerequisite for success for these carrier-dependent drugs.

The development of drug targeting concepts is a typical multidisciplinary activity. Reviewing the literature dealing with site-specific drug delivery, it is clear that key articles are published in scientific journals with a widely varying signature. Molecular biologists, anatomists, physiologists, pathologists, cellular biologists, immunologists and pharmacists all contributed and will continue to contribute to the fine tuning of drug delivery in the future to fight life-threatening diseases. Only through input from all these disciplines will the dream of drug targeting come true.

7 References

Allen, T.M. (1989). In: G. Lopez-Berenstein and I.J. Fidler (Eds.), *Liposomes in the Therapy of Infectious Diseases and Cancer*, pp. 405–13, Alan, R. Liss, New York.

Baumler, E. (1984). *Paul Erhlich, Scientist for Life*, Holms & Meir, New York.

Boddy, A., Aarons, L. & Petrak, K. (1989). Efficiency of drug targeting: Steady state considerations using a three compartment model. *Pharm. Res.*, **6**, 367–72.

Bodor, N. & Loftsson, T. (1987). In: J.R. Robinson and V.H.L. Lee (Eds.), *Controlled Drug Delivery*, pp. 337–71, Marcel Dekker, New York.

Bos, R., Otter, M. & Nieuwenhuizen, W. (1990). In: D.J.A., Crommelin and H. Schellekens (Eds.), *From Clone to Clinic*, pp. 167–74, Kluwer Academic, Dordrecht.

Buchsbaum, D.J. & Lawrence, T.S. (1990). In: P. Tyle and B.P. Ram (Eds.), *Targeted Therapeutic Systems*, pp. 215–55, Marcel Dekker, New York.

Bundgaard, H. (1983). In: D.D. Breimer and P. Speiser (Eds.), *Topics in Pharmaceutical Sciences*, pp. 329–43, Elsevier Science Publishers, Amsterdam.

Burton, D.R., Persson, M.A., Huse, W.D. & Golub, E.S. (1990). In: D.J.A. Crommelin and H. Schellekens (Eds.), *From Clone to Clinic*, pp. 67–72, Kluwer Academic, Dordrecht.

Crommelin, D.J.A. & Storm, G. (1990). In: J.B. Taylor (Ed.), *Comprehensive Medicinal Chemistry*, Vol. 5, *Biopharmaceutics*, pp. 661–701, Pergamon Press, Oxford.

Davis, S.S. & Illum, L. (1986). In: E. Tomlinson and S.S. Davis (Eds.), *Site Specific Drug Delivery*, pp. 93–111, J. Wiley, Chichester.

De Leij, L., De Jonge, M.W.A., Ter Haar, J., Spakman, H., De Vries, E., Willemse, P., Mulder, N.H., Berendsen, H., Elias, M., Smit Sibinga, C., De Lau, W., Tax, W. & The, T.H. (1990). In: D.J.A. Crommelin and H. Schellekens (Eds.), *From Clone to Clinic*, pp. 159–65, Kluwer Academic, Dordrecht.

Emmen, F. & Storm, G. (1987). Liposomes in the treatment of infectious diseases. *Pharm. Weekbl. Sci. Ed.*, **9**, 162–71.

Felgner, P.L. (1990). Particulate systems and polymers for *in vitro* and *in vivo* delivery of polynucleotides. *Adv. Drug Del. Rev.*, **5**, 163–87.

Fendler, J.H. (1980). In: G. Gregoriadis and A.C. Allison (Eds.), *Liposomes in Biological Systems*, p. 87, J. Wiley, Chichester.

Fidler, I.J. (1986). In: E. Tomlinson and S.S. Davis (Eds.), *Site-Specific Drug Delivery,* pp. 111–34, J. Wiley, Chichester.

Fujimori, K., Covell, D.G., Fletcher, J.E. & Weinstein, J.N. (1989). Modeling of the global and microscopic distribution of immunoglobulin G, F(ab′)2 and Fab in tumours. *Cancer Res.*, **49**, 5656–63.

Gabizon, A. & Papahadjopoulos, D. (1988). Liposome formation with prolonged circulation time in blood and enhanced uptake by tumours. *Proc. Natl. Acad. Sci. USA*, **85**, 6949–53.

Goldstein, G. (1987). Monoclonal antibody specificity: Orthoclone OKT3 T-cell blocker. *Nephron*, **46**, Suppl. 1, 5–11.

Green, R. & Widder, K. (Eds.) (1987). *Methods in Enzymology*, Vol. 149, part B: *Drug and Enzyme Targeting*, Academic Press, London.

Gregoriadis, G. (Ed.) (1987). *Liposome Technology,* Vols. I,II and III, CRC Press, Boca Raton, FL.

Gregoriadis, G. (Ed.) (1988). *Liposomes as Drug Carriers, Recent Trends and Progress,* J. Wiley, New York.

Gupta, P.K. & Hung, C.T. (1989). Quantitative evaluation of targeted drug delivery systems. *Int. J. Pharm.*, **56**, 217–26.

Heath, T.D., Lopez, N.G., Stern, W.H. & Papahadjopoulos, D. (1985a). 5-Fluoroorotate: a new liposome-dependent cytotoxic agent. *FEBS Lett.*, **187**, 73–5.

Heath, T.D., Lopez, N.G. & Papahadjopoulos, D. (1985b). The effect of liposome size and surface charge on liposome-mediated delivery of methotrexate-γ-aspartate to cells *in vitro. Biochim. Biophys. Acta*, **820**, 74–84.

Hellström K.E., Hellström, I. & Goodman, G.E. (1987). In: J.R. Robinson and V.H.L. Lee (Eds.), *Controlled Drug Delivery*, pp. 623–53, Marcel Dekker, New York.

Hughes, B.J., Kennel, S., Lee, R. & Huang, L. (1989). Monoclonal antibody targeting of liposomes to mouse lung *in vivo. Cancer Res.*, **49**, 6214–20.

Hunt, C.A., MacGregor, R.D. & Siegel, R.A. (1986). Engineering targeted *in vivo* drug delivery. I. The physiological and physicochemical principles governing opportunities and limitations. *Pharm. Res.*, **3**, 333–44.

Jain, R.K. (1987a). Transport of molecules across tumor vasculature. *Cancer Metastasis Rev.*, **6**, 559–93.

Jain, R.K. (1987b). Transport of molecules in the tumor interstitium: a review. *Cancer Res.*, **47**, 3039–51.

Klibanov, A., Maruyama, K., Torchilin, V.P. & Huang, L. (1990). Amphipatic polyethyleneglycol effectively prolongs the circulation time of liposomes. *FEBS Lett.*, **268**, 235–43.

Los, G., Kop, W.J. & Deurloo, M.J.M. (1991). Antitumor response and nephrotoxicity after intraperitoneal administration of a slow release formulation with cisplatin in cancers restricted to the peritoneal cavity of rats. *Br. J. Cancer*. In press.

Los, G., Mutsaerts, P.H.A., Van der Vijgh, W.J.F., Baldew, G.S., Van de Graaf, P.W. & McVie, J.G. (1989). Direct diffusion of *cis*-diamminedichloroplatinum (II) in intraperitoneal chemotherapy: a comparison with systemic chemotherapy. *Cancer Res.*, **49**, 3380–4.

Lowder, J.N. (1986). *Curr. Probl. Cancer,* **10**, 487–55.

Matzku, S., Krempel, H., Weckenmann, H.-P., Schirrmacher, V., Sinn, H. & Stricker, H. (1990).

Tumour targeting with antibody-coupled liposomes: failure to achieve accumulation in xenografts and spontaneous liver metastases. *Cancer Immunol. Immunother.*, **31**, 285–91.

Mitra A.K. & Ghosh, M.K. (1990). In: P. Tyle and B.P. Ram (Eds.), *Targeted Therapeutic Systems,* pp. 141–87, Marcel Dekker, New York.

Nässander, U.K., Storm, G., Steerenberg, P.A., De Jong, W.H., De Groot, G., Van Hoesel, Q.G.C.M. & Crommelin, D.J.A. (1990). In: D.J.A. Crommelin and H. Schellekens (Eds.), *From Clone to Clinic,* pp. 357–65, Kluwer Academic, Dordrecht.

Nitta, T., Sato, K., Yagita, H., Okumura, K. & Ishii, S. (1990). Preliminary trial of specific targeting therapy against malignant myeloma. *Lancet*, **335**, 368–71.

Peeters, P.A.M., Storm, G. & Crommelin, D.J.A. (1987). Immunoliposomes *in vivo*: the state of the art. *Adv. Drug Del. Rev.*, **1**, 249–66.

Peeters, P.A.M., Brunink, B.G., Eling, W.M.C. & Crommelin, D.J.A. (1989). Therapeutic effect of chloroquine (CQ) containing liposomes or combinations of antibodies and CQ or liposomal CQ in rats infected with *Plasmodium berghei* parasitized mouse red blood cells. *Biochim. Biophys. Acta*, **981**, 269–76.

Pietersz, G.A., Smyth, M.J., Kanellos, J., Cunningham, Z., Sacks, N.P.M. & McKenzie, I.F.C. (1988). *Antibody, Immunoconjugates and Radiopharmaceuticals*, **1**, 79–103.

Poste, G. (1985). In: G. Gregoriadis, G. Poste, J. Senior and A. Trouet (Eds.), *Receptor-mediated Targeting of Drugs*, pp. 427–74, Plenum Press, New York.

Ramakrishnan, S. (1990). In: P. Tyle and B.P. Ram (Eds.), *Targeted Therapeutic Systems,* pp. 189–213, Marcel Dekker, New York.

Roitt, I., Brostoff, J. & Male, D. (1987). *Immunology*, pp. 11.6/7, Churchill-Livingstone, Edinburgh.

Senior, J. (1987). Fate and behaviour of liposomes *in vivo*: a review of controlling factors. *Crit. Rev. Ther. Drug Carrier Syst.*, **3**, 123–93.

Shield, C.F., Norman, D.J., Marlett, P., Fucello, A.J. & Goldstein, G. (1987). Comparison of antimouse and antihorse antibody production during the treatment of allograft rejection with OKT3 or antithymocyte globulin. *Nephron*, **46**, 48–51.

Storm, G., Wilms, H.P. & Crommelin, D.J.A. (1990). Liposomes and biotherapeutics. *Biotherapy*, **3**, 25–42.

Storm, G., Oussoren C. & Peeters, P.A.M. (1991). Safety of liposomes. In: C. Vigo-Pelfrey (Ed.), *Membrane Lipid Oxidation,* Vol. 3, CRC Press, Boca Raton, FL.

Straubinger, R.M., Lopez, N.G., Debs, R.J., Hong, K. & Papahadjopoulos, D. (1988). Liposome-based therapy of human ovarian cancer: parameters determining potency of negatively charged and antibody-targeted liposomes. *Cancer Res.*, **48**, 5237–45.

Takakura, Y., Fujita, T., Hashida, M. & Sezaki, H. (1990). Disposition characteristics of macromolecules in tumor-bearing mice. *Pharm. Res.*, **7**, 339–46.

Tomlinson, E. (1987). Theory and practice of site-specific drug delivery. *Adv. Drug Del. Rev.*, **1**, 87–198.

Toonen, P.A.H.M. & Crommelin, D.J.A. (1983). Immunoglobulins as targeting agents for liposome-encapsulated drugs. *Pharm. Weekbl. Sci. Ed.*, **5**, 269–80.

Weinstein, J.N., Black, C.D.V., Barbet, J., Eger, R.R., Parker, R.J., Holton, O.D., Mulshine, J.L., Keenan, A.M., Larson, S.M., Carrasquillo, J.A., Sieber, S.M. & Covell, D.G. (1986). In: E. Tomlinson and S.S. Davis (Eds.), *Site-specific Drug Delivery,* pp. 81–91, J. Wiley, Chichester.

Woodle, M.C., Newman, M., Collins, L., Redemann, C. & Martin, F. (1990). Improved long-circulating (Stealth®) liposomes using synthetic lipids. *Proc. Int. Symp. Control. Rel. Bioact. Mater.*, **17**, pp.77–8. Controlled Release Soc.

Wright, S. & Huang, L. (1989). Antibody-directed liposomes as drug delivery vehicles. *Adv. Drug Del. Rev.*, **3**, 343–89.

Zonneveld G. & Crommelin, D.J.A. (1988). In: G. Gregoriadis (Ed.), *Liposomes as Drug Carriers; Recent Trends and Progress,* pp. 795–817, J. Wiley, New York.

23 Novel Drug-delivery Systems

R. KIRSH and G. WILSON

Department of Drug Delivery, SmithKline Beecham Pharmaceuticals, 709 Swedeland Road, King of Prussia, PA 19406, USA

1 Introduction

Drug-delivery systems are designed to deliver therapeutic agents into the body to improve the efficacy, safety, convenience and cost of therapy. Traditionally, these systems have been used to deliver conventional low-molecular-weight molecules into the bloodstream at a controlled rate. To meet these requirements a broad range of dosage forms and delivery systems has achieved widespread use within the pharmaceutical industry. These include a wide range of controlled-release oral or injectable systems, and specialized devices such as transdermal patches and implantable pumps (for reviews see Robinson & Lee, 1987; Tyle, 1990). Most drug-delivery systems currently in clinical practice are, however, not suitable for new delivery approaches involving new classes of pharmacological agents (e.g. peptides, proteins, nucleotides, genes and cells), or new therapeutic strategies (e.g. drug targeting and chronotherapeutics) that are based on an understanding of the cellular and molecular basis of disease pathogenesis and which require more complex dosing regimens. Novel approaches include the following:

1 The design of soluble receptor analogues of the cellular receptors involved in viral attachment and penetration for pathogenic viruses such as HIV and rhinoviruses (e.g. CD4, ICAM-1) in order to block interaction of these pathogenic viruses in the relevant target cells (Deen *et al.*, 1988; Greve *et al.*, 1989).

2 Synthesis of a variety of peptidomimetic growth factor analogues for treatment of hormone deficiencies (Walker *et al.*, 1988).

3 Synthesis of peptidomimetic aspartyl protease inhibitors for enzymes such as renin (Greenfield *et al.*, 1989) or the HIV protease (McQuade *et al.*, 1990; Meek *et al.*, 1990) for use in hypertension and AIDS therapeutics respectively.

4 Transfection of human genes into mammalian cells for gene-replacement therapy (Anderson, 1985; Friedmann, 1989).

Although significant progress has been made in designing strategies for the discovery of novel therapeutic agents as well as novel strategies for their utilization, translation of many of these novel compounds into commercially viable drugs will depend on the development of new delivery systems. For these reasons, considerable efforts have been devoted to the design of drug-delivery systems and approaches capable of ensuring:

- reproducible absorption of new molecules that do not naturally penetrate cellular barriers and whose absorption cannot be reproducibly predicted from physicochemical properties (e.g. log *P*), and
- selective localization at relevant target sites.

2 Drug-delivery approaches

A number of approaches have been used in an attempt to achieve delivery of new molecular classes. These can be defined as:

1 Synthesis of simple analogues of the native drug molecule.

2 Drug carrier systems and polymeric prodrugs (drug conjugates) used to improve the absorption and biodistribution of the compound.

3 Use of a specialized drug-delivery device to deliver drug into the systemic circulation.

The majority of attention has focused on drug-carrier or novel-formulation strategies. Accordingly, a wide variety of macromolecular, microparticulate and cell-based delivery systems has been evaluated. These systems include antibodies, lectins and hormones, soluble and microparticulate carriers prepared from a diverse array of naturally occurring and synthetic polymeric matrices as well as lipid-based systems (including liposomes, emulsions, microemulsions and mixed micelles; for reviews see Chien, 1982; Levy & Miller, 1983; Tomlinson & Davis, 1986; Juliano, 1988). Similarly a number of specialized novel drug-delivery devices are under investigation.

Although traditionally it has been possible to match conventional drugs with existing delivery systems, a number of new criteria that relate either to the drug molecule, the required dosing regimen (i.e. acute versus chronic therapy), or the destination in the body are being used to design novel drug-delivery systems. These systems must be able to achieve the following for maximum efficacy and patient compliance.

1 Suitable pharmacokinetic/pharmacodynamic profiles.

2 An acceptable route of administration in consideration of the anticipated dose, dosing frequency and chronicity of the disease.

3 Access to, and retention of, the pharmacological agent at the site of action.

4 Exclusion of the compound from non-target organ, tissue and cells as well as defining the potential for adverse effects due to interaction of the drug with non-target tissues.

3 Pharmacokinetic/pharmacodynamic considerations

In principle, drug-delivery systems are designed to meet the most desired pharmacokinetic and pharmacodynamic properties relevant to the drug and the disease (for reviews see Stella & Himmelstein, 1985; Boddy & Aarons, 1989). Initial advances in maintaining continuous blood levels (e.g. with osmotic pumps and transdermal devices), within a narrow therapeutic window, have been superseded by considerations of targeting drugs to the desired cell types, and the need to achieve complex dosing regimens to meet the chronobiology of specific diseases. For site-specific drug targeting, the pharmacokinetic profile at any target site is dependent on the following:

- the input rate of drug into the systemic circulation;
- the rate at which the drug gains access to the relevant target site;
- the rate at which the drug is lost from the target site;
- the rate of elimination from the body.

Moreover, in regard to prodrugs and macromolecular or microparticulate carriers, the rate of release of free compound from the targeting species and the pharmacokinetics of the free compound released following cleavage at the target site are also of paramount importance. *In vivo*, the rate of systemic elimination of the drug–carrier complex is the most critical factor influencing the efficiency of any targeting system. For example, if the elimination rate is very rapid, it will be difficult to achieve and maintain sufficient concentrations at the target site to produce a pharmacological effect. For this reason, it is important to design drug carriers that are not rapidly eliminated from the body.

The concentration of drug–carrier complex at the target site is also dependent upon the rate of delivery to the target site and the rate of removal of targeted complex from the target site. If access to the target site is low (i.e. due to inability to penetrate microvascular endothelium), the probability of achieving effective drug concentrations at the target site, regardless of the specificity of the targeting ligand, is low. As the targeting ligand serves to ensure retention of the complex at the relevant site, clearance of the complex should not be a major issue. However, if the rate of elimination of the free drug from the target site is rapid, the concentration at the target site may be sufficiently low, such that a therapeutic response may not be observed irrespective of the specificity of the targeting ligand. As such, delivery-system design requires a thorough understanding of the pharmacokinetic behaviour of the parent compound in relation to its pharmacology in order to establish an effective dose, dose frequency and duration of therapy in designing effective therapeutic protocols.

4 Absorption of peptidergic drugs

The development of drug-delivery systems for the non-intravenous administration of peptidergic molecules has become a priority in order to translate the therapeutic potential of these molecules into commercial drug products. Whereas the absorption of peptidergic molecules from almost every conceivable anatomical site has been examined (Davis, 1986), approaches to the non-injectable administration of these molecules have progressed without a clear understanding of the permeability and metabolic properties of the biological barriers at these sites. Recent progress in maintaining epithelial and endothelial barriers *in vitro* (Audus *et al.*, 1990) has provided new test systems to examine the transport properties of specific biological barriers, to distinguish between paracellular and transcellular pathways, and to monitor the effects of delivery approaches on the integrity of the cellular barrier. In addition there is increasing evidence that such systems can be conveniently used for screening new compounds for their ability to cross biological barriers, and may be used to predict absorption *in vivo.*

Of the many routes of administration that have been investigated, delivery of peptidergic molecules across the skin and the intestinal and respiratory epithelia continues to receive considerable attention. Currently, commercial non-injectable systems in clinical practice are restricted to those that deliver small quantities of potent peptides (e.g. DDAVP, leutinizing hormone-releasing hormone (LH-RH) antagonists, growth hormone releasing peptides) across the nasal mucosa, or into the systemic circulation from implanted biodegradable polymeric delivery systems. The nasal route is, however, limited due to generally low bioavailability and in some cases poor reproducibility. The mechanism of action and toxicity of absorption enhancers still being proposed for the nasal route remain to be established. Implantable polymeric systems are limited by the amount of drug that can be 'loaded' and by their use only as sustained-release (rather than pulsatile) systems. The potential of delivering peptidergic molecules via the oral route is currently the focus of much attention and speculation particularly in view of our generally poor understanding of the permeability properties of different regions of the gastrointestinal tract. A number of peptidergic molecules (e.g. insulin and erythropoietin) give a pharmacological response when given orally in a variety of systems including lipid emulsions and other particulate and macromolecular carriers; however, the potential of all systems currently under investigation awaits demonstration of acceptable bioavailability, reproducibility, safety and pharmaceutics. Until such evidence is published, claims that these systems can significantly increase transport across the intestinal epithelium (rather than offer protection against proteolysis which may result in increased bioavailability) must be viewed against the existing literature which documents the generally poor absorption of macromolecules and particles, evidence which is entirely consistent with the physiological function of the intestinal epithelium and the transport mechanisms that are present. The potential of exploiting a naturally occurring endocytic mechanism, i.e. the uptake of particles by M-cells in the gut for oral vaccination, has recently been demonstrated (Muecbrock & Tice, 1989). Based on the low capacity of this system and the proximity of M-cells to gut-associated lymphoid tissue, it is unlikely that it could be used for the systemic absorption of peptidergic molecules. Potential exploitation of physiological pathways that transport nutrients and structural mimetics of nutrients (e.g. the transport of some orally active cephalosporins that mimic dietary tripeptides (Kramer *et al.*, 1990); and cobalamin conjugates (Russell-Jones & Aizpurna, 1988)) will depend on whether these carrier systems can transport (and in some cases release) sufficient quantities of a wide range of peptidergic structures. Although few systematic studies have been performed, the applicability of prodrug strategies to enhance the stability and delivery of peptidergic molecules has been demonstrated by the work of Bundgaard *et al.* (Moss *et al.*, 1990). This approach is most applicable to small peptides (2–8 amino acids) which can passively diffuse through membranes.

Interest in exploiting the pulmonary route for the absorption of peptides and proteins has gained momentum following the recent demonstration that leuprolide, a nonapeptide LH-RH antagonist, has an absolute bioavailability of up to 18% when administered to human volunteers as an aerosal (Adjei & Garren, 1990). Similarly, human growth hormone ($M_r = 22\,000$) shows high pharmacological availability when administered to rats as an aerosol (Patton *et al.*, 1990). Although there is little information on the permeability of the pulmonary epithelium to a wide range of peptidergic molecules, current evidence indicates that absorption takes place predominantly across the alveolar epithelium. The key challenge in the future design of inhalation devices is thus to achieve optimal delivery of either aqueous or dry-powder formulations to the lower airways. Exploitation of the pulmonary route in rats for the delivery of an adenovirus vector containing the

α_1-antitrypsin gene was recently reported (Crystal, 1991). In an approach that may be applicable to a number of lung diseases, including cystic fibrosis, it was demonstrated that α_1-antitrypsin was synthesized and secreted into the lung for at least 1 week.

While the iontophoretic transdermal delivery of peptidergic molecules is at an early stage in its development, the potential of this approach has been demonstrated by the transport of therapeutic doses of insulin and LH-RH across the skin of experimental animals or human volunteers, respectively (Lattin *et al.*, 1991). Although a complete understanding of each of the system components and the interactions that occur at the electrode/reservoir and the reservoir/skin interfaces is required for analysis of drug stability and skin irritation, it is anticipated that commercial devices will be available in the next 2–3 years. Potential advantages of iontophoretic devices are the precise control of drug delivery with the rate of delivery proportional to current, and the capability of programmed release to suit ultradian rhythms. The possibility of using 'reverse iontophoresis' for the sampling of body fluids raises the hope that the delivery of some peptidergic hormones (e.g. insulin) could be adjusted in a closed-loop fashion by sensing the amount of glucose pulled from the body.

5 Site-specific drug targeting

Once absorbed into the vascular compartment, drugs and/or the drug–carrier complex must be able to reach their target site. In many disease states (e.g. neoplastic, neurodegenerative and chronic inflammatory diseases), relevant target cells reside outside of the microvasculature within both tissue parenchyma as well as connective tissues. Therefore, the drug candidates must be able to escape from the vasculature in order to interact with their target tissues and cells. Appropriately designed conventional lipophilic low-molecular-weight molecules have capability to permeate the cellular barrier of the microvasculature in most tissues including the brain (for review see Chien *et al.*, 1982; Graybill *et al.*, 1982; Levy & Miller, 1983). Macromolecules, on the other hand, do not have the capability of freely permeating the microvascular barrier. Therefore, their utilization will be limited by the anatomical and physiological constraints to peptide and protein permeability.

In normal tissues, the microvasculature is composed of three distinct histological types: (i) continuous, (ii) fenestrated and (iii) discontinuous (for review of anatomy, see Weiss & Greep, 1977). Simply from a mechanical standpoint, continuous and fenestrated capillaries represent a major barrier to the escape of macromolecules from the circulation. In contrast, macromolecular drug and microparticulate drug-carrier extravasation might be expected to occur by simple percolation through the pore structure of sinusoidal capillaries in the liver, spleen and bone marrow, allowing access to the tissue parenchyma in these organs (for review, see Weiss & Greep, 1977; Wolff, 1977; Poste *et al.*, 1984). On the other hand, delivery of macromolecular drugs to organs with a microvascular supply composed of a tight continuous endothelial layer, such as the brain, represents a significant problem for the design of effective delivery approaches (Greig, 1989). This becomes particularly important for the utilization of novel peptidergic drugs for the treatment of neurodegenerative disorders such as Alzheimer's and Huntington's diseases where new agents show tremendous promise *in vitro*, but lack activity *in vivo* because they do not have sufficient access to the diseased site for an appropriate time to exert pharmacological activity. Several approaches have emerged in the past several years aimed at subverting the blood–brain barrier, albeit transiently, in order to facilitate uptake of compounds from the systemic circulation into the CNS. These include osmotic opening and metrazol shock (Greig, 1989). Although these approaches already increase brain permeability, substantial issues exist for the chronic utilization of such drastic measures.

Drug modifications either through the design of new drug analogues or prodrugs that are more lipophilic than the parental compound have also been extensively studied to augment the penetration of compounds into the CNS. However, as discussed above, unless there is some form of retention, the compound will reach the target site, then rapidly redistribute. Prodrugs offer the added advantage that, once at the target site, they are cleaved from the transport-facilitating ligand thereby trapping the drug (Bodor & Brewster, 1982; Greig & Rapoport, 1988). Similarly, there have been several studies during the past several years demonstrating the transport of oligopeptides across the blood–brain barrier *in vitro* and *in vivo*, presumably via receptor/carrier-mediated processes. These include amino acids (Greig *et al*., 1987), cationized albumin (Pardridge, 1986), insulin (Pardridge *et al*., 1985) and transferrin (Pardridge *et al*., 1987). Whether transport of drugs into the brain can be facilitated via conjugation to molecules effectively transported remains to be determined.

It is now well established that capillary permeability increases significantly during the inflammatory process (for references see Arfors *et al*., 1979; Gabbiani & Majno, 1980; Simionescu, 1980). Many inflammatory mediators appear to increase permeability by opening gaps between adjacent endothelial cells at the level of the post-capillary venule. Anatomically, these gaps appear sufficiently large enough to allow efficient extravasation of macromolecules and microparticulates in the size range of 0.2 μm. However, using a granuloma pouch assay to monitor permeability of liposomes in this size range has revealed that extravasation does not occur even under conditions in which extensive macromolecular leakage occurs (Poste *et al*., 1983). This highlights the difficulties in correlating morphological evidence of large pores with functional transport studies.

Although apparent difficulties are currently present for selective targeting to extravascular sites, potential opportunities exist, however, for targeting of macromolecular drug carriers to sites of inflammation by exploiting the surface expression of adhesion molecules such as intercellular adhesion molecule (ICAM) and endothelial leukocyte adhesion molecule (ELAM) and GMP-140 on the surface of inflammatory endothelial cells (for reviews, see Bevilacqua *et al*., 1989; Wawryk *et al*., 1989; McEver, 1991). For example, early in inflammation, histamine or thrombin induces the transient expression GMP-140 which increases the adhesiveness of endothelial cells for polymorphonucleocytes. Release of inflammatory cytokines interleukin-1 and tumour necrosis factor results in stimulation of *de novo* synthesis of ELAM and additional ICAM (McEver, 1991). The kinetics of this are such that ELAM peaks at 4–6 h which coincides with the peak neutrophil influx. ICAM levels remain high presumably to facilitate late phase influx of mononuclear phagocytes. If suitable ligands with high affinity for these endothelial cell receptors can be identified for attachment to macromolecular drug carrier, it may become possible selectively to target drugs to inflammatory lesions *in vivo*. For example, the sialyl-Le^X glycolipid appears to be the natural neutrophil ligand that binds to ELAM-1 on inflammatory endothelial cells (Walz *et al.*, 1990). As such, incorporation of this molecule into a liposome membrane should serve to target liposomes to these inflammatory endothelial cells thereby delivering liposome-encapsulated drugs to sites of inflammation.

However theoretically promising and clinically attractive this approach may be, significant scientific issues still remain. These include:

- release and retention of associated drug within the lesion—this is critical for drugs such as extracellular enzyme inhibitors;
- internalization of the carrier–ligand complex, stability of the carrier-associated drug in the endosomal/lysosomal compartment;
- and release of active drug from this compartment into the cell cytoplasm for compounds such as antisense oligonucleosides or enzyme inhibitors designed to interrupt synthesis and/or processing and activity of inflammatory mediators.

5.1 *Intravascular targeting*

The issues outlined above highlight the problems associated with rationally designing targetable drug-delivery systems that will have access to—and be retained in—relevant extravascular sites. However, realistic opportunities exist for targeting of specific drugs by exploiting the cell-surface differentiation antigens expressed on subpopulations of circulating lymphoid cells. Similarly, the phagocytic activity of monocytes as well as fixed elements of the reticuloendothelial system in the liver, spleen and bone marrow represent realistic targets for the selective delivery of microparticulate-associated drugs.

Tremendous progress has been made during the past 10 years using immunological and molecular probes to classify the major subsets of circulating lymphocytes based on expression of well-defined differentiation antigens (for reviews see Clark & Ledbetter, 1986; Ledbetter & Clark, 1988). These surface markers represent useful targets for ligand-specific targeting. Although monoclonal antibodies currently represent the most prevalent source of targeting moiety, as further biochemical characterization of these cell-surface molecules delineates their physiological binding specificities, it is likely that additional targeting ligands will become available.

The ability to target drugs selectively to circulating blood cells has significant clinical applications. One obvious example for this approach would be to exploit the abnormal expression of lymphocyte differentiation antigens on leukaemia cells as target molecules for ligand-directed targeting of liposomes containing anticancer drugs. The feasibility of using these molecules for selective drug targeting has already been demonstrated in studies using immunotoxins (for reviews see Blakey & Thorpe, 1986; Lord *et al.*, 1988). A related application would be immunosuppression by exploiting T-lymphocyte antigens as targets for liposomes containing cytotoxic drugs to destroy the T-cell subsets involved in allograft rejection (Cosimi *et al.*, 1981) and autoimmune diseases (Rnages *et al.*, 1985; Rose *et al.*, 1988). However, cytotoxic ablating of specific T-cell subsets raises the potential of increasing the potential for opportunistic infections. For example, anti-CD4 treatment of mice has been found to reactivate CNS toxoplasmosis (Vollmer *et al.*, 1987) as well as increase susceptibility to *Mycobacterium bovis* infections (Pedrazzini *et al.*, 1987).

5.2 *Microparticulate carriers*

The majority of research efforts on microparticles *in vivo* has been on analysis of the disposition following intravenous administration. This route will provide access to the microvasculature of all organs; however, successful targeting to extravascular locations will require at a minimum that the microparticle be capable of exiting from the vasculature. At present there is little evidence to support this possibility, although numerous classes of drug carrier have been studied (for review see Juliano, 1988).

Clearance of microparticles by Kupffer cells lining the hepatic sinusoids represents the major route of elimination of microparticulates including liposomes from the circulation following intravenous administration (for reviews, see Poste *et al.*, 1984; Juliano, 1988). The other major sites of particulate retention after intravenous injection are in the mononuclear phagocytes of the spleen and the bone marrow as well as circulating blood monocytes (Poste *et al.*, 1984). Accumulation in the spleen is enhanced under conditions where hepatic uptake of liposomes is saturated (Abra & Hunt, 1981; Ellens *et al.*, 1983). Splenic retention thus represents clearance of the 'spillover' fraction from the liver. Saturation of the ability of splenic macrophages to remove particulates results, in turn, in 'spillover' to bone-marrow macrophages. As such, multiple injections of particulates can eventually 'exhaust' the capabilities of macrophages in the liver, spleen and bone marrow to clear liposomes and other particles with resulting toxicity due to impaired reticuloendothelial function (Poste *et al.*, 1984; Allen, 1988).

Localization of intravenously administered microparticulates including liposomes within the macrophages and monocytes can be exploited to deliver drugs selectively to these cells to enhance their intrinsic host defence functions. For example, the systemic administration of liposomes containing immunostimulants has been shown to activate macrophage-mediated tumoricidal and virucidal activity *in vitro* and *in vivo*. Similarly, delivery of anti-infective agents to these cells can be exploited in therapy of intracellular infections that are not responsive to conventional therapy due to limited access of the antibiotic to intracellular sites. For example, liposomes containing immunostimulants injected intravenously result in significant destruction of established metastases produced by murine tumours of diverse histological origins (for reviews see Poste & Fidler, 1981; Fidler, 1988). The therapeutic efficacy of liposome-encapsulated immunostimulants in rendering pulmonary macrophages tumoricidal in these experimental tumour models is encouraging since the lung is a major site of metastatic disease. Similar results have also been observed in treatment of hepatic metastases (Thombre & Deodhar, 1984). Similarly, a lipophilic muramyl dipeptide analogue muramyl tripeptide phosphatidylethanolamine (MTP-PE) encapsulated in liposomes is significantly more effective than free MTP-PE in protecting experimental animals against systemic lethal HSV-2 infections (Koff *et al.*, 1985). In this particular case, the therapeutic molecule has been chemically modified to allow retention in the delivery system. This approach has recently been extended to include therapy of pulmonary viral infections due to HSV-1 and influenza virus (Gangemi *et al.*, 1987).

As mentioned above, mononuclear phagocytes play an important role in the clearance and destruction of pathogenic micro-organisms. However, a number of bacteria, fungi, viruses and pathogenic protozoa can also replicate intracellularly within macrophages. Intracellular infections caused by these micro-organisms are difficult to manage clinically and are often refractory to conventional chemotherapeutic treatment protocols because of poor drug penetration into cells. Administration of antimicrobial agents in association with liposomes offers a potential solution to this problem. The merits of this approach for the chemotherapy of intracellular infections of the mononuclear phagocyte system caused by bacteria (Fountain *et al.*, 1981; Desiderio & Campbell, 1983; Emmen & Storm, 1987), parasites (Alving & Steck, 1979, Alving, 1986; Croft *et al.*, 1988), fungi (Graybill *et al.*, 1982; Lopez-Berestein *et al.*, 1983, 1985) and viruses (Smolin *et al.*, 1981; Kende *et al.*, 1985; Popescu *et al.*, 1987) have been demonstrated over the last few years. Particularly encouraging are the clinical studies with liposomal formulations of amphotericin-B (Lopez-Berestein *et al.*, 1985, 1987; Sculier *et al.*, 1988; Wiebe & DeGregorio, 1988). The liposomal formulation of this drug shows significant reduction in toxicity without change in efficacy thereby improving the therapeutic index of amphotericin-B resulting in improved clinical management of these life-threatening infections.

Systems for specific, ligand-directed targeting of microcarrier-associated drugs to circulating blood cells have been proposed. The critical issue associated with the development of targeted microparticles is whether these carriers can be prevented from localizing within mononuclear phagocytes. It is possible, of course, that equipping carriers with ligands that have a high affinity for cells other than the mononuclear phagocytes including blood monocytes may be sufficient to achieve this goal. However, it is uncertain whether insertion of a cell-recognition ligand with any microcarrier would alter their clearance kinetics. This must certainly be considered when antibodies (monoclonal or heterospecific) are used as targeting ligands. Any exposure of F_c regions on the antibody molecules would result in rapid clearance via their bindings to F_c receptors on circulating monocytes and tissue macrophages.

Prolongation of the clearance times of circulating microparticles by imposing a reticuloendothelial blockade would be expected to increase the likelihood of interaction with cells other than mononuclear phagocytes. Similarly, approaches to designing microparticles with prolonged circulation half-lives such as the utilization of hydrophilic block copolymers or inclusion of sialyl-containing gangliosides to alter liposome surface properties, thereby masking features recognized by mononuclear phagocytes, has been recently proposed (Davis & Illum, 1986; Gabizon & Papahadjopoulos, 1988). If localization in mononuclear phagocytes can be prevented, or at least reduced substantially, ligand-directed targeting of liposomes to other cells within the vasculature becomes more realistic.

6 Chronopharmacological considerations

Increasing evidence that a number of diseases exhibit strong circadian rhythms in the occurrence or exacerbation of symptoms is leading to attempts (chronotherapeutics) to optimize therapies that will both improve efficacy and reduce side-effects (Houshesky *et al.*, 1991). Many chronic disorders, such as allergic rhinitis, angina, asthma, epilepsy, rheumatoid and osteoarthritis, and ulcer disease, follow this pattern. In addition, endogenous release of a number of hormones (e.g. growth hormone, luteinizing hormones) is also subject to circadian rhythms. Adjustment of the mode and timing of delivery (i.e. continuous vs. pulsatile vs. self-regulated) to meet the biological rhythms is thus becoming a major factor in the design of delivery systems. The delivery approach must thus take into account: (i) time patterns in the expression or intensity of disease, (ii) administration time-dependent effects of drugs in terms of their kinetics/dynamics, and (iii) individual differences between patients in the time(s) when medication is most required to reduce/prevent symptoms or acute episodes of disease. Optimization of therapeutic protocols to account for circadian rhythms has been reported for a number of drugs including theophylline, H_2-receptor antagonists, non-steroidal anti-inflammatory drugs, Adriamycin® and cisplatin, and gonadorelin acetate.

In the past, the clinical use of novel delivery systems capable of meeting the requirements of chronotherapeutic regimens has been largely limited to programmable pumps. A number of new delivery approaches that can respond to environmental stimuli have been described (Kort & Langer, 1991). These include externally regulated systems (modulated by magnetic fields, ultrasound, heat, electric current) or self-regulated systems (modulated by pH and ionic strength), including glucose-responsive insulin-delivery systems. In addition, recent advances in iontophoretic systems having the potential for feedback control provide an alternative technology for adjustable/programmable/stimuli-responsive dosing. Many of these systems are still at an early stage of development. Critical issues surrounding biocompatibility, toxicology, responsiveness to stimuli, ability to contain practical levels of drug, and pharmaceutics still need to be resolved before the potential for commercialization can be assessed. Perhaps the ultimate in feedback control will be the use of genetically engineered cellular-delivery systems, implanted into the body to produce protein in response to physiological demand, although this approach is clearly in its infancy (Cornetta *et al.*, 1991).

7 Implications for medicinal chemistry

New classes of drug molecules together with new therapeutic strategies have led to a variety of novel approaches for the design of drug-delivery systems. A major trend is to utilize information on the site of disease pathogenesis and the biological barriers (cellular, metabolic and pharmacokinetic) that separate the site of drug administration from drug action, in the design of delivery approaches. In particular, new information on the cellular and molecular biology of cellular barriers, and the systems that transport molecules and particles across such barriers, has provided new tools and approaches for evaluating the intrinsic properties of new drug molecules and for designing delivery strategies. As the knowledge of transport pathways is still poorly understood, our ability to deliver new molecular classes is likely to progress hand-in-hand with advances in this area.

Traditionally, drug delivery and formulation approaches are considered when compounds reach development status; however, new biological approaches to drug delivery are more appropriately incorporated at an earlier stage in the discovery process. In principle, such approaches provide valuable input into medicinal chemistry strategies in all the therapeutic areas, to provide drug molecules that have improved bioavailability and selectivity, in addition to pharmacological activity. In practice, close cooperation between those involved in compound synthesis, pharmacology, drug delivery and pharmaceutics is required to address delivery and disposition issues at the earliest possible stage, and to ensure the selection of the most attractive lead compounds that meet these additional criteria. Although trends in medicinal chemistry (e.g. synthesis of smaller and more stable peptidergic molecules) may provide some compounds with more intrinsic ability to penetrate membranes, the successful administration of a variety of new pharmacological agents on the therapeutic horizon, coupled with the desirability of matching their delivery mode and frequency to treat diseases in the most efficacious and patient compliant manner, will depend to a large extent on our ability to design novel delivery approaches and systems.

8 References

Abra, R.M. & Hunt, C.A. (1981). *Biochim. Biophys. Acta*, **666**, 493–503.

Adjei, A. & Garren, J. (1990). *Pharm. Res.*, **7**, 565–9.

Allen, T.M. (1988). *Adv. Drug Del. Rev.*, **2**, 55–67.

Alving, C.R. (1986). *Parasitol. Today*, **2**, 101–7.

Alving, C.R. & E.A. Steck (1979). *Trends Biochem. Sci.*, **4**, N175–7.

Anderson, W.F. (1985). *Science*, **226**, 401–9.

Arfors, E.-E., Rutili, G. & Svensjo, E. (1979). *Acta Physiol. Scand. (Suppl.)*, **463**, 93–103.

Audus, K., Bartel, R.L., Hidalgo, I. & Borchardt, R.T. (1990). *Pharm. Res.*, **7**, 435–51.

Bevilacqua, M.D., Stengelin, S., Gimbrone, M.A. & Seed, B. (1989). *Science*, **243**, 1160–5.

Blakey, D.C. & Thorpe, P.E. (1986). *Bioessays*, **4**, 292–7.

Boddy, A. & Aarons, L. (1989). *Adv. Drug Del. Rev.*, **3**, 155–64.

Bodor, N. & Brewster, M. (1982). *Pharmacol. Therap.*, **19**, 337–86.

Chien, Y.W. (Ed.) (1982). *Novel Drug Delivery Systems*, Marcel Dekker, New York.

Clark, E.A. & Ledbetter, J.A. (1986). *Immunol. Today*, **7**, 267–70.

Cornetta, K., Morgan, R.A. & Anderson, W.F. (1991). *Human Gene Ther.*, **2**, 5–14.

Cosimi, A.B., Colvin, R.B., Burton, R.C., Rubin, R.H., Goldstein, G., Kung, P.C., Hansen, W.P., Delmonico, F.L. & Russell, P.S. (1981). *N. Engl. J. Med.*, **305**, 308–11.

Croft, S.L., Neal, R.A. & Rao, L.S. (1988). In: D.T. Hart (Ed.), *Leishmaniasis: The First Centenary (1885–1985). New Strategies for Control*, Plenum Publishing, New York, in press.

Crystal, R. (1991). *Science*, **252**, 400–3.

Davis, S.S. (1986). In: S.S. Davis, L. Illum and E. Tomlinson (Eds.) *Delivery Systems for Peptide Drugs*, p.21, Plenum Press, New York.

Davis S.S. & Illum L. (1986). In: E., Tomlinson, S.S. Davis (Eds.), *Site-Specific Drug Delivery*, pp. 93–110, Wiley, Chichester.

Deen, K.C., McDougal, J.A., Inacker, R., Folena-Wasserman, G., Arthos, J., Rosenberg, M.J., Maddon, P.J., Axel, R. & Sweet, R.W. (1988). *Nature (London)*, **331**, 82–4.

Desiderio, J.V. & Campbell, S.G., (1983). *J. Reticuloendothelial Soc.*, **34**, 279–87.

Ellens, H., Morself, H., Dontje, B., Kalicharan, D., Hulstaert, C. & Scherphof, G. (1983). *Cancer Res.*, **43**, 2927.

Emmen, F. & Storm, G. (1987). *Pharm. Weekbl. (Sci).*, **9**, 162–71.

Fidler, I.J. (1988). In: E. Tomlinson and S.S. Davis (Eds.), *Drug Delivery*, pp. 111–35, John Wiley, Chichester.

Fountain, M.W., Dees, C. & Schultz, R.D. (1981). *Curr. Microbiol.*, **6**, 373–6.

Friedmann, T. (1989). *Science*, **244**, 1275–80.

Gabbiani, G. & Majno, G. (1980). In: G. Kaley and B.M. Altura (Eds.), *Microcirculation*, Vol. III, pp. 143–63, University Park Press, Baltimore, Maryland.

Gabizon, A. & Papahadjopoulos, D. (1988). *Proc. Natl. Acad. Sci. USA*, **85**, 6949–53.

Gangemi, J.D., Nachtigal, M., Barnhart, D., Krech, L. & Jani, P. (1987). *J. Infect. Dis.*, **155**, 510–16.

Graybill, J.E., Craven, P.C., Taylor, R.L., Williams, D.M. & Magee, W.E. (1982). *J. Infect. Dis.*, **145**, 748–51.

Greenfield, J.C., Cook, K.J. & O'Leary, I.A. (1989). *Drug Met. Disp.*, **17**, 518–25.

Greig, N.H. (1989). In: E.A. Newelt (Ed.), *Implications of the Blood Brain Barrier and its Manipulation*, pp. 311–67, Plenum Press, New York.

Greig, N. & Rapoport, S.I. (1988). *Cancer Chemother. Pharmacol.*, **31**, 1–8.

Greig, N., Momma, S. & Sweeny, D. (1987). *Cancer Res.*, **47**, 1571–6.

Greve, J.M., Davis, G., Meyer, A.M., Forte, C.P., Connolly Yost, S., Marlor, C.W., Kamarck, M.E. & McClelland, A. (1989). *Cell*, **56**, 839–47.

Houshesky, W.J.M., Laugh, R. & Theeuwes, F. (1991). *Anal. N.Y. Acad. Sci.*, **618**.

Juliano, R.L. (1988). *Adv. Drug Del. Rev.*, **2**, 31–54.

Kende, M., Alving, C.R., Rill, W.L., Swatz, Jr., G.M. & Canonico, P.G. (1985). *Antimicrob. Agents Chemother.*, **27**, 903–7.

Koff, W.C., Showalter, S.D., Hampar, B. & Fidler, I.J. (1985). *Science*, **228**, 495–7.

Kort, J. & Langer, R. (1991). *Adv. Drug Del. Rev.*, **6**, 19–50.

Kramer, W., Girbig, F., Gutjahr, U., Kleemann, H.-W., Leipe, I., Urbach, H. & Wagner, A. (1990). *Biochim Biophys. Acta*, **1027**, 25–30.

Lattin, G.A., Padmanabhan, R.V. & Phipps, B.J. (1991). *Anal. N.Y. Acad. Sci.*, **618**, 450–64.

Ledbetter, J.A. & Clark, E.A. (1988). *Adv. Drug Del. Rev.*, **2**, 319–42.

Levy, R. & Miller, R.A. (1983). *Fed. Proc., Fed. Am. Soc. Exp. Biol.*, **42**, 2650–6.

Lopez-Berestein, G., Mehta, R., Hopfer, R.L., Mills, K., Kasi, L., Mehta, K., Fainstein, V., Luna, M., Hersh, E.M.J. & Juliano, R. (1983). *J. Infect. Dis.*, **5**, 939–45.

Lopez-Berestein, G., Fainstein, V., Hopfer, R., Mehta, K., Sullivan, M.P., Keating, M.J., Rosenblum, M.G., Mehta, R., Luna, M., Hersh, E.M., Reuben, J., Juliano, R.L. & Bodey, G.P. (1985). *J. Infect. Dis.*, **151**(4), 704–10.

Lopez-Berestein, G., Bodey, G.P., Frankel, L.S. & Mehta, K. (1987). *Br. J. Clin. Oncol.*, **5**, 310–17.

Lord, J.M., Spooner, M.A., Hussain, K. & Roberts, L.M. (1988). *Adv. Drug. Del. Rev.*, **2**, 297–319.

McEver, R.P. (1991). *J. Cell. Biochem.*, **45**, 156–61.

McQuade, T.J., Tomasselli, A.G., Liu, L., Karacostas, V., Moss, B., Sawyer, T.K., Heinrikson, R.L. & Tarpley, W.G. (1990). *Science*, **247**, 454–6.

Meek, T.D., Lambert, D.M., Dreyer, G.B., Carr, T.J., Tomasek, Jr, T.A., Moore, M.L., Strickler, J.E., Debouck, C., Hyland, L.J., Matthews, T.J., Metcalf, B.W. & Petteway, S.R. (1990). *Nature (London)*, **343**, 90–2.

Moss, J., Burr, A. & Bundgaard, H. (1990). *Int. J. Pharmaceut.*, **66**, 183–9.

Muecbrock, J.A. & Tice, T.R. (1989). *Curr. Topics Micro. Immunol.*, **146**, 59–66.

Pardridge, W.M. (1986). *Endocrine Rev.*, **1**, 314–30.

Pardridge, W.M., Eisenberg, J. & Yang, J. (1985). *J. Neurochem.*, **44**, 1771–8.

Pardridge, W.M., Eisenberg, J. & Yang, J. (1987). *Metabolism*, **36**, 892–5.

Patton, S.J., McGabe, J.G., Hansen, S.E. & Dougherty, A.L. (1990). *Biotechnol. Therapeut.*, **1**, 213–28.

Pedrazzini, T., Hug, K. & Louis, J.A. (1987). *J. Immunol.*, **128**, 2032–7.

Popescu, M.C., Swenson, C.E. & Ginsberg, R.S. (1987). In: M.J. Ostro (Ed.), *Liposomes from Biophysics to Therapeutics*, p. 219, Marcel Dekker, New York.

Poste, G., & Fidler, I.J. (1981). In: C. Nicolau and A. Paraf. (Eds.), *Liposomes, Drugs and Immunocompetent Cell Functions*, pp. 147–62, Academic Press, New York.

Poste, G., Kirsh, R. & Koestler, T. (1983). In: C. Gregoriadis (Ed.), *Liposome Technology*, Vol. III, pp. 1–29, CRC Press, Boca Raton, Florida.

Poste, G., Kirsh, R. & Bugelski, P.J. (1984). In: G. Gregoriadis (Ed.), *Liposome Technology*, pp. 1–28, CRC Press, Boca Raton, Florida.

Rnages, G.E., Sriram, S. & Cooper, S.M. (1985). *J. Exp. Med.*, **162**, 1105–10.

Robinson, J.R. & Lee, V.H.L. (Eds.) (1987). *Controlled Drug Delivery*, Marcel Dekker, New York.

Rose, L.M., Alvord, Jr, E.C., Hruby, S., Jackvicius, S., Peterson, R., Warner, N.L. & Clark, E.A. (1988). *Clin. Immunol. Immunopath.*, **45**, 405–23.

Russell-Jones, G.J. & Aizpurna, H.J. (1988). *Proc. Int. Symp. Contr. Release Bioact. Mater.*, **15**, 142–3.

Sculier, J.P., Coune, A., Meunier, F., Brassinne, C., Laduron, C., Hollaert, C., Collette, N., Heyman, C. & Klastersky, J. (1988). *Eur. J. Cancer Clin. Oncol.*, **24**, 527–38.

Simionescu, M. (1980). Blood cells and vessel wall: functional interactions, *Ciba Found. Symp.*, **71**.

Smolin, G., Okumoot, M., Feiler, S. & Condon, D. (1981). *Am. J. Ophthalmol.*, **91**, 220–5.

Stella, V.J. & Himmelstein, K.J. (1985). In: R.T. Borchardt, A.J. Repta and V.J. Stella (Eds.), *Directed Drug Delivery*, pp. 247–67, Humana Press, New York.

Thombre, P. & Deodhar, S.D. (1984). *Cancer Immunol. Immunother.*, **16**, 1984–91.

Tomlinson, E. & Davis, S.S. (Eds.) (1986). *Site-specific Drug Delivery*, John Wiley, Chichester.

Tyle, P. (Ed.) (1990). *Specialized Drug Delivery Systems*, Marcel Dekker, New York.

Vollmer, T.L., Waldor, M.K., Steinman, L. & Conley, F.K. (1987). *J. Immunol.*, **138**, 3737–41.

Walker, R.F., Codd, E.E., Barone, F.C., Nelson, A.H., Goodwin, T. & Campbell, S.A. (1988). *Life Sci.*, **47**, 29–36.

Walz, G., Aruffo, A., Kolzenus, W., Bevilacqua, M. & Seed, B. (1990). *Science*, **250**, 1132–5.

Wawryk, S.O., Novotny, J.R., Wicks, I.P., Wilkinson, D., Maher, D., Salvaris, E., Welch, K., Fecondo, J. & Boyd, A.W. (1989). *Immunol. Rev.*, **108**, 135–61.

Weiss, L. & Greep, R.O. (1977). *Histology*, 4th edn., McGraw-Hill, New York.

Wiebe, V.J. & DeGregorio, M.W. (1988). *Rev. Infect. Dis.*, **10**, 1097–101.

Wolff, J.R. (1977). In: G. Kaley and B.M. Altura (Eds.), *Microcirculation*, Vol. I, pp. 95–130, University Park Press, Baltimore, Maryland.

24 Lipid Microsphere Preparations and Lecithinized Peptides for Drug-delivery Systems

Y. MIZUSHIMA and R. IGARASHI

Institute of Medical Science, St Marianna University, 2-16-1 Sugao, Miyamae-ku, Kawasaki 216, Japan

1 Introduction

Progress in drug development has been amazing in recent years. One of the areas where progress has been particularly rapid is drug-delivery systems. These consist of three major categories:

1 targeting therapy;
2 controlled release;
3 biological barrier modification.

Here we deal with lipid microsphere (LM) preparations made by the authors for targeting therapy, as well as chemical modification of bioactive peptides whose value seems to be comparable to that of LM. The former is reviewed by Mizushima and the latter by Igarashi.

In addition to malignant tumours, a great variety of diseases such as infectious and inflammatory diseases, autoimmune diseases and atherosclerosis can be treated with targeting therapy. Carriers for targeting therapy which have been studied are liposomes. Liposomes have been found to be deposited in inflamed tissues and the reticuloendothelial system as well as in malignant tumours. However, liposomes are relatively unstable and are not easily mass produced. LM with an average diameter of 0.2–0.3 μm and consisting of soybean oil and lecithin (Fig. 24.1) are widely used in clinical medicine for parenteral nutrition (Intralipid®, etc.). LM are very stable and can be stored for 2 years at room temperature. Their distribution in the body is similar to that of liposomes and they have no significant side-effects, even at doses of 500 ml. Based on the properties of these LM, the authors studied first their use in targeting therapy for certain drugs such as steroids and non-steroidal anti-inflammatory drugs (Mizushima *et al.*, 1982, 1983a).

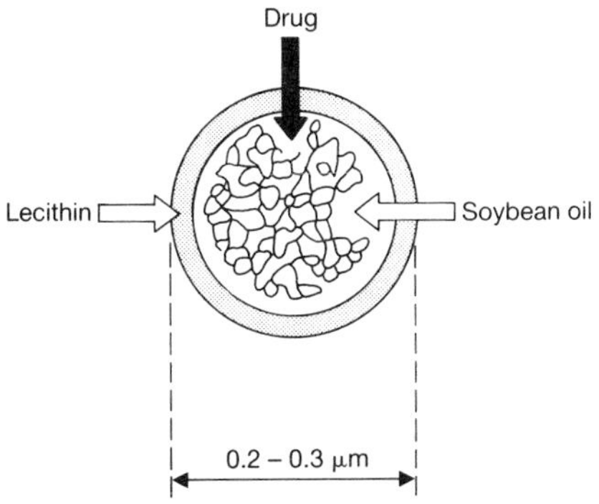

Figure 24.1. Lipid microsphere preparation.

Our previous studies have shown that LM are depositd in vascular lesions (e.g. atherosclerotic blood vessels) in larger quantities than in inflamed tissues. Based on this finding, we conducted studies on prostaglandin (PG)-related substances and biologically active proteins such as superoxide dismutase which are believed necessary for delivering to vascular lesions. Findings of these studies are also presented.

2 Lipid microspheres (LM)

2.1 *Accumulation of LM in inflamed tissues and vascular lesions*

When LM were injected intravenously into rats with carrageenan-induced inflammation, an accumulation of LM at the inflamed sites was observed by the radioisotope technique or by electron microscopy. First, ^{3}H-labelled dexamethasone palmitate was incorporated into lipid microspheres; a high concentration of ^{3}H was found in the liver, spleen and inflamed paws (Mizushima *et al.*, 1982). An electron microscopic study showed that LM were entrapped by Kupffer cells in the liver and by macrophages in the spleen. In the inflamed tissues, they were partly phagocytosed by the endothelial cells and partly penetrated the basement membranes of vessels (Mizushima, 1985). No LM were found in the tissues of non-inflamed rat paws.

Because of a similarity in the tissue distribution of LM and liposomes (Post *et al.*, 1982), many studies performed with liposomes can be applied to those with LM (Gregoriadis, 1984). Several studies indicate that liposomes are accumulated in some tumours and in myocardial and intestinal infarction lesions of animals (Palmer *et al.*, 1984). The slow release of drugs from liposomes increases the plasma half-life of the drug, decreases toxicological and allergic side-effects and alters immunological responses.

Earlier studies (Hallberg, 1965; Davis *et al.*, 1984) of lipid emulsions demonstrated also that LM had an affinity for vascular walls (including capillaries) like chylomicrons. Shaw *et al.* (1979) reported that they had even more affinity for vascular walls at inflamed sites. As described below, our study showed that LM accumulated, particularly at high concentrations, in atherosclerotic vessels.

Accumulation of LM at the site of vascular lesions can be evaluated by three methods:

1 Electron microscopy, to examine the site of tissue damage.
2 By following the delivery of LM incorporating a radiolabelled compound.
3 Scintigraphic clinical assessment of technetium-labelled LM.

Hamano *et al.* (1990) injected LM intravenously into SHR (spontaneously hypertensive rats), and then examined them by electron microscopy. Many LM accumulated below the vascular endothelium in the arterial lesions of SHR which are similar to those of arteriosclerosis. They speculated that LM passed through the gaps between endothelial cells or through endothelial cells, and accumulated below them. In contrast, LM do not accumulate in subendothelial spaces in normal rats or SHR before the development of vascular lesions.

We investigated the delivery of LM to atherosclerotic lesions induced by cannulation in rabbits. Many LM which passed through the gaps between the

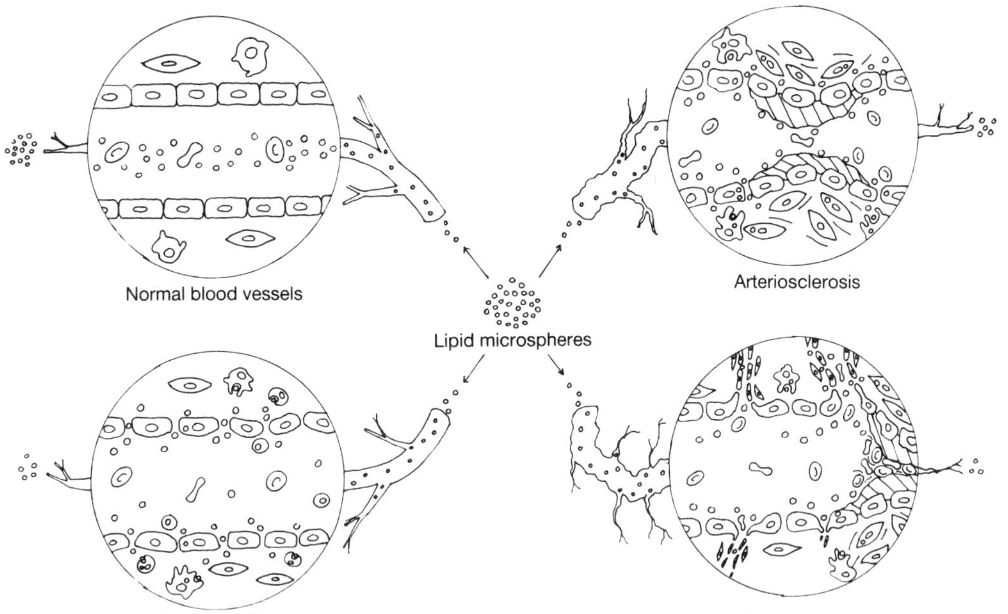

Figure 24.2. Distribution and accumulation of LM in the vascular lesions.

endothelial cells accumulated below the endothelium at the site of atherosclerotic lesions (Mizushima *et al.*, 1990). When LM incorporating radiolabelled dexamethasone palmitate ester were injected, radioactivity in the atherosclerotic vascular wall was about double the level in normal sites (Mizushima *et al.*, 1990). Nakura *et al.* (1986) similarly studied the distribution of radioactivity in SHR, and found that lipo-PGE_1 accumulated at the site of pathological lesions at higher concentrations than free PGE_1. Technetium-labelled LM were given to patients with arteriosclerosis obliterans by intravenous injection, followed by scintigraphy. Technetium accumulated at the sites corresponding to the atherosclerotic lesions (Kiyokawa *et al.*, 1987).

All these results show that LM are taken up at high concentrations in the vascular walls with atherosclerotic or other similar lesions, while they are not taken up to the same extent by normal vessels.

Of the various pharmacological actions of PGE_1 and PGI_2, one action which has not yet been studied intensively, but seems to be important for the treatment of arteriosclerotic diseases, is their effect on vascularization. This suggests the need to determine whether LM accumulate at the site of vascularization. We are now conducting a study to investigate this possibility. Since the space between endothelial cells in new blood vessels is as large as that of vessels with sclerosis or inflammation, LM are very likely to accumulate in new blood vessels. Figure 24.2, shows a diagrammatic representation of the distribution and accumulation of LM at the site of vascular lesions.

2.2 *Lipid microsphere preparations (lipo-preparations)*

Since LM are deposited in inflamed tissues and atherosclerotic lesions as mentioned above, they are most suitable as carriers of drugs such as steroid hormones,

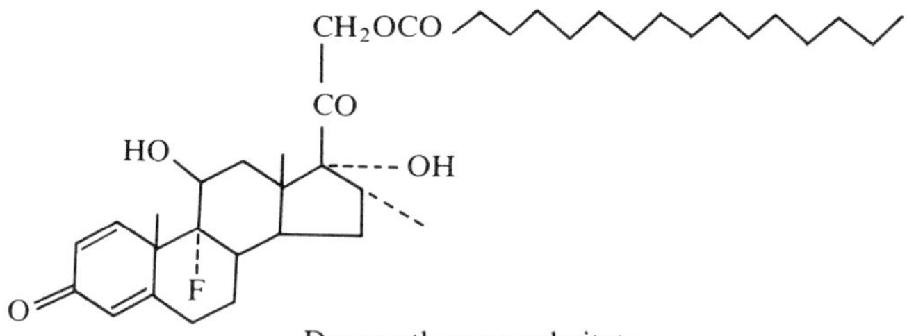

Dexamethasone palmitate

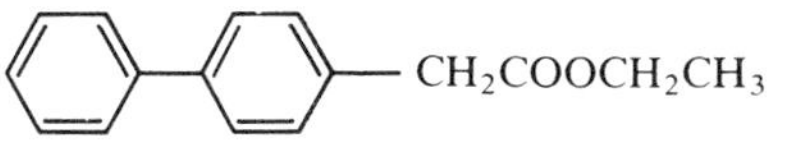

Biphenylacetic acid ethyl ester (BPAA Et)

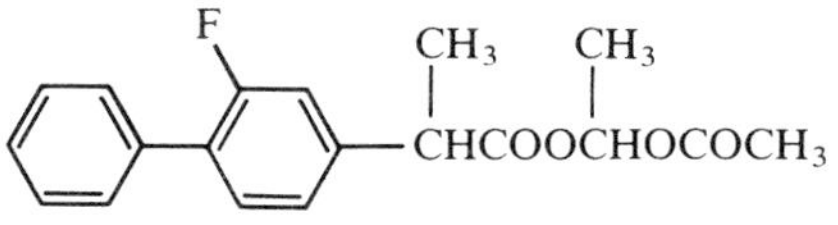

Acetoxyethyl flurbiprofen (FP-AcO-αEt)

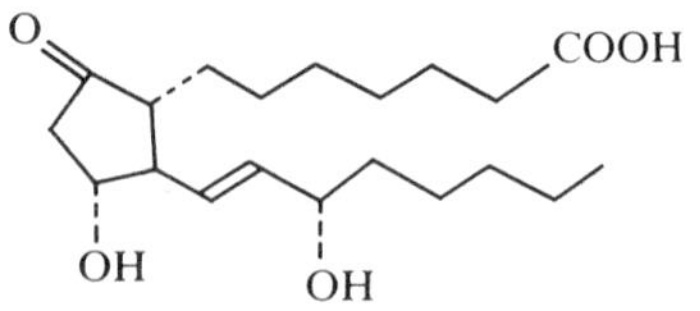

Prostaglandin E_1 (PGE_1)

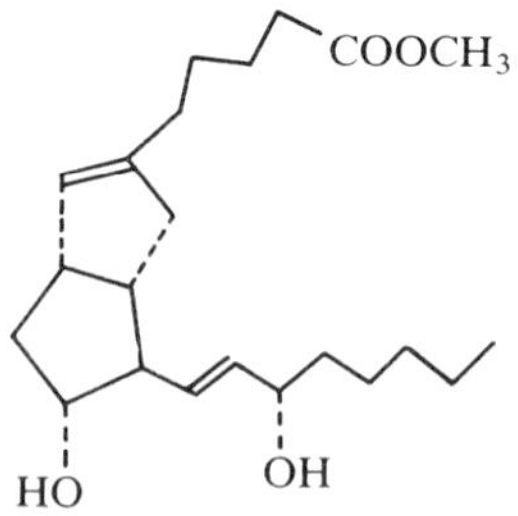

Isocarbacyclin methyl ester (TEI 9090)

Figure 24.3. Chemical structure of drugs used for LM preparation.

non-steroidal anti-inflammatory drugs and prostaglandin-related substances. We incorporated some of these drugs into LM. Since most of these drugs were not sufficently lipo-soluble, alkyl esters were synthesized to prepare lipo-preparations. The chemical structures of major drugs which have been incorporated in lipo-preparations so far are shown in Fig. 24.3.

In order to prepare lipo-preparations, lipo-soluble drugs are first dissolved in soybean oil and emulsified with lecithin using a Manton–Gaulin homogenizer (Mizushima *et al.*, 1982). The composition of various lipo-preparations is shown in Table 24.1. All lipo-preparations contain different amounts of drug.

Table 24.1. Composition of lipo-preparations

Drug	*x*	μg
Soybean oil	100	mg
Egg-yolk phospholipids	18	mg
(Oleic acid	2.4	mg)
Glycerol	22.1	mg
Water for injection	q.s.	

Total volume = 1 ml; pH = 4.5–6.0; osmolarity = 280–300 mOsm; particle diameter = 200–300 nm; shelf life (cold room) = 12 months.

2.3 *Liposteroid*

Dexamethasone palmitate incorporated in LM is known as a liposteroid. When the anti-inflammatory activity of this liposteroid was compared with free dexamethasone sodium phosphate, it was 5.6-fold as potent as an equivalent amount of the latter using the carrageenan granuloma pouch method (Mizushima *et al.*, 1982) and 3.3-fold as potent in adjuvant arthritis. Liposteroids have been tested clinically against rheumatoid arthritis. First, intravenous injections of 1 ml of liposteroid emulsion (2.5 mg as dexamethasone) and 1 ml of dexamethasone sodium phosphate (3.3 mg as dexamethasone) were evaluated in a double-blind crossover test. Judging from patient preference, the liposteroid was significantly superior to dexamethasone sodium phosphate (Mizushima *et al.*, 1983a). Next, either 1 ml of liposteroid or 1 ml of dexamethasone sodium phosphate was injected intravenously every 2 weeks for 8 weeks and the two regimens were evaluated in terms of effectiveness and side-effects in a multicentre, double-blind clinical trial. The results indicated a significant superiority of liposteroid over free dexamethasone (Hoshi *et al.*, 1984).

Shaw *et al.* (1979) and Phillips *et al.* (1979) had reported that hydrocortisone palmitate incorporated into liposomes had a much stronger anti-inflammatory effect than free hydrocortisone given by systemic administration in animals and by local administration in humans.

Dexamethasone palmitate was entrapped in LM by the present authors. It was found that the dexamethasone palmitate ester is cleaved by rat serum, human liver and human monocytes but not by human serum. The relatively weak anti-inflammatory activity of liposteroids in humans compared with that in rats may be due to a difference in susceptibility to hydrolysis of the ester. A search must be made for steroid esters which are more susceptible to human esterases in order to develop a better liposteroid for clinical use.

2.4 *Lipo-non-steroidal anti-inflammatory drugs (lipo-NSAIDs)*

Indomethacin ethoxycarbonylmethyl ester (lipo-indomethacin) (Mizushima *et al.*, 1983c), biphenylacetic acid ethyl ester (lipo-BPAA) (Shoji *et al.*, 1986) and flurbiprofen acetoxy α-ethyl ester (lipo-flurbiprofen) have been incorporated into LM by the present authors. The anti-inflammatory activity of lipo-indomethacin and sodium indomethacin was assessed in the rat carrageenan paw oedema test. When the drugs were given intravenously before the carrageenan injection, there was no difference in activity between the two products. However, when they were injected after the induction of inflammation, i.e. 2 h after the carrageenan injection, the anti-inflammatory activity of lipo-indomethacin was fivefold more potent than that of free indomethacin. These results can easily be explained by the fact that LM accumulate in the inflamed paw, as mentioned above. Furthermore, LM are phagocytosed by cells which produce prostaglandins, such as macrophages and endothelial cells. Thus lipo-NSAIDs, whose mechanism of action is inhibition of prostaglandin synthesis, are delivered to the sites of inflammation and then delivered to the target cells in the inflamed tissues.

Lipo-BPAA and lipo-flurbiprofen also have a much stronger anti-inflammatory activity than the sodium salts of the parent drugs. Moreover, the latter two NSAID esters are more lipophilic than the indomethacin ester. Therefore, a lipo-preparation containing a high dose of NSAID can be obtained. For example, 1 ml of lipo-BPAA contains 20 mg of BPAA, which is strong enough for clinical application. Several blind clinical trials have shown that intravenous injection of 3 ml of lipo-BPAA is very effective and safe in relieving pain induced by tumours, operation, urinary stones, gout and other acute rheumatic conditions.

2.5 *Lipo-prostaglandins*

2.5.1 LIPO-PROSTAGLANDIN E_1 (LIPO-PGE_1)

As described above, LM incorporating PGE_1 (lipo-PGE_1) (Mizushima *et al.*, 1983b; Otomo *et al.*, 1985) are delivered preferentially to the site of vascular lesions, and PGE_1 becomes less irritant because it is incorporated into the microspheres. In addition, its inactivation in the lungs is slightly reduced for the same reason. All these features of the preparation strongly suggest that it would be very valuable not only in animal models but also in humans. Thus, Sim *et al.* (1986) found that in the hamster's cheek pouch test lipo-PGE_1 was better for thrombolysis than free PGE_1. In an animal model of lauric acid-induced peripheral vascular disease, better effectiveness was obtained with lipo-PGE_1 than with free PGE_1 (Otoma *et al.*, 1985).

Hamano *et al.* (1986) compared the hypotensive response to lipo-PGE_1 with that to free PGE_1-cyclodextrin (CD) in diabetic and SHR rats. The reactivity to vasoactive substances decreased as diabetes progressed. This phenomenon was found in both PGE_1-CD and isoproterenol, while response to lipo-PGE_1 was enhanced as diabetes progressed (Hamano *et al.*, 1986). The most striking difference between lipo-PGE_1 and PGE_1-CD was found in diabetic rats aged 10 weeks. In these animals, the hypotensive effect of lipo-PGE_1 was about 25-fold the level of PGE_1-CD. In the SHR model, lipo-PGE_1 was shown to be much more potent than PGE_1-CD.

After several Phase 2 clinical studies were completed (Mizushima *et al.*, 1983b; Hoshi *et al.*, 1986), controlled multicentred trials of lipo-PGE_1 were conducted for the treatment of arterial duct-dependent congenital heart disease, Buerger's disease plus arteriosclerosis obliterans, diabetes-associated peripheral vascular and nervous disorders, collagen disease-associated peripheral vascular disturbances, and vibration disease. In all diseases tested, lipo-PGE_1 was significantly more beneficial than PGE_1-CD or other reference standard. A cold feeling in the extremities, numbness, paraesthesia, and pain were very responsive to lipo-PGE_1. The size of the ulcers which could be assessed objectively was measured in a blind manner. Ulcer lesions associated with collagen diseases regressed significantly within a week of daily treatment with lipo-PGE_1. After 4 weeks' treatment, the difference from the control group was significant (Mizushima *et al.*, 1987a). One double-blind study demonstrated a better safety profile for lipo-PGE_1 than for placebo.

Lipo-PGE_1 was introduced onto the Japanese market in October 1988, and since then it has been widely used in clinical practice. There have been many sporadic reports that it is effective in fulminant hepatitis, neuralgia associated with herpes zoster, multiple spinal canal stenosis, cerebral infarction, myocardial infarction, chronic renal failure, and bed sores as well as for its registered indications.

Lipo-PGE_1 was reported to be 10–20-fold more effective than PGE_1 in the treatment of ductus-dependent congenital heart diseases. It can be used at much lower doses, and, accordingly, adverse reactions are reduced (Momma, 1984). The drug is approved for this indication, and has already been used safely in many patients.

2.5.2 LIPO-PGI_2

Prostacyclin (PGI_2) is more potent than PGE_1 in its antiplatelet and antithrombotic effects, and is expected to be very useful in the treatment of various thrombotic diseases (Grose *et al.*, 1985; Gryglewski, 1985). However, it causes more adverse reactions, such as hypotension and facial flushes, because of its vasodilatory effect and possible suppression of feedback mechanisms. Many attempts at clinical applications have failed. We have conducted studies on isocarbacyclin methyl ester (TEI 9090) (Fig. 24.3), a chemically stable and lipophilic PGI_2 derivative (Shibasaki *et al.*, 1983), incorporated in LM (lipo-PGI_2) similar to those carried out with lipo-PGE_1. TEI 9090 could be incorporated into LM well and was chemically and physically stable (Mizushima *et al.*, 1987b), and so its clinical application seemed to be feasible.

Sim and his co-workers evaluated the antithrombotic effect of lipo-PGI_2 in the hamster's cheek pouch model. Lipo-PGI_2 was 500-fold stronger than free TEI 9090 in its antithrombotic effect (refer to Mizushima *et al.*, 1987b). Therefore, the clinical effect of lipo-PGI_2 seemed to appear with doses $< 1\%$ those recommended for PGI_2. Ohtsu *et al.* (1988) demonstrated that TEI 9090 produced more potent vascularization than natural PGI_2. As stated above, LM are expected to be preferentially delivered to the site of vascularization. These results suggest that lipo-PGI_2 may be very valuable in the treatment of peripheral vascular diseases or cerebral infarction.

We carried out a double-blind comparison of lipo-PGI_2 at a low dose of 2 μg daily (compared with placebo) in the treatment of chronic cerebral infarction. The duration of the treatment was short (2 weeks), but lipo-PGI_2 was obviously and significantly more effective (Hoshi & Mizushima, 1990). Both mental and neurological symptoms responded and the patients were better motivated for further therapy. In Japan, lipo-PGI_2 is being evaluated in a late Phase 2 clinical study (a dose-finding study). It is expected to be marketed in 1996. The results of completed studies showed that lipo-PGI_2 was virtually free from adverse reactions, suggesting that intravenous prostacyclin may be given safely probably only when it is incorporated in LM.

2.5.3 LIPO-PROSTAGLANDINS: FUTURE PROSPECTS

TEI 9090, which was used to prepare lipo-PGI_2, is chemically stable and lipophilic. It can be incorporated efficiently into LM (Mizushima *et al.*, 1987b). PGE_1 itself is relatively unstable, and is degraded to produce PGA_1 in some part when stored at room temperature. It is not very lipophilic. Consequently, when serum is mixed with lipo-PGE_1, a certain amount of PGE_1 will be released from the microspheres (Igarashi *et al.*, 1988). Therefore, there are two approaches to improvement of the lipo-PGE_1 preparation. One is the improvement of storage. The other is that only a small amount of PGE_1 is released before it reaches the target organ when it is given by intravenous injection. It is recommended that a new lipo-PGE_1 preparation should be developed to overcome the problems in the present lipo-PGE_1. To prevent degradation to PGA_1 and increase the lipophilicity of the drug, certain prodrugs of PGE_1 should be prepared. When these new PGE_1 derivatives are incorporated into LM, it is expected that very stable preparations can be obtained. Response might increase, because release of the prostaglandin is expected to decrease in serum.

2.6 *Lipo-preparations: conclusions*

Lipid microspheres with an average diameter of 0.2–0.3 μm resemble liposomes in terms of tissue distribution. They accumulate in the reticuloendothelial organs, inflamed tissues and vascular lesions and are taken up by phagocytic cells such as macrophages and endothelial cells. LM are more stable than liposomes and can be mass produced for commercial use. Moreover, there is no toxicity associated with the use of small amounts of LM. PGE_1, isocarbacyclin methyl ester, some NSAID esters and dexamethasone palmitate were incorporated in LM, and it was found that these lipo-preparations were significantly superior to the each free drug in terms of efficacy and safety. One problem encountered with LM is their incompatibility with lipid-insoluble drugs.

3 Chemical modification of bioactive peptides

In recent years, rapid progress in biotechnology and its introduction into medical fields have made it possible to mass produce bioactive peptides such as insulin, calcitonin, erythropoietin, tumour necrosis factor and superoxide dismutase (SOD) for clinical applications. However, most bioactive peptides have common disadvantages as drugs, including short half-life, poor cellular affinity and absence of selectivity for target organs and lesions. Therefore, the use of drug delivery systems is essential in order to obtain significant clinical efficacy with these substances, such as the following:

1 Lecithinization of bioactive peptides (which has been studied by the present authors).

2 Some chemical modifications of SOD.

3 Chemical modifications of other bioactive peptides.

These will be discussed in turn.

3.1 *Lecithinization of peptides*

As mentioned above, the present authors have successfully used LM of 0.2–0.3 μm in diameter as targeting carriers for low-molecular-weight drugs such as PGE_1, PGI_2 and steroids. Since LM are deposited in atherosclerotic lesions, thrombi, bone marrow and inflamed tissues, they were expected to be useful as carriers not only for low-molecular-weight drugs, but also for larger molecular-weight bioactive peptides such as urokinase, tissue plasminogen activator, erythropoietin, colony stimulating factors and SOD. Initially attempts were made to prepare lipo-bioactive peptides through covalent linkage of LM and bioactive peptides. However, problems encountered included the limited amount of peptide which can be bound to LM due to the 3-D size of peptide molecules, and the unstable activity of bioactive peptides in aqueous glycerin (Igarashi, 1989). As an alternative approach, efforts were made to fix lecithin around the peptide molecules by covalent linkage. Peptide molecules with covalently bound lecithin around them are smaller than LM, but their pharmacokinetics were expected to be similar to those of LM because their structure with an outer layer of lecithin is similar to that of LM. In addition, this approach seems to overcome the problem related to the stability of bioactive peptide activity because it makes possible freeze-drying of lecithinized bioactive peptides. IgG was selected as a model peptide and covalently bound with lecithin in order to evaluate pharmacokinetics and cellular affinity of the bioactive peptide–lecithin complex (Igarashi & Mizushima, 1989).

3.1.1 PREPARATION, DISTRIBUTION IN THE BODY AND CELLULAR AFFINITY OF LECITHINIZED IgG

An aminolecithin (shown below) and IgG were coupled using water-soluble carbodiimide. Lecithinized IgG in which about 60 lecithin molecules are coupled to one IgG molecule was obtained (Igarashi & Mizushima, 1989; Mizushima & Igarashi, 1991b).

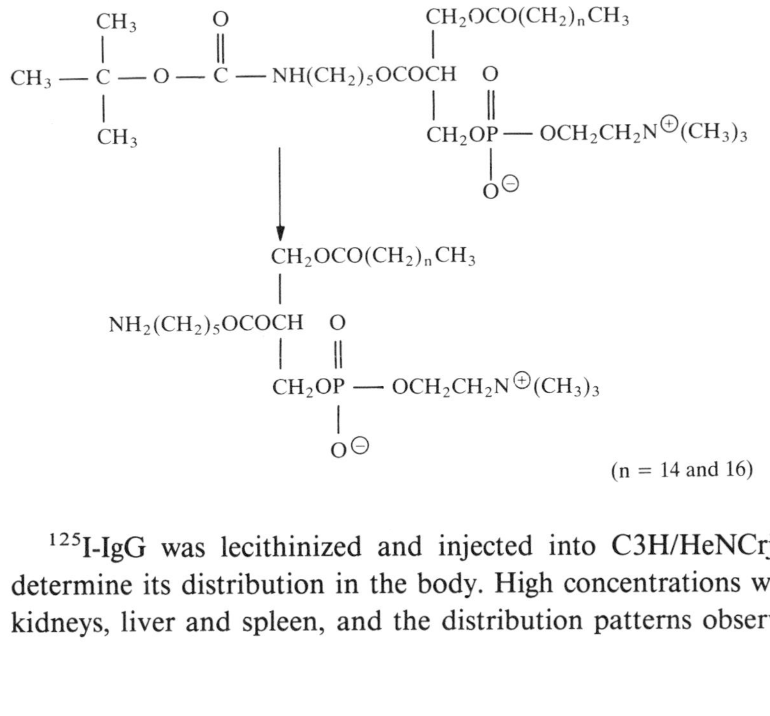

(n = 14 and 16)

^{125}I-IgG was lecithinized and injected into C3H/HeNCrj mice in order to determine its distribution in the body. High concentrations were observed in the kidneys, liver and spleen, and the distribution patterns observed were similar to those observed with LM. Lecithinized IgG was added to MM46 cancer cells, sarcoma 180 cancer cells and human lymphocytes suspended in RPMI 1640 medium in order to evaluate the affinity of lecithinized IgG to these cells. The lecithinized IgG showed much greater affinity than IgG alone (Igarashi & Mizushima, 1989; Mizushima & Igarashi, 1991).

Thus, lecithinized IgG showed high concentrations in organs and great cellular affinity. Affinity with MM46 cancer cells and sarcoma 180 cancer cells was particularly great probably because these cancer cells actively take up lecithin, which serves as a nutrient. This finding suggests that the lecithinization of monoclonal antibodies and anticancer drugs is promising. However, our studies have shown that γ-interferon and calcitonin are largely inactivated when they are lecithinized by our method, although the activity of SOD and anti-TSH receptor antibodies was preserved almost intact.

Questions which must be answered in the future include whether lecithin should be introduced into the carboxylic terminal (–COOH) or amino terminal (–NH_2) of peptides and how the activity, immunogenicity (ability to produce antibodies) and antigenicity of peptides are altered when lecithin is introduced. The fact that IgG covalently coupled with lecithin showed distribution patterns and cellular affinity similar to those of LM suggests that lecithinization techniques can become effective drug-delivery system for bioactive peptides.

3.2 *Chemical modification of superoxide dismutase (SOD)*

O_2^-, an active form of oxygen in the body, exerts bactericidal activity in appropriate amounts. In excessive amounts, however, it damages tissues, aggravates most inflammatory diseases and serves as a carcinogenesis promoter. SOD is an enzyme which is able to scavenge O_2^- and has been highlighted as a therapeutic agent for inflammatory diseases in recent years. Five types of SOD have been reported: CuZu-SOD, Mn-SOD, Fe-SOD, FeZn-SOD and EC-SOD. CuZu-SOD (molecular weight ca. 32 000) and Mn-SOD (molecular weights ca. 40 000 and 80 000) have been studied for development as therapeutic agents.

CuZu-SOD is rapidly eliminated from the blood, with a half-life of about 6 min. Therefore, efforts have been made to increase its half-life by chemical modifications. High-molecular-weight substances were chemically coupled with SOD.

3.2.1 PEG-SOD (POLYETHYLENE GLYCOL-COUPLED SOD)

Several groups have studied PEG-SOD (Pyatak *et al.*, 1980; Malaisse, 1982; Veronese *et al.*, 1983; Hatherill *et al.*, 1986; Conforti *et al.*, 1987; Miyata *et al.*, 1988). The half-life of PEG-SOD is 10–200-fold longer than that of free SOD depending on the molecular weight of the PEG used. Another major advantage of PEG-SOD is that it has less immunogenic potential than SOD because SOD is covered with PEG which has no immunogenic potential.

Veronese *et al.* (1983) coupled SOD with PEG of various molecular weights and found that PEG3-SOD and PEG18-SOD, in which 3 and 18 molecules, respectively, of PEG with a molecular weight of 5000 are coupled with 1 SOD molecule, have blood half-lives of 4 and 20 h, respectively. Free SOD, PEG3-SOD and PEG18-SOD inhibited carrageenan-induced oedema in rats by 18%, 22% and 33%, respectively. Miyata *et al.* (1988) coupled PEG of molecular weight 5000 with *Serratia*-derived Mn-SOD (molecular weight 48 000) and found that 52% of SOD activity was retained at an amino group modification rate of 24%; the blood half-life increased tenfold compared to that of free SOD, while the antigenic and immunogenic potentials decreased to 10%. In addition, this PEG-SOD dose-dependently inhibited carrageenan-induced abscess in rats at 0.5–4 mg/kg (inhibition rate 17–30%), while free SOD was almost ineffective. This PEG–SOD also effectively prevented (31–61%) ovalbumin-induced delayed type hypersensitivity in mice at 1.6–8 mg/kg, while free SOD was ineffective at 8 mg/kg. However, major disadvantages of PEG-SOD include low cellular affinity and absence of lesion selectivity.

The binding of PEG to a peptide can be illustrated as follows:

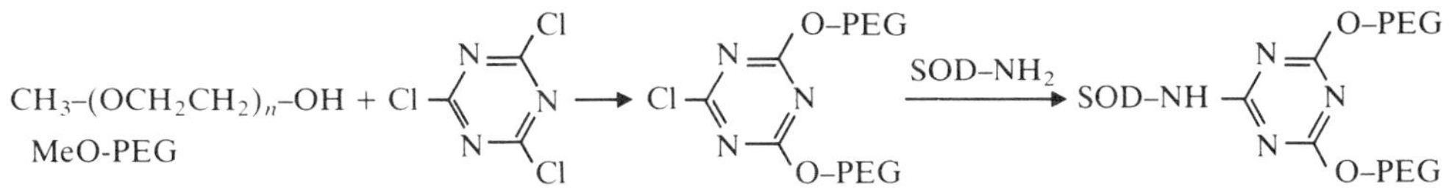

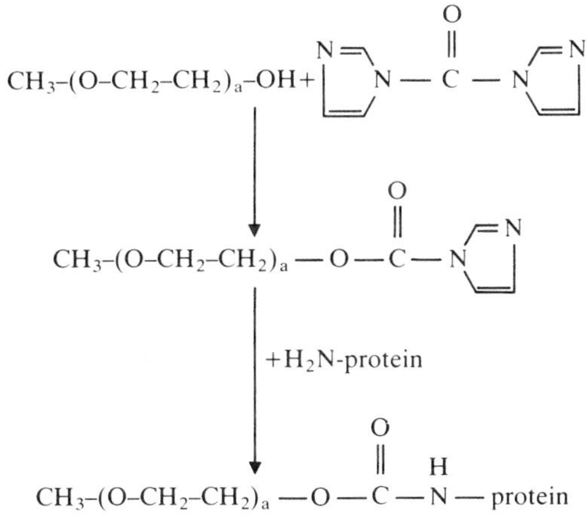

3.2.2 PYRAN-SOD AND SMA-SOD

Pyran-copolymer is a polymer with the molecular structure:

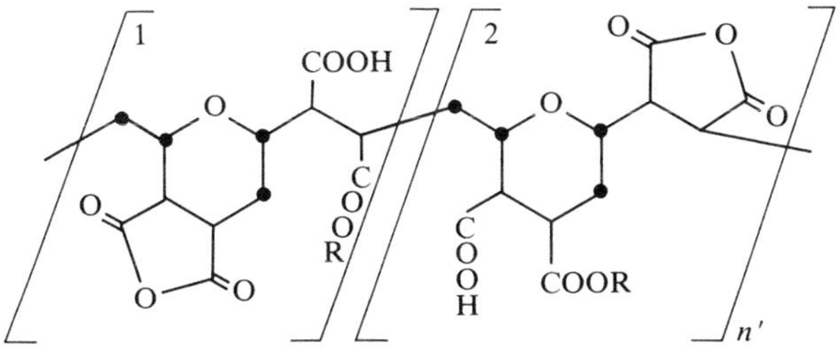

(R:H, C_2H_5, C_4H_9, ...)

Pyran-copolymer
m.w. 5600

It was discovered as a synthetic high-molecular-weight substance which induces production of interferon. It has been used to increase the molecular weight of Adriamycin. Oda *et al.* (1989) coupled pyran-copolymer with a molecular weight of 5600 with SOD and evaluated the efficacy for influenza virus infection.

The pyran-SOD used, in which 10% of lysine residues were coupled with pyran, retained 80% of SOD activity. In *in vivo* experiments, while free SOD in the plasma was almost completely inactivated in 30 min, pyran-SOD remained active up to 5 h after it came in contact with the plasma. The survival rate of mice treated with 200 U of pyran-SOD after infection with twice the LD_{50} of influenza virus was 90%, indicating that its therapeutic effects are significantly greater than those of free SOD. This enhancement of efficacy is thought to be attributable to the longer presence in the blood and greater selectivity for lesions achieved by coupling with pyran-copolymer. It is also interesting that O_2^- is involved in influenza virus infection. Ogino *et al.* (1988) also studied SMA-SOD obtained by coupling stylene comaleic acid butyl ester

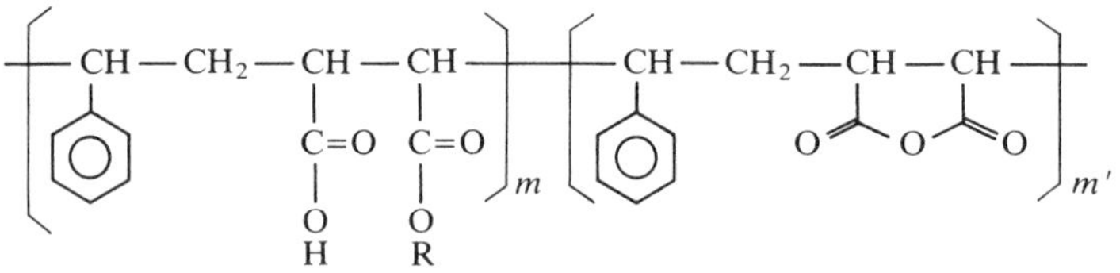

$m, m' \geq 1$ [R=H, CH_3, C_2H_5, C_4H_9, ...]

Stylene comaleic acid butyl ester
(SMA) m.w. 1600

and SOD, and found that SMA-SOD, as well as pyran-SOD, had much longer half-lives than free SOD and that it showed remarkable therapeutic effects on various types of inflammation and ischaemic diseases.

3.2.3 LIPOSOMAL-SOD AND OTHER COMBINATIONS

Michelson *et al.* (1981a,b) developed liposomal-SOD by incorporating SOD in liposome. Liposomal-SOD was found to have longer blood half-life and greater cellular affinity than free SOD and to exert anti-inflammatory effects (Michelson *et al.*, 1981a,b; Baillet *et al.*, 1986; Somiya *et al.*, 1986; Michelson *et al.*, 1988). It has been suggested that the penetration of this compound into organs can be regulated by the selection of liposomal electric charges (Michelson *et al.*, 1981b). Various clinical effects of this compound have been reported. We have also shown that it is effective for diseases such as scleroderma. In recent years, incorporation of human recombinant SOD into liposomes has also been evaluated.

Other high-molecular-weight substances which have been coupled with SOD include albumin (Wong *et al.*, 1980), polylysin (Schalkwijk *et al.*, 1985), complexes of several C_{10}–C_{14} fatty acid molecules, basic peptides with great affinity to heparin sulphate and inulin polysaccharides. Although SOD is expected to have various clinical applications, its development as a drug has been delayed due to great difficulties encountered in its evaluation.

3.3 *Chemical modifications of other bioactive peptides*

3.3.1 INSULIN

Insulin, a drug indispensable for diabetic patients, can be administered only by injection, which is a major inconvenience for patients. Therefore, many studies have been conducted to develop oral and transdermal insulin preparations. Kim *et al.* (1990) have developed a controlled-release system with a sensor using concanavalin A (ConA)-coupled high-molecular-weight substances coupled with sugar-chain insulin. This system makes it possible to release ConA-coupled sugar-chain insulin and take up glucose when blood glucose levels increase. Cho & Ito (1989) administered oral insulin preparations prepared using carrier complexes composed of triglyceride, cholesterinester and phospholipid (ODDS insulin) to insulin-dependent and insulin-independent diabetic patients over a long period of time and reported that fasting high blood-glucose levels could be reduced. It was also reported that insulin-receptor binding potency was altered when insulin was coupled with cholera toxin (Roth *et al.*, 1983) and that the half-life of insulin was increased when it was coupled with PEG (Ehrat & Luisi, 1983). Hashimoto *et al.* (1989) tried to increase the potency and duration of action of insulin by increasing its lipo-solubility through palmitoylation. They found that monopalmitoylinsulin was longer-acting than native insulin.

3.3.2 HAEMOGLOBIN (Hb)

Modified Hb has also been actively studied as a blood substitute. Erythrocytic Hb plays a major role in blood oxygen transport. Exogenous administration of Hb is believed to be useful in blood substitution following intraoperative blood loss, treatment of anaemia, and elimination of viruses and toxic substances during plasmapheresis. Although stroma free-Hb (SFH) obtained after elimination of erythrocyte membrane and blood-group substances causes no adverse reactions such as renal failure (Rabiner *et al.*, 1967), it has several disadvantages: (i) 2,3-diphosphoglycerate is metabolized during storage, resulting in increases in the affinity of Hb and oxygen molecules and a decrease in the ability of Hb to deliver the oxygen to tissues; and (ii) in the blood, Hb, a tetramer, is transformed into a dimer, coupled with haptoglobin, and is rapidly excreted. Benesch *et al.* (1972) found that the oxygen affinity of Hb could be reduced by coupling Hb with pyridoxal 5-phosphate (PLP), a B_6 vitamin. Efforts have been made to increase the blood half-life of Hb by coupling it with high-molecular-weight substances such as PEG, dextran, hydroxyethyl starch and inulin. The duration of Hb in the blood is directly related to the molecular weight of the PEG and the number of binding sites (Ajsaka & Iwashita, 1980). Iwasaki & Iwashita (1986) prepared PEG–PLP–Hb by coupling PEG of a molecular weight of 4000 to Hb previously coupled with PLP. They found that the oxygen affinity of this compound was comparable to that of erythrocytes and that it remained in the blood between four- and sevenfold longer than SFH. Although PEG–PLP–Hb is expected to be an effective erythrocyte substitute, problems yet to be solved include how to ensure a supply of Hb in large quantities and how to eliminate viral contamination completely. Studies on artificial haem are also ongoing.

3.3.3 OTHER COMPOUNDS

In addition to SOD, compounds which are under study with the objective of increasing blood half-life and decreasing antigenic potential by chemical modifications using PEG include interleukin-2 (Katre *et al.*, 1987), urokinase (Sakuragawa *et al.*, 1986), uricase (Nishimura *et al.*, 1981) and L-asparaginase (Matsushima *et al.*, 1980; Yoshimoto *et al.*, 1986). Lipase, chymotrypsin, papain, catalase and peroxidase are soluble in organic solvents when they are coupled with PEG. These substances then efficiently catalyse various reactions in organic solvents, such as hydrolysis, acid amide synthesis, esterification, and decomposition of H_2O_2 and oxidation by H_2O_2. In addition to PEG, IgG, dextran and albumin have been used to increase blood half-life.

Anti-AIDS drugs have been highlighted in recent years. A Japanese company is developing a drug which selectively kills HIV by coupling antigen CD4, a membrane protein which is able to recognize HIV, with liposomes which contain diphtheria toxin. Genentech, an American company, is developing an antibody in which the antigen binding site is replaced with CD4 as an anti-AIDS drug.

4 Conclusions

Various types of chemical modifications which have been attempted for drug-delivery systems of bioactive peptides have been reviewed. Major objectives of these in relation to bioactive peptides include increases in half-life, cellular affinity and tissue selectivity. Bioactive peptides should not be inactivated by chemical modification. It is also important to minimize adverse effects which may be produced by chemical modification. In chemical modifications of bioactive peptides for drug-delivery systems, therefore, it is desirable that substances introduced be able to increase tissue affinity of peptides, increase concentrations of peptides in target organs, not be toxic and be readily decomposed in the body other than at the target sites. At present, various outstanding intelligent biomaterials are available to achieve major objectives of drug-delivery systems such as targeting, controlled release and sensing. Technological breakthroughs have been remarkable in this field in recent years. Various drug-delivery systems can be used not only alone but also in combination. A more multidisciplinary approach, which uses technologies from a wide range of fields, will be required for drug-delivery systems of bioactive peptides in the future.

5 References

Ajisaka, K. & Iwashita, Y. (1980). Modification of human hemoglobin with polyethylene glycol: a new candidate for blood substitute. *Biochem. Biophys. Res. Commun.*, **97**, 1076.

Baillet, F., Housset, M. & Michelson, A.M. (1986). Treatment of radiofibrosis with liposomal superoxide dismutase. Preliminary results of 50 cases. *Free Rad. Res. Commun.*, **1**, 387.

Benesch, R.E., Benesch, R., Renthal R. & Maeda, N. (1972). Affinity labeling of the polyphosphate binding site of hemoglobin. *Biochemistry*, **11**, 3576.
Cho, Y.W. & Ito, H. (1989). Oral DDS for insulin. *Pharma Medica*, **7**, 67.
Conforti, A., Franco, L., Milanio, R. & Velo, G.P. (1987). PEG superoxide dismutase derivatives: anti-inflammatory activity in carrageenan pleurisy in rats. *Pharmacol. Res. Commun.*, **19**, 28.
Davis, S.S., Hadgraft, J. & Palin, K.J. (1985). In: P. Becher (Ed.), *Encyclopedia of Emulsion Technology 3*, p. 159, Marcel Dekker, New York.
Ehrat, M. & Luisi, P. (1983). Synthesis and spectroscopic characterization of insulin derivatives containing one or two poly(ethylene oxide) chains at specific positions. *Biopolymer*, **22**, 569.
Gregoriadis, G. (1984). In: G. Gregoriadis & A.C. Allison (Eds.), *Liposomes in Biological Systems*, Vols I–III, John Wiley, New Jersey.
Grose, R., Greenberg, M., Strain, J. & Mueller, H. (1985). Intracoronary prostacyclin in evolving acute myocardial infarction. *Am. J. Cardiol.*, **55**, 1625.
Gryglewski, R.J. (1985). Effects of prostacyclin in atherosclerotic vascular disease. *Adv. Prostaglandin Thromboxane Leukot. Res.*, **15**, 539.
Hallberg, D. (1965). Elimination of exogenous lipids from the blood stream. *Acta Physiol. Scand.*, **65** (Suppl.), 254.
Hamano, T., Shintome, M. & Watanabe, M. (1986). Vasodilation activity of lipo-PEG_1. *Basic Clin. Res.*, **20**, 93.
Hashimoto, M., Takada, K. & Kiso Y. (1989). Synthesis of palmitoyl derivatives of insulin and their biological activities. *Pharm. Res.*, **6**, 171.
Hatherill, J.R., Till, G.O., Bruner, L.H. & Ward, P.A. (1986). Thermal injury, intravascular hemolysis, and toxic oxygen products. *J. Clin. Invest.*, **78**, 629.
Hoshi, K. & Mizushima, Y. (1990). A preliminary double-blind cross-over trial of lipo-PGI_2, a prostacyclin derivative incorporated in lipid microspheres, in cerebral infarction. *Prostaglandins*, **40**, 155.
Hoshi, K., Kaneko, K. & Mizushima, Y. (1984). Clinical study on liposteroid. *First World Conference on Inflammation, Venice, 16 April, 1984.*
Hoshi, K., Mizushima, Y., Kiyokawa, S. & Yanagawa, A. (1986). Prostaglandin E_1 incorporated in lipid microspheres in the treatment of peripheral vascular disease and diabetic neuropathy. *Drugs Exp. Clin. Res.*, **12**, 681.
Igarashi, R. (1989). Planning of LM conjugated BRM. *Pharma Medica*, **7**, 91.
Igarashi, R. & Mizushima, Y. (1989). Biological active peptide conjugated with lecithin for DDS. *Jpn. J. Drug Del. Syst.*, **4**, 257.
Igarashi, R. Nakagawa, M. & Mizushima, Y. (1988). Biological activity of several esterified PGE_1 and the application of them to lipid microsphere (LM) preparations. *Jpn. J. Inflam.*, **8**, 243.
Iwasaki, K. & Iwashita, Y. (1986). Preparation and evaluation of hemoglobin–polyethylene glycol conjugate (pyridoxalated PEG hemoglobin) as an oxygen-carrying resuscitation fluid. *Artif. Organs*, **10**, 411.
Katre, N.V., Knauf, M.J. & Laird, W.J. (1987). Chemical modification of recombinant interleukin-2 by polyethylene glycol increases its potency in the murine Meth A sarcoma model. *Proc. Natl. Acad. Sci. USA*, **84**, 1487.
Kim, S.W., Pai, C.M. & Makino, K. (1990). Self-regulated glycosylated insulin delivery. *J. Cont. Rel.*, **11**, 193.
Kiyokawa, S., Igarashi, R., Iwayama, T., Haramoto, S., Matsuda, T., Hoshi, K. & Mizushima, Y. (1987). ^{99m}Tc labeled lipid microspheres (LM) would be useful for an imaging study of those diseases. *Jpn. J. Inflam.*, **7**, 551.
McCord, J.M. & Wong, K. (1989). In: *Ciba Foundation Symposium 65*, p. 257, Excerpta Medica, Amsterdam.
Malaisse, W.J. (1982). Alloxan toxicity to the pancreatic cell: a new hypothesis. *Biochem. Pharmacol.*, **31**, 3527.
Matsushima, A., Nishimura, H. & Ashihara, Y. (1980). Modification of *E. coli* asparaginase with 2,4-bis(O-methoxypolyethyleneglycol)-6-chloro-*s*-triazine (activated PEG_2); disappearance of

binding ability towards anti-serum and retention of enzymic activity. *Chem. Lett.*, **7**, 773.
Michelson, A.M., Puget, K., Perdereau, B. & Barbaroux, C. (1981a). Scintigraphic studies on the localization of liposomal superoxide dismutase injected into rabbits. *Mol. Physiol.*, **1**, 71.
Michelson, A.M., Puget, K. & Durosay, P. (1981b). Studies of liposomal superoxide dismutase in rats. *Mol. Physiol.*, **1**, 85.
Michelson, A.M., Jodot, G. & Puget, K. (1988). Treatment of brain trauma with liposomal superoxide dismutase. *Free Rad. Res. Commun.*, **4**, 209.
Miyata, K., Nakagawa, Y., Nakamura, M., Ito, T., Sugo, K., Fujita, T. & Tomoda, K. (1988). Altered properties of *Serratia* superoxide dismutase by chemical modification. *Agric. Biol. Chem.*, **52**, 1575.
Mizushima, Y. (1985). Lipid microspheres as novel drug carriers. *Drugs Exp. Clin. Res.*, **9**, 595.
Mizushima, Y. (1989). Clinical advantages of new drug delivery systems in arthritis—an overview of current trends. *JAMA*, **5** (Suppl.), 13.
Mizushima, Y. & Igarashi, Y. (1991a). Lipid microspheres and lecithinized polymer for drug delivery systems. In: *Third International Symposium on Drug Delivery Systems*, p. 266, ACS Books, Washington D.C.
Mizushima, Y. & Igarashi, R. (1991b). Studies on polypeptide drug delivery systems: tissue distribution of immunoglobulin G conjugated with lecithin. *J. Contr. Rel.*, **17**, 99–104.
Mizushima, Y., Hamano, T. & Yokoyama, K. (1982). Tissue distribution and anti-inflammatory activity of corticosteroids incorporated in lipid emulsion. *Ann. Rheum. Dis.*, **41**, 263.
Mizushima, Y., Kaneko, K. & Hoshi, K. (1983a). Targeting steroid therapy in rheumatoid arthritis. *Ann. Rheum. Dis.*, **42**, 479.
Mizushima, Y., Yanagawa, A. & Hoshi, K. (1983b). Prostaglandin E_1 is more effective, when incorporated in lipid microspheres, for treatment of peripheral vascular diseases in man. *J. Pharm. Pharmacol.*, **35**, 666.
Mizushima, Y., Hamano, T. & Haramoto, S. (1990). Distribution of lipid microspheres incorporating prostaglandin E_1 to vascular lesions. *Prostaglandins Leukot. Essent. Fatty Acids*, **41**, 269.
Mizushima, Y., Wada, Y., Etoh, Y. & Watanabe, K. (1983c). Anti-inflammatory effects of indomethacin ester incorporated in a lipid microsphere. *J. Pharm. Pharmacol.*, **35**, 398.
Mizushima, Y., Shiokawa, Y., Homma, M., Kashiwazaki, S., Ichikawa, Y., Hashimoto, H. & Sakuma, A. (1987a). A multicenter double blind controlled study of lipo-PGE_1, PGE_1 incorporated in lipid microspheres, in peripheral vascular disease secondary to connective tissue disorders. *J. Rheumatol.*, **14**, 97.
Mizushima, Y., Igarashi, R., Hoshi, K., Sim, A.K., Cleland, M.E., Hayashi, H.& Goto, J. (1987b). Marked enhancement in antithrombotic activity of isocarbacyclin following its incorporation into lipid microspheres. *Prostaglandins*, **33**, 161.
Momma, K. (1984). Prostaglandin E_1 treatment of ductus dependent infants with congenital heart disease. *Int. Angiol.*, **3** (Suppl.), 33.
Nakura, K., Hamano, T., Shintome, M. & Watanabe, M. (1986). Location of lipo-PGE_1 in blood vessel. *Basic Clin. Res.* (in Japanese), **20**, 143.
Nishimura, H., Matsushima, A. & Inaba, Y. (1981). Improved modification of yeast uricase with polyethylene glycol, accompanied with non-immunoreactivity towards anti-uricase serum and high enzymic activity. *Enzyme*, **26**, 49.
Oda, T. Akaike, T., Hamamoto, T., Suzuki, F., Hirano, T. & Maeda, H. (1989). Oxygen radicals in influenza-induced pathogenesis and treatment with pyran polymer-conjugated SOD. *Science*, **244**, 974.
Ogino, T., Inoue, M., Ando, Y., Awai, M., Maeda, H. & Morino, Y. (1988). Chemical modification of superoxide dismutase. Extension of plasma half life of the enzyme through its reversible binding to the circulating albumin. *Int. J. Pept. Protein Res.*, **32**, 153.
Ohtsu, A., Fujii, K. & Kurozumi, S. (1988). Induction of angiogenic response by chemically stable prostacyclin analogs. *Prostaglandins Leukot. Essent. Fatty Acids*, **33**, 35.
Otomo, S., Mizushima, Y., Aihara, H., Yokoyama, K., Watanabe, M. & Yanagawa, A. (1985).

Prostaglandin E_1 incorporated in lipid microspheres (lipo-PGE_1). *Drug Exp. Clin. Res.*, **11**, 627.

Palmer, T.N., Caride, V.J., Caldecourt, M.A., Twickler, J. & Abdullah, V. (1984). The mechanism of liposome accumulation in infarction. *Biochim. Biophys. Acta*, **797**, 363.

Phillips, N.C., Thomas, D.P.P., Night, C.G. & Dingle, J.T. (1979). Liposome-incorporated corticosteroids. II. Therapeutic activity in experimental arthritis. *Ann. Rheum. Dis.*, **38**, 553.

Poste, G., Bucana, C., Raz, A., Bugelski, P., Kirsh, R. & Fidler, I.J. (1982). Analysis of the fate of systemically administered liposomes and implications for their use in drug delivery. *Cancer Res.*, **42**, 1412.

Pyatak, P.S., Abuchowski, A. & Davis, F.F. (1980). Preparation of a polyethylene glycol: superoxide dismutase adduct, and an examination of its blood circulating life and anti-inflammatory activity. *Res. Commun. Chem. Pathol. Pharmacol.*, **29**, 113.

Rabiner, S., Helbert, J., Lopas, H. & Friedman, L. (1967). Evaluation of a stroma-free hemoglobin solution for use as a plasma expander. *J. Exp. Med.*, **126**, 1127.

Roth, R.A. & Maddux, B. (1983). Insulin–cholera toxin binding unit conjugate: a hybrid molecule with insulin biological activity and cholera toxin binding specificity. *J. Cell. Physiol.*, **115**, 151.

Sakuragawa, N., Shimizu, K., Kondo, K., Kondo, S. & Niwa, M. (1986). Studies of the effect of PEG-modified urokinase on coagulation–fibrinolysis using beagles. *Thromb. Res.*, **41**, 627.

Schalkwikj, J., van den Berg, W.B., van de Putte, L.B.A. *et al.*, (1985). Cationization of catalase, peroxidase, and superoxide dismutase. *J. Clin. Invest.*, **76**, 198.

Shaw, I.H., Knight, C.G., Thomas, D.P.P., Phillips, N.C. & Dingle, J.T. (1979). Liposome-incorporated corticosteroids: 1. The interaction of liposomal cortisol palmitate with inflammatory synovial membrane. *Br. J. Exp. Pathol.*, **60**, 142.

Shibasaki, M., Torisawa, Y. & Ikegami, S. (1983). Synthesis of 9(O)-methano-PGI_1; the highly potent carbon analog of prostacyclin. *Tetrahed. Lett.*, **24**, 3493.

Shoji, Y., Mizushima, Y., Yanagawa, A., Shiba, T., Takei, H., Fujii, M. & Amino, M. (1986). Enhancement of anti-inflammatory effect of biphenylyl-acetic acid ester by incorporation into lipid microspheres. *J. Pharm. Pharmacol.*, **38**, 118.

Sim, A.K., McCraw, A.P., Cleland, M.E., Aihara, H., Otomo, S. & Hosoda, K. (1986). The effect of thrombus formation and disaggregation and its potential to target to the site of vascular lesions. *Arzneimittel-Forschung/Drug Res.*, **36**, 1206.

Somiya, K., Niwa, Y. & Michelson, A.M. (1986). Effects of liposomal superoxide dismutase on human neutrophil activity. *Free Rad. Res. Commun.*, **1**, 329.

Veronese, F.M., Boccu, E., Schiavon, O., Velo, G.P., Conforti, A., Franco, L. & Milanino, R. (1983). Anti-inflammatory and pharmacokinetic properties of superoxide dismutase derivatized with polyethyleneglycol via active esters. *J. Pharm. Pharmacol.*, **35**, 757.

Wong, K., Cleland, L.G., Poznansky, M.J. (1980). Enhanced anti-inflammatory effect and reduced immunogenicity of bovine liver superoxide dismutase by conjugation with homologous albumin. *Agents Actions*, **10**, 231.

Yoshimoto, Y., Nishimura, H., Saigo, Y. & Sakurai, K. (1986). Characterization of polyethylene glycol-modified L-asparaginase from *Escherichia coli* and its application to therapy of leukemia. *Jpn. J. Cancer Res.*, **77**, 1264.

Index

Gr

1-800-658-3567.

Graduct study.

financial

A-3. SAS.

4-6.

renewal.

federal staff loan.

Free-yoyo 8,500.

7.43% 12.1册.

10 end: 6周.